Lecture Notes in Mathematics

Edited by A. Dold and B. Eckmann

Series: Institute of Mathematical Sciences. Madras
Advisor: K. R. Unni

399

Functional Analysis and its Applications

International Conference, Madras, 1973

Edited by H. G. Garnir, K. R. Unni and J. H. Williamson

Springer-Verlag
Berlin · Heidelberg · New York 1974

H. G. Garnir
Université de Liège/Belgium

K. R. Unni
Matscience, Madras/India

J. H. Williamson
University of York/England

Library of Congress Cataloging in Publication Data
Main entry under title:

Functional analysis and its applications.

(Lecture notes in mathematics, 399)
1. Functional analysis--Congresses. I. Garnir, Henri G., ed. II. Unni, K. R., 1933- ed.
III. Williamson, John Hunter, 1926- ed.
IV. Series: Lecture notes in mathematics (Berlin) 399.
QA3.L28 no. 399 [QA320] 510'.8s [515'.7] 74-13998

AMS Subject Classifications (1970): 46-XX

ISBN 3-540-06869-4 Springer-Verlag Berlin · Heidelberg · New York
ISBN 0-387-06869-4 Springer-Verlag New York · Heidelberg · Berlin

Offsetdruck: Julius Beltz, Hemsbach/Bergstr.

This volume is dedicated to Professor ALLADI RAMAKRISHNAN, Founder-Director of MATSCIENCE, The Institute of Mathematical Sciences, Madras, who has completed twenty-five years of active research in the realm of mathematical sciences.

PREFACE

This International Conference on Functional Analysis was held at Madras from January 1 to January 7, 1973 on the occasion of the Eleventh Anniversary of Matscience, The Institute of Mathematical Sciences, Madras.

Functional Analysis is not only a very important branch of pure mathematics but perhaps the most fundamental for various applications. It is also a subject in which the members of Matscience have actively contributed.

In this conference, the emphasis was made on different aspects of Functional Analysis, namely Topological Vector Spaces, Harmonic Analysis, Approximation Theory and Partial Differential Equations.

This meeting was unique in the sense that it was the biggest international conference in pure mathematics so far held in India. There were more than 80 participants including 50 eminent mathematicians from more than 20 countries outside India. There were one hour invited lectures, half hour invited lectures and ten minutes communications. In addition to the papers presented at the conference, this volume also contains three papers submitted in writing to the conference by the authors who could not participate at the last minute.

This conference was inaugurated by the Patron of the Institute, Hon'ble Sri C.Subramaniam, Minister of Science and Technology, Government of India, who is also the founding father of Matscience. It is with pleasure that we record our sincere gratitude to him and to the Chairman of the Board of Governors, Hon'ble Thiru V.R.Nedunchezhiyan, Minister of Education, Government of Tamil Nadu, for their unfailing interest in the growth and development of the Institute.

For the conduct of this conference, the Institute has received full support and cooperation from the Government of Tamil Nadu and the Government of India and the good will and cooperation from the international mathematical community and we wish to express our warmest gratitude to them. Our thanks are due to the citizens of Madras who have shown a keen interest in the programmes of the Institute and in particular to Dr. Malcolm S.Adiseshiah, former Acting Deputy Director General of the UNESCO, who presided over the inaugural function.

Our particular thanks are due to Miss Vimala Walter who has undertaken the tedious task of filling up all the mathematical equations for the entire proceedings and to the staff of the Institute namely Messrs R.Jayaraman, S.Krishnan, N.S.Sampath and A.R.Balakrishnan for typing various manuscripts.

H.G.G.

K.R.U.

J.H.W.

CONTENTS

LIST OF PARTICIPANTS

C.A.AKEMANN	University of California, Santa Barbara, California 93106, U.S.A. and University of Copenhagen, Denmark.
W.A.AL SALAAM	University of Alberta, Edmonton, Canada.
R.A.ALO(*)	Carnegie Mellon University, Pittsburgh, Pennsylvania 15213, U.S.A.
C.BERGE	C.E.N., Saclay, Paris, France.
J.T. BURNHAM(*)	University of York, Heslington, York, England and University of Iowa, Iowa City 52240, U.S.A.
L.COLLATZ	Institut für angewandte Mathematik, Universität Hamburg, Hamburg, West Germany.
T.R. DHANAPALAN	Annamalai University, Annamalainagar, India
J.G.DHOMBRES	Université de Nantes, 44 Nantes, France.
PH.A.DIONNE	University of Auckland, Auckland, New Zealand.
M.M.DZHRBASHYAN	Institute of Mathematics, Armenian Academy of Sciences, Erevan, U.S.S.R.
LE BARON O.FERGUSON	University of California, Riverside, California 92502, U.S.A.
J.P.FERRIER	Universite de Nancy, Nancy, France.
S.H.FESMIRE(*)	Carnegie-Mellon University, Pittsburgh, Pennsylvania 15232, U.S.A.
G.FREUD	Mathematical Institute of the Hungarian Academy of Sciences, Budapest, Hungary.
D.GAIER	Justus Liebig Universität, Giessen, West Germany.
H.G.GARNIR	Institut de Mathématique, Université de Liège, 4000 Liège, Belgium.
P.K.GEETHA	Matscience, Madras, India.

B.R. GELBAUM	SUNY at Buffalo, Buffalo, New York 14214, U.S.A.
A.GIROUX	Université de Montreal, Montreal, Canada.
J.GOBERT	Institut de Mathématique, Université de Liège, 4000 Liège, Belgium.
R.R.GOLDBERG	University of Iowa, Iowa City, Iowa 52240, U.S.A.
K.GOPALASAMY	The Flinders University of South Australia, Bedford Park, South Australia 5042.
J.P.GOSSEZ	Free University of Brussels, Brussels, Belgium.
K.N.GOWRISANKARAN	McGill University, Montreal, Canada.
A.K.GUPTA	Indian Institute of Technology, Kanpur, India.
S.C.GUPTA	Aligarh Muslim University, Aligarh, India.
A.GUSSEINOV	Mathematical Institute of Academy of Sciences of Azerbaidjan, Baku, U.S.S.R.
G.HARDY	University of Queensland, St.Lucia, Brisbane, Australia.
P.HESS	Universität Zürich, 8032 Zürich, Switzer-land.
T.HUSAIN	McMaster University, Hamilton, Ontario, Canada.
S.IGARI	Tohoku University, Sendai, Japan.
R.JAGANNATHAN	Matscience, Madras, India.
H.JOHNEN	Techn. Hochschule Aachen, 51 Aachen, West Germany.
J.P.KAHANE	Université de Paris, Orsay, France.
G.K.KALISCH	University of California, Irvine, California 92664, U.S.A.
N.KASTHURI	Matscience, Madras, India.
G.N.KESHAVA MURTHY	Matscience, Madras, India.
KRISHNASWAMI ALLADI	Vivekananda College, Madras, India.

L.D.KUDRYAVTSEV	Steklov Mathematical Institute, Moscow, U.S.S.R.
P.LEONARD	Institut de Mathématique, Université de Liège, 4000 Liège, Belgium.
F.E.MACSUDOV	Mathematical Institute of the Academy of Sciences of Azerbaidjan, Baku, U.S.S.R.
A.MALLIOS	Mathematical Institute, University of Athens, Athens, Greece.
S.MANI	Matscience, Madras, India.
K.H.MARIWALLA	Matscience, Madras, India.
B.MONTADOR	Université de Sherbrooke, Sherbrooke, Canada.
C.W.MULLINS	Pahlavi University, Shiraz, Iran.
G.S.N.MURTHY	Matscience, Madras, India.
B.SZ.-NAGY	University of Szeged,Szeged, Hungary.
L.NIRENBERG	Courant Institute of Mathematical Sciences, New York University, New York 10012, U.S.A.
F.J.NORONHA	University of Bangalore, Bangalore, India.
D.V.PAI	Indian Institute of Technology, Bombay, India.
A.PELCZYNSKI	Institute of Mathematics, Polish Academy of Sciences, Warsaw, Poland.
E.T.POULSEN	Aarhus University, Aarhus, Denmark.
Y.S.PRAHLAD	Matscience, Madras, India.
V.RADHAKRISHNAN	Matscience, Madras, India.
ALLADI RAMAKRISHNAN	Matscience, Madras, India.
N.R.RANGANATHAN	Matscience, Madras, India.
T.J.RIVLIN	Thomas J.Watson Research Centre, York Town Heights, New York 10598, U.S.A.
T.SATYANARAYANA	Matscience, Madras, India.
K.SCHERER	Techn. Hochschule Aachen, 51 Aachen, West Germany.

J.SCHMETS	Institut de Mathématique, Université de Liège, 4000 Liège, Belgium.
I.J.SCHOENBERG	Mathematics Research Centre, University of Wisconsin, Madison, Wisconsin 53705, U.S.A.
B. SENDOV	University of Sofia, Sofia 26, Bulgaria.
A.SHARMA	University of Alberta, Edmonton, Canada.
R.N.SEN	Kalyani University, West Bengal, India.
A.H.SIDDIQI	University of Tabriz, Iran.
K.R.SONAVANE	Shivaji University, Kolhapur, India.
R.SRIDHAR	Matscience, Madras, India.
K.SRINIVASA RAO	Matscience, Madras, India.
R.P.SRIVATSAV	SUNY at Stony Brooke, Stony Brooke, New York, U.S.A.
V.SUBBA RAO	Indian Institute of Technology, Madras, India.
M.R.SUBRAHMANYA	Matscience, Madras, India.
A.R.TEKUMALLA	Matscience, Madras, India .
K.R.UNNI	Matscience, Madras, India.
A.VIJAYAKUMAR	Matscience, Madras, India.
VIMALA WALTER	Matscience, Madras, India.
S.G.WAYMENT	Utah State University, Logan, Utah, U.S.A.
J.H.WILLIAMSON	University of York, Heslington, York, England.
B.H.YOON	Sogang University, Seoul, Korea.

(*) indicates persons whose papers are included in this volume, but could not participate.

LECTURE PROGRAMME

Monday, 1st January, 1973

10.00 -	Registration of the participants. The registration continuous for the whole day.
16.00 -	Inanguration of the conference.

Tuesday, 2nd January, 1973

FIRST SESSION

Chairman : Alladi Ramakrishnan

9.00 - 9.50	I.J. Schoenberg, 'On cardinal spline interpolation'.
9.55 - 10.45	B.R.Gelbaum, 'Operator theory and Q-uniform Banach spaces'.

SECOND SESSION

Section A. Chairman : T.Husain

11.15 - 11.40	J.Schmets, 'Generalization of the theorem of Nachbin-Shirota about the spaces of continuous functions'.
11.45 - 12.10	C.A.Akemann, 'Non-abelian Pontryagin Duality'.
12.15 - 12.40	A.Pelczynski, 'Subspaces of L_p-spaces'.

Section B. Chairman : J.Gobert

11.15 - 11.40	K.Scherer, 'Approximation on compact homogeneous spaces'.
11.45 - 12.10	B.Ferguson, 'Müntz-Szasz theorem with integral coefficients'.
12.15 - 12.40	Ph.A.Dionne, 'Wave front sets and hypoellipticity'.

THIRD SESSION

Chairman : B.R.Gelbaum

15.10 - 16.00	A.Mallios, 'Topological algebras and several complex variables'.
16.30 - 17.20	L.D.Kudryavtsev, 'On properties of traces of functions belonging to weight spaces'.

Wednesday, 3rd January 1973

FOURTH SESSION

Chairman : J.P.Kahane

9.00 - 9.50 H.G.Garnir, 'The axiom of Solovay and functional analysis'.

9.55 - 10.45 B.Sz.-Nagy, 'A general view on the theory of unitary dilations of operators'.

FIFTH SESSION

Section A. Chairman : S.Igari

11.15 - 11.40 T.Husain, 'Quasicomplemented Banach spaces'.

11.45 - 12.10 S.G.Wayment, 'v-integrals in topological vector spaces'.

12.15 - 12.40 A.H.Siddiqi, 'Invariant means and convergence in non-Archimedian analysis'.

12.40 - 12.50 G.N.Keshava Murthy, 'On a class of functions satisfying Lipschitz condition'.

Section B. Chairman : P.Hess

11.15 - 11.40 J.P.Gossez, 'Non-linear elliptic boundary value problem in Orlicz - Sobolev spaces'.

11.45 - 12.10 J.Gobert, 'A priori inequalities for systems of partial differential equations'.

12.15 - 12.40 G.Freud, 'On weighted polynomial approximation'.

12.40 - 12.50 A.Sharma, 'Averaging interpolation'.

SIXTH SESSION

Chairman : B.Sz.- Nagy

15.10 - 16.00 L.Nirenberg, 'Remarks on linear partial differential equations'.

16.30 - 17.20 D.Gaier, 'Approximation of modules of quadrilaterals and of ring domains'.

Thursday, 4th January 1973

SEVENTH SESSION

Chairman : L.D.Kudryavtsev

9.00 - 9.50 J.P.Kahane, 'Some problems in metric projection'.

9.55 - 10.45 Bl.Sendov, 'Some problems in Hausdorff approximation'.

11.15 - 12.05 L.Collatz, 'Field approximation in partial differential equations'.

EIGHTH SESSION

Chairman : A.Mallios

12.15 - 13.05 J.H.Williamson, 'Isotone measures'.

13.10 - New and unsolved problems.

Afternoon : Excursion

<u>Friday, 5th January 1973</u>

NINTH SESSION

Chairman : H.G.Garnir

9.00 - 9.50 E.T.Poulsen, 'Quantization in Hamiltonian Particle mechanics'.

9.55 - 10.45 R.R.Goldberg, 'Segal algebras'.

TENTH SESSION

Section A. Chairman : R.R.Goldberg

11.15 - 11.40 J.G.Dhombres, 'Linear interpolation of functions on a compact space'.

11.45 - 12.10 G.N.Keshava Murthy, 'Multipliers on weighted spaces'

12.15 - 12.40 J.P.Ferrier, 'Inductive limits of Banach spaces and complex analysis'.

12.40 - 12.50 M.Montador, 'The spectrum of composition operators in $C[0,1]$'.

Section B. Chairman : G.Freud

11.15 - 11.40 P.Leonard, 'Fundamental solutions of hyperbolic differential equations'.

11.45 - 12.10 H.Johnen, 'Approximation by splines'.

12.15 - 12.40 Vimala Walter, 'Splines in Hilbert spaces'.

12.40 - 12.50 A.Geroux, 'Polynomials with a prescribed Zero'.

ELEVENTH SESSION

Chairman : L.Collatz

15.10 - 16.00 G.K.Kalisch, 'On unitary equivalence classes of certain normal operators and their unitary invariants'.

16.30 - 17.20 P.Hess, 'Non Coercive non linear boundary value problems and homotopy arguments'.

Saturday, 6th January 1973

TWEELFTH SESSION

Chairman : E.T.Poulsen

9.00 - 9.50 T.J.Rivlin, 'The Lebesgue constant for polynomial interpolation'.

9.55 - 10.45 S.Igari, 'On the (L^p, L^p) multipliers'.

THIRTEENTH SESSION

Chairman : T.J.Rivlin

11.15 - 12.05 M.M.Dzhrbashyan, 'Harmonic analysis in the complex domain and its application in the theory of infinitely differentiable functions'.

12.10 - 13.00 K.R.Unni, 'Multipliers of Segal algebras'.

Afternoon : Excursion

Sunday, 7th January 1973

FOURTEENTH SESSION

Section A. Chairman : J.H.Williamson

10.00 - 10.25 K.N.Gowrisankaran, 'Measurability of lattice of operations on a cone and an application to potential theory'.

10.30 - 10.55 R.P.Srivatsav, 'A Hilbert space technique for the solution of dual integral equations and dual series equations'.

11.25 - 11.35 T.R.Dhanapalan, 'Matrix Transformations of spaces of sequences in a Banach algebra'.

11.35 - 11.45 K.R.Sonavane, 'Generalized functions depending on a parameter'.

11.45 - 11.55 C.W.Mullins, 'Linear functionals on vector valued Köthe spaces'.

12.00 - 12.25 F.Macsudov.

Section B. Chairman : K.R.Unni

10.00 - 10.25 D.V.Pai, 'Approximation of vector valued functions'.

10.30 - 10.55 M.R.Subrahmanya, 'On simultaneous best approximation'.

11.00 - 11.45 R.N.Sen, 'On the iterative process of solving non-linear operator equations'.

11.45 - 12.00 P.K.Geetha, 'On Bernstein approximation problem'.

NON-ABELIAN PONTRYAGIN DUALITY

Charles A. Akemann

In this lecture I shall indicate how the Pontryagin duality theorem could be approached from a non-abelian point of view. That is, suppose your overall goal is to study general groups, but at some point along the way you wish to prove the Pontryagin duality theorem as an application of the general theory. There are undoubtedly many ways to do this, but since my own area of competence lies in the direction of C*-algebras, this point of view will predominate.

Let G be a locally compact group. I shall assume knowledge of Haar measure (which I take on the left), the convolution algebras $L^1(G)$ and $M(G)$ (the measure algebra), the basic properties of positive definite functions and their relation to the representation theory of $L^1(G)$ and G. This material can be found in Naimark's book [4] .

Now comes a somewhat larger pill to swallow. I shall assume that we have moved (selectively) through Eymard's thesis [3] so that the following notation and facts are established. The completion of $L^1(G)$ in its largest C*-algebra norm is called $C^*(G)$. The dual space of $C^*(G)$ (as a Banach space) is the space $B(G)$ which consists of all linear combinations of continuous positive definite functions on G, the positive linear functionals on $C^*(G)$ being precisely the positive definite functions in $B(G)$. Now $B(G)$ is an abelian Banach algebra with a closed ideal $A(G)$ generated by the positive definite functions with compact support. The spectrum of $A(G)$ is exactly G. This is perhaps the most important theorem in Eymard's thesis. It means precisely that every non-zero multiplicative linear functional on A(G) is given by evaluation at a point of G and that A(G) separates the point of G. Further if the set of such functionals is given the topology of point wise convergence on A(G), then it is homomorphic to G. Almost all the necessary material from C^*-algebra theory is also discussed by Eymard at the beginning of his thesis. The dual $W^*(G)$ of B(G) is a W*-algebra.

Now we are ready to look at the case in which G is abelian. Our goal, however, is to use this additional assumption only when it is necessary. When G is abelian, $C^*(G) = C_o(\Gamma)$, the continuous functions from the dual group Γ to the complex numbers which vanish at infinity [5, p. 9]. Also Γ is exactly the set of non-zero extreme points of the set $P(G)_1$ of positive definite functions in B(G) of norm less than or equal to one. We now borrow the proof from Rudin [5, p.10] that Γ is a locally compact abelian group with the topology of pointwise convergence on $C^*(G)$ coinciding with the compact open topology as functions on G. It is at this point that the duality theorem makes sense. There is now seen to be a natural group isomorphism from G into the group $\hat{\Gamma}$ of continuous characters on Γ. The theorem asserts that this map is onto and bicontinuous, where $\hat{\Gamma}$ has the compact open topology as functions on Γ.

To see that it is onto, let $x \in \hat{\Gamma}$. Since x is a continuous bounded function on Γ, we may consider x as a multiplier of $C^*(G) = C_o(\Gamma)$ and hence as lying in $W^*(G)$. Now G also lies in $W^*(G)$ under the natural injection which takes g into the linear functional on B(G) which takes $F \in B(G)$ into F(g). Other natural injections of G into $W^*(G)$ coincide with this one, and it is shown by Ernest [2] that G can be identified with its image in $W^*(G)$, the topology being that of pointwise convergence on $B(G)$ (or in fact on $A(G)$). Further as a subset of $W^*(G)$, G consists of unitary operators.

If we let z be the smallest central projection in $W^*(G)$ such that $F(z) = 1$ for every extreme point F of $P(G)_1$ (the existence and the properties of z have nothing to do with the fact that $C^*(G)$ comes from a group), then the map $M(C^*(G)) \longrightarrow zM(C^*(G))$ is a * isomorphism. Since $zM(C^*(G)) \cong C^b(\Gamma)$, the algebra of bounded continuous functions on Γ, we see that zx is unitary, so x is unitary. Further there is another central projection y in $W^*(G)$ such that $A(G) = \{ F \in B(G): F(yb) = 0 \text{ for all } b \in W^*(G)\}$. Thus x is non-zero on $A(G)$. Suppose we can now show that x is multiplicative on $B(G)$. Then by Eymard's theorem there would be a $g \in G$ with $g|_{A(G)} = x|_{A(G)}$. This implies $g = x$, since both are multiplicative.

Given any F and H in $P(G)_1$ of norm 1 we may approximate them in the weak* topology by convex combinations of elements of Γ. Since $x \in M(C^*(G))$, it follows from general C*-algebra theory that this approximation is valid for x as well. Since x is multiplicative on the elements of Γ, $x(FH) = x(F)\, x(H)$. Since all of $B(G)$ consists of linear combinations of elements of $P(G)_1$ of norm 1, we get that the mapping $G \longrightarrow \hat{\Gamma}$ is onto.

To see that G and $\hat{\Gamma}$ have the same topologies is straightforward. If $\{g_\alpha\} \subset G$ and $g_\alpha \longrightarrow g$, then for all $F \in L^1(G)$, $\| g_\alpha F - gF \|_1 \longrightarrow 0$, hence for all $a \in C^*(G)$, $\| g_\alpha a - ga \| \longrightarrow 0$. But if $K \subset \Gamma$ is compact there exists $a \in C^*(G) = C_0(\Gamma)$ with $a(r) = 1 \;\forall\, r \in K$. Thus $| g_\alpha(r) - g(r) | = | g_\alpha a(r) - ga(r) | \leq$ $\leq \| g_\alpha a - ga \| \longrightarrow 0$. For the other direction, by the result of

Ernest mentioned above, if $g_\alpha \longrightarrow g$ uniformly on compact subsets of Γ we need only show that $g_\alpha \longrightarrow g$ for each $F \in B(G)$. Since $G \subset M(C^*(G))$, this follows from general C^*-algebra theory by a standard argument using the Krein-Milman theorem.

References

1. Akemann, C.A. and Walter, M.E., Non-abelian Pontryagin duality, to appear in Duke Math. J.

2. Ernest, J., A new group algebra for locally compact groups, Amer. J. Math. 86 (1964), 467-492.

3. Eymard, P., L'algèbra de Fourier d'un groupe localement compact, Ball. Soc. Math. France. 92 (1964), 181-236.

4. Naimark, N.A., Normed Rings, P.Noordhoof, N.V.Groningen, Netherlands (1964).

5. Rudin, W., Fourier Analysis on Groups, Interscience, New York (1962).

THE TOPOLOGICAL DUAL, THE ALGEBRAIC DUAL AND RADON-NIKODYM DERIVATIVES

Richard A. Alò and Andre de Korvin

In this paper we will consider a Nikodym-type theorem for a Pettis-type integral. As compared with the generalized Nikodym theorem given in N. Dinculeanu [2], our query is how the result will be affected when this new integral is defined. In this respect for the Pettis-type integral, the derivative has values in a space of bounded linear operators, the operators now having their range in an appropriate algebraic dual as opposed to the topological dual given in [2]. This Pettis integral may be used to represent weakly compact operators (with weak c. a. property) on a space of continuous functions whose ranges are totally bounded subsets of some Banach space.

In beginning the discussion let us remark that a traditional approach to integration theory is to define the integral of a simple function and then to extend that integral by some limit process to a more general class of functions. Such an approach is followed in [2] and [3] for example. Other approaches have been taken to construct an integral. In [4], for example, the integral of a scalar valued function relative to an X-valued measure, where X is a locally convex space, is defined as follows.

Let Σ denote a σ-algebra of subsets of a non-empty set T and let m be a set function defined on Σ with values in the locally convex space X. A scalar-valued function f is called <u>m-integrable</u> if f is x^*m-integrable for all x^* in the dual, X^*, of X and if for each A in Σ there exists a $x \in X$ such that

$$x^*(x) = \int_A fd(x^*m).$$

In that case we write $x = \int_A fdm$.

Several convergence theorems are obtained in [4] for this integral. It is also used to obtain a representation theorem for weakly compact operators from the space $C(S)$ (all continuous real-valued functions on a compact Hausdorff space S) into X. Here two things utilized in the development should be stressed. Firstly, it was assumed that x^*m be countably additive. This implies of course that m is countably additive (see [3]). Secondly it was necessary to have S as a compact Hausdorff space.

More genrally one could define the integral of a function f relative to m (f not necessarily scalar valued) by requiring that for each A in Σ there exists an x in X such that $x^*(x)$ is a pre-assigned scalar defined via a traditional approach to integration. Such an approach is taken in [1].

In particular, let f now be a function from a set T into a Banach space E, let Σ denote a σ-ring of subsets of T and let m be a finitely additive set function from Σ into $L(E,F)$ (the set of bounded linear operators from E into another Banach space F). Let σ^* denote the unit sphere of F^* and let σ denote the unit sphere of E. For y^* in σ^*, define the function m_{y^*} from Σ into E^* by

$$\langle m_{y^*}(A), x \rangle = \langle y^*, m(A)x \rangle .$$

Let us assume that the semi-variation $\tilde{m}$ of m is finite and that the m_{y^*} are countably additive. Then the function f is said to be m-<u>integrable</u> if, for y^* in σ^*, f is m_{y^*}-integrable in the sense of [2], and if for each A in Σ there exists y in F such that

$$y^*(y) = \int_A f\,dm_{y^*}.$$

If f is m-integrable, we will denote y by $\int_A f\,dm$.

It should be noted immediately, that all simple functions are m-integrable.

In [5], m is defined to be <u>variationally semi-regular</u>, or v.s.r. if $\tilde{m}$ is finite and if for every decreasing sequence $(A_n)_{n\in N}$ of sets in Σ such that $\cap A_n = \emptyset$ the sequence $(\tilde{m}(A_n))_{n\in N}$ converges to 0.

In [5], conditions were studied which insured the variational semi-regularity of a set function. Under certain conditions it is shown that m is v.s.r. if and only if for y* in σ^*, m_{y^*} is countably additive .

With regard to the presently defined integral, it is now possible to develop an integration theory of E-valued functions relative to L(E,F)-valued measures. If H is a normal topological space and if $C_{TB}(H,E)$ represents all continuous functions on H with totally bounded range in E, then the integral may be used to represent certain weakly compact operators from $C_{TB}(H,E)$ into F (see [1]). This is rather different than the approach taken in [4]. For example, here, m may be finitely additive.

The specific weakly compact operators considered above are those with the weak c.a. property (see [1]). It develops that if H is a compact Hausdorff space then every weakly compact operator has the weak c.a. property. More generally for a normal space H, it appears that this property is closely related to the property of representing linear functionals on some space of real-valued functions by a countably additive measure.

Some convergence theorems have been developed in [1] for this Pettis-type integral presently being defined. The property of m being variationally semi-regular is the right one needed to obtain these theorems. In the present work the concept also plays an important role as shown by the Lemma, the theorem and our remarks at the end. Loosely speaking it means that the semi-variation is continuous from above.

Let us recall that a map ρ from $\mathcal{L}^{\infty}(\gamma)$ into $\mathcal{L}^{\infty}(\gamma)$ is called a <u>linear lifting</u> if ρ is a linear map, $\rho(f) \sim f$, $f \sim g$ implies $\rho(f) = \rho(g)$, $f \geq 0$ implies $\rho(f) \geq 0$ and $f = \alpha$ (α a constant) implies $\rho(f) = \alpha$.

Some facts on linear liftings may be found in [2]. An important result is that if γ has the direct sum property and if $\gamma \neq 0$ then $\mathcal{L}^{\infty}(\gamma)$ always has a linear lifting.

As stated above a Nikodym-type theorem will be obtained for the m-integral presently being defined. Again as stated above, the main difference between our theorem and the generalized Nikodym theorem given in [2], is that for the latter the operator had values in $L(E,F^{**})$, whereas here the operator has values in $L(E,F^{*A})$ where F^{*A} denotes the algebraic dual of F^*. Some technical details contained in [1] will be used. In particular, the general setting is that m is absolutely continuous with respect to the positive measure γ and that γ satisfies some growth condition relative to m, that is, as $\{m(A_n)\}_{n\in N}$ converges to 0, $\{\gamma(A_n)\}_{n\in N}$ may not converge to 0 too fast. The functions to which the theorem applies are scalar-valued and in some sense are limits of a uniformly integrable family.

<u>The Nikodym Theorem</u>

In this section we will maintain x as an element in the unit sphere σ of E, γ as a positive finite measure on the σ-ring Σ and m as a measure from Σ into $L(E,F)$, where E is assumed to be of countable type (see [2]),

First a convergence property for m-integrability is needed.

LEMMA. <u>Assume</u> $\{f_n\}_{n\in N}$ <u>is a sequence of scalar-valued m-integrable functions defined on the Banach space</u> E <u>such that</u> $\{f_n\}_{n\in N}$ <u>converges a.e. (relative to m) to the function</u> f. <u>Assume also that</u>

(1) <u>there exists a scalar-valued function</u> g <u>defined on</u> E <u>which is</u> m_{y^*} <u>integrable for all</u> $y^* \in \sigma^*$ <u>and</u> $|f_n| < g$ <u>for all</u> n <u>in</u> N,

(2) <u>the measure</u> m <u>is</u> v.s.r.,

(3) <u>for every</u> $\varepsilon > 0$ <u>there exists</u> $\delta > 0$ <u>such that</u> $\tilde{m}(A) < \delta$ <u>implies</u> $\int_A |f_n| \overline{dm_{y^*}} < \varepsilon$ <u>for all</u> n <u>in</u> N.

<u>Then</u> f <u>is m-integrable and</u> $\{\int_A f_n \cdot x\, dm\}_{n\in N}$ <u>converges to</u> $\int_A f.x dm$ <u>uniformly for</u> $x \in \sigma$.

<u>Proof</u>. The function f is m_{y^*}- integrable by the classical dominated convergence theorem. So let $E_n = \{ t \in T : |f_n(t) - f(t)| \geq \varepsilon \}$. Since the functions f_n converge to f a.e., by [2] it follows that the sequence $(\tilde{m}(E_n))_{n \in N}$ converges to 0 . Then

$$| \langle \int_A f\, dm_{y^*}, x \rangle - y^* \int_A f_n \cdot x\, dm | \leq$$

$$\leq | \langle \int_{A-E_n} f\, dm_{y^*}, x \rangle - y^* \int_{A-E_n} f_n \cdot x\, dm | +$$

$$+ | \langle \int_{A\cap E_n} f\, dm_{y^*}, x \rangle - y^* \int_{A \cap E_n} f_n \cdot x\, dm | .$$

The first term of the second member is dominated by $\varepsilon \overline{m_{y^*}}(A-E_n)$ which in turn is dominated by $\varepsilon \tilde{m}(A)$. Now

$$| \int_{A\cap E_n} (f - f_n)\, dm_{y^*} | \leq \liminf_m \int_{A\cap E_n} |f_m - f_n| \overline{dm_{y^*}} .$$

By picking n large enough so that $\tilde{m}(A \cap E_n) \leq \delta$, the second term is dominated by 2ε . Thus $\int_A f_n \cdot x\, dm$ converges in F to some element, say $y_{x,A}$ and it is seen

$$\langle \int_A f dm_{y^*}, x \rangle = y^*(y_{x,A}).$$

Thus

$$y_{x,A} = \int_A fxdm.$$

THEOREM. *Assume the measure* m *is absolutely continuous with respect to* γ *and that the* m_{y*} *are measures of finite variation* $\overline{m_{y*}}$. *Assume moreover that there exists for each* y^* *in* σ^* *a function* φ_{y*} *from the positive integers into the positive integers such that for every sequence* $(A_n)_{n\in N}$ *of sets in* Σ, *for which* $\overline{m_{y*}}(A_n)$ *converges to* 0,

$$\overline{m_{y*}}(A_n) \leq \varphi_{y*}(n)\, \gamma\,(A_n)$$

for $n \geqslant n_o$ (n_o *depends on the sequence* $(A_n)_{n\in N}$). *Let* φ *be a real-valued function such that:*

(1) *There exists a sequence of simple functions* $(s_n)_{n\in N}$ *converging to* φ (γ a.e.).

(2) *There is a* γ*-integrable function* g *for which* $|s_n| \leq g$, n *in* N.

(3) *For every* $\varepsilon > 0$ *there exists a* $\delta > 0$ *such that* $\tilde{m}(A) < \delta$ *implies* $\int_A |s_n|\, \overline{dm_{y*}} < \varepsilon$ *for all* n *in* N *and* $y^*\in\sigma^*$.

Assume also that m *is* v.s.r.

Then there exists a mapping U *from* T *into* $L(E,F^{*A})$ *such that*

$$\langle \int_A \varphi\cdot xdm, y^* \rangle = \int_A \varphi \langle Ux, y^* \rangle\, d\gamma\ .$$

(The integral on the left is in the sense of [1]. On the right $\varphi\langle Ux,y^*\rangle$ denotes the pointwise product of two functions.)

Proof. If m is 0 then pick U = 0 and the result holds. Thus let us assume that $m \neq 0$ and without loss of generality we may assume that φ_{y*} is increasing. Now the $\|m(A)\| = \sup_{y^*\in\sigma^*} \|m_{y*}(A)\|$, so m_{y*} is absolutely continuous with respect to γ. Since m_{y*} is E^*-valued, by page 271 of [2] one obtains a γ-integrable function f_{y*} mapping T into E^* such that

$$m_{y*}(A) = \int_A f_{y*} d\gamma \quad \text{for } A\in\Sigma.$$

Moreover

$$\overline{m}_{y*}(A) = \int_A |f_{y*}| d\gamma .$$

For convenience let us write $m_{y*} = f_{y*}\gamma$. Then $\overline{m}_{y*} = |f_{y*}|\gamma$.

We now show that $\langle x, f_{y*}\rangle \in \mathcal{L}^{\infty}(\gamma)$. Assume that $\langle x, f_{y*}\rangle$ does not belong to $\mathcal{L}^{\infty}(\gamma)$ and let $A_n = \{t : |\langle x, f_{y*}(t)\rangle| > n\}$. Then clearly $\gamma(A_n) > 0$. Also $\bigcap_{n=1}^{\infty} A_n$ has zero measure since $\langle x, f_{y*}\rangle$ is finite. Thus $\{\gamma(A_n)\}_{n\in N}$ converges to 0. Let us pick a B_2 in Σ, $B_2 \subset A_1$ such that $|\langle x, f_{y*}\rangle| \geq \frac{\varphi_{y*}(2)}{\gamma(A_1)}$ on B_2. Since it is assumed that $\langle x, f_{y*}\rangle$ does not belong to $\mathcal{L}^{\infty}(\gamma)$, it follows that $\gamma(B_2) > 0$. Now by induction pick B_{n+1} in Σ, $B_{n+1} \subset B_n$ such that on B_{n+1}, $|\langle x, f_{y*}\rangle| \geq \frac{\varphi_{y*}(n+1)}{\gamma(B_n)}$. Then $B_2 - B_3$, $B_3 - B_4, \ldots$, and $B_{n-1} - B_n$ are disjoint sets in Σ. Since m_{y*} has finite variation, $\{\overline{m_{y*}}(B_{n-1} - B_n)\}_{n\in N}$ converges to 0. Also

$$m_{y*}(B_{n-1} - B_n) \geq \int_{B_{n-1} - B_n} |\langle x, f_{y*}\rangle| \, d\gamma \geq \frac{\varphi_{y*}(n-1)}{\gamma(B_{n-2})} \gamma(B_{n-1} - B_n)$$

for all n. This contradicts the growth condition, $\overline{m_{y*}}(A_n) \leq \varphi_{y*}(n) \cdot \gamma(A_n)$ for $n \geq n_0$. Thus $\langle x, f_{y*}\rangle$ is an element of $\mathcal{L}^{\infty}(\gamma)$.

Now we shall construct the map U from T into $L(E,F^{*A})$ where F^{*A} denotes the algebraic dual of F^{*} Let ρ be a linear lifting of $\mathcal{L}^{\infty}(\gamma)$and define $\bar{\rho}$ by

$$\langle \bar{\rho}(f_{y*}), x \rangle = \rho \langle f_{y*}, x \rangle .$$

Now the maps $y^* \to m_{y*}$ and $y^* \to \rho\langle f_{y*}, x\rangle$, are linear. Let $\pi_{x,t}$ denote the map $y^* \to \rho\langle f_{y*}(t), x\rangle$ where t is fixed. Let U(t) denote the map $x \to \pi_{x,t}$ where again t is fixed. Then U(t) is a linear map from E into F^{*A}. Thus U maps T into $L(E,F^{*A})$.

If f is an m-integrable function in the sense of [1], then

$$\langle \int_A f dm, y^* \rangle = \int_A f dm_{y*}$$

and $m_{y*} = f_{y*}\gamma$. By page 169 of [2] it is nec... sary to check that f is m-integrable with respect to $|f_{y*}||\gamma| = |f_{y*}| = m_{y*}$. This is clearly the case by the very definition of m-integrability. So

$$\int_A f dm_{y*} = \int_A \langle f_{y*}, f \rangle d\gamma .$$

Now consider remark 3, page 212 of [2]. For a fixed y^* in σ^* $\langle f_{y*}, x \rangle \sim \langle \bar{\rho}(f_{y*}), x \rangle$, so $f_{y*} \sim \bar{\rho}(f_{y*})$ (for E is assumed to be of countable type). Thus

$$\int_A \langle \bar{\rho}(f_{y*}), f \rangle d\gamma = \int_A \langle f_{y*}, f \rangle d\gamma .$$

Now let φ be a function satisfying (1), (2) and (3) of the theorem. The argument used to show that $\langle x, f_{y*}\rangle$ belongs to $\mathcal{L}^{\infty}(\gamma)$. Then for every simple function s

$$\langle \int_A s{\cdot}x dm, y^* \rangle = \int_A \langle U(s x), y^* \rangle d\gamma .$$

Indeed consider the case when s is a characteristic function $s = \chi_B$,

$$\langle \int_A \chi_B . x dm, y^* \rangle = \langle y^*, m(A \cap B) x \rangle$$

(see [1]). On the other hand

$$\int_A \langle U(\chi_B \cdot x), y^* \rangle \, d\gamma = \int_{A \cap B} \langle Ux, y^* \rangle \, d\gamma$$

$$= \int_{A \cap B} \langle f_{y^*}, x \rangle \, d\gamma = \int_{A \cap B} \langle f_{y^*}, x \rangle \, d\gamma$$

$$= \langle \int_{A \cap B} f_{y^*} \, d\gamma, x \rangle = \langle m_{y^*}(A \cap B), x \rangle .$$

This extends by linear combination to the desired equality.

Since $\| f_{y^*} \|$ belongs to $\mathcal{L}^{\infty}(\gamma)$ there exists a constant, depending on y*, say K_{y^*} such that $\overline{m}_{y^*} < K_{y^*}\gamma$. Now statement (1) implies that the sequence $(s_n)_{n \in N}$ converges to φ a.e. (relative to m). The statement (2) implies that g is m_{y^*}-integrable. Since $\varphi . x$ is m-integrable, the above lemma then implies that the sequence of terms $\langle \int_A s_n . x dm, y^* \rangle$, for n in N, converges to the term $\langle \int_A \varphi \cdot x dm, y^* \rangle$.

By the classical dominated convergence theorem then, the sequence of terms $\int_A s_n \langle Ux, y^* \rangle \, d\gamma$ converges to $\int_A \varphi \langle Ux, y^* \rangle \, d\gamma$. This proves the theorem.

Remarks.

(1) Assume there exists a function H from the positive integers into the positive integers such that $\tilde{m}(A_n) \leq H(n) \, \gamma \, (A_n)$ for $n \geq n_o$ whenever $(A_n)_{n \in N}$ is a sequence of disjoint sets in Σ. Since $m(A) = \sup_{y^* \in \sigma^*} | m_{y^*}(A) |$, one has the existence of the φ_{y^*}.

(2) If $\sum_{i \in N} m_{y_i^*}(F_i) < \infty$ for every sequence $\{y_i^*\}_{i \in N}$ in σ^* and sequence $\{F_i\}_{i \in N}$ of disjoint sets in Σ, it is shown in [1] that m is v.s.r.

(3) Assume m is dominated by γ , that is, for some constant K, $|m(A)| < K\gamma(A)$ for all $A \in \Sigma$. Then there exists a function U from T into L(E,F**) with the stated properties. One notes then that $\|f_{y^*}\| < K$ for all y* in σ^* . The proof proceeds as above but this time $\|\pi_{x,t}\| \leq K$. Thus U maps T into L(E,F**). In the notations of [2] this corresponds to the case $m_\infty(T) < \infty$.

(4) Conditions (1), (2) and (3) may be interpreted as stating that in some sense φ is the limit of a uniformly integrable family.

(5) In the scalar case a relation between m-integrability and Bochner integrability is pointed out in [4]. Specifically f is m-integrable if and only if there exists a sequence $\{f_n\}_{n \in N}$ of bounded measurable functions which converge pointwise to f such that for A in Σ, $\{\int_A f_n dm\}_{n \in N}$ is a Cauchy sequence. The following proposition is valid when the functions are E-valued.

PROPOSITION.

(1) *If f is m-integrable and if $|f|$ is m_{y^*}-integrable* ($y^* \in \sigma^*$) *then f is the pointwise limit of bounded measurable functions* f_n *such that for every A in Σ and for every y*,* $\{\int_A f_n dm_{y^*}\}_{n \in N}$ *is a Cauchy sequence*.

(2) *If there exists a sequence of simple functions* $\{f_n\}_{n \in N}$ *which converge to f pointwise such that for A in Σ ,* $\{\int_A f_n dm\}_{n \in N}$ *is a Cauchy sequence and such that for every* $y^* \in \sigma^*$, $\{\int_A f_n dm_{y^*}\}_{n \in N}$ *converges to* $\int_A f dm_{y^*}$, *then f is m-integrable*.

Proof. Assuming (1), it is clear that f is m_{y^*}-integrable. Thus let $f_n = f\chi_{E_n}$ where χ_{E_n} is the characteristic function of $E_n = \{t : |f(t)| \leq n\}$. The rest follows. Assuming (2), let $y_A = \lim \int_A f_n dm$. Then $y^*(y_A) = \lim \int_A f_n dm_{y^*} = \int_A f dm_{y^*}$. Thus f is m-integrable.

Bibliography

1 de Korvin, A. and Kunes, L., 'Some non-weak integrals defined by linear functionals', (submitted).

2 Dinculeanu, N., Vector Measures, Pergamon Press, Berlin, 1967.

3 Dunford, N. and Schwartz, J., Linear operators, Vol. 1, Interscience Publishers, Inc., New York, 1958.

4 Lewis, D.,'Integration with respect to vector measures', Pacific Journal of Mathematics 33 (1970), 157-165.

5 Lewis, P., Vector measures and topology (submitted).

The authors are indebted to N.Dinculeanu for his many helpful suggestions and comments. The work of the first author was done as a recipient of a Scaife Research Grant.

SEGAL ALGEBRAS AND DENSE IDEALS IN BANACH ALGEBRAS

James T. Burnham

I should like very much to thank Professor K.R.Unni, and other members of the Institute, for inviting me to this conference and affording me the opportunity to present this paper. Also, it gives me great pleasure to record here my thanks to Professor Hans Reiter for words of encouragement, some years ago during a time at which I had nearly abandoned mathematics. Furthermore, Professor Reiter's published works are directly responsible for the course of my own investigations.

As a matter of fact the things I have chosen to talk about today have for their origin some results announced by Hans Reiter in 1965 [38].

The notion of Segal algebra, introduced by Professor Reiter in 1965 is now fairly familiar to mathematicians working in modern harmonic analysis. The idea of introducing *via* axioms classes of Banach algebras that have properties common to a 'well known' Banach algebra is not new. However in many respects, Segal algebras and their generalizations appear to belong to a class all their own. The (historical) origin of Segal algebras is deeply rooted in classical analysis. Indeed, as is the case with many of the seminal ideas of harmonic analysis, Norbert Wiener defined and studied the 'first' Segal algebra [52] , [53] . Subsequently, T.Carleman [10] looked at Wiener's algebra (along with a great deal of Wiener's work in general) from a fresh point of view. As noted, and expounded, by Reiter in his book [39] , Carleman's 'fresh point of view' permeates the whole of modern Fourier analysis. This particular example, Wiener's algebra, of a Segal algebra, all continuous f on the line $\mathbb{R}$ for which $\sum_{q \in \mathbb{Z}} \max_{x \in [q, q+1]} |f(x)|$ is finite, has been studied by a number of mathematicians: [1], [10], [12], [14], [18], [21], [22], [27], [37], [38], [44], [48], [51], [52], [53], [2], to cite a few. Wiener's algebra is evidently a very popular Banach algebra.

There is no question that I.Segal [44, theorem 3.1] recognised a general algebraic structure underlying Wiener's algebra. However, Segal did not exploit this algebraic structure in a systematic fashion. It remained for Reiter [38] (and subsequently [39], [40]) to recognise the true significance of Segal's 'axioms' and their relationship to analysis on locally compact groups. Presumably Reiter was in possession of his ideal theorem (theorem 3 below) independently of the analogous result in [31] where a specialised version of the ideal theorem (for the algebras $A_p(G) = \{f \in L^1 G \mid \hat{f} \in L^p(\hat{G})\}$) was presented. The reason for this presumption is because Segal [44] had already observed the 'ideal theorem' for certain closed ideals (in algebras satisfying the hypotheses of theorem 3.1 of [44]) and with some experience it would not have been difficult to 'guess' the general result.

The Segal algebras $L^1(G) \cap L^p(G)$, $1 \leq p < \infty$ (G not necessarily abelian) have been around for some years. Indeed, as early as 1944 K.Iwasawa [23] studied $L^1(G) \cap L^p(G)$ and proposed that $L^1(G) \cap L^p(G)$ (with the norm $\|f\| = \mathrm{MAX}(\|f\|_1, \|f\|_p)$) be called the group algebra of G. (This is in contrast with Segal [43] terming $L^1(G)$ with a formally adjoined identity the group algebra of G). Iwasawa[1] [23]

1 The contents of this paper indicate quite clearly that Iwasawa was in possession of Gelfand's theory as announced in the early 1940's. International situations of these times makes Iwasawa's knowledge of Gelfand's work of some historical interest. In general Iwasawa's paper is of interest both for its mathematical content and facts concerning the early history of group representations and harmonic analysis.

worked out many properties of $L^1(G) \cap L^p(G)$ including representation theory thereby extending many of the results obtained by Segal [43]. Subsequently, though the authors appear to be unaware of Iwasawa's paper [23], a couple of papers [33], [54] (and [50] which deals with $L^1(G) \cap L^2(G)$) have appeared treating various properties of $L^1(G) \cap L^p(G)$ for LCAG's.

Another class of Segal algebras, $A_p(G) = \{f \in L^1(G) \mid \hat{f} \in L^p(\hat{G})\}$, one for each $p \in [1,\infty)$ were introduced and subject to systematic analysis by Larsen, Liu and Wang [30]. It was in this paper, [30] that the full statement (specialized of course) of Reiter's ideal theorem first appeared[1]. In 1969 H-C Lai [28] corrected an error in the proof of Reiter's ideal theorem for $A_p(G)$ appearing in [28] and observed the consequence that every closed primary ideal of $A_p(G)$ is maximal. It appears as though Lai was not aware of Iwasawa's paper [23] or Reiter's work [38], [39] during the course of his research on $A_p(G)$. Some interesting properties of both $A_p(G)$ and $L^1(G) \cap L^p(G)$ are studied by Lai in [29][2].

---

1. For compact (not necessarily abelian) G, Loomis [34, p.161] sketches the details of a bijection between the closed ideals of $L^1(G)$ and those of (the Segal algebra) $L^2(G)$. (Though the full content of Reiter's ideal theorem is not written down). This observation was repeated (almost verbatum) by R.L.Lipeman in his Yale lecture notes [32]. Loomis indicates that the same result(s) hold for the $L^p(G)$ algebras.

2. For $1 < p < 2$ and $1/p + 1/q = 1$ Lai shows that $L^1(G) \cap L^p(G)$ is a dense set of the 1st category in $A_q(G)$ (G is of course non-discrete) . As a matter of fact one can prove: If $S_i(G), i = 1,2$ are commutative Segal algebras on the non-discrete l.c. G with $S_1(G) \subseteq S_2(G)$ then $S_1(G)$ is an $S_2(G)$-Segal algebra (definition 12 of the present paper). Furthermore, either $S_1(G) = S_2(G)$ or $S_1(G)$ is a dense set of the 1st category in $S_2(G)$.

Generalizations of Segal algebras were given by Cigler (1969) [7] , and Burnham (1972)[1] [4] . I was lucky enough to extend Segal algebras to a class of general Banach algebras and find a proof of Reiter's ideal theorem that applied at once to non-commutative Segal algebras. In this talk I shall describe yet a further generalization of Segal algebra.

Intended as a generalization of classical Fourier series it should be noted that G.Shilov (1951) [47] introduced a class of Banach algebras (of functions) on compact abelian groups called Homogeneous Banach spaces. Shilov's paper has subsequently been translated into English[2]. H.Mirkil presents a summary of the essential results of Shilov's work on Homogeneous Banach spaces in the last chapter of [36] . Generalizations of Shilov's work to arbitrary compact groups are given by De Leeuw (1958) [30] , one might care to look at Kitchen's paper [26]. Undoutedly influenced by Shilov's work on homogeneous Banach spaces, Katznelson, in his 1968 text book [25] on harmonic analysis carried out much of the classical Fourier series theory known for $L^1(T)$, [T = circle] on

---

1. As a matter of fact the details of [4] were worked out late in 1968 and rougn notes (c.f. the bibliography in Reiter [40]) were (privately) circulated in 1969 and then in revised form in 1970. My notes we prepared on the basis of Reiter's paper [38] without knowledge of either [11] or [39] . Professor L.Y.H.Yap kindly brought Reiter's book [39] to my attention.

2. In view of the number of Russian papers on Banach algebras that have appeared in English translation it is a mystery to me why Ditkin's paper [12] has not yet been 'officially' translated into English.

general homogeneous Banach spaces[1]. Katznelson indicated (presenting selected results and suggested exercises) that homogeneous Banach spaces can be defined on the line $\mathbb{R}$ and even more generally on LCAG's. (Katznelson goes a step further and defines homogeneous Banach spaces on $\mathbb{R}$ that are $L^1(\mathbb{R})$ -modules which are <u>not</u> necessarily subalgebras (even subsets) of $L^1(\mathbb{R})$). R.E.Edwards [15], independently of Katznelson, exploited the common properties shared by the classical Banach algebras on the circle, $C(T)$, $L^p(T)$, $1 \leq p < \infty$ and $L^\infty(T)$ with the weak topology, and presented much of the classical Fourier series theory for these algebras by studying a generic algebra. For general infinite compact groups, Edwards has carried the results of [15] to their full generality in a recent book [16]. The recent text of Butzer and Nessel [9][2] follows the trend set down by Katznelson and Edwards. However, amongst these authors, only Katznelson works things out explicitly for abstractly defined algebras. By the way, the main difference, as observed by H-C Wang [49], between homogeneous Banach spaces, as subalgebras of L^1, and Segal algebras is that the former need not be dense in L^1.

---

1. It appears as though Katznelson [25] discovered the Banach module factorization theorem (specialized to homogeneous Banach spaces on the circle) quite independently of other mathematicians (e.g. [22], and the appropriate references therein) working on the problem. Katznelson isolated the essential ingredient (convex sequences and their intimate relationship to Fourier-cosine series) used by R.Salem [42] in his proof of the relation $L^1(T) * C(T) = C(T)$. Edwards [15,1.5.1] has given the same result with the same proof specialized to the <u>classical</u> homogeneous Banach spaces: $C(T)$, $L^p(T)$, $1 < p < \infty$. Evidently Edwards was unaware of Katznelson's discovery.

2. The authors promise a second volume intending to deal with more 'abstract matters'.

With respect to the general Segal algebras (definition 12 below) introduced by the author in [4] there is a closely related (and historically earlier) concept, this is the notion of <u>normed ideal</u> as introduced by J. von Neumann and R.Schatten - see [45] and [46]. (The normed ideals of J.Cigler [11] had for their origin the work of von Neumann and Schatten). These normed ideals were used by Schatten in his investigations of certain problems in the theory of Hilbert space - more precisely, the theory of certain linear operators acting on Hilbert space. There is a vast Russian literature on this topic for which I refer to the monograph of Gohberg and Krein [17] and the references therein. One can also consult the treatise of Dunford and Schwartz [13] (especially section 6 of chapter 6) for relevant analysis on these matters.

If one consults some of the references we have thus far cited it will become apparent that the general Segal algebras of [4] include a variety of objects occurring in analysis. There is, to be sure, much work yet to be done. In the course of the skeleton analysis to follow I shall point out a few new results. My main interest is in the closed ideal theory of (generalized) Segal algebras and I will show that in a certain sense, among the various types of subalgebras of Banach algebras, Reiter's ideal theorem obtains only for Segal algebras. I conclude with an extension of the ideal theorem presented in [4] .

We begin our analysis.

Unless something to the contrary is explicitly stated, G denotes a locally compact Hausdorff group equipped with a left Haar measure dm. To avoid trivialities we further suppose that G is non-discrete; for the only Segal algebra on a discrete group is $L^1(G)$ - see [5] . For a complex valued function f on G, the map $y \to f_y$ is denoted by $\mathcal{L}_f$, (for fixed f we have $\mathcal{L}_f(y) = f_y$) where f_y is defined by $f_y(x) = f(y^{-1}x)$. Note that $\mathcal{L}_f^y$ is defined on G whereas $L_y f = f_y$ (L_y for fixed $y \in G$) is a function with a set of functions as domain. $\mathcal{L}_f$ and L_y are, of course, related but they do not have 'all' properties in common. Indeed, (take the domain of L_y to be $L^{\infty}(G)$ for infinite compact G) L_y can be continuous while $\mathcal{L}_f$ fails to be continuous. As usual, the Lebesgue spaces (with respect to dm) are denoted by $L^p(G)$, $1 \leq p < \infty$ with norms $\| \ \|_p$. For p = 1 (all p in case G is compact) $L^1(G)$ is, of course, the Banach convolution algebra of absolutely dm - integrable functions in G. The Banach space of (Borel) measurable essentially bounded functions is denoted by $L^{\infty}(G)$ with norm $\| \ \|_{\infty}$. For $1 \leq p < \infty$ ($p = \infty$) we adopt the usual practice of identifying functions that are equal almost everywhere (locally almost everywhere). The Banach convolution algebra of bounded Radon measures on G is denoted by M(G) with the total variation norm $\| \ \|_M$. Finally, $C_o(G)$ is the (pointwise)[1] Banach sup-norm algebra of continuous functions on G that vanish at infinity.

We recall the definition of Segal algebra.

Definition1. Reiter [40]. The subalgebra S(G) of $L^1(G)$ is a left Segal algebra in case

S1. S(G) is dense in $L^1(G)$.

S2. S(G) is a Banach space with respect to a norm $\| \ \|_S$.

S3. $L_y f \in S(G)$ for all $y \in G$ and all $f \in S(G)$.

S4. $\| L_y f \|_S = \| f \|_S$ for all $y \in G$ and all $f \in S(G)$.

S5. $\mathcal{L}_f$ is continuous from G to $(S(G), \| \ \|_S)$ for each $f \in S(G)$.

S6. There is a constant $C > 0$ so that $\|f\|_1 \leq C \| f \|_S$ for all $f \in S(G)$.

---

1. As it turns out $C_o(G)$ has a convolution structure. Indeed $C_o(G)$ is but a special case of a Banach (convolution) $L^1(G)$-module.

Remarks. We may, and do, take C = 1 in S6 without any loss of generality. There is, of course, an analogous definition for right Segal algebras. Shilov [47] constructs interesting examples of algebras where S3 holds but S4 and S5 fail, S3, S4 hold but S5 fails. As we have noted, an easy example illustrating that S3 and S4 need not imply S5 is $L^{\infty}(G)$ for infinite compact G.

The fundamental 'working definition' of left Segal algebras is summarized in

THEOREM 2. (**Reiter** [39], [40], **Cigler** [11]). If S(G) is a left Segal algebra then S(G) is a left ideal in M(G), $(S(G), \| \|_S)$ is a Banach algebra and $\| \mu * f \|_S \leq \| \mu \|_M \| f \|_S$ for all $\mu \in M(G)$ and all $f \in S(G)$. In particular, S(G) is a left ideal in $L^1(G)$ and $\| f * g \|_S \leq \| f \|_1 \| g \|_S$ for all $f \in L^1(G)$ and all $g \in S(G)$.

Henceforth I'll drop the adjective 'left' and simply say that S(G) is a Segal algebra. By the way, in case S(G) is right[1] 'translation invariant' (i.e., S3, S4 and S5 hold for right as well as left translations) then the Segal algebra S(G) is said to be symmetric. We refer to Reiter [40] for this concept as well as other species of Segal algebras. A Segal algebra need not be symmetric. A Segal algebra stable under the natural involution for $L^1(G)$ is said to be star symmetric [40]. As B.E.Johnson [24] has shown even for G = $\mathbb{R}$, Segal algebras may fail to be star symmetric. Johnson's example is this: fix $f \in L^1(\mathbb{R})$ and set $S_f = \{ g \in L^1(\mathbb{R}) \mid f * g \in C_0(\mathbb{R}) \}$ with norm $\| g \|_1 + \| f * g \|_{\infty}$.

---

1. The right translation operators $R_y f, R_f(y)$ are given by the common value $f(xy^{-1}) \Delta_G(y^{-1})$ where Δ_G is the modular function for G.

Reiter's ideal theorem entails that the closed left ideal theory of any Segal algebra is the same as that of $L^1(G)$. Indeed,

THEOREM 3. (Burnham [4], Reiter [40]). Let $S(G)$ be a Segal algebra. Then every closed left ideal I_S of $S(G)$ is of the form $I \cap S(G)$ where I is a unique closed left of $L^1(G)$. Indeed, I is the closure in $L^1(G)$ of I_S. Conversely, if I is a closed left ideal of $L^1(G)$, then $I \cap S(G)$ is a closed left ideal of $S(G)$. Thus the map $I_S \to$ closure of I_S in $L^1(G)$ is a bijection between the closed left ideals of $S(G)$ and the closed left ideals of $L^1(G)$.

Some Observations. For the present discussion 'ideal' means left ideal.

(1) We first note that if $(B, \| \|_B)$ is any normed subalgebra of the Banach algebra $(A, \| \|_A)$ with $\|f\|_A \leq C\|f\|_B$ and J is a closed ideal of A then $J \cap B$ is a closed ideal of B.

(2) Clearly $I \subseteq I^A \cap B$ where I^A is the closure in A of the closed ideal, I, of B (A and B are as in (1)).

(3) As a matter of fact the more substantial assertions of Theorem 3 are:

(i) I^{L^1} is an ideal of L^1; (ii) the inclusion $I^{L^1} \cap S \subseteq I$ and (iii) uniqueness.

(4) Concerning uniqueness we need only establish the following: Let J be a closed ideal of $L^1(G)$ and set $J_S = J \cap S$. Then J is the closure of J_S in $L^1(G)$. As a close examination of the verification of this assertion given by Reiter [39, (1), (ii) p.40] reveals the full defining characteristics of Segal algebras comes to the front.

(5) The inequality $\|f * g\|_S \leq \|f\|_1 \|g\|_S$ plays a crucial role in the proof of (3) (i) and (ii). The existence of an approximate identity is used (and is crucial) to establish both (ii) and (iii). See the details in [4].

The inequality $\|f*g\|_S \leq \|f\|_1 \|g\|_S$ plays an active role in the analysis of Segal algebras and their generalizations. As it turns out, this inequality can only be satisfied by Banach subalgebras[1] of $L^1(G)$. A more precise statement follows.

THEOREM 4. *Let* $(A, \|\ \|_A)$ *be a Banach algebra. Suppose* B *is a subalgebra of* A *that is a Banach space with respect to a norm* $\|\ \|_B$. *Furthermore, suppose there exists* $M > 0$ *so that* $\|f\|_A \leq M\|f\|_B$ *for all* $f \in B$. *If there exists* $K > 0$ *so that* $\|fg\|_B \leq K\|f\|_A \|g\|_B$ *for all* $f, g \in B$, *then* B *is a left ideal in the* A-*closure of* B.

Proof. Let f belong to the closure of B in A. Choose a sequence $\langle f_n \rangle$ in B with $\|f_n - f\|_A \to 0$. By hypothesis

$\|f_n g - f_m g\|_B \leq K\|f_n - f_m\|_A \|g\|_B$ for all $g \in B$ since $f_n - f_m \in B$.

Thus $\langle f_n g \rangle$ is Cauchy in B so $\exists$ $h \in B$ with $\|f_n g - h\|_B \to 0$. Since $f_n g - h \in B$ we have $\|f_n g - h\|_A \leq M\|f_n g - h\|_B$ so that $f_n g \to h$ in A. But A is a Banach algebra so $f_n g \to fg$ in A and hence $fg = h \in B$. Thus, B is a left ideal in its A-closure as was to be shown.

Here is a 'comparison' to Theorem 4. Actually we present a slight restatement of Theorem F appearing in [8].

THEOREM 5. **Burnham and Goldberg [8]**. *Let* $(B, \|\ \|_B)$ *be a Banach algebra which is left ideal in the* A-*closure of* B *in the Banach algebra* $(A, \|\ \|_A)$. *If there exists* $C > 0$ *so that* $\|f\|_A \leq C\|f\|_B$ *for all* $f \in B$ *then there exists* $K > 0$ *so that* $\|fg\|_B \leq K\|f\|_A \|g\|_B$ *for all* $g \in B$ *and all* f *in the* A-*closure of* B.

---

1. We are ignoring the general theory of Banach $L^1(G)$-modules.

The arguments involved in verifying the observations following the statement of Theorem 3 together with the contents of Theorem 4 and 5 entail the full content of Theorem 3 can only hold for ideals (rather than just subalgebras) of $L^1(G)$.

Putting the preceding discussion together we can answer, in part (though in more generality) a 'project' suggested by Reiter [39, p.138].

THEOREM 6. *Let $(A, \| \|_A)$ be a Banach algebra. Let $(B, \| \|_B)$ be a Banach algebra with B a subalgebra of A. If B contains an approximate identity of A, and $\exists\, C > 0$ so that $\|f\|_A \le C \|f\|_B (f \in B)$ then the map $I \to I^A$ is a bijection between the closed ideals of B and the closed ideals of A if and only if B is a dense ideal of A.*

Proof. If B is a dense ideal of A then the theorem follows from Theorem 5 and Theorem 13 (our generalized ideal theorem of Reiter). On the other hand I^A can only be an ideal in A provided B is a dense ideal of A (this can be shown based on the techniques in our proof of Theorem 4).

Specializing in Reiter's project we have in particular,

COROLLARY 7. *If $L^1_\omega(G)$ is a proper Beurling algebra in $L^1(G)$ then the map $I \to I^{L^1}$ is not surjective.*

Proof. No proper Beurling algebra is an ideal of $L^1(G)$ - [4].

Remark. As illustrated by Definition 12 and Theorem 13, Corollary 7 does not preclude the analysis of subalgebras of $L^1_\omega(G)$ where now $L^1_\omega(G)$ plays the role of $L^1(G)$.

As a matter of fact, Reiter's ideal theorem provides rather complete information as regards the structure of Segal algebras on compact groups. Indeed, all of section 2.12 pages 129-137 in Edwards little book [16] applies at once to any Segal algebra on an infinite compact group[1]. To verify this latter assertion one needs to make use of the representation theory of compact groups. Incidentally, Hans Reiter [40] has made use of the representation theory of compact groups in his analysis of the structure of closed ideals in Segal algebras on compact groups.

It is interesting to note that as early as 1944, Iwasawa [23] had generalized the then embryonic representation theory of $L^1(G)$ (due to I.Segal [43]) in terms of the Segal algebras $L^1(G) \cap L^p(G)$, $1 \leq p \leq \infty$ for possibly non-commutative and non-compact G! Iwasawa noted that certain properties of $L^1(G)$ could be deduced from those of $L^1(G) \cap L^p(G)$ and vice-versa. As a matter of fact this seems to be a general feature of Segal algebras that is apparently not shared by other classes of subalgebras of $L^1(G)$. However, we have already noted that the theory is not entirely symmetrical. (i.e. the example we cited due to B.E.Johnson). For further comment on this point of view we refer to Reiter's lecture notes [40].

Closely related to Segal algebras are J.Cigler's normed ideals [11].

Definition 8. The subalgebra N(G) of $L^1(G)$ is a normed ideal in case

N1. N(G) is a dense left ideal of $L^1(G)$

N2. N(G) is a Banach space with respect to a norm $\| \|_N$.

N3. There is a constant $M > 0$ so that $\|f\|_1 \leq M \|f\|_N$ $(f \in N(G))$.

N4. There is a constant $C > 0$ so that $\|f * g\|_N \leq C \|f\|_1 \|g\|_N$ for all $f \in L^1(G)$ and all $g \in N(G)$.

We see at once that a normed ideal is a Banach algebra. The main difference between normed ideals and Segal algebras is the

---

1. We hope to elaborate on this matter in a future communication.

possible absence of a continuous translation in the former. Indeed, $L^{\infty}(G)$(G infinite compact) is a normed ideal but $L^{\infty}(G)$ is _not_ a Segal algebra. As it turns out, the ideal theorem (Theorem 13) is not valid for general normed ideals. Again $L^{\infty}(G)$ serves as an example. For C(G) is a closed ideal of $L^1(G)$ which is _not_ the intersection with $L^{\infty}(G)$ of any closed ideal of $L^1(G)$. As the ideal theorem is valid for a large class of Banach algebras (Theorem 13 below) where the notion of translation may have no meaning we must look a bit closer at $L^{\infty}(G)$ to see why the ideal theorem fails. The culprit is that $L^{\infty}(G)$ does _not_ have an approximate identity! It is not difficult to show that a normed ideal has a _continuous translation_ if and only if it has an approximate identity. This is an equivalent characterization of Cigler's characterization of _commutative_ normed ideals [11] : N(G) is a Segal algebra if and only if the ideal of all $L^1(G)$ functions with compactly supported Fourier transforms is dense in N(G). The 'approximate identity' assertion can be made to resemble a non-commutative version (infinite compact G) of Cigler's characterization of commutative normed ideals. The details have for their ingredients approximate identities made up of _central_ $L^1(G)$ functions together with the 'completeness' of coordinate functions in every _Segal_ algebra on a compact group.

J.Cigler [11] raises the interesting question: Can the inequality N3 be proved, instead of assumed, if instead one assumes in N2 that N(G) is a Banach algebra? In [8] Richard Goldberg and I gave an affirmative answer to Cigler's question within the more general context of A-Segal algebras (Definition 12 below). The solution R.Goldberg and I presented takes on added significance in view of the following comments. J.Cigler [11] provides an affirmative answer to his problem under the added assumption that N(G) is stable under the natural involution of $L^1(G)$. However, we have already noted B.E.Johnson's example of a <u>commutative</u> Segal algebra (and so a normed ideal) that fails to be stable under the involution of $L^1(G)$.

Our solution in [8] to Cigler's problem together with Theorem 5 (also in [8]) as stated in the present paper are in fact results asserting the continuity (in an appropriate topology) of certain linear operators ('the' inclusion map in fact). Here is another 'continuity' result whose proof is similar to the one R.Goldberg and I gave for Theorem D in [8]. We don't present things in maximum generality but we do include a proof.

<u>THEOREM 9</u>. <u>Let</u> $(N(G), \|\ \|_N)$ <u>be a normed ideal. Let</u> T <u>be a linear operator on</u> $L^1(G)$ <u>into</u> $L^1(G)$. <u>If there exists</u> $K > 0$ <u>so that the restriction</u>, T^N, <u>of</u> T <u>to</u> N(G) <u>satisfies</u> $\|T^N f\|_1 \leq K\|f\|_N$ (<u>in particular</u>[1], <u>by</u> N3, <u>if</u> $\|T^N f\|_N \leq K\|f\|_N$) <u>then there exists</u> $K' > 0$ <u>so that</u> $\|Tf\|_1 \leq K'\|f\|_1$ <u>for all</u> $f \in L^1(G)$ <u>and hence</u> T <u>is a continuous operator on</u> $L^1(G)$.

<u>Proof</u>. Let $g \in L^1(G)$, $\langle g_n \rangle$ a sequence in N(G) with $\|g_n - g\|_1 \to 0$. Furthermore, suppose $Tg_n \to h$ in $L^1(G)$. By the Closed Graph Theorem it suffices to show that $Tg = h$. Let $f \in N(G)$ and note that

$$\|(Tg - h) * f\|_1 \leq \|T\{(g - g_n) * f\}\|_1 + \|(Tg_n - h) * f\|_1 . \tag{1}$$

But $(g-g_n) * f \in N(G)$ since N(G) is a left ideal of $L^1(G)$ and by hypothesis, $\|T\{(g-g_n) * f\}\|_1 \leq K\|(g-g_n)*f\|_N$. Also, $\|g - g_n * f\|_N \leq \|g-g_n\|_1 \|f\|_N$ and $\|(Tg_n - h) * f\|_1 \leq \|Tg_n - h\|_1 \|f\|_1$. These estimates together with (1) obtain

$$\|(Tg - h) * f\|_1 \leq K\|g - g_n\|_1 \|f\|_N + \|Tg_n - h\|_1 . \|f\|_1 .$$

---

1. Of course here we must have $TN(G) \subseteq N(G)$.

Thus $(Tg-h) * f = 0$ a.e. and so, [8,C] $Tg-h = 0$ a.e. Hence, $Tg = h$ a.e. and the proof is complete.

Note that Theorem 9 does not entail that bounded operators on N(G) are bounded on $L^1(G)$. The Hilbert (transform) distribution defined on $N(G) = L^2(G)$, G = circle, illustrates the hypothesis that T be defined on all of $L^1(G)$ is essential.

We note, for example, we can deduce the continuity of $f \to f_y$ on $L^1(G)$ from the corresponding continuity property on any proper Segal algebra.

We hope to treat consequences of Theorem 9 (with applications to representation theory) in some detail in a future communication. We state here, without proof, a selected sample of results where Theorem 9 fits in.

THEOREM 10. *Let* G *be a locally compact group. Let* $T: L^1(G) \to L^1(G)$ *be linear and further suppose that* T *commutes with left translations. If* $TI \subseteq I$ *for every dense ideal of* $L^1(G)$ *then* T *is bounded on* $L^1(G)$.

Idea of proof. As T fixed all dense ideals, in particular $T(L^1(G) \cap L^p(G)) \subseteq L^1(G) \cap L^p(G)$ for all $p \in [1,\infty)$. But (a classical result for G = circle) it can be shown that any linear operator on $L^1(G) \cap L^2(G)$ commuting with translations is bounded (with respect to $\| \ \|_1 + \| \ \|_2$). Now apply Theorem 9.

With a little analysis Theorem 9, teamed up with 3, can be used to obtain

THEOREM 11. *Let* G *be an infinite compact group. Let* T *be a linear operator on* $L^1(G)$ *that leaves invariant all closed left ideals of* $L^1(G)$. *Then, the restriction of* T *to any Segal algebra* S(G) *leaves invariant the closed left ideals of* S(G). *Furthermore* *the restriction of T to S(G) is S(G)-continuous and hence T is continuous on all of $L^1(G)$.*

By the way, it is not known whether or not a linear operator commuting with translations on $L^1(G)$ is necessarily bounded for arbitrary locally compact G.

We now take up some 'generalizations'. The types of algebras we have been discussing thus far are but special cases of the following class of algebras.

Definition 12. Let $(A, \| \|_A)$ be a Banach algebra. The subalgebra B of A is an A-Segal algebra in case

A1. B is a dense left ideal of A

A2. B is a Banach space with respect to a norm $\| \|_B$.

A3. $\|f\|_A \leq C \|f\|_B$ for all $f \in B$.

A4. $\|fg\|_B \leq K\|f\|_A \|g\|_B$ for all $f, g \in B$.

Many of the fundamental properties of A-Segal algebras have been worked out in some detail in [4], [5], [8]. In particular, [4], we have a generalization of Reiter's ideal theorem.

THEOREM 13. Let $(B, \| \|_B)$ be an A-Segal algebra. If $(B, \| \|_B)$ has a (weak) right approximate identity, then

(1) If J is a closed left ideal in A, then $J \cap B$ is a closed left ideal in B.

(2) If I is a closed left ideal in B, then I^A (= closure of I in A) is a closed left ideal in A and $I = I^A \cap B$.

As a consequence of a step in the proof of (2) we further have

(3) If I is a closed left ideal in B, then I is a left ideal in A.

The 'observations' following the statement of Theorem 3 apply here. As a matter of fact, as a little thought illustrates,

Theorem 13 is <u>not</u> dependent on a <u>Banach</u> algebra structure. The crucial property is the fact that 'multiplication' is <u>continuous</u> from AxB to B in an appropriate product topology. We shall consider <u>only</u> locally convex Fréchet topological algebras. This a fixed assumption and topological entails these conditions.

<u>Definition 14</u>. Let $(A, \mathcal{T}_A)$ be a topological algebra. The subalgebra B of A is an A-Segal algebra in case

A1. B is a dense left ideal of A.

A2. B is a topological algebra with respect to some topology $\mathcal{T}_B$.

A3. The identity map from $(B, \mathcal{T}_B)$ to $(B, \mathcal{T}_A)$ is continuous

A4. Multiplication is continuous from AxB into B with respect to the $\mathcal{T}_A \times \mathcal{T}_B$ product topology.

We say that a topological algebra $(X, \mathcal{T}_X)$ has a right approximate unit in case for every $x \in X$ and every neighbourhood $\mathcal{V}_0$ of zero in X there is an $e(x) \in X$ with $xe(x) \in \{x\} + \mathcal{V}_0$.

Happily enough, with only trifling changes in the proof, the ideal theorem (Theorem 13) carries over verbatim for this new class of A-Segal algebras . Reiter constructed his ideal theorem (following the announcement in [38]) for a class of <u>commutative</u> algebras called Wiener algebras [39,2.4 p.22 then 2.5 pp.129-130] . These are considerably more general than commutative <u>Banach</u> algebras so that even though our ideal theorem vastly generalizes Reiter's result specialized to Segal algebras, it does not contain his Wiener algebra result. However, the generalization to the class of algebras covered by Definition 14, we have just mentioned completely generalizes Reiter's theorem. To the preceding observations we add that many of the results of [5] have valid analogues for the A-Segal algebras of Definition 14.

I should like to conclude by presenting a few examples and stating some (apparently) open problems. There are many well known examples of Segal algebras but here are a few 'new' ones. The first two examples are due to Richard R. Goldberg (personal communication).

Example 15. Let G = circle group and define $S(T) = \{f \in L^1(T) \mid \|f - D_N * f\|_1 \to 0\}$ with norm $\|f\|_S = \sup_N \|D_N * f\|_1$. Here D_N is the Dirichlet kernel of order N. The slightly larger algebra (with the same norm) $N(T) = \{f \in L^1(T) \mid \sup_N \|D_N * f\|_1 < \infty\}$ is a normed ideal which is not a Segal algebra. (These examples are readily generalized to $G = \mathbb{R}$ = real line).

Example 16. $S(T) = \{f \in L^1(T) \mid \hat{f}(n) = o(1/\ell n(|n|))\}$ with norm $\|f\|_S = \|f\|_1 + \max_{1 \leq n \leq \infty} |\ell n(n)\hat{f}(n)|$ is a Segal algebra.

The next two examples are taken from [7].

Example 17. Let G be a locally compact abelian group supporting a measure $\mu \in M(G)$ with $\hat{\mu}$ free of zeros on $\hat{G}$. Furthermore, suppose μ^{-1} does not exist in M(G). $S(G) = \{\mu * f \mid f \in L^1(G)\}$ with norm $\|\mu * f\|_S = \|f\|_1 + \|\hat{\mu}\hat{f}\|_\infty$ is a Segal algebra.

Example 18. Let G be a locally compact abelian group. Let $S_i(G)$, i = 1,2, be Segal algebras. Let T be an (S_1, S_2) multiplier with $\hat{T}$ free of zeros on $\hat{G}$. Then $S(G) = T(S_1(G))$ is an $S_2(G)$-Segal algebra. $T(S_1(G))$ has an approximate identity and so the ideal theorem (Theorem 13) applies here.

Example 19. R.Larsen [30]. Let $(A, \|\ \|_A)$ be a commutative Banach algebra with maximal ideal space $\mathfrak{M}$. Let μ be a positive regular Borel measure on $\mathfrak{M}$. $A_p(\mu) = \{a \in A \mid \hat{a} \in L^P(\mathfrak{M}, \mu)\}$ is an A-Segal algebra with norm $\|a\|_{A_p} = \|a\|_A + \|\hat{a}\|_p$.

If $(A, \| \|_A)$ has an approximate identity then so does $(A_p, \| \|_A)$ and hence Theorem 13 is applicable.

Example 20. Let A be the ideal of all compact operators on some (separable) Hilbert space. Let $B \subseteq A$ be the ideal of Hilbert-Schmidt operators with the Hilbert-Schmidt norm. Then B is an A-Segal algebra (A is equipped with the operator norm). Furthermore the ideal of operators of finite rank constitute an approximate identity for B, so Theorem 13 is applicable. See Schatten [46].

Example 21. Let $(B, \| \|_B)$ be an A-Segal algebra. $B^2 = \{a.b \mid a,b \in B\}$ is a B-Segal algebra with norm $\|h\|_{B^2} = \mathrm{INF}\{\|a\|_B \|b\|_B \mid \text{all } a,b \text{ with } h = a.b\}$. Applying this to example 20, B^2 turns out to be the trace class which is a proper subset of B. Furthermore the operators of finite rank are an approximate identity for B^2 so Theorem 13 is applicable.

Example 22. See Edwards [16, 2.14.2] and the references to Hewitt and Ross [22] for discussion and definitions. Let G be an infinite compact group. For $1 \le p \le \infty$, $A_p(G) = \{f \in L^1(G) \mid \hat{f} \in L^p(\hat{G})\}$ with norm $\|f\|_{A_p} = \|f\|_1 + \|\hat{f}\|_{L^p}$ is a Segal algebra.

Here are a couple of examples of A-Segal algebras as described by Definition 14.

Example 23. Reiter [39]. Let G be a locally compact abelian group. $S(G) = \{f \in L^1(G) \mid f \in L^p(G)$ for all $p \in (1,\infty)\}$ is an $L^1(G)$-Segal algebra with the topology induced by the family of norms $\| \|_1 + \| \|_p$, $1 < p < \infty$. Here S(G) has an approximate identity and so the generalization of Theorem 13 alluded to in the paragraph following the statement of Definition 14 is applicable. By the way the results of [5] carry over to this setting.

Incidentally, L. Yap [55] has vastly generalized Example 23 and noted the validity of the 'ideal' theorem.

__Example 24__. Reiter [39]. Let $G = \mathbb{R}$. Let $S(\mathbb{R})$ be the algebra of all $f \in L^1(\mathbb{R})$ such that f is indefinitely differentiable and $f^{(n)} \in L^1(\mathbb{R})$ for all $n \geq 1$, with the topology defined by the norms $\|f\|_1 + \|f^{(n)}\|_1$. $S(\mathbb{R})$ with this topology is an $L^1(\mathbb{R})$-Segal algebra. The remarks stated in Example 23 apply equally as well here.

The following 'framework' (see Edwards [16, 2.13.2-3 pp.138-143]) provides one with numerous capabilities of constructing A-Segal algebras as covered by Definition 14. Let S be an index set and a family of algebras $\{A_s\}_{s \in S}$ given. Form the product algebra PA_s. One selects various subalgebras B of PA_s and by appropriately constructed topologies for B, B becomes PA_s-Segal algebras.

__Some problems.__

(1) What makes a dense (proper) ideal of a Banach algebra a Banach algebra in its own right?

(2) In Example 21 there appears an example of a Banach algebra B with $B^2 \subsetneq B$. Most of the known Segal algebras $S(G)$ $(\subseteq L^1(G))$ have the property $S^2 \subsetneq S$, [6], [35], [49], [54], [55]. Does every Segal algebra fail to factor?

(3) Determine the multiplier theory of Segal algebras. See [7] and the references therein.

(4) Can 'Sidonicity' be characterized in terms of __arbitrary__ Segal algebras on compact groups? The only result I know of is summarized by K.A.Ross in [41], see reference 1 therein - here also a particular type of Beurling algebra appears.

(5) What Segal algebras on the circle have the 'almost everywhere convergence property'? The only results (other than obvious ones) I know of are the famous results of Carleson ($L^2(T)$) and Hunt ($L^p(T)$).

(6) Let G be a locally compact (non-discrete) abelian group. Let $M_o(G)$ be the closed ideal of measures in $M(G)$ whose Fourier-Stieltjes transforms vanish at infinity. Are there any non-trivial $M_o(G)$-Segal

algebras?

(7) What is the structure (in terms of equivalence relations and idempotents) of closed subalgebras in Segal algebras on the circle? See the discussion in Edwards [15,11.3,pp.11-17].

(8) Let S(G) be a Segal algebra on the locally compact abelian group G. Let $\mu,\eta \in M(G)$. Suppose $\mu * S(G) \subseteq \eta * S(G)$. What is the relationship between μ and η? In case $S = L^1$, then it can be shown (Brainerd and Edwards [3], Glicksberg [20]) that $\exists\, \lambda \in M(G)$ so that $\mu = \eta * \lambda$. By the way, this latter result can be used to give a simple proof of J.Williamson's 'Kawada into problem' [56] for locally compact abelian G. That is, if T is a bipositive ($Tf \geq 0 \Leftrightarrow f \geq 0$) multiplier of $L^1(G)$ then $T = \varepsilon_a$ = point mass supported at some $a \in G$. In particular the range of T is all of $L^1(G)$.

I should like to express my gratitude to the Science Research Council (United Kingdom) for supporting portions of the research presented herein and for complete support of the preparation of this paper.

References

1. Bochner, S., Uber factorfolgen für Fouriersche reihen, Acta. Sci. Math. (Szeged) 4 (1928) 125-129.
2. Bourbaki, N., Elements de Mathematique Fasc. 32 theories Spectrales, Hermann, Paris 1967.
3. Brainerd, B. and Edwards, R.E., Linear operators which commute with translations, I and II. J. Austr. Math. Soc. VI (1966) 289-350.
4. Burnham, J.T., Closed ideals in subalgebras of Banach algebras I, Proc. Amer. Math. Soc. 32 (1972) 551-555.
5. Burnham, J.T., Closed ideals in subalgebras of Banach algebras II: Ditkin's condition, to appear in Monatsh. für Math.
6. Burnham, J.T., Nonfactorization in Subsets of the measure algebra, Proc. Amer. Math. Soc. 35 (1972) 104-106.
7. Burnham, J.T., Topics in the multiplier theory of commutative Segal algebras, to be submitted for publication.
8. Burnham, J.T. and Goldberg, Richard, R., Basic properties of Segal algebras, to appear in J. for Math. Analysis and Applications.
9. Butzer, P.L. and Nessel, R.J., Fourier Analysis and Approximation, Vol. I : One-Dimensional theory, Academic Press, New York 1971.
10. Carleman, T. L'Integral de Fourier et questions qui s'y rattachent, Inst. Mittag-Leffler Publ. Scient., Uppsala 1944 (Lectures given at the Mittag-Leffler Institute in 1935).
11. Cigler, J., Normed ideas in $L^1(G)$, Nederl. Akad. Wetensch. Proc. Ser. A 72 = Indag. Math. 31 (1969) 273-282.
12. Ditkin, V.A., Study of the structure of ideals in certain normed rings, Ucenye Zapiski Moskov. Gos. Unive. Matematika 30 (1939) 83-130 (in Russian).

13. Dunford, N. and Schwartz, J.T., Linear Operators, Part II: Spectral theory, self adjoint operators in Hilbert Space, Interscience Publishers, Inc., New York, 1963.

14. Edwards, R.E., Comments on Wiener's Tauberian theorems, J. London Math. Soc. 33 (1958) 462-466.

15. Edwards, R.E., Fourier Series: A Modern Introduction, Vols. I and II, Holt Reinhart and Winston, New York, 1967.

16. Edwards, R.E., Integration and Harmonic Analysis on Compact Groups, London Mathematical Society Lecture Notes Series 8, Cambridge University Press, 1972.

17. Gohberg, I.C. and Krein, M.G., Introduction to the theory of linear nonselfadjoint operators, Translations of Math. Monographs Vol.18 Amer. Math. Soc. 1969.

18. Goldberg, R.R., On a space of functions of Wiener, Duke Math. J. 34 (1967) 683-691.

20. Glicksberg, I., When is $\mu * L$ closed, Trans. Amer. Math. Soc. 1972 .

21. Hardy, G.H., Divergent Series, Oxford University Press, New York, 1949.

22. Hewitt, E. and Ross, K.A., Abstract Harmonic Analysis, Vol.II: Structure and Analysis for Compact Groups and Locally Compact Abelian Groups, Springer-Verlag: Berlin-Gottingen-Heidelberg 1970.

23. Iwasawa, K., On group rings of topological rings, Proc. Imp. Acad. Japan, Tokyo 20 (1944) 67-70.

24. Johnson, B.E., Some examples in harmonic analysis Preprint 1972.

25. Katznelson, Y., An Introduction to Harmonic Analysis, John Wiley and Sons, Inc., 1968.

26. Kitchen, J.W. Jr., Normed modules and almost periodicity, Monatsh. Für Math. 70 (1966) 233-243.

27. Korenblyum, B.I., On certain special commutative normed rings, Doklady Akad. Nauk, SSSR (N.W.) 64 (1949) 281-287 (In Russian).

28. Lai, H-C., On some properties of $A^p(G)$-algebras, Proc. Japan Acad. 45 (1969) 572-576.

29. Lai, H-C., On the category of $L^1(G) \cap L^p(G)$ in $A^q(G)$, Proc. Japan Acad. 45 (1969) 577-581.

30. Larsen, R., Liu , T-S. and Wang, J-K., On functions with Fourier transforms in Lp, Michigan Math. J. 11 (1964) 369-378.

31. de Leeuw, K., Homogeneous algebras on compact groups, Trans. Amer. Math. Soc. 87 (1958) 372-386.

32. Lipsman, R., Abstract Harmonic Analysis, Yale University, 1968.

33. Liu, T-S., Sums and intersections of normed linear spaces, Math. Nachr. 42 (1969) 29-42.

34. Loomis, L., Abstract Harmonic Analysis, D. Van Nostrand and Company 1952.

35. Martin, J.C. and Yap, L.Y.H., The algebra of functions with Fourier transforms in L^P, Proc. Amer. Math. Soc. 24 (1970) 217-219.

36. Mirkil, H., The Work of Shilov on Commutative Semi-Simple Banach Algebras, Fasciculo Pelo Instituto de Mathematical Pura e Aplicada, Rio de Janeiro, 1966.

37. Pitt, H.R., Tauberian Theorems, Oxford University Press 1958.

38. Reiter, H., Subalgebras of $L^1(G)$, Nederl. Akad. Wetensch. Proc. Ser. A 68 (1965) 691-696.

39. Reiter, H., Classical Harmonic Analysis and Locally Compact Groups, Oxford 1968.

40. Reiter, H., L^1-Algebras and Segal Algebras, Lecture Notes in Mathematics, No. 231 Springer-Verlag: Berlin-Heidelberg-New York 1971.

41. Ross, K.A., Some new characterizations of Sidon sets (for details consult reference 1 therein) in Conference on Harmonic Analysis, Lecture Notes in Mathematics, No.266 Springer-Verlag, Berlin-Heidelberg-New York 1972.

42. Salem, R., Sur les transformations des séries de Fourier, Fund. Math. 33 (1939).

43. Segal, I., The group ring of a locally compact group I, Proc. Nat. Acad. Sci., U.S.A. 27 (1940) 348-352.

44. Segal, I., The group algebra of a locally compact group, Trans. Amer. Math. Soc. 61 (1947) 69-105.

45. Schatten, R., A Theory of Cross Spaces, Ann. of Math. Stud. No.26, Princeton University Press, 1950.

46. Schatten, R., Norm Ideals of Completely Continuous Operators, Springer-Verlag: Berlin-Gottingen-Heidelberg, 1960.

47. Shilov, G.E., Homogeneous rings of functions, Uspehi Mat. Nauk. (N.S.) 41 (1951) 91-137 (In Russian) English translation: Translations of the Amer. Math. Soc. Series 1 Vol.8 (1962) 392-455.

48. Wang, J-K., Lectures on Banach Algebras, Lecture notes, Department of Mathematics, Yale University, 1965.

49. Wang, H-C., Non-factorization in Group Algebras, Ph.D. thesis, The University of Iowa, Iowa City, Iowa, 1971. To appear in revised form in Studia Math.

50. Warner, C.R., Closed ideals in the group algebra $L^1(G) \cap L^2(G)$, Trans. Amer. Math. Soc. 121 (1966) 408-423.

51. Widder, D.V., The Laplace Transform, Princeton University Press, Princeton, 1941.

52. Wiener, N., Tauberian Theorems, MIT Press 1964, 143-242 (Reprinting of Wiener's 1932 Acta. Paper).

53. Wiener, N., The Fourier Integral and Certain of Its Applications, Cambridge University Press 1933.

54. Yap, L.Y.H., Ideals in subalgebras of the group algebras, Studia Math. 35 (1970) 165-175.

55. Yap, L.Y.H., Non-factorization in Fréchet subalgebras of $L^1(G)$, Preprint 1971.

56. Williamson, J.H., On theorems of Kawada and Wendel, Proc. Edinburgh Math. Soc. 11 (1958-59).

FIELD APPROXIMATION AND FREE APPROXIMATION FOR DIFFERENTIAL EQUATIONS

L. Collatz

Summary

In applying approximation theory to differential and integral equations one gets very often unusual types of approximation problems even in simple cases, especially one gets free approximation problems and field approximation of different levels.

1. Introduction

For linear boundary value problems one can often use the boundary maximum principle (shortly:maximum principle) and the monotonicity principle (shortly: monotonicity). If both principles are applicable, the maximum principle is easier to handle than the monotonicity and gives a Tschebyscheff Approximation (T.A.) problem. The monotonicity gives an onesided Tschebyscheff Approximation (T_1.A.) problem and has great advantages compared with the maximum principle. The monotonicity gives in many cases better bounds and is applicable often in nonlinear cases (Collatz [71] , [73]). Therefore let us consider which types of approximation problems occur by applying the Monotonicity.

Let B be a given closed domain in the $(x_1,\dots,x_n)$-space R^n and $\mathscr{C}(B)$ the linear space of continuous functions $h(x)$ defined in B, where x is the vector $x = (x_1,\dots,x_n)$. Let $f(x)$ be a given element of $\mathscr{C}(B)$ and $W = \{w(x,a) = w(x_1,\dots,x_n,a_1,\dots,a_p)\}$ a given class of functions of $\mathscr{C}(B)$, which depends on a parameter vector $a = (a_1,\dots,a_p)$. Then we can start the optimization problem

$$-\delta_1 \leq w(x,a) - f(x) \leq \delta_2 \;, \text{ for all } x \in B,\ \delta_1 \geq 0 \;,\ \delta_2 \geq 0,$$
$$Q = \delta_1 + \delta_2 = \text{Infimum}$$

for the unknowns δ_1 , δ_2, $a_1,\dots,a_p$ with infinitely many restrictions (compare Krabs [71], Lempio [72], Collatz-Wetterling [71]). We have the T.A. for $\delta_1 = \delta_2$ and the T_1.A. from above for $\delta_1 = 0$, and the T_1.A. from below for $\delta_2 = 0$.

2. Free Approximation

We consider the linear boundary value problem for a function y(x) in the interval J = [-1, 1]:

$$Ly = -y'' - \frac{1+x^2}{2} y = 0, \quad y(\pm 1) = 1. \tag{2.1}$$

We can use the iteration procedure

$$v_{n+1} = T v_n, \quad n = 0, 1, 2, \ldots \tag{2.2}$$

where the operator T is defined by

$$Tg = h \text{ is equivalent to } \quad -h'' = \frac{1+x^2}{2} g, \quad h(\pm 1) = 1.$$

Starting with

$$v_0(x) = 1 + \sum_{\nu=1}^{p} a_\nu (x^{2\nu-2} - x^{2\nu}) \tag{2.3}$$

which satisfies the boundary conditions for all (real) parameters a_ν, one gets

$$v_1(x) = \frac{37 - 12x^2 - x^4}{24} + \sum_{\nu=1}^{p} \left[\frac{a_\nu}{2} \frac{(1-x^{2\nu})}{(2\nu-1)2\nu} + \frac{x^{2\nu+4} - 1}{(2\nu+3)(2\nu+4)} \right] \tag{2.4}$$

For instance, let us ask for a lower bound for the solution y(x). If $v_1 \geq v_0$ in J, this has the consequence $y \geq v_1$ in J. But y is not known and we have the optimization

$$0 \leq v_1 - v_0 \leq \delta, \qquad \delta = \text{Min}. \tag{2.5}$$

(2.5) describes a typical situation in boundary value problem and we introduce for this type of approximations the name "Free Approximation Problem" (abbreviated by "Free Appr. Pr."). We interpret the situation geometrically : To every

$$v_0 = v_0(x, a_1, \dots, a_p) \tag{2.6}$$

is associated a certain

$$v_1(x, a_1, \dots, a_p) . \tag{2.7}$$

Only the parameter vectors $a = (a_1, \dots, a_p)$ are admitted, for which $v_1 \geqslant v_0$ in J. With every admissible parameter vector a distance $\delta(a_1, \dots, a_p)$ is associated; the minimum of these distances is sought; fig.1.

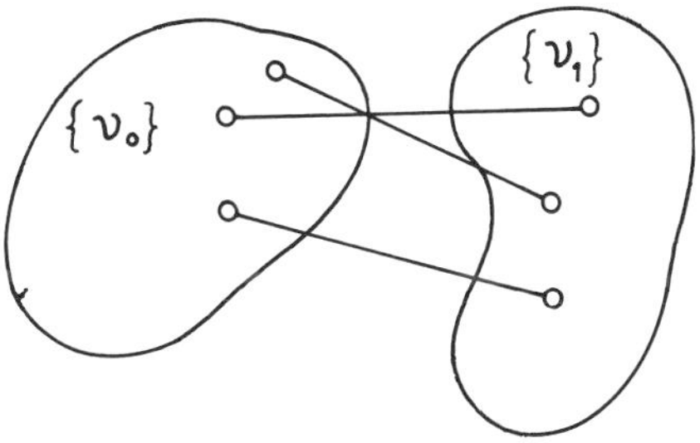

FIG. 1

Of course, this free approximation problem can be considered as a special case of the generalized onesided T_1.A.

Another complication, compared with the classical T.A. or T_1.A. is the fact, that the parameters a_ν occur at several places, and that this cannot be avoided by using a parameter transformation. We call this a "Chained Approximation". This phenomenon occurs also often in applying approximation theory to boundary value problems.

3. Field Approximation of stage R.

Let us consider the nonlinear boundary value problem for a function $u(x_1, x_2)$ in the circle (distribution of temperature u in a plate with influence of chemical reactions)

$$B = \{(x_1, x_2) , r^2 = x_1^2 + x_2^2 \leq 1\} \tag{3.1}$$

$$-\sum_{j=1}^{2} \frac{\partial^2 u}{\partial x_j^2} = -\Delta u = 1 + r^2 + \frac{1}{2} e^u \quad \text{in} \quad r < 1, \qquad (3.2)$$

$$u = 0 \quad \text{on} \quad r = 1 .$$

We use the iteration procedure (2.2) with

$$-\Delta v_{n+1} = 1 + r^2 + \frac{1}{2} \exp(v_n) \quad \text{in} \quad r < 1, \; v_{n+1} = 0 \quad \text{on } r = 1.$$

Calculation with only one parameter a gives

$$v_1 = (\tfrac{5}{8} - 4a)(1 - r^2) + a(1 - r^4)$$

$$v_0 = \ell_n [3 - 32a + (32a - 2) r^2] .$$

Again we have with (2.5) chained Free Appr. Pr. and similarly for more parameters $a_1, \ldots, a_p$.

For instance we get for $a = \frac{1}{16}$

$$v_0 = 0, \qquad v_1 = \frac{1}{16}(7 - 6r^2 - r^4) .$$

Analogously we have Free Approximation Pr. from above by substituting a, v_0, v_1 by b, w_0, w_1. We get the best Free Approximation from above with

$$b = 0.0423, \qquad w_0 = \ell_n (1.646 - 0.646 r^2),$$

$$w_1 = 0.4558(1 - r^2) + 0.0423(1 - r^4) .$$

Let us mention, that here the theory of Nr. 4 is applicable; the conditions (4.3) are satisfied and therefore exists a solution u of (3.2) in the strip $v_1 \leq u \leq w_1$; compare fig. 2.

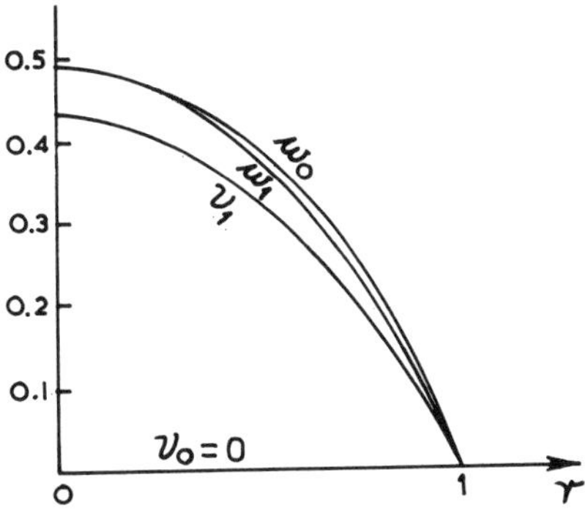

FIG. 2

In more complicated cases we have several, say (k-1) function-inequalities as additional conditions and we call this a "Field Approximation of Level k". Often different formulations with different values of k are possible; compare Bredendiek [61] .

4. Operators in ordered linear spaces

Let R_1, R_2 be two linear partially ordered spaces with $\prec$ as symbol for the ordering and T an operator, which maps the domain D of definition (a subset of R_1) into R_2. The operator T is called syntone, if $f \prec g$ has the consequence $Tf \prec Tg$ for all $f, g \in D$, and T is called antitone, if $f \prec g$ has the consequence $Tf \succ Tg$ for all $f, g \in D$; the operator T is called a MDO (Monotonically Decomposible Operator, see J.Schröder [60]), if there exist a syntone operator T_1 and an antitone operator T_2 with the same domain D of definition and $T = T_1 + T_2$.

Let us look for a solution u of the equation

$$u = Tu + r \tag{4.1}$$

with a M D O operator T and a given element $r \in R_2$.

If the spaces R_1, R_2 are partially ordered Banach spaces and if T_1, T_2 are completely continuous, one can apply the iteration procedure for two sequences of elements (J. Schröder [60]) v_n, w_n corresponding to

$$\left.\begin{aligned} v_{n+1} &= T_1 v_n + T_2 w_n + r \\ w_{n+1} &= T_1 w_n + T_2 v_n + r \end{aligned}\right\} (n = 0, 1, 2, \dots) . \qquad (4.2)$$

starting with two elements $v_o, w_o \in D$. If then the initial conditions

$$v_o \prec v_1 \prec w_1 \prec w_o \qquad (4.3)$$

are satisfied, the application of Schauders fix-point theorem (compare for instance Collatz [66] , p.352-376) gives the existence of at least one solution u of (4.1) in the interval $\mathcal{M}_n = \langle v_n, w_n \rangle$ for every $n = 0,1,2,\dots$, that means

$$v_n \prec u \prec w_n \quad \text{for every} \quad n = 0,1,\dots, \qquad (4.4)$$

Let us consider some special cases.

We look in the case that R_1 is the space of continuous functions $h(x)$, defined in a domain B of the $(x_1,\dots,x_n)$-space for functions

$$\begin{cases} v_o = v_o(x_j, a_1, \dots, a_p) \\ w_o = w_o(x_j, b_1, \dots, b_q) \end{cases} \qquad (4.5)$$

with using (4.2)

$$\begin{cases} v_1 = v_1(x_j, a_1, \dots, a_p, b_1, \dots, b_q) \\ w_1 = w_1(x_j, a_1, \dots, a_p, b_1, \dots, b_q) \end{cases} \qquad (4.6)$$

and for such values of the parameters a_ν, b_μ , that the initial conditions are satisfied as closely as possible; this gives the chained field approximation of level $k = 3$;

$$\left.\begin{aligned} &0 \le w_1(x_j, a_\nu, b_\mu) - v_1(x_j, a_\nu, b_\mu) \le \delta, \\ &\qquad\qquad\qquad\qquad \delta = \text{Min} \\ &0 \le v_1(x_j, a_\nu, b_\mu) - v_o(x_j, a_\nu, b_\mu), \\ &0 \le w_o(x_j, a_\nu, b_\mu) - w_1(x_j, a_\nu, b_\mu) \end{aligned}\right\} \text{for all } x \in B. \quad (4.7)$$

Let us consider some special cases.

5. Free Approximation with syntone operators

If T is syntone, then we can put $T_2 = 0$ = zero-operator, and (4.2) reduces to

$$\left.\begin{aligned} v_{n+1} &= T_1 v_n + r \\ w_{n+1} &= T_1 w_n + r \end{aligned}\right\} \quad (n = 0, 1, \ldots) \qquad (5.1)$$

First method: One can try to use the simple way with separation of the parameters a_ν from the b_μ; then one has Free Appr. Pr. as in (2.6),(2.7)

$$0 \leq v_1(x_j, a_\nu) - v_0(x_j, a_\nu) \leq \delta, \quad \delta = \text{Min} \quad \text{for } \forall x \in B \qquad (5.2)$$

and another Free Appr. Pr. for the b_μ:

$$0 \leq w_0(x_j, b_\mu) - w_1(x_j, b_\mu) \leq \hat{\delta}, \quad \hat{\delta} = \text{Min} \quad \text{for all } x \in B \qquad (5.3)$$

These are Field Approximations of level k=1 with no further additional functional inequality.

If one has sets of parameters a_ν, b_μ, which are "admissable" (so that the restrictions in (5.2),(5.3) are satisfied for all $x \in B$; it is not necessary to solve the optimization problems exactly), one has to test, whether $w_1 \succ v_1$ is satisfied. If $w_1 \succ v_1$ holds in B, one has the existence of a solution u of (4.1) with the error estimation

$$v_1 \prec u \prec w_1 \, . \qquad (5.4)$$

If $w_1 \succ v_1$ is not satisfied in the whole domain B, one has to use the

Second method : One has to look for the field approximation (4.7) of level 3 with the only simplification, that v_1 does not depend on the b_μ and w_1 does not depend on the a_ν. But it is a chained approximation.

6. Examples for the First Method

I. The first method works : We have a syntone operator T in the linear integral equation

$$u(x) = Tu(x) \qquad \text{with} \qquad Tu(x) = 1 + \int_0^1 \frac{1+x^2+s^2}{3} u(s)\,ds. \tag{6.1}$$

Starting with

$$v_0 = a_1 + a_2 x \quad , \qquad w_0 = b_1 + b_2 x \tag{6.2}$$

we get

$$v_1 = Tv_0 = 1 + \frac{4}{9} a_1 + \frac{1}{4} a_2 + x^2\left(\frac{a_1}{3} + \frac{a_2}{6}\right) \tag{6.3}$$

and analogously $w_1 = Tw_0$ (substitute a_ν by b_ν). The Free Appr. Pr. (5.2) is here

$$0 \le v_1 - v_0 = 1 - \frac{5}{9} a_1 + \frac{1}{4} a_2 - a_2 x + x^2\left(\frac{a_1}{3} + \frac{a_2}{6}\right) \le \delta \,, \quad \delta = \text{Min}. \tag{6.4}$$

and has the solution $a_1 = \frac{9}{5}$, $a_2 = \frac{18}{25}$.

The corresponding Free Appr. Pr. (5.3) has the solution

$b_1 = \frac{90}{41}$, $b_2 = \frac{36}{41}$.

Fig.3a shows the graphs of v_0, v_1, w_1, w_0. The initial conditions (4.3) are satisfied and therefore exists a solution $u(x)$ of (6.1) in the strip $v_1 \le u \le w_1$

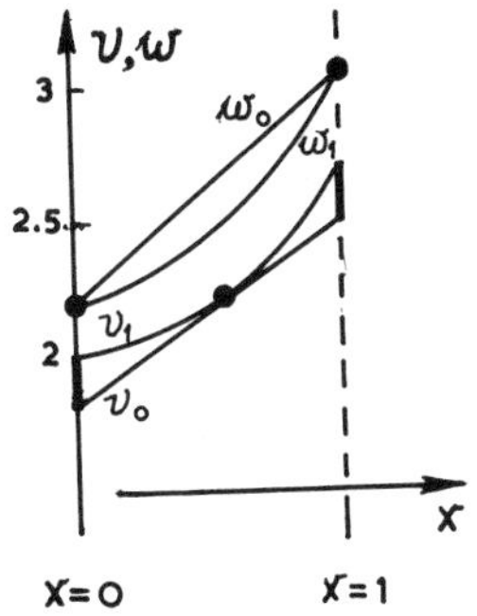

FIG.3a

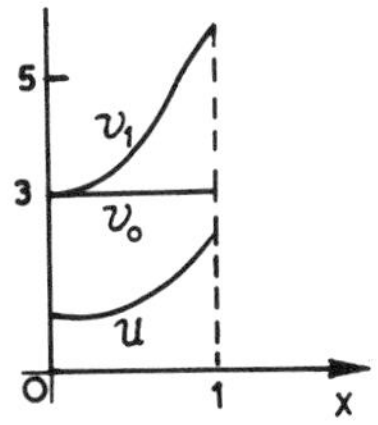

FIG.3b

In this case the kernel in (6.1) is degenerate so that one has easily the exact solution

$$u(x) = \frac{360 + 135\,x^2}{176}$$

which is in fact contained in the strip $v_1 \leq u \leq w_1$. We observe that we have for the solutions v_o, v_1, w_1, w_o characteristic common points (for w_o, w_1), a touching point (for v_o, v_1) and equal extrema for $v_1 - v_o$ at the endpoints.

II. The first method fails : We consider an example which is very similar to the foregoing example:

$$u(x) = Tu(x) \qquad \text{with} \qquad Tu(x) = -1 + \int_0^1 (1+x^2+s^2)u(s)\,ds. \tag{6.5}$$

We solve for

$$v_o = a_1 + a_2 x\,, \quad v_1 = Tv_o = -1 + \frac{4}{3}a_1 + \frac{3}{4}a_2 + x^2\left(a_1 + \frac{a_2}{2}\right)$$

the Free Appr. Pr., corresponding to (6.4),

$$0 \leq v_1 - v_o \leq \delta\,, \qquad \delta = \text{Min} \tag{6.6}$$

and get $v_o = 3$, $v_1 = 3 + 3x^2$; the condition $v_o \leq v_1$ and (6.6) is satisfied, and it seems, that v_1 is a lower bound for the solution $u(x)$. But this is not true, the exact solution is

$$u(x) = \frac{30 + 45x^2}{34}$$

see fig.3b. In this case we do not get functions w_o, w_1, which satisfy (4.3).

7. Field Approximation of level 3 for a nonlinear problem with antitone operator

With the same notations as in (3.1),(3.2) we consider the nonlinear boundary value problem

$$-\Delta u = 1 - u^2 \quad \text{for} \quad r < 1, \quad u = 0 \quad \text{on} \quad r = 1. \tag{7.1}$$

The operator is here antitone; starting with

$$v_o = a(1-r^2), \quad w_o = b(1-r^2)$$

we get

$$w_1 = Tv_o = \frac{1}{4} - \frac{11}{72}a^2 + \frac{r^2}{4}(a^2-1) - \frac{r^4}{8}a^2 + \frac{r^6}{36}a^2$$

and analogously $v_1 = Tw_o$ (change a,b).

We have

$$w_o - w_1 = \frac{1}{72}(1-r^2)(72b - 18 + a^2\varphi(r))$$

$$w_1 - v_1 = \frac{1}{72}(1-r^2)(b^2 - a^2)\varphi(r)$$

$$v_1 - v_o = \frac{1}{72}(1-r^2)(18 - 72a - b^2\varphi(r)) \text{ with } \varphi(r) = 11 - 7r^2 + 2r^4.$$

The chained field approximation (4.7) of level 3 is solved by solution of the equations

$$18 - 72a - 11b^2 = -3 + a^2 + 12b = 0$$

with

$$a = 0.2408, \quad b = 0.2452$$

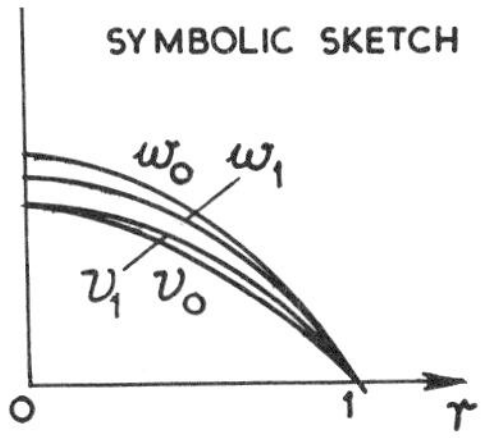

FIG. 4

The functions v_0, v_1, w_1, w_0, fig. 4, satisfy (4.3) and therefore exists at least one solution in the strip $v_1 \leq u \leq w_1$, especially we have the inclusion

$$v_1(0) = 0.2408 \leq u(0,0) \leq 0.2412 = w_1(0).$$

Again we observe characteristic touchings for v_0 and v_1 at r=0 and for w_1 and w_0 at r=1.

8. Field Approximation of level 2 for an antitone operator

For an antitone operator T one can start with v_0 and put $w_1 = w_0$; then (4.7) reduces to the field approximation of level 2:

$$0 \leq w_1 - v_1 \leq \delta, \qquad \delta = \text{Min}, \qquad 0 \leq v_1 - v_0. \tag{8.1}$$

But for the calculation it can happen, that the determination of $v_1 = T(Tv_0)$ causes more work than using (4.7), which avoids the operator T^2.

Sometimes one can avoid even the field approximation of level 2 and use the free Appr. Pr.

$$0 \leq w_1 - v_0 \leq \delta \qquad \delta = \text{Min}; \tag{8.2}$$

but then one has to test, whether

$$v_0 \leq v_1 \leq w_1 \tag{8.3}$$

is satisfied; if (8.3) does not hold, then (8.2) is useless and (8.1) is necessary.

Example for this phenomenon: $y'' = y$ in $-1 \leq x \leq 1$, $y(\pm 1) = 1$. Starting with

$v_o = 1+a(1-x^2)$ one gets $w_1 = \frac{1}{2} - \frac{5a}{12} + (1+a)\frac{x^2}{2} - \frac{a}{12}x^4$.

(8.2) gives $a = -\frac{3}{8}$, $w_1 = \frac{1}{32}(21 + 10x^2 + x^4) = w_o$,

$$v_1 = \frac{1}{960}(619 + 315x^2 + 25x^4 + x^6)$$

but $v_1 - v_o = \frac{1}{960}(1-x^2)(19 - 26x^2 - x^4)$ changes the sign in $-1 < x < 1$; therefore (8.2) is not allowed in this example.

9. Field approximation in more independent variables

Consider the linear integral equation with a syntone operator T:

$$u(x_1,x_2) = Tu(x_1,x_2) \text{ with} \tag{9.1}$$

$$Tu(x_1,x_2) = x_1x_2 + \int_0^1\int_0^1 \frac{1}{4}(1+x_1x_2s_1s_2)\, u(s_1,s_2)ds_1ds_2.$$

For $v_o = a_1 + a_2(x_1+x_2)$ one gets

$$v_1 = Tv_o = \frac{1}{4}(a_1+a_2) + xy(1+\frac{1}{16}a_1+\frac{1}{12}a_2).$$

The Free Appr. Pr. (5.2) gives

$$-a_1 = \frac{48}{95} = a_2$$

and $v_o = \frac{48}{95}(-1 + x_1 + x_2)$, $v_1 = \frac{96}{95}x_1x_2$.

Fig.5 gives a graphical representation of v_o, v_1 with the characteristic touchings (at $x_1=1$, $x_2=0$ and $x_1=0$, $x_2=1$) and the equal extrema of the difference v_1-v_o at $x_1 = x_2 = 0$ and $x_1 = x_2 = 1$.

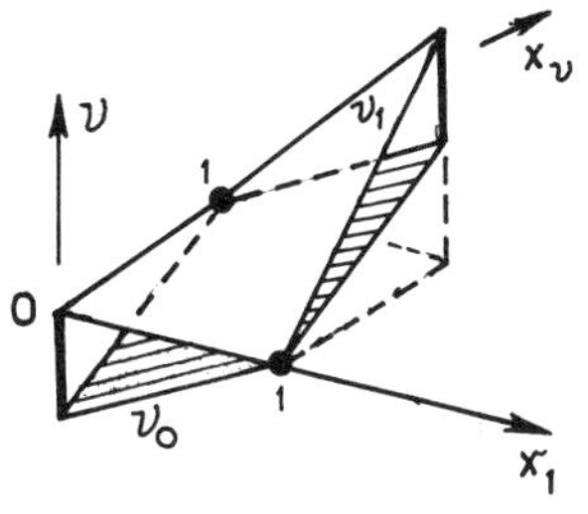

FIG. 5

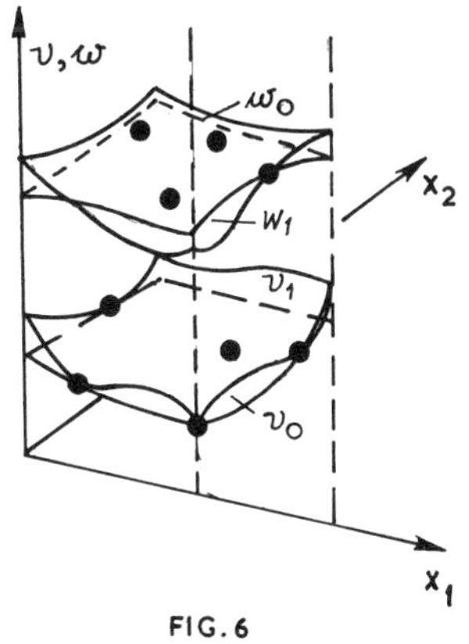

FIG. 6

By using more parameters one gets the typical picture, that one has more touching points for v_o and v_1, and for w_1 and w_o, but not for v_1 and w_1 (except for homogeneous boundary conditions) and more extremal points for the differences v_1-v_o and w_o-w_1; a symbolic sketch is shown in fig.6.

10. Comparison: Field Approximation and classical iteration

In the following very simple example the field approximation gives better results than the classical iteration procedure:

$$y'' = y, \quad y(\pm 1) = 1, \quad \text{exact solution } y(x) = \frac{\cosh x}{\cosh 1} .$$

Iteration procedure: $y''_{n+1} = y_n, \; y_{n+1}(\pm 1) = 1, \; n=0,1,2,\ldots$

$$y_o = 1, \; y_1 = \frac{1}{2}(1+x^2), \; y_2 = \frac{1}{24}(17+6x^2+x^4),$$

$$y_3 = \frac{1}{720}(434+255x^2+30x^4+x^6)$$

values at $x = 0$:

n	0	1	2	3
$y_n(0)$	1	0.5	0.70833	0.6028

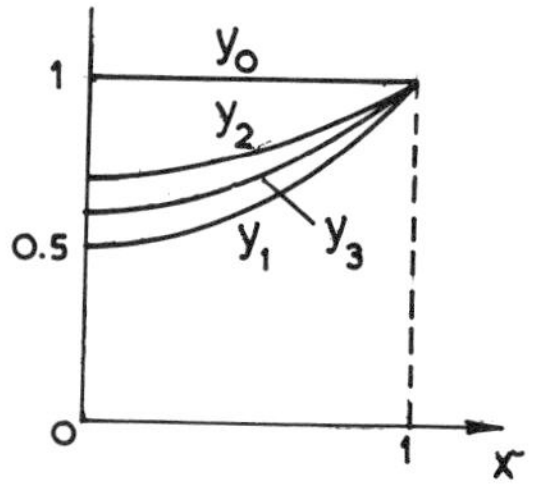

FIG. 7

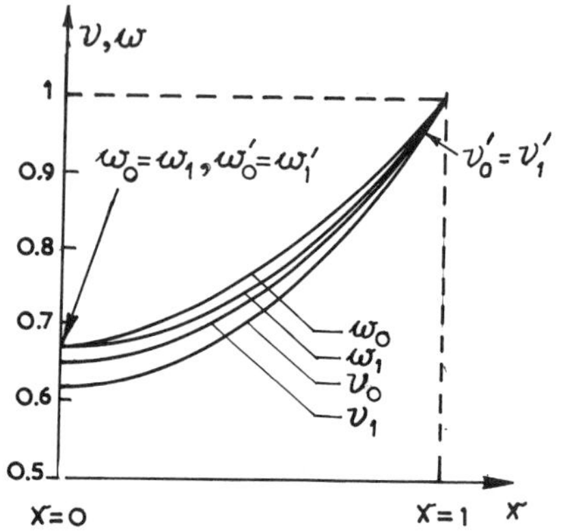

FIG. 8

The convergence is slow, fig. 7; the field approximation of level 3 with

$$v_o = a + (1-a)x^2, \quad w_o = b + (1-b)x^2,$$

$$w_1 = \frac{1}{12}[(11-5a) + 6ax^2 + (1-a)x^4],$$

$$v_1 = \frac{1}{12}[(11-5b) + 6bx^2 + (1-b)x^4]$$

gives with (4.7) $a = \frac{19}{31}; \quad b = \frac{41}{62}$,

$$v_1(0) = \frac{477}{744}, \; w_1(0) = \frac{492}{744}, \; 0 \leq w_1(0) - v_1(0) = \frac{5}{248} \approx 0.02$$

compared with $\quad y_2(0) - y_3(0) \approx 0.105$

Fig. 8 shows again the typical touchings.

11. Application to nonlinear vibrations

More complicated problems cause field approximations with greater values of the level k. We consider a simple example of forced nonlinear vibrations:

$$Ty = -y'' - g(y) + \cos x = 0 \quad \text{with} \quad g(y) = 2y - y^3. \tag{11.1}$$

In the domain

$$\frac{dg}{dy} = 2 - 3y^2 \leq 0 \tag{11.2}$$

or

$$|y| \geq \sqrt{\frac{2}{3}} = \mu \approx 0{\cdot}816 \tag{11.3}$$

we expect forced periodic solutions $y(x)$ and the theory of J.Werner [70] holds.

We start with functions $v_o = a_1+a_2 \cos x$, $w_o = b_1+b_2 \cos x$ and get

$$\left.\begin{aligned} Tv_o &= -2a_1-a_2\cos x+(a_1+a_2\cos x)^3+\cos x \leq 0 \\ Tw_o &= -2b_1-b_2\cos x+(b_1+b_2 \cos x)^3+ \cos x \geq 0 \end{aligned}\right\} \tag{11.4}$$

Then we have the Field **Approximation**:

$$0 \leq w_o-v_o \leq \delta \quad , \quad \delta = \text{Min} \tag{11.5}$$

with $Tv_o \leq 0$, $Tw_o \leq 0$. Sometimes further restrictions on the range of y occur (compare for instance Werner [70]), and one has field approximations of level $k > 3$. If we take v_o, w_o as constants, we get $a_1 = 1$, $b_1 = \gamma$ where $\gamma \approx \frac{1}{2}(1 +\sqrt{5})$ is the positive root of $\gamma^3 = 1 + 2\gamma$; better values are

$$a_1 = 1,\ a_2 = -0.5,\ b_1 = \gamma\ ,\ b_2 = -0.245$$

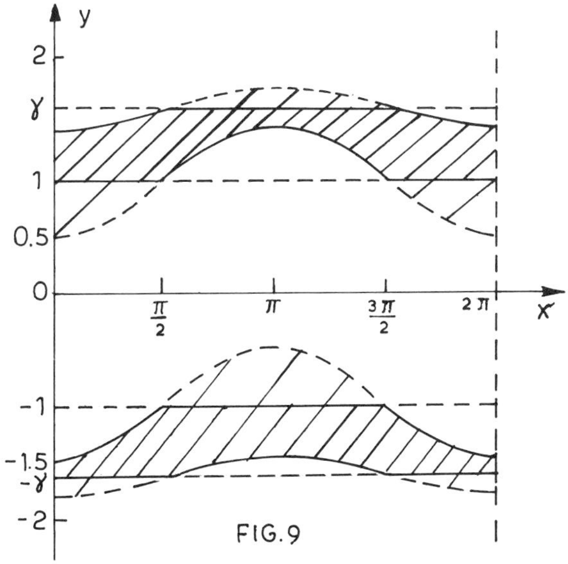

FIG.9

Fig.9 gives the hatched strips, in which the existence of at least one periodic answer $y(x)$ is proved. With $y(x)$ is $z(x) = -y(x+\pi)$ also a periodic solution and therefore fig.9 gives the strips for two periodic solutions.

References

Bredendiek, E. [69] Simultan-Approximation, Arch. Rat. Mech. Anal. 33 (1969) 307-330.

Collatz, L. [66] Functional Analysis and Numerical Mathematics, Academic Press 1966, 473 p.

Collatz, L. [71] Some applications of functional analysis to analysis, particularly to nonlinear integral equations, Proc. Symp. nonlinear functional analysis and applications, edited by Rall, Academic Press 1971, 1-43.

Collatz, L. [73] Finite und infinite Optimierung mit einigen Anwendungen; Vortrag 6. I.K.M., Weimar 1972, to appear in wissenschaftlicher Zeitschrift der Hochschule für Architektur und Bauwesen Weimar, 1973.

Collatz, L. and Wetterling, W. [71] Optimierungsaufgaben, Springer Verlag, 2.ed. 1971, 222 p.

Krabs, W. [71] Nichtlineare Optimierung mit unendlich vielen Nebenbedingungen, Computing 7, 204-214 (1971).

Lempio, F. [72] Lineare Optimierung in unendlich dimensionalen Vektorräumen, Computing (1972).

Schröder, J. [60] Anwendung von Fixpunktsatzen bei der numerischen Behandlung nichtlinearer Gleichungen in halbgeordneten Räumen. Arch. Rat. Mech. Anal. 4 (1960) 177-192.

Werner, J. [70] Einschließungssätze für periodische Lösungen der Liénard'schen Differentialgleichung. Computing 5 (1970), 246-252.

LINEAR INTERPOLATION AND LINEAR EXTENSION OF FUNCTIONS

J. G. Dhombres

Our aim is to relate linear interpolation and linear extension of functions in a rather general setting in order to gather some results and raise up some problems.

Let us start with a non-empty topological space X. With each closed subset Y of X we associate a locally convex topological algebra A(Y). As usual, we denote by C(Y) the algebra of all continuous, bounded and complex-valued functions over Y, equipped with the sup-norm. We shall always suppose that A(Y) is algebraically and topologically embedded into C(Y) for each closed subset Y. By the expression locally convex topological algebra, we simply mean that, for each Y, A(Y) is locally convex and for any given g in A(Y), $f \longrightarrow fg$ is continuous from A(Y) into A(Y).

Definition 1. An operator $Q : A(Y) \longrightarrow A(X)$ is a linear extension operator if Q is a linear operator satisfying $R_Y(Qf) = f$ for any f in A(Y) where R_Y is the restriction operator from X to the subspace Y.

Related to such an operator is another one according to $P = QR_Y$ as soon as $R_Y(A(X)) \subset A(Y)$. Such an operator P is defined over A(X) and takes its values in A(X). It satisfies a very simple functional equation :

$$\begin{aligned} P(f.Pg) &= Q\,R_Y\,(f.QR_Y(g)) \\ &= Q\,(R_Yf\,.\,R_Y\,Q\,R_Y(g)) = Q(R_Y(f)\,.\,R_Y(g)) \\ &= Q\,R_Y\,(fg) = P(fg), \end{aligned}$$

that is $P(fPg) = P(fg)$ for all f and g in A(X).

Definition 2. A linear operator $P : A(X) \longrightarrow A(X)$ which satisfies the functional equation $P(fPg) = P(fg)$ is called an interpolating operator.

Conversely, does an interpolating operator originate from a linear extension operator? Some more properties of A(X) seem to be required in order to obtain a positive result : we ought to know the structure of closed ideals of A(X). We introduce some convenient definitions.

Definition 3. The topological algebra A(X) is a global algebra if the following conditions are satisfied :

(a) X is a paracompact space,

(b) For any locally finite open covering $\{\Theta_i\}_{i\in J}$ of X, there exist functions $\{f_i\}_{i\in J}$ belonging to A(X) such that for any f in A(X) and in the sense of A(X), we get $\sum_{i\in J} f f_i = f$ and $\mathrm{Supp}\, f_i \subset \Theta_i$ $(\forall\, i\in J)$ where $\mathrm{Supp}\, f_i$ denotes the support of f_i.

(c) Let L be a linear and continuous form on A(X) such that L(fg) = 0 for all f in A(X), then g(x) = 0 for any x in X having the following property : for any neighbourhood $\mathcal{V}(x)$ of x, there exists an h in A(X), the support of which is included in $\mathcal{V}(x)$ and for which $L(h) \neq 0$.

For a given non-zero linear and continuous form L on A(X), let us consider the subset of all x in X having the property stated in (c) of Definition 3. This subset, called the support of L, is obviously closed. It is not empty. In fact, let us suppose that for each x in X, there exists an open neighbourhood $\mathcal{V}(x)$ of x such that whatever h might be in A(X), with $\mathrm{Supp}\, h \subset \mathcal{V}(x)$, then L(h) = 0. As X is paracompact, it is possible to build a locally finite open sub-covering $\{\mathcal{V}(x_i)\}_{i\in J}$. There exist $\{f_i\}_{i\in J}$ in A(X) such that $\sum_{i\in J} f_i f = f$ for all f in A(X) and $\mathrm{Supp}\, f_i \subset \mathcal{V}(x_i)$ $(\forall\, i \in J)$. As $L(f_i f) = 0$, we deduce L(f) = 0. As $L \neq 0$, this contradiction proves that Supp L $\neq 0$.

Example 1. If $X = \mathbb{R}^n$, a locally convex algebra $A(\mathbb{R}^n)$ containing the space D of all complex-valued functions having derivatives of all order and compact support, as a dense subset is a global algebra.

Definition 4. Let A(X) be a global algebra. A closed subset Y of X is said to have the continuous interpolation property (towards A(Y) and A(X)) if for each f in A(Y) there exists an $\tilde{f}$ in A(X) such that $f \longrightarrow L(\tilde{f})$ is a linear and continuous form on A(Y) for any linear and continuous form L on A(X), the support of which is included in Y.

If A(X) is a normed algebra, Y has the bounded interpolation property if there exists a constant A such that for all L in $(A(X))'$:

$$| L(\tilde{f}) | \leq A \, \| L \| \, \| f \|_{A(Y)}$$

For any linear interpolating operator $P : A(X) \longrightarrow A(X)$, we call interpolator the subset of all x in X such that $Pf(x) = f(x)$ for all f in A(X). We may now give a converse result under some strong conditions which could certainly be weakened.

THEOREM 1. Let A(X) be a global algebra such that every closed subset has the continuous interpolation property. Let P be a linear and continuous interpolating operator on A(X). Then, there exists a linear and weakly continuous extension operator $Q : A(Y) \longrightarrow A(X)$ where Y is the interpolator of P. This extension operator Q is continuous if A(X) is a normed algebra and if the bounded interpolation property is assumed for any closed subset Y in X.

The interpolator Y of P is obviously closed and it is not an empty subset according to the following reasoning. For any linear and continuous form L on A(X), the application $P^*(L) : f \longrightarrow \langle Pf, L \rangle$ is also linear and continuous. We then get $\langle f(g - Pg), P^*(L) \rangle = 0$ and due to the definition of a global algebra, we may deduce $g(x) = Pg(x)$ for all g in A(X) and for every x in the support of P*(L). Let us now define $Qf = P(\tilde{f})$ for any f belonging to A(Y), which is possible as Y has the continuous interpolation property. Qf is an element of A(X) and from $Qf(x) = \langle \tilde{f}, P^*(\delta_x) \rangle$ where δ_x is the application $f \longrightarrow f(x)$, we get $Qf(x) = f(x)$ for any x in Y and so Q is a linear extension operator from A(Y) into A(X). As $\langle Qf, L \rangle = \langle \tilde{f}, P^*(L) \rangle$, it is easy to see that Q is a continuous operator from A(Y) equipped with the topology $\sigma(A(Y), A'(Y))$ into A(X) equipped with $\sigma(A(X), A'(X))$. In the same way, if A(X) is a normed algebra and if Y has the bounded interpolation property, then the following estimation holds:

$$|\langle Qf, L \rangle| \leq A \|f\|_{A(Y)} \|P^* L\|_{A'(X)} \leq A \|f\|_{A(Y)} \|P\| \|L\|_{A'(X)}$$

and so

$$\|Qf\|_{A(X)} \leq A \|P\| \|f\|_{A(Y)} .$$

<u>Example 2</u>. Let X be a paracompact space and take A(Y) to be C(Y). The C(X) is a global normed algebra. Tietze's theorem asserts that any closed subset of X has the bounded interpolation property. Then, to construct a continuous interpolating operator is equivalent to the construction of a linear extension operator according to $P = QR_Y$.

<u>Example 3</u>. Let us consider $X = \mathbb{R}^n$, a positive integer k and $A(\mathbb{R}^n)$ the Banach algebra of all complex-valued functions of class C^{k-1}, which tend to zero with their derivatives of all order less than or equal to k-1 as soon as x tends to infinity. Let Y be a closed subset of $\mathbb{R}^n$ and consider the family of applications $f^{(h)} : x \in Y \longrightarrow f^{(h)}(x) \in \mathbb{C}$ where (h) is any n-tuple of integers such that $|h| = h_1 + \dots + h_n < k$. We define, for an n-tuple j with $0 \leq |j| \leq k-1$ and a, b in Y

$$R_j(a,b) = f^{(j)}(b) - \sum_{0 \leq |j+h| \leq k-1} \frac{1}{h!} f^{(j+h)}(a) \, (b-a)^h.$$

Then A(Y) is the set of all such families for which there exists a constant M and, for all $0 \leq |j| \leq k-1$, for all a and b in Y, we have :

$$|f^{(j)}(a)| \leq M \; ; \quad |R_j(a,b)| \leq M(d(a,b))^{k-|j|}$$

where d(a,b) is for instance the Euclidean distance in $\mathbb{R}^n$. With the use of an extension theorem of Whitney (cf. [7], where we take only a weak part of the main theorem), and due to known structure properties of distributions with compact supports, we see that any compact subset Y has the bounded interpolation property. We may then apply Theorem 1 which gives the structure of any interpolating and continuous operator. Precisely such an operator can always be written as $P = QR_Y$ where $Q : A(Y) \longrightarrow A(X)$ is an extension operator, R_Y the restriction operator from C(X) onto C(Y) and Y the interpolator of P.

Example 4. Take $X = T = \mathbb{R}/\mathbb{Z}$ and for A(T) we take the Banach algebra of all continuous and complex-valued functions over T having Fourier coefficients in $\ell^1(\mathbb{Z})$ and equipped with the norm inherited from $\ell^1(\mathbb{Z})$. When Y is a closed subset of T, A(Y) shall be the Banach algebra quotient of A(T) by the closed ideal of all functions in A(T) which are equal to zero on Y. Then, any linear and continuous interpolating operator $P : A(T) \longrightarrow A(T)$ having an interpolator which satisfies the harmonic synthesis can be written as $P = QR_Y$ where Q is a linear and continuous extension operator from A(Y) into A(X) and R_Y is as before the restriction operator.

Within the present frame, some questions appear quite naturally.

(1) How to construct non-trivial interpolating operators on a given algebra?

(2) Which conditions are sufficient for a closed subset Y of X

to be an interpolator related to a continuous interpolating operator on a topological algebra A(X)?

(3) What is a characterisation of those topological spaces Y which are always interpolators as soon as they are considered as closed subspaces of a topological space X?
(Here we suppose A(Z) = C(Z) for any closed subset Z in X).

(4) Which conditions ensure that X possesses a non-empty perfect closed subset which is an interpolator related to a continuous interpolating operator on A(X) ?

(5) Which conditions can be found for the union of two interpolators to be an interpolator ?
Naturally such problems have already been studied in the case where C(Y) = A(Y) (with the language of continuous extension operators).

A result of Dugundji (cf. [4]) proves that a metrisable compact space is of the kind of topological spaces which are looked for in Problem 3. By taking the topological product of any such spaces we still get an interpolator (cf. [2]).

On an opposite way, without metrisability assumptions over X, the norm of an interpolating operator plays an interesting part. For example, using simple and elegant combinatorial properties, Corson and Lindenstrauss (cf. [1]) have shown that with a compact space X and a closed subset Y homeomorphic to the one-point compactification of an uncountable discrete set, then either Y is not an interpolator, or Y in an interpolator for an interpolating operator with a norm equal to an odd integer and this norm cannot be lessened. If we require the norm of the interpolating operator to be one and P to preserve constants, then Problem 2 is solved at least for an hyperstonian space X but this could certainly be improved. The necessary and sufficient condition then is that Y is a retract of X (cf. [2]).

For algebras which are different from C(X), linear extension

operators Q have been studied with $Q : C(Y) \longrightarrow A(X)$ (cf. [5]).
We shall end with an extension of Example 4. Let G be a locally compact abelian group and $G^\wedge$ its dual group equipped with the open-compact topology. We choose Haar measures on each of these two locally compact groups. The algebra A(G) is the set of all continuous and complex-valued functions on G which are Fourier transforms of functions belonging to $L^1(G^\wedge)$. It is a Banach algebra for the norm it inherits from $L^1(G^\wedge)$. In the same way, B(G) is the Banach algebra of all continuous functions on G which are Fourier transforms of bounded Radon measures on $G^\wedge$. We also use T(G), the set of all trigonometric polynomials on G, i.e. functions like

$$x \longrightarrow t(x) = \sum_{i=1}^{n} a_i \langle x, x_i^\wedge \rangle .$$

PROPOSITION 1. *A linear and continuous operator* $P : A(G) \longrightarrow A(G)$, *such that* $P(tf) = tPf$ *for any* t *in* T(G), *is an interpolating operator if and only if* P *is an homomorphism. Moreover,* P *can be written as a finite sum of products like* $P_1 P_2 \dots P_n$ *and* P_i *is either a multiplication by* χ_i *or a multiplication by* $(1 - \chi_i)$ *where* χ_i *denotes the characteristic function of an open sub-group of* G *after possibly a translation.*

The idea of the proof is first to establish that P is a multiplication operator : $Pf = \chi f$ for all f in A(G). Then χ must be a characteristic function and the result of Cohen (cf. [6]) on idempotent Radon measures furnishes the conclusion. Using operator $P_i : f \longrightarrow \chi_i f$, we deduce that the translate of an open subgroup of G is an interpolator associated with an interpolating operator of norm one. This last property characterises such operators P_i among all interpolating operators satisfying the hypotheses of Proposition 1. All the elements of the Boolean algebra generated by translates of open subgroups are interpolators, which paves the way to Problem 5. Concerning the norm, it is interesting to note that there exists a universal constant α ($\alpha > 1$) such that the norm of an interpolating operator satisfying the hypotheses of Proposition 1 is greater than α as soon as P is not of the kind P_i (cf. [6]).

Let us try to find other continuous and interpolating operators on B(G). The problem is the same as finding bounded interpolating operators Q on the convolution algebra $M(G^\wedge)$:

$$Q(\mu * Q\nu) = Q(\mu * \nu)$$

Let us suppose that for all μ , the Radon measure $Q\mu$ is the image of μ through a given $\Phi : G^\wedge \longrightarrow G^\wedge$. There Φ necessarily satisfies the following functional equation on $G^\wedge$:

$$\Phi(\hat{x} + \Phi\hat{y}) = \Phi(\hat{x} + \hat{y}) . \qquad (1)$$

All such functions can be described (for analogous functional equations, cf. [3]).

PROPOSITION 2. *A function* $\Phi : G^\wedge \longrightarrow G^\wedge$ *satisfies* (1) *if and only if there exists a subgroup* H *of* $G^\wedge$ *and a function* $\varphi : G/H \rightarrow H$ *such that, if we write* $x \sim \dot{x} + x'$ *where* $\dot{x} \in G/H$ *and* $x' \in H$, *then*

$$\Phi(x) \sim \dot{x} + \varphi(\dot{x}) .$$

For the present problem, certain regularity assumptions have to be made over Φ (like being a Borel function or even an universally measurable function) and this somewhat restricts the choice of H and of φ . Instead of general results, let us take two simple examples. For $G = T$ and $\varphi \equiv 0$, we have $H = h\mathbb{Z}$ and get when h is an odd integer, $h = 2n+1$:

$$Pf(x) = \frac{1}{2n+1} \sum_{\ell=0}^{\ell=2n} f\left(\frac{2\pi \ell}{2n+1}\right) D_n\left(x - \frac{2\pi \ell}{2n+1}\right)$$

where $D_\ell(x) = \dfrac{\sin \frac{2\ell+1}{2} x}{\sin \frac{x}{2}}$, that is, the Dirichlet kernel of order ℓ. Then Pf is the trigonometric interpolating polynomial of f with 2n+1 equidistant nodes $x_\ell = \frac{2\pi\ell}{2n+1}$, $\ell = 0, \ldots, 2n$. The norm of P over B(T) is one and P(1) = 1 (P is also a positive operator).

For $G = \mathbb{R}$ and $H = \beta\mathbb{Z}$, we find

$$Pf(x) = \sum_{n=-\infty}^{+\infty} \int_{\beta(n-1)}^{\beta n} e^{i(\lambda - \beta(n-1))x} e^{i\varphi[\lambda - \beta(n-1)]x} d\mu(\lambda).$$

We get $\|P\| = 1$ and P(1) = 1. Moreover $(2\pi/\beta)\mathbb{Z}$ is the interpolator of P and P commutes with translation operators τ_h as soon as h belongs to $(2\pi/\beta)\mathbb{Z}$.

Bibliography.

1 Corson, H.H. and Lindenstrauss, J., On simultaneous extensions of continuous functions. Bull. Amer. Math. Soc. 71 (1965) 542-545

2 Dhombres, J., Sur les opérateurs multiplicativement liés. Mémoires de la Soc. Math. de France, No.57, 1971 (Thèse).

3 Dhombres, J., Functional equations on semi-groups arising from the theory of means. Nanta Mathematica No.1, 6, (1972) (to appear)

4 Dugundji, J., An extension of Tietze's theorem. Pacific J. Math. 1, (1951), 353-367

5 Michael, E. and Pelczynski, A., A linear extension theorem. Illinois J. Math., 11, (1967), 555-562

6 Rudin, W., Fourier analysis in groups. Interscience Pub., New York, 1962

7 Whitney, H., Analytic extensions of differentiable functions defined in closed sets. Trans. Amer. Math. Soc., 36, (1934), 63-89

WAVE FRONT SETS AND HYPOELLIPTIC OPERATORS

Ph.A. Dionne

The theory of hypoelliptic operators was initiated in 1950 by L.Schwartz [13] in order to study the interior regularity of the solutions of linear partial differential equations. On the other hand, L.Hörmander [4], [5] has characterized the hypoelliptic operators with constant coefficients in terms of singular supports of distributions and recently [6] has noticed that a stronger description of these operators is obtained in using a refinement of the notion of singular support, what he has called wave front set of a distribution.

Hypoelliptic operators and singular supports.

Let M be a real $\mathcal{C}^\infty$ paracompact manifold of dimension n,

$$D = \left(-i\frac{\partial}{\partial x_1}, \ldots, -i\frac{\partial}{\partial x_n}\right), \quad \alpha = (\alpha_1, \ldots, \alpha_n) \in \mathbb{N}^n,$$

$$|\alpha| = \alpha_1 + \ldots + \alpha_n, \quad D^\alpha = \frac{i^{-|\alpha|}\,\partial^{|\alpha|}}{\partial x_1^{\alpha_1} \ldots \partial x_n^{\alpha_n}}, \quad x \in M,$$

$$a = a(x, D) = \sum_{|\alpha| \leq m} a_\alpha(x)\, D^\alpha, \quad a_\alpha : M \to \mathbb{C}$$

of class $\mathcal{C}^\infty$, Ω an open set of M.

By definition, a is hypoelliptic, iff

$$\forall\, \Omega \subset M,\ \forall\, h \in \mathcal{D}'(\Omega),\ a h \in \mathcal{C}^\infty(\Omega) \Rightarrow h \in \mathcal{C}^\infty(\Omega). \tag{1}$$

If a = a (D) is with constant coefficients, then a is hypoelliptic iff

$$\forall\, \Omega \subset M,\ \forall\, h \in \mathcal{D}'(\Omega),\ \text{sing supp } h = \text{sing supp } ah, \tag{2}$$

where sing supp h means the least closed set F of Ω such that $h \in \mathcal{C}^{\infty}(\Omega \setminus F)$.

However if a is with variable coefficients, then for a fixed Ω, (1) $\not\Leftrightarrow$ (2). Here is an example.

Let

$$M = \{(x, y) \in \mathbb{R}^2 \mid 1 < \sqrt{x^2 + y^2} < 2\},$$

$$a = x \frac{\partial}{\partial y} - y \frac{\partial}{\partial x} + 1,$$

then for $\Omega = M$, (1) holds but (2) does not hold.

To obtain a stronger description of hypoelliptic operators, Hörmander [6] used the notion of wave front set of a distribution suggested to him by M. Sato [11] .

Let M be a real analytic paracompact manifold of dimension n. Sato has defined on M a sheaf $\mathcal{C}$ over $S^*(M)$, the cosphere bundle of M, (i.e. the quotient space $(T^*(M) \setminus \{0\})/R^*_+$ where $T^*(M)$ denotes the cotangent bundle of M) such that there exists a surjective linear map u from the hyperfunctions in M into the sections of $\mathcal{C}$, the kernel of u being the analytic functions in M. Let h be a hyperfunction in M. Then the support of u(h) is a closed cone in $T^*(M) \setminus \{0\}$. The projection $\pi : T^*(M) \setminus \{0\} \to M$ verifies

$$\pi\,(\text{supp } u(h)) = \{x \in M \mid h(x) \text{ is not analytic}\}$$

Let p be the principal part of a linear differential operator in M of order m with analytic coefficients, i.e. $p(x, D) = \sum_{|\alpha|=m} a_{\alpha}(x) D^{\alpha}$; then

$$ah \text{ analytic} \Rightarrow \text{supp } u(h) \subset \{(x, \xi) \in T^*(M) \setminus \{0\} \mid p(x,\xi) = 0\}$$

and the singularities of h propagate along the bicharacteristics, i.e. supp u(h) is invariant under the flow defined by the Hamiltonian vector field of p (cf. Sato [12]).

If h is a distribution then a set denoted by $WF(h)$ will be defined having similar properties to $u(h)$ when h is a hyperfunction, i.e.

$WF(h)$ will be a closed cone in $T^*(M) \setminus \{0\}$ such that $\pi(WF(h)) = \text{sing supp } h$,
and $ah \in \mathcal{C}^\infty \Longrightarrow (WF(h) \setminus WF(ah)) \subset p^{-1}(0)$

$WF(h)$ can be defined in using pseudo-differential operators.

Pseudo-differential Operators and Wave Front Sets.

Let M be an open set of $\mathbb{R}^n$, B a relatively compact subset of M, $p = p(x, \xi) \in \mathcal{C}^\infty(M \times \mathbb{R}^n)$, such that

$$\exists\, m \in \mathbb{R},\ \forall\, x \in B,\ \forall \text{ multi-indices } \alpha, \beta,$$

$$| D_\xi^\alpha D_x^\beta p(x, \xi) | \leq C(\alpha, \beta, B)(1 + \|\xi\|)^{m - \rho|\alpha| + \delta|\beta|}$$

$$\text{with } 0 \leq \delta < \rho \leq 1 \text{ and } 1 - \rho \leq \delta . \tag{3}$$

Let P be the set of all functions p verifying (3). An element of P is called a symbol. If p can be written in the form $p = p_0 + p_1$, where $p_0(x, \xi)$ is homogeneous with respect to ξ of degree m and p_1 is of degree $m-1$, then p_0 is called a principal symbol.

Let $\mathcal{D}(M)$ be the vector space of all $\mathcal{C}^\infty$ functions defined in M and with compact supports. A map

$$A : \mathcal{D}(M) \longrightarrow \mathcal{C}^\infty(M)$$

is said to be a pseudo-differential operator of order m, if A can be written in the form

$$A\varphi(x) = \frac{1}{(2\pi)^n} \int e^{\langle x, \xi \rangle i} p(x, \xi) \hat{\varphi}(\xi)\, d\xi$$

when $p \in P$ and $\hat{\varphi}(\xi) = \int e^{-\langle x, \xi \rangle i} \varphi(x)\, dx$

If p is a polynomial in ξ, then A is a differential operator:

$$A = a(x, D) = \sum_{|\alpha| \le m} a_\alpha(x)\, D^\alpha \varphi(x)$$

$$= \frac{1}{(2\pi)^n} \sum_{|\alpha| \le m} a_\alpha(x) \int \widehat{D^\alpha \varphi}(\xi)\, e^{\langle x, \xi \rangle i}\, d\xi$$

$$= \frac{1}{(2\pi)^n} \sum_{|\alpha| \le m} a_\alpha(x) \int \xi^\alpha\, \hat{\varphi}(\xi)\, d\xi$$

$$= \frac{1}{(2\pi)^n} \int a(x, \xi)\, e^{\langle x, \xi \rangle i}\, \hat{\varphi}(\xi)\, d\xi .$$

In fact a pseudo-differential operator A is a continuous map and can be extended to a continuous map $A : \mathcal{E}'(M) \longrightarrow \mathcal{D}'(M)$ ($\mathcal{E}'$: the vector space of distributions with compact supports, $\mathcal{E}' \subset \mathcal{D}'$).

We have for $\varphi \in \mathcal{D}(M)$

$$A\, \varphi(x) = \frac{1}{(2\pi)^n} \int_{\mathbb{R}^n} p(x, \xi) \int_M \varphi(y)\, e^{-\langle \xi , y - x \rangle i}\, dy\, d\xi$$

$$= \int_M \hat{p}_2(x, x - y)\, \varphi(y)\, dy$$

where $\hat{p}_2(x, \cdot)$ means the Fourier transform of $p(x, \xi)$ with respect to ξ:

$$\hat{p}_2(x, x - y) : (x, y) \longrightarrow \frac{1}{(2\pi)^n} \int_{\mathbb{R}^n} e^{\langle x - y, \xi \rangle i}\, p(x, \xi)\, d\xi .$$

$\hat{p}_2(x, x-y)$ is the kernel of A; it is a distribution which in general is not a function, it coincides with a $\mathcal{C}^\infty$ function in the complement of the diagonal $y = x$, i.e. the diagonal contains all the singularities of the kernel; hence

$$\forall \varphi \in \mathcal{E}'(M), \quad \text{sing supp } A\varphi \quad \subset \quad \text{sing supp } \varphi .$$

The pseudo-differential operators form an algebra, invariant under change of variables, which allows an extension to M, real $\mathcal{C}^\infty$ paracompact manifold; their principal symbols are invariantly defined on $T^*(M)$ [1] .

A pseudo-differential operator on the manifold M is said to be <u>properly supported</u> if both projections

$$\text{supp } \hat{p}_2 \, (x, x-y) \longrightarrow M$$

are proper, i.e. such that the set

$$\{(x, y) \in \text{supp } \hat{p}_2(x, x-y), \, x \in K \text{ or } y \in K\}$$

is compact for every compact set $K \subset M$.

Let $Z^{(m)}(M)$ be the vector space of all properly supported pseudo-differential operators of order m.

A pseudo-differential operator A of order m with symbol p is said to be <u>characteristic</u> at $(x,\xi) \in T^*(M) \setminus \{0\}$ if, for $\lambda \in \mathbb{R}$,

$$\lim_{\lambda \to +\infty} \frac{|p(x, \lambda\xi)|}{\lambda^m} = 0 .$$

A is said to be <u>elliptic</u> if it has no characteristic points, $(A \text{ elliptic}, h \in \mathcal{D}', Ah \in \mathcal{C}^\infty) \Rightarrow h \in \mathcal{C}^\infty$ and

$$\text{sing supp } Ah = \text{sing supp } h.$$

In the algebra over the ring $\mathcal{C}^\infty(M)$ of pseudo-differential operators of order 0, with $\delta = 0$ and $\rho = 1$ the invertible elements are the elliptic ones.

<u>Wave Front Sets</u>.

Let $h \in \mathcal{D}'(M)$, the wave front set of h, denoted by WF(h) is the set of points in $T^*(M) \setminus \{0\}$ which are characteristic for every pseudo-differential operator A such that $Ah \in \mathcal{E}^\infty$; A runs over $Z^{(m)}(M)$.

$WF(h)$ is a closed cone in $T^*(M) \setminus \{0\}$. One advantage of considering $WF(h)$ instead of sing supp h is to eliminate the difference between local and global results.

THEOREM. <u>The projection</u> $\pi : T^*(M) \setminus \{0\} \longrightarrow M$ <u>verifies</u>

$$\pi\,(W F\,(h)\,) = \text{sing supp } h.$$

<u>Proof</u>. 1) $\pi\,(W F\,(h)\,) \subset$ sing supp h. Indeed, let $x \notin$ sing supp h, then there exists $\varphi \in \mathcal{E}^\infty$ such that $\varphi(x) \neq 0$ and $\varphi h \in \mathcal{E}^\infty$; hence $A = \varphi$ is not characteristic at (x,ξ).

2) sing supp $h \subset \pi\,(W F\,(h)\,)$. Indeed let $(x,\xi) \notin \pi(W F\,(h)\,)$ for a fixed x and all ξ. To show $x \notin$ sing supp h. For every ξ, there exists a pseudo-differential operator A such that

$$A\,h \in \mathcal{E}^\infty \quad \text{and} \quad \underline{\lim}_{\lambda \to +\infty}\, p\,(x,\lambda\,\xi) \neq 0\,;$$

hence A is elliptic at x and therefore $h \in \mathcal{E}^\infty$ at x. This result holds globally; therefore $h \in \mathcal{E}^\infty(M)$.

<u>Hypoelliptic Operators and Wave Front Sets</u>.

A linear differential operator a will be said to be <u>hypoelliptic</u> if

$$\forall\, h \in \mathcal{D}'(M)\,, \quad WF(h) = WF(a\,h)\;.$$

If a is hypoelliptic, then

$$W F\,(h) = W F\,(ah) \Longrightarrow \text{sing supp } h = \text{sing supp } ah;$$

the converse is true if $a = a\,(\,D\,)$.

Let $s \in \mathbb{R}$ and $W^s(M)$ be the vector space of all the distributions $h \in \mathscr{E}'(M)$ the Fourier transform $\hat{h}$ of which is a function verifying for $\xi \in \mathbb{R}^n$

$$\int_{\mathbb{R}^n} (1 + \|\xi\|^2)^{s/2} \, |\hat{h}(\xi)|^2 \, d\xi < +\infty .$$

$A \in Z^{(m)}(M)$ is said to be <u>subelliptic</u> if $\exists \; s \in \mathbb{R}$ and $d \in \mathbb{R}^*_+$,

$$h \in W^s(M) \cap \mathscr{E}'(M) \; , \; Ah \in W^{s+1-m} \Longrightarrow h \in W^{s+d}(M)$$

A subelliptic $\Longrightarrow$ A hypoelliptic .

<u>Hypoelliptic Operators of Principal Type</u>.

Let M be a real $\mathscr{C}^\infty$ paracompact manifold, $A \in Z^{(m)}(M)$ with a principal symbol p such that

$$(x,\xi) \in T^*(M) \setminus \{0\} \text{ and } p(x,\xi)=0 \Longrightarrow dp(x,\xi) \neq 0 \; ; \qquad (4)$$

in that case A is said to be of <u>principal type</u> if it is a differential operator.

$A \in Z^{(m)}(M)$ verifying (4), $h \in \mathscr{D}'(M)$ and $Ah \in \mathscr{C}^\infty \Longrightarrow WF(h) \subset p^{-1}(0)$.

Two cases are considered.

1. p is <u>real valued</u>. Then

$$h \in \mathscr{D}'(M) \Longrightarrow WF(h) \setminus WF(Ah) = S \subset p^{-1}(0)$$

and S is invariant under the flow defined by the Hamiltonian field of p.

$Ah \in \mathscr{C}^\infty \Longrightarrow WF(h)$ is invariant under the flow defined by the Hamiltonian field of p and therefore it can be said in that case that the singularities of h propagate along the bicharacteristics, i.e. the integral curves of the equations

$$\frac{\partial x_j}{\partial t} = \frac{\partial p}{\partial \xi_j} , \qquad \frac{\partial \xi_j}{\partial t} = - \frac{\partial p}{\partial x_j} .$$

Hence if $x \in$ sing supp h then there exists a bicharacteristic curve staying in the singular support, W F (h) is the union of bicharacteristic strips (integral curves of Hamiltonian field contained in $p^{-1}(0)$) for p.

2. p is <u>complex valued</u>.

Let $T = \{ (x,\xi) \in T^*(M) \setminus \{0\} \mid \exists\, q \in \mathcal{C}^\infty$ in a neighbourhood of (x,ξ) , $q(x,\xi) \neq 0$ and $\operatorname{Im} qp \geq 0 \}$. T is an open set and contains $T^*(M) \setminus p^{-1}(0)$. Suppose Γ is an open cone contained in T and containing $p^{-1}(0)$ but containing no bicharacteristic strips, then

$$h \in \mathcal{D}'(M) \implies \text{W F (h)} = \text{WF(ah) i.e. A is hypoelliptic.}$$

<u>Local Solvable and Hypoelliptic Operators</u>.

A pseudo-differential operator A is said to be <u>solvable at</u> $x \in M$, if there exists an open neighbourhood N of x such that

$$\forall f \in \mathcal{C}^\infty(M), \ \exists\, k \in \mathcal{D}'(M), \quad Ak = f \quad \text{in } N .$$

Let $(x,\xi) \in T^*(M) \setminus \{0\}$ such that $p(x,\xi) = 0$, then A is said to be <u>solvable</u> at (x,ξ) if

$$\forall f \in \mathcal{D}'(M), \ \exists\, k \in \mathcal{D}'(M), (x,\xi) \notin WF(Ak - f) .$$

Let $(x,\xi) \in T^*(M) \setminus \{0\}$ such that $p(x,\xi) = 0$ and $d \operatorname{Re} p(x,\xi) \neq 0$. Let H be the smooth hypersurface defined by $\operatorname{Re} p = 0$ and containing (x,ξ) and let $s \in H$. Through s there exists an oriented bicharacteristic of the Hamiltonian field of $\operatorname{Re} p$ which stays in H.

Let A be a partial differential operator of principal type.

Then

A hypoelliptic $\Longrightarrow$

(4) in a neighbourhood of $(x,\xi) \in H$, the restriction of $\operatorname{Im} p$ to bicharacteristics of Hamiltonian field of $\operatorname{Re} p$ never has a zero of finite order where the sign changes from positive to negative.

Let A^* be the adjoint of A and suppose that the Hamiltonian field of p does not have the radial direction. In that case Nirenberg and Trèves [9] , [10] have conjectured

A^* is solvable at $(x,\xi) \Longleftrightarrow$ the sign does not change from positive to negative at any zero either of finite order or infinite order.

Suppose the last equivalence true, then A hypoelliptic partial differential operator of principal type $\Longleftrightarrow$ (4) (Trèves [14])

References

1 Bokolza-Haggiag, J., Opérateurs differentials sur une variété différentiable. Ann.Inst.Fourier, Grenoble 19 (1969) pp. 125 - 177.

2 Duistermaat, J.J. and Hörmander, L., Fourier Integral Operators II, Acta Math. 128 (1972) pp. 183 - 269.

3 Hörmander, L., Fourier integral operators I. Acta Math. 127 (1971) pp. 79 - 183.

4 ------------, On the theory of general partial differential operators. Acta Math. 94 (1955) pp. 161 - 248.

5 ------------, Linear partial differential operators, Grundlehren d. Math. Wiss. 116, Springer-Verlag, 1963.

6 ------------, On the existence and the regularity of solutions of linear pseudo-differential equations. L'Ensignement Math. 17 (1971) pp.671-704.

7 Hörmander, L., Linear differential operators. Actes Cong.Inst. Math. Nice 1970, 1, pp. 121-133.

8 Nirenberg, L., Pseudo-differential operators. Amer.Math.Soc., Symp. Pure Math. 16 (1970), pp.149-167.

9 Nirenberg, L. and Trèves, F., Solvability of a first order partial differential equation . Comm. Pure Appl. Math. 14(1963) pp.331-351.

10 ----------, On local solvability of linear partial differential equations. Part I. Necessary conditions. Comm. Pure Appl. Math. 23 (1970) pp. 1-38. Part II. Sufficient conditions. Comm. Pure Appl. Math. 23 (1970) pp. 459-510.

11 Sato, M., Hyperfunctions and partial differential equations. Proc. Int. Conf. on Functional Analysis, Tokyo, 1969 pp.91-94

12 -------, Regularity of hyperfunction solutions of partial differential equations. Actes Cong. Int. Math. Nice, 1970, 2, pp. 785 - 794.

13 Schwartz, L., Théorie des distributions (3rd édition) 1966, Hermann, Paris.

14 Trèves, F., Hypoelliptic partial differential equations of principal type. Sufficient conditions and necessary conditions. Comm. Pure Appl. Math. 24 (1971) pp. 631-670.

HARMONIC ANALYSIS IN THE COMPLEX DOMAIN AND APPLICATIONS IN THE THEORY OF ANALYTICAL AND INFINITELY DIFFERENTIABLE FUNCTIONS

M.M.Dzhrbashyan

The present report briefly reviews the results of a big series of investigations by the author and his students dealing with the problem of harmonic analysis in the complex domain and the most important applications of the developed theory to some fundamental questions of the classical theory of functions.

The major part of these investigations has been published in different mathematical journals and also in a monograph by the author Nevertheless, I consider it appropriate to present this report to such a representative forum of contemporary analysis as this conference sponsored by the Madras Institute of Mathematical Sciences.

The report consists of three parts:

I. The theory of integral transforms with Mittag-Leffler and Volterra kernels (Plancherel-type theorems in the complex domain).

II. The classes of entire, quasi-entire and analytical functions and their integral representations. (Wiener-Paley-type theorems).

III. The uniqueness of certain general classes of infinitely differentiable functions (Denjoy-Carleman-type theorems).

I

(a) An important landmark in the development of functional analysis in general and of the theory of harmonic analysis in particular was the fundamental theorem on Fourier transforms in the classes L_2 established in 1910 by M.Plancherel.

It is well known that for complex form of Fourier transforms this theorem reads:

1^{0} For any function $f(x) \in L_2(-\infty,\infty)$ the following limit in the mean exists

$$F(u) = \underset{\sigma \to +\infty}{\mathrm{l.i.m}} \frac{1}{\sqrt{2\pi}} \int_{-\sigma}^{\sigma} f(t)\, e^{-iut}\, dt \in L_2(-\infty, \infty)$$

defining the Fourier transform, $\mathfrak{F}[f] = F(u)$

2^{0} The reverse limit relation

$$f(x) = \underset{\sigma \to +\infty}{\mathrm{l.i.m}} \frac{1}{\sqrt{2\pi}} \int_{-\sigma}^{\sigma} F(u)\, e^{ixu}\, du \equiv \mathfrak{F}^{-1}[f]$$

and also the equality

$$\int_{-\infty}^{\infty} |f(x)|^2\, dx = \int_{-\infty}^{\infty} |F(u)|^2\, du$$

hold.

Thus, this theorem has for the first time constructed a Fourier operator in L_2 and established a complete equivalence between the function and its Fourier transform giving a unitary mapping of the whole space L_2 upon itself.

At a considerably later date G. Watson (1933) constructed the general theory of the Fourier-type transforms in $L_2(0,+\infty)$ performing unitary or quasi-unitary mapping of L_2 upon itself with the aid of formulae of the form

$$g(x) = \frac{d}{dx} \int_{0}^{+\infty} \frac{K(x,y)}{y}\, f(y)\, dy \ , \qquad f(x) = \frac{d}{dx} \int_{0}^{+\infty} \frac{h(x,y)}{y}\, g(y)\, dy$$

A note should be also taken of the significant generalizations of Plancherel's theorems discovered within the theory of singular boundary problems of the Sturm-Liouville-type on the half-axis $(0,+\infty)$.

(b) The theory of integral transforms developed in our investigations is very substantially supported by the remarkable asymptotic properties of the two families of functions; the functions of the Mittag-Leffler-type

$$E_\rho(z;\mu) = \sum_{k=0}^{\infty} \frac{z^k}{\Gamma(\mu+\frac{k}{\rho})} \qquad (\mu > 0,\ \rho > 0)$$

and their continuous analogs-the functions of Volterra

$$\nu_\rho(z;\mu) = \int_0^{+\infty} \frac{z^t}{\Gamma(\mu+\frac{t}{\rho})}\,dt \qquad (\mu > -1,\ \rho > 0)$$

Note that the function $E_\rho(z;\mu)$ with $(\mu = 1)$ was first introduced into analysis by Mittag-Leffler at the beginning of this century (1903-1905) in connection with his discovery of a new rather strong method of summing divergent series. In this connection he was also the first to discover asymptotic formulae for this function when $|z| \to \infty$ in the domain

$$\Delta_\rho = \left\{ z : |\arg z| < \frac{\pi}{2\rho} \right\} \qquad (\rho > \tfrac{1}{2})$$

and in its complement $\Delta_\rho^* = C\Delta_\rho$.

As to the function $\nu_\rho(z;\mu)$ (for $\rho = 1$, $\mu = 0$) it was introduced into the analysis by V.Volterra (1916) in connection with the solution of a special integral equation occuring in the theory of heredity.

The considerable period of time that has elapsed since then has not enriched the mathematical literature with any more or less serious investigations on the properties, or, all the more so, on applications of these functions to the contemporary analysis.

The asymptotic properties of the functions $E_\rho(z;\mu)$ and $\nu_\rho(z;\mu)$ are given by the following theorem:

THEOREM 1. 1^0 For each $\mu \in (-\infty,\infty)$ and $\rho > \frac{1}{2}$ the function $E_\rho(z;\mu)$ is entire of order ρ and type $\sigma = 1$, and, if α is any number satisfying the condition

$$\frac{\pi}{2\rho} < \alpha < \min\left\{\pi, \frac{\pi}{\rho}\right\}$$

then for $|z| \longrightarrow \infty$

$$\text{a)} \quad E_\rho(z;\mu) = \rho z^{\rho(1-\mu)} e^{z^\rho} + O\left(\frac{1}{z}\right) \quad \text{for } |\arg z| \leq \alpha \tag{I.1}$$

$$\text{b)} \quad E_\rho(z;\mu) = O\left(\frac{1}{z}\right) \quad \text{for } \alpha \leq |\arg z| \leq \pi \tag{I.2}$$

2^0 For each $\mu \in (-\infty,\infty)$ and $\rho > 0$ the function $\nu_\rho(z;\mu)$ is analytical on the whole Riemann surface

$$G_\infty = \left\{z : |\operatorname{Arg} z| < \infty, \quad 0 < |z| < \infty\right\}$$

(we call such functions quasi-entire) has order ρ and type $\sigma = 1$, and, for any $\alpha \in \left(\frac{\pi}{2\rho}, \frac{\pi}{\rho}\right)$ when $|z| \longrightarrow \infty$

$$\text{a)} \quad \nu_\rho(z;\mu) = \rho z^{\rho(1-\mu)} e^{z^\rho} + O\left(\frac{1}{\log|z|}\right) \quad \text{for } |\operatorname{Arg} z| \leq \alpha \tag{I.3}$$

$$\text{b)} \quad \nu_\rho(z;\mu) = O\left(\frac{1}{\log|z|}\right) \quad \text{for } \alpha \leq |\operatorname{Arg} z| < \infty \tag{I.4}$$

The asymptotic properties of $E_\rho(z;\mu)$ for $0 < \rho \leq \frac{1}{2}$ have a different formulation. For us it was particularly important to reveal these properties for $\rho = \frac{1}{2}$.

THEOREM 2. If $0 < \mu < 3$ then

a) For $0 \leq \arg z \leq \pi$ or $-\pi \leq \arg z \leq 0$ respectively, when $|z| \longrightarrow \infty$ we have

$$E_{\frac{1}{2}}(z;\mu) = \frac{1}{2} z^{\frac{1}{2}(1-\mu)} \left\{e^{z^{1/2}} + e^{\mp i\pi(1-\mu)} e^{-z^{1/2}}\right\} + O\left(\frac{1}{z}\right) \tag{I.5}$$

b) <u>For</u> $0 < x < +\infty$ <u>when</u> $x \to \infty$ <u>we have</u>

$$E_{\frac{1}{2}}(-x,\mu) = x^{\frac{1}{2}(1-\mu)} \cos\left[\sqrt{x} + \frac{\pi}{2}(1-\mu)\right] + O\left(\frac{1}{x}\right) \qquad (I.5^{I})$$

c) It will be assumed below that by a given arbitrary $\rho \geqslant \frac{1}{2}$ the parameter μ is subject to the condition

$$\frac{1}{2} < \mu < \frac{1}{2} + \frac{1}{\rho} \qquad (I.6)$$

Henceforth we shall write $g(y) \in L_{2,\mu}(0,\infty)$ if $g(y)y^{\mu-1} \in L_2(0,\infty)$. Finally, if $g(y) \in L_{2,\mu}(0,\infty)$ and the family of functions $\{g_\sigma(y)\}$, $g_\sigma(y) \in L_{2,\mu}(0,\infty)$ depending on the parameter $\sigma(0<\sigma<\infty)$ is such that

$$\lim_{\sigma \to +\infty} \int_0^{+\infty} |g(y) - g_\sigma(y)|^2 y^{2(\mu-1)} dy = 0$$

then we shall write

$$g(y) = \underset{\sigma \to +\infty}{\overset{(\mu)}{\text{l.i.m}}}\; g_\sigma(y)$$

The first basic theorem on the transforms with Mittag-Leffler kernels reads:

<u>THEOREM 3.</u> <u>Let</u> $g(y) \in L_{2,\mu}(0,\infty)$,

$$f(x) = \frac{1}{\sqrt{2\pi\rho}} \frac{d}{dx} \int_0^{+\infty} \frac{e^{-ixy} - 1}{-iy} g(y)\, y^{\mu-1} dy, \quad x \in (-\infty,\infty) \qquad (I.7)$$

<u>and thus</u>

$$\int_0^{\infty} |g(y)|^2 y^{2(\mu-1)} dy = \rho \int_{-\infty}^{\infty} |f(x)|^2 dx \qquad (I.8)$$

I°. <u>Putting</u>

$$g(y;\varphi) \equiv \frac{y^{1-\mu}}{\sqrt{2\pi\rho}} \frac{d}{dy}\left\{ y^{\mu-1} \int_{-\infty}^{\infty} E_\rho\{(ix)^{\frac{1}{\rho}} y^{\frac{1}{\rho}} e^{i\varphi};\mu+1\}(ix)^{\mu-1} f(x)\,dx \right\} \quad (I.9)$$

<u>we shall have almost everywhere on the half-axis</u>

$$g(y;\varphi) = \begin{cases} g(y) & \rho \geq \frac{1}{2}, \ \varphi = 0 \\ 0 & \rho \geq 1, \ \frac{\pi}{\rho} \leq |\varphi| \leq \pi \end{cases} \quad (I.10)$$

2°. <u>Besides, putting</u>

$$g(y;\varphi;\sigma) = \frac{1}{\sqrt{2\pi\rho}} \int_{-\sigma}^{\sigma} E_\rho\{(ix)^{\frac{1}{\rho}} y^{\frac{1}{\rho}} e^{i\varphi};\mu\}(ix)^{\mu-1} f(x)\,dx \quad (I.11)$$

<u>with respect to the conditions (I.10) we shall also have</u>

$$g(y;\varphi) = \underset{\sigma \to +\infty}{\overset{(\mu)}{\mathrm{l.i.m}}}\; g(y;\varphi;\sigma) \quad (I.10^{I})$$

Thus, this theorem contains two substantially new statements of harmonic analysis:

1) For the usual Fourier transform of functions from $L_{2,\mu}(0,\infty)$ there is not one but a whole family of inversion formulae with $E_\rho(z;\mu)$-type kernels, where $\rho \geq \frac{1}{2}$ is any number. Only the special selection of parameters ρ and μ (when $\rho = \frac{1}{2}$, $\mu = 1,2$ or $\rho = \mu = 1$) leads to the inversion formulae well known in the Fourier - Plancherel theory.

2) When $\rho \geq 1$, the general inversion formula (I.10) has an important property: representing the function $g(y)$ on the ray $\varphi = 0$. It simultaneously represents zero identically in the angular domain $\frac{\pi}{\rho} \leq |\varphi| \leq \pi$.

This permits the construction of the apparatus of integral transforms and their inversions for an arbitrary finite system of rays proceeding from one point of the complex plane.

Prior to formulating the theorem we shall introduce some notations.

Denote by $L\{\varphi_1, \varphi_2, \dots, \varphi_p\}$ the set of rays

$$\ell_k : \arg z = \varphi_k \qquad k = 1, 2, \dots, p$$

$$0 \leq \varphi_1 < \varphi_2 < \dots < \varphi_p < \varphi_{p+1} = 2\pi + \varphi_1$$

emerging from the origin. The system of these rays divides the z-plane into p angular domains with a common vertex at z = 0.

Denote

$$\omega = \max_{1 \leq k \leq p} \left\{ \frac{\pi}{\varphi_{k+1} - \varphi_k} \right\}$$

noting that $\frac{\pi}{\omega}$ is the value of the smallest of these angles.

<u>THEOREM 4. Let</u> $\rho \geqslant \omega$ <u>and</u> $g(z)$ <u>be an arbitrary function defined on the set of rays</u> $L\{\varphi_1, \varphi_2, \dots, \varphi_p\}$ <u>and such that</u>

$$\int_{L\{\varphi_1, \varphi_2, \dots, \varphi_p\}} |g(z)|^2 \, |z|^{2(\mu-1)} |dz| < +\infty \tag{I.12}$$

1°. <u>Putting</u>

$$f_k(x) = \frac{1}{\sqrt{2\pi\rho}} \frac{d}{dx} \int_0^{+\infty} \frac{e^{-ixr} - 1}{-ir} g(r e^{i\varphi_k}) r^{\mu-1} dr \in L_2(0, +\infty) \qquad (k = 1, 2, \dots, p) \tag{I.13}$$

<u>we shall have for almost any</u> $re^{i\varphi} \in L\{\varphi_1, \varphi_2, \dots, \varphi_p\}$

$$g(re^{i\varphi}) = \frac{r^{1-\mu}}{\sqrt{2\pi\rho}} \sum_{k=1}^{p} \frac{d}{dr} \left\{ r^{\mu} \int_{-\infty}^{\infty} E_\rho\left\{(ix)^{\frac{1}{\rho}} r^{\frac{1}{\rho}} e^{i(\varphi-\varphi_k)} ; \mu+1\right\} (ix)^{\mu-1} f(x) \, dx \right\} \tag{I.14}$$

2°. The Parseval type equality

$$\int\limits_{L\{\varphi_1,\varphi_2,\dots\varphi_p\}} |g(z)|^2 |z|^{2(\mu-1)} |dz| = p \sum_{k=1}^{p} \int_{-\infty}^{\infty} |f_k(x)|^2 dx \tag{I.15}$$

holds.

It is easy to see, that the basic theorem of Plancherel is a special case of this theorem when a system of two rays $L\{0,\pi\}$ is considered and $\mu = 1$.

Finally we shall note that the inverse of Theorem 3 holds:

THEOREM 5. 1° For any function $f(x) \in L_2(0,\infty)$ the transforms

$$g^{(\pm)}(y) = \frac{y^{1-\mu}}{\sqrt{2\pi p}} \frac{d}{dy}\left\{ y^{\mu} \int_0^{\infty} E_p\{e^{\pm i\frac{\pi}{2p}} y^{\frac{1}{p}} x^{\frac{1}{p}};\mu+1\} x^{\mu-1} f(x)\,dx \right\} \tag{I.16}$$

determine functions $g^{(\pm)}(y) \in L_{2,\mu}(0,\infty)$ and the inversion formula

$$f(x) = \frac{1}{\sqrt{2\pi p}} \left\{ e^{-i\frac{\pi}{2}(1-\mu)} \frac{d}{dx} \int_0^{+\infty} \frac{e^{-ixy}}{iy} g^{(+)}(y)\, y^{\mu-1}\, dy + \right.$$

$$\left. + e^{i\frac{\pi}{2}(1-\mu)} \frac{d}{dx} \int_0^{+\infty} \frac{e^{ixy}-1}{iy} g^{(-)}(y)\, y^{\mu-1} dy \right\}, \quad x \in (0,\infty) \tag{I.17}$$

holds.

2°. The inequalities

$$\int_0^{\infty} |g^{(\pm)}(y)|^2 y^{2(\mu-1)} dy \leq \frac{M\mu^2}{2\pi p} \int_0^{+\infty} |f(x)|^2 dx \tag{I.18^{I}}$$

$$\int_0^{\infty} |f(x)|^2 dx \leq p^{-1} \left\{ \int_0^{+\infty} |g^{(+)}(y)|^2 y^{2(\mu-1)} dy + \right.$$

$$\left. + \int_0^{+\infty} |g^{(-)}(y)|^2 y^{2(\mu-1)} dy \right\} \tag{I.18^{II}}$$

hold.

d) The results of the theorems 3,4 and 5 remain valid if the function $E_\rho(z;\mu)$ is replaced by $\mathcal{V}_\rho(z;\mu)$ putting everywhere $\mu = \frac{1}{2}$ and $\rho = 0$.

For this reason we do not present here the reformulations of these theorems, but only give the analogue of the Theorem 4 which is of rather extensive generality.

Let $\mathcal{L}\{v_1,\dots,v_p\}$ denote the set of parallel straight lines

$$w = u + i\,v_k, \quad -\infty < u < \infty \quad (k = 1,2,\dots,p)$$

$$-\infty < v_1 < v_2 < \dots < v_p < +\infty.$$

Denote

$$\omega = \max_{1 \leqslant k \leqslant p-1} \left\{ \frac{\pi}{v_{k+1} - v_k} \right\}.$$

We see that the value $\frac{\pi}{\omega}$ is equal to the width of the smallest of the strips generated on the z plane by the system of lines $\mathcal{L}\{v_1,v_2,\dots,v_p\}$.

THEOREM 6. *Let the function $G(w)$ be defined on the system of straight lines $\mathcal{L}\{v_1,v_2,\dots,v_p\}$ and satisfy the condition*

$$\int_{\mathcal{L}\{v_1,v_2,\dots,v_p\}} |G(w)|^2 \, d(\mathrm{Re}\, w) < \infty \tag{I.19}$$

Then for any $\rho \geqslant \omega$ the following statements hold:

1°. *For almost any* $w \in \mathcal{L}\{v_1,v_2,\dots,v_p\}$

$$G(w) = e^{-\frac{\rho w}{2}} \sum_{k=1}^{p} \frac{1}{\sqrt{2\pi\rho}} \frac{d}{dw} \left\{ e^{\frac{\rho w}{2}} \int_{-\infty}^{\infty} \left[\mathcal{V}_\rho \left(e^{i(\frac{\pi}{2\rho} - v_k) + w} x^{\frac{1}{\rho}}; \frac{3}{2} \right) \times \right. \right.$$

$$\left. \left. \times\, x^{-\frac{1}{2}} f_k(x) \right] dx \right\} \tag{I.20}$$

where

$$f_k(x) = \frac{e^{-i\frac{\pi}{4}}}{\sqrt{2\pi\rho}} \frac{d}{dx} \int_{-\infty}^{\infty} (e^{-ixe^{\rho u}} - 1)\, e^{-\frac{u\rho}{2}}\, G(u+iv_k)\,du$$

$$(k = 1, 2, \dots, p) \qquad (I.21)$$

2^0. The equality

$$\int_{\mathcal{L}\{v_1, v_2, \dots, v_p\}} |G(w)|^2\, d(\operatorname{Re} w) = \rho \sum_{k=1}^{p} \int_{-\infty}^{\infty} |f_k(x)|^2\, dx \qquad (I.22)$$

holds.

In conclusion we would like to note that the Theorem 3,4 and 6 may be considered as theorems of approximation by entire functions on the half-axis $(0,+\infty)$ or on the system of rays $L\{\varphi_1, \varphi_2, \dots, \varphi_p\}$ or on the system of parallel straight lines $\mathcal{L}\{v_1, \dots, v_p\}$.

II

(a) In their famous monograph "Fourier transforms in the complex domain" (1934) Wiener and Paley established two fundamental theorems: on representations of exponential type entire functions from $L_2(-\infty,\infty)$ and functions from H_2 analytical in the half-plane. We shall formulate these famous theorems for they served as a starting point of our investigations in this section.

The first of these theorems reads:

The class of entire functions from $L_2(-\infty,+\infty)$ of exponential type $\leqslant \sigma$ coincides with the set of functions admitting a representation of the form

$$f(z) = \int_{-\sigma}^{\sigma} e^{itz}\, \varphi(t)\,dt$$

where $\varphi(t)$ is an arbitrary function from $L_2(-\sigma,\sigma)$.

The second theorem reads:

The class H_2 *of functions* $F(z)$ *analytical in the half-plane* $\operatorname{Re} z > 0$ *and satisfying the condition*

$$\sup_{0 < x < +\infty} \left\{ \int_{-\infty}^{\infty} |F(x+iy)|^2 \, dy \right\} < +\infty$$

coincides with the set of functions admitting a representation of the form

$$F(z) = \int_0^{\infty} e^{-tz} \varphi(t) dt$$

where

$$\varphi(t) \in L_2(0, \infty).$$

Our results on representations of analytical functions are far-reaching substantial generalizations of these two theorems.

(b) Let us first give the formulation of the most general theorem on entire functions. To this end we shall first introduce some preliminary notations.

We shall assume that the natural number $\varkappa = \varkappa(\rho) > 0$ satisfies the condition

$$\varkappa \geqslant [2\rho] - 1 \tag{II.1}$$

for any $\rho \geqslant \frac{1}{2}$.

Then, for a given $\rho \geqslant \frac{1}{2}$ we shall assume that the set of numbers $\{\vartheta_0, \vartheta_1, \ldots, \vartheta_{\varkappa+1}\}$ satisfies the conditions

$$-\pi < \vartheta_0 < \vartheta_1 < \ldots < \vartheta_\varkappa \leq \pi < \vartheta_{\varkappa+1} = \vartheta_0 + 2\pi,$$
$$\max_{0 \leq k \leq \varkappa} \{\vartheta_{k+1} - \vartheta_k\} = \frac{\pi}{\rho} \tag{II.2}$$

Starting with the set $\{\vartheta_0, \vartheta_1, \ldots, \vartheta_{\varkappa+1}\}$ we form a sequence of pairs

$$(\vartheta_k, \vartheta_{k+1})_1^{\varkappa} = \{(\vartheta_0, \vartheta_1), (\vartheta_1, \vartheta_2), \ldots, (\vartheta_\varkappa, \vartheta_{\varkappa+1})\} \tag{II.3}$$

and then, preserving the mutual order of their succession, we shall isolate all pairs for which the equality

$$\vartheta_{r_{k+1}} - \vartheta_{r_k} = \frac{\pi}{\rho} \quad (k = 0, 1, \dots, p \le \varkappa) \tag{II.4}$$

holds.

Here, if $p < \varkappa$, let us denote the remaining pairs of (II.3) by $(\vartheta_{s_k}, \vartheta_{s_{k+1}})_1^q \quad (q = \varkappa - p)$

Denote further

$$\theta_k = \frac{1}{2}\left\{\vartheta_{r_k} + \vartheta_{r_{k+1}}\right\} \quad (k = 0, 1, \dots, p)$$

and assuming that $-1 < \omega < 1$, $\sigma_k \geqslant 0$ $(k = 0,1,2,\dots,p)$ associate with the set of numbers $\{\vartheta_0, \vartheta_1, \dots, \vartheta_{\varkappa+1}\}$ the class $W_\sigma^{(\rho)}(\omega; \{\vartheta_k\}; \{\sigma_k\})$ of entire functions of order ρ and normal type $\leqslant \sigma$ satisfying the conditions

$$\int_0^{+\infty} \left| f(te^{-i\vartheta_k})\right|^2 t^\omega dt < +\infty \quad (k = 0,1,\dots,\varkappa) \tag{II.6}$$

$$\overline{\lim_{r \to +\infty}} \frac{\log |f(re^{-i\theta_k})|}{r^\rho} = h(-\theta_k; f) \leqslant \sigma_k \leqslant \sigma \quad (k=0,1,2,\dots p)$$

Basing substantially on our theory of integral transforms with Mittag-Leffler-type kernels we have established the following general theorem.

<u>THEOREM 7.</u> <u>The class</u> $W_\sigma^{(\rho)}(\omega; \{\vartheta_k\}; \{\sigma_k\})$ <u>coincides with the set of functions</u> $f(z)$ <u>admitting the representation</u>

$$f(z) = \sum_{k=0}^{p} \int_0^{\sigma_k} E_\rho\left\{e^{i\theta_k} z\tau^{\frac{1}{\rho}}; \mu\right\} \varphi_k(\tau)\, \tau^{\mu-1} d\tau, \tag{II.7}$$

<u>where</u>

$$\mu = \frac{\omega+\rho+1}{2\rho}, \quad \varphi_k(\tau) \in L_2(0, \sigma_k) \quad (k = 0,1,2,\dots,p)$$

Here if

$$\Phi_k(\tau) = \frac{1}{\sqrt{2\pi}} \frac{d}{d\tau} \int_0^{+\infty} f(e^{-i\vartheta_k} v^{\frac{1}{\rho}}) \frac{e^{-i\tau v} - 1}{-iv} v^{\mu-1} dv \tag{II.8}$$

then almost everywhere we have

$$\frac{i}{\sqrt{2\pi}\,\rho} \left\{ e^{-i\frac{\pi}{2}\mu} \Phi_{r_{k+1}}(-\tau) - e^{i\frac{\pi}{2}\mu} \Phi_{r_k}(\tau) \right\} =$$

$$= \begin{cases} \varphi_k(\tau), & \tau \in (0, \sigma_k) \\ 0, & \tau \in (\sigma_k, +\infty) \end{cases} \qquad (k = 0, 1, 2, \ldots, p) \tag{II.9}$$

By special selection of the system of rays $\{\arg z = \vartheta_k\}_0^{\varkappa}$ along which the conditions of the form (II.6) are imposed, some corollaries of a more special character, including the Wiener-Paley theorem itself may be obtained.

For example, the following statement is true

THEOREM 8. 1°. The class of entire functions $f(z)$ of order $\rho \geq \frac{1}{2}$ and type $\leq \sigma$ subject to the condition

$$\sup_{\frac{\pi}{2\rho} < \vartheta \leq \pi} \left\{ \int_0^{+\infty} |f(te^{-i\vartheta})|^2 t^{\omega} dt \right\} < +\infty \quad (-1 < \omega < 1)$$

coincides with the set of functions of the form

$$f(z) = \int_0^{\sigma} E_\rho\{ z\tau^{\frac{1}{\rho}}; \mu\} \varphi(\tau) \tau^{\mu-1} d\tau$$

where $\mu = \frac{\omega+1+\rho}{2\rho}$ and $\varphi(t) \in L_2(0,\sigma)$.

2°. The class of entire functions $f(z)$ of order one and type $\leq \sigma$ for which

$$\int_{-\infty}^{\infty} |f(x)|^2 |x|^{\omega} dx < +\infty$$

coincides with the set of functions of the form

$$f(z) = \int_{-\sigma}^{\sigma} E_1\{ i\tau z; \mu\} \varphi(\tau) |\tau|^{\mu-1} d\tau$$

<u>where</u> $\mu = 1 + \frac{\omega}{2}$ <u>and</u> $\varphi(\tau) \in L_2(-\sigma,\sigma)$.

Remark, that the theorem of Wiener-Paley is contained in the statement 2° of this theorem as a special case when $\omega = 0$ because then $\mu = 1$ and $E_1(z,1) = e^z$.

(c) A function analytical on the whole Riemann surface G_∞ with

1) $\widetilde{M}_f(r) = \sup_{\varphi} |f(r e^{i\varphi})| < \infty \qquad (0 < r < +\infty)$

2) $\lim_{r \to +0} \widetilde{M}_f(r) < \infty$

is called quasi-entire.

The order ρ and type σ of quasi-entire functions are determined in a way analogous to that in the theory of entire functions.

A result similar to Theorem 7 is also established for the quasi-entire functions with the function $E_\rho(z;\mu)$ replaced by $\nu_\rho(z;\frac{1}{2})$ and $\omega = -1$, $\rho > 0$ may be arbitrary.

In order to state the result some new notations are needed.

For the given value of $\rho (0 < \rho < \infty)$ assume that the set of numbers $\{\vartheta_k\}_{-p}^{q}$ $(p \geq 1,\ q \geq 0)$ satisfies the following conditions:

$$-\alpha_0 \leq \vartheta_{-p} < \vartheta_{-p+1} < \dots < \vartheta_{-1} < \vartheta_0 < \vartheta_1 < \dots < \vartheta_q \leq \alpha_0$$

where $\alpha_0 > 0$ and

$$\max_{-(p+1) \leq k \leq q-1} \{\vartheta_{k+1} - \vartheta_k\} = \frac{\pi}{\rho}$$

Further, forming the successive pairs $\{(\vartheta_k, \vartheta_{k+1})\}_{-p}^{q-1}$ we shall isolate, retaining their mutual order of succession, all pairs $(\vartheta_{r_k}, \vartheta_{r_{k+1}})_1^m$ for which

$$\vartheta_{r_{k+1}} - \vartheta_{r_k} = \frac{\pi}{\rho} \qquad (k = 1,2,\dots,m \leq p+q)$$

Finally denote by $\tilde{W}_{\sigma}^{(\rho)}(\{\vartheta_k\};\{\sigma_k\})$ the set of quasi-entire functions $f(z)$ of order ρ $(0 < \rho < +\infty)$ and type $\leqslant \sigma$ subject to the conditions

1) $$\int_0^{+\infty} |f(r e^{-\vartheta_k})|^2 r^{-1} dr < +\infty \qquad (-p \leq k \leq q)$$

2) $$\sup_{|\varphi| \geqslant \alpha_0} \left\{ \int_0^{+\infty} |f(r e^{i\varphi})|^2 r^{-1} dr \right\} < +\infty$$

3) $$h(-\theta_k;f) = \overline{\lim_{r \to +\infty}} \frac{\log |f(r e^{-i\theta_k})|}{r^\rho} \leqslant \sigma_k \leqslant \sigma$$
$$(k = 1,2,\ldots,m)$$

where
$$\theta_k = \frac{1}{2} \{\vartheta_{r_k} + \vartheta_{r_{k+1}}\}$$

THEOREM 9. The class $\tilde{W}_{\sigma}^{(\rho)}(\{\vartheta_k\}, \{\sigma_k\})$ coincides with the set of functions $f(z)$ representable in the form

$$f(z) = \sum_{k=1}^{m} \int_0^{\sigma_k} \nu_\rho \left\{ e^{i\theta_k} z \tau^{\frac{1}{\rho}}; \frac{1}{2} \right\} \tau^{-\frac{1}{2}} \varphi_k(\tau) d\tau \qquad (\text{II.10})$$

where $\varphi_k(\tau) \in L_2(0,\sigma_k)$ $(1 \leqslant k \leqslant m)$.

(d) As to the results of the type of Wiener-Paley second basic theorem here we have

THEOREM 10. 1°. The class $\mathcal{H}_2(\alpha,\omega)$ $(\frac{1}{2} < \alpha < +\infty; -1 < \omega < 1)$ of functions $F(z)$ analytical in the angle

$$\Delta_\alpha = \left\{ z: |\arg z| < \frac{\pi}{2\alpha}, \; 0 < |z| < \infty \right\}$$

for which

$$\sup_{|\varphi| < \frac{\pi}{2\alpha}} \left\{ \int_0^{+\infty} |F(re^{i\varphi})|^2 r^\omega dr \right\} < +\infty$$

coincides with the set of functions of the form

$$F(z) = \int_0^{+\infty} E_\rho\left(e^{i\frac{\pi}{2\gamma}} z\, t^{\frac{1}{\rho}}; \mu\right) V_{(-)}(t)\, t^{\mu-1}\, dt \; +$$

$$+ \int_0^{+\infty} E_\rho\left(e^{-i\frac{\pi}{2\gamma}} z\, t^{\frac{1}{\rho}}; \mu\right) V_{(+)}(t)\, t^{\mu-1}\, dt, \qquad z \in \Delta_\alpha \tag{II.11}$$

where $V_{(\pm)}(t) \in L(0,\infty)$, $\rho \geqslant \dfrac{\alpha}{2\alpha - 1}$,

$$\gamma = \frac{\rho\,\alpha}{\rho+\alpha}, \quad \mu = \frac{1+\omega+\rho}{2\,\rho} \tag{II.12}$$

2^{0}. The class $\mathcal{H}_2[\alpha]$ $(0 < \alpha < +\infty)$ of functions analytical in the domain $\Delta_\alpha \in G_\infty$ and subject to the condition

$$\sup_{|\varphi| < \frac{\pi}{2\alpha}} \left\{ \int_0^{+\infty} |F(re^{i\varphi})|^2\, dr \right\} < +\infty$$

coincides with the set of functions representatable in the form

$$F(z) = z^{-\frac{1}{2}} \int_0^{+\infty} \nu_\rho\left(e^{i\frac{\pi}{2\gamma}} z\, t^{\frac{1}{\rho}}; \tfrac{1}{2}\right) t^{-\frac{1}{2}} V_{(-)}(t)\, dt \; +$$

$$+ z^{-\frac{1}{2}} \int_0^{+\infty} \nu_\rho\left(e^{-i\frac{\pi}{2\gamma}} z\, t^{\frac{1}{\rho}}; \tfrac{1}{2}\right) t^{-\frac{1}{2}} V_{(+)}(t)\, dt, \qquad z \in \Delta_\alpha \tag{II.13}$$

where $\rho > 0$ is any number,

$$\gamma = \frac{\alpha\rho}{\alpha+\rho}, \quad V_{(\pm)}(t) \in L_2(0,\infty)$$

An important feature of the integral formulae (II.11) and (II.13) is that they also represent zero identically in the domain complementary to $\Delta(\alpha)$.

Namely, for (II.11) this is the case when $\rho \geqslant \dfrac{2\alpha}{2\alpha - 1}$ in the domain

$$|\pi - \arg z| < \frac{\pi}{2\varkappa}, \qquad \varkappa = \frac{\alpha\,\rho}{(2\alpha-1)\rho - 2\alpha}$$

and for (II.13) this is the case in the domain of values

$$\frac{\pi}{2\alpha} + \frac{\pi}{\rho} < |\operatorname{Arg} z| < +\infty, \qquad 0 < |z| < +\infty .$$

This remarkable fact permits the construction of the apparatus and the development of the theory of Fourier - Plancherel-type integrals for the sets consisting of a finite number of rays and angular domains lying on the z-plane or the Riemann surface G_∞.

Precise formulations of the theorems in question shall not be dwelt upon.

III

As early as 1912 J.Hadamard posed the problem of determining conditions for a sequence of positive numbers $\{M_n\}_1^\infty$ ensuring the uniqueness of the class $C\{M_n\}$ of functions infinitely differentiable on some interval $\mathfrak{I} = (a,b)$ and for values $\{\varphi^{(n)}(x_0)\}_0^\infty$, $x_0 \in \mathfrak{I}$ satisfying the conditions

$$|\varphi^{(n)}(x)| \leq A.B^n . M_n \quad (n = 1,2,3,\ldots).$$

In 1921 A.Denjoy for the first time established the existence of such classes substantially wide (as compared with the usual classes $C\{n!\}$) of analytical functions. In particular, he proved that the class $C\{M_n\}$ is quasi-analytical, if, e.g.

$$M_n = (n \log n \ldots \log_p n)^n , \qquad n \geqslant N_p .$$

T.Carleman (1923-1926) gave a comprehensive solution of Hadamard's problem by laying down the necessary and sufficient condition for the class $C\{M_n\}$ to possess the property of uniqueness or, in other words for its quasi-analyticity.

In the formulation by A.Ostrowsky the Carleman's result usually called the theorem of Denjoy-Carleman reads:

The condition

$$\int_1^{+\infty} \frac{\log T(r)}{r^2}\,dr = +\infty, \quad T(r) = \sup_{n \geqslant 1} \frac{r^n}{M_n} \qquad \text{(III.1)}$$

is necessary and sufficient for the class $C\{M_n\}$ to be quasi-analytical.

In the five decades that followed, many original investigations have come into being dealing with the theory of quasi-analytical functions. Here a special note should be given to the study by S.Mandelbrojt on the generalized quasi-analyticity in the sense of Denjoy-Carleman where, however, the condition (III.1) is always assumed to be fulfilled.

(b) According to the theorem of Denjoy-Carleman by the condition

$$\int_1^{+\infty} \frac{\log T(r)}{r^2}\,dr < +\infty \qquad \text{(III.2)}$$

the class $C\{M_n\}$ of functions infinitely differentiable on the half-axis $[0,\infty)$ or on a segment $[0,\ell]$ will certainly not be quasi-analytical.

Namely, it is well known that by the condition (III.2), say, in the case of the half-axis $[0,+\infty)$ there exist non-trivial functions $\varphi(x) \in C\{M_n\}$ satisfying, moreover the conditions:

$$|\varphi^{(n)}(x)| \leqslant A.B^n M_n e^{-\nu x} \quad (\nu > 0, n \geqslant 1, x \in [0,\infty)) \qquad \text{(III.3)}$$

$$\varphi^{(n)}(0) = 0 \quad (n = 0,1,2,\ldots.) \qquad \text{(III.3}^{I}\text{)}$$

In this connection it would be natural to put the following question:

If the class $C\{M_n\}$ is not quasi-analytical on $[0,+\infty)$ or on $[0,\ell]$ then which are the functionals $\{L_n(\varphi)\}_0^{\infty}$ that can determine the functions of this class in a unique way instead of the values $\{\varphi^{(n)}(0)\}_0^{\infty}$?

It turned out that it is possible to introduce a new general notion of α-quasi-analyticity covering also the notion of classical quasi-analyticity and to obtain a complete solution of this problem in the spirit of the classical theorem of Denjoy-Carleman.

(c) Preliminarily we introduce some notations and definitions.

Let us consider a set $C_\alpha^{(\infty)}$ $(0 \leq \alpha < 1)$ of functions $\varphi(x)$ infinitely differentiable on $[0, +\infty)$ and satisfying the conditions:

$$\sup_{0 \leq x < +\infty} |(1 + x^{\alpha m})\ \varphi^{(n)}(x)| < +\infty \quad (n,m = 0,1,2,3,\ldots)$$

Assuming that $\varphi(x) \in C_\alpha^{(\infty)}$ $(0 \leq \alpha < 1)$ and putting $\frac{1}{\rho} = 1 - \alpha$ $(\rho \geq 1)$ let us consider the operator of successive differentiation of the function $\varphi(x)$ in the sense of Weyl of orders $\frac{n}{\rho}$ $(n = 0,1,2,\ldots.)$.

$$D_\infty^{\frac{0}{\rho}} \varphi(x) \equiv \varphi(x), \quad D_\infty^{\frac{n}{\rho}} \varphi(x) = D_\infty^{\frac{1}{\rho}} D_\infty^{\frac{n-1}{\rho}} \varphi(x) \qquad (n \geq 1)$$

where

$$D_\infty^{\frac{1}{\rho}} \varphi(x) = \frac{d}{dx} D_\infty^{-\alpha} \varphi(x)$$

$$D_\infty^{-\alpha} \varphi(x) = \frac{1}{\Gamma(\alpha)} \int_x^\infty (t - x)^{\alpha - 1} \varphi(t)\, dt$$

Note that in a special case when $\alpha = 0$ $(\rho = 1)$ we shall have

$$D_\infty^{n} \varphi(x) \equiv \varphi^{(n)}(x) \qquad (n = 0,1,2,\ldots.)$$

Finally, for an arbitrary sequence of positive numbers $\{M_n\}_1^\infty$ we introduce two classes of infinitely differentiable functions:

The class $C_\alpha^*\{[0,+\infty); M_n\}$ is the set of functions $\varphi(x)$ from $C_\alpha^{(\infty)}$ for which

$$\sup_{0 \leq x < +\infty} |D_\infty^{\frac{n}{\rho}} \varphi(x)| \leq A.B^n.M_n \qquad (n = 1,2,3,\ldots) \qquad \text{(III.4)}$$

and

The class $C_\alpha\{[0,\infty) ; M_n\}$ is the set of functions $\varphi(x)$ from $C_\alpha^{(\infty)}$ for which

$$\sup_{0 \leq x < +\infty} (1 + \alpha x^2) \, |\varphi^{(n)}(x)| \leq A.B^n M_n \quad (n=1,2,\ldots) \qquad (III.5)$$

For both these classes a question is put similar to Hadamard's problem and reducible to this problem when the parameter $\alpha = 0$.

<u>What is the sequence of numbers</u> $\{M_n\}_1^\infty$ <u>to be in order that for each function</u> $\varphi(x)$ <u>from the corresponding class the equalities</u>

$$D_\infty^{\frac{n}{\rho}} \varphi(0) = \frac{1}{\Gamma(\alpha n)} \int_0^\infty x^{\alpha n-1} \varphi^{(n)}(x)\, dx = 0 \qquad (n=0,1,2,\ldots) \qquad (III.6)$$

<u>yield the identity</u>

$$\varphi(x) \equiv 0, \qquad 0 \leq x < +\infty$$

Classes of this kind are called "α-quasi-analytical", and it is easily seen that the 0-quasi-analytical classes $C_0^*\{[0;\infty); M_n\}$ or $C_0\{[0,+\infty); M_n\}$ are identical to the classically quasi-analytical class $C\{M_n\}$.

The following basic theorem has been established:

<u>THEOREM 11</u>. 1°. <u>The class</u> $C_\alpha^*\{[0,+\infty); M_n\}$ <u>is α-quasi-analytical if and only if</u>

$$\int_1^\infty \frac{\log T(r)}{r^{1+\frac{1}{1+\alpha}}}\, dr = +\infty \qquad (III.7)$$

2°. <u>The class</u> $C_\alpha\{[0,+\infty); M_n\}$ <u>is α-quasi-analytical, if and only if</u>

$$\int_1^\infty \frac{\log T(r)}{r^{1+\frac{1-\alpha}{1+\alpha}}}\, dr = +\infty \qquad (III.8)$$

In both these statements

$$T(r) = \sup_{n \geq 1} \frac{r^n}{M_n}$$

is the function of Carleman-Ostrowsky.

Each of the statements 1° and 2° is reduced to the classical theorem of Denjoy-Carleman when $\alpha = 0$.

Elementary estimations show that if

$$M_n = (n^{\frac{1+\alpha}{1-\alpha}} \log n \dots \log_p n)^n, \quad n \geq N_p ,$$

where $p \geq 1$ is any integer, then the condition (III.8) is fulfille

This example shows that in the α-quasi-analytical class $C_\alpha \{ [0,+\infty); M_n \}$ $(0 < \alpha < 1)$ the successive derivatives of functions may have a substantially faster growth (as $\frac{1+\alpha}{1-\alpha} > 1$ when $0 < \alpha < 1$) than it is possible for 0-quasi-analytical classes; this can be observed from the original results of Denjoy.

The notion of α-quasi-analyticity is also introduced for the classes of functions infinitely differentiable on a finite segment $[0, \ell]$. Here the classes $C^*_\alpha \{ [0,\ell] ; M_n \}$ and $C_\alpha \{ [0 , \ell] , M_n \}$ are determined as subclasses of functions $\varphi(x)$ from the corresponding classes on $[0,+\infty)$ satisfying the additional condition $\varphi(x) \equiv 0$, $\ell \leq x < +\infty$.

The corresponding theorem on α-quasi-analyticity of these classes has exactly the same formulation as statements 1° and 2° of the Theorem 11.

(d) Let us consider briefly the method of proving these theore

As it is also the case with the original proof of the Denjoy-Carleman theorem, the problem of α-quasi-analyticity is solved by reducing it to the problem of Watson. In our case such reduction is possible only by making use of the apparatus of integral transforms and representations with Mittag-Leffler kernels $E_\rho(z;\mu)$.

Here a substantial role is played by the following basic theorem:

THEOREM 12. 1°. Let the function f(z) be analytical in the interior and continuous on the closed angular domain

$$\Delta_\rho^* = \left\{ z : \frac{\pi}{2\rho} < |\arg z| \leq \pi , \quad 0 < |z| < +\infty \right\}$$

and in the neighbourhood of $z = \infty$,

$$\max_{\frac{\pi}{2\rho} \leq |\varphi| \leq \pi} \left\{ |f(re^{i\varphi})| \right\} = O(r^{-\omega}) \qquad (\omega > 1) \tag{III.9}$$

Then we have an integral representation of the form

$$f(z) = \int_0^{+\infty} E_\rho(zt^{\frac{1}{\rho}} ; \frac{1}{\rho}) \, t^{\frac{1}{\rho} - 1} \varphi(t)dt, \quad z \in \Delta_\rho^* \tag{III.10}$$

where

$$\varphi(t) = \frac{1}{2\pi i} \int_{L_\rho} e^{-t \zeta^\rho} f(\zeta) \, d\zeta , \quad 0 < t < +\infty \tag{III.11}$$

and L_ρ is the boundary of the domain Δ_ρ^* advanced in the positive direction.

2°. If f(z) in an entire function of the order $\rho > \frac{1}{2}$ and type ℓ $(0 < \ell < +\infty)$ satisfying the condition (III.9), then in the representation (III.10) we have

$$\varphi(t) \equiv 0, \quad \ell < t < +\infty$$

These statements permit to establish the necessity of conditions (III.7) and (III.8) of the Theorem 11. Here we rely substantially upon our discovery of an important property of the function

$$\mathcal{E}_\rho(x ; \lambda) = E_\rho(\lambda x^{\frac{1}{\rho}} ; \frac{1}{\rho}) \, x^{\frac{1}{\rho} - 1}$$

to be a solution of the Cauchy-type problem for a differential operator of fractional order on the half-axis $[0,+\infty)$:

$$\begin{aligned} & D_0^{\frac{1}{\rho}} \mathcal{E}_\rho(x;\lambda) - \lambda \, \mathcal{E}_\rho(x;\lambda) = 0 \\ & D_0^{-\alpha} \mathcal{E}_\rho(x;\lambda) \Big|_{x=0} = 1 \end{aligned} \tag{III.12}$$

where

$$D_0^{-\alpha} f(x) \equiv \frac{1}{\Gamma(\alpha)} \int_0^x (x-t)^{\alpha-1} f(t)dt$$

and

$$D_0^{\frac{1}{\rho}} f(x) \equiv \frac{d}{dx} D_0^{-\alpha} f(x)$$

is the derivative in the sense of Riemann-Liouville of the order $\frac{1}{\rho}$.

Also important is that the function

$$e_\rho(x;\lambda) = e^{-\lambda^\rho x} \qquad (|\arg \lambda| \leq \frac{\pi}{2\rho},\ \rho \geq 1)$$

is also a solution of the Cauchy problem, but this time of another kind:

$$D_\infty^{\frac{1}{\rho}} e_\rho(x;\lambda) + \lambda e_\rho(x;\lambda) = 0$$

$$e_\rho(0;\lambda) = 1 \qquad \text{(III.13)}$$

As to the sufficiency of the conditions in Theorem 11, here an essential role is played by the fact that if

$$\varphi(x) \in C_\alpha^* \{[0,\infty); M_n\} , \quad D_\infty^{\frac{n}{\rho}} \varphi(o) = 0 \qquad (n \geq 0)$$

then the function

$$f(z) = \int_0^{+\infty} E_\rho(zt^{\frac{1}{\rho}}; \frac{1}{\rho}) t^{\frac{1}{\rho}-1} \varphi(t)dt, \quad z \in \Delta_\rho^*$$

admits the representation

$$f(z) = \frac{(-1)^n}{z^n} \int_0^{+\infty} E_\rho(zt^{\frac{1}{\rho}}; \frac{1}{\rho}) t^{\frac{1}{\rho}-1} D_\infty^{\frac{n}{\rho}} \varphi(t)dt, \quad z \in \Delta_\rho$$

for any $n \geq 1$.

This representation permits to establish that the fulfilment of the conditions of theorem 11 results in $f(z) \equiv 0$. Then using the theorem on inverse transforms with the Mittag-Leffler kernel which has been cited above, we come to the conclusion that $\varphi(t) \equiv 0$.

It should be noted in conclusion that within the last two years the reporter and his co-workers have been continuing their investigations in this direction.

First, we succeeded in obtaining an analogous theorem of uniqueness in the case when the function $\varphi(x)$ is analytical in the domain of arbitrary angle

$$\Delta_{\gamma} = \left\{ z : |\mathrm{Arg}\, z| < \frac{\pi}{2\gamma}, \quad 0 < |z| < +\infty \right\}$$

of the span $\frac{\pi}{\gamma}$ $(0 < \gamma < +\infty)$ on the Riemann surface of the logarithm.

Secondly, we have extended the Theorem 11 to the case when the parameter α lies within the limits of $-1 < \alpha < 0$. This case yields results of a completely new quality this time for the classes of Denjoy-Carleman.

Formulations of the theorems in question will not be dwelt upon in view of the lack of time.

References.

1. M.M. Dzhrbashyan, Integral transforms and representations of functions in the complex domain. Moscow, "Nauka", 1966, Chapters III-VIII (Russian).

2. M.M. Dzhrabashyan, Extension of quasi-analytical classes of Denjoy-Carleman, Izvestia AN Arm. SSR, ser. Matem. IV, N 4, 1969, 225-243 (Russian).

3. M.M. Dzhrbashyan and H.S. Kocharian, Uniequeness theorems for certain classes of analytical functions. Izvestia AN SSSR, Ser. Matem. 37 (1973), 98-134, (Russian).

MÜNTZ-SZASZ THEOREM WITH INTEGRAL COEFFICIENTS I

Le Baron O. Ferguson

Let $C[a,b]$ be the continuous real valued functions defined on an interval $[a,b]$ and $\|\cdot\|$ the supremum norm on $C[a,b]$ ($\|f\| = \sup \{|f(x)| : a \le x \le b\}$). Let $\Lambda = \{\lambda_i\}$ be a sequence of real numbers satisfying

$$0 < \lambda_1 < \lambda_2 < \ldots .$$

A Λ-polynomial p is a function of the form

$$p(x) = a_0 + \sum_{i=1}^{\infty} a_i x^{\lambda_i} \qquad (*)$$

where the a_i's are real numbers. The classical Müntz-Szasz theorem, in one of its slight variations, reads as follows (c.f.Feller [1]).

THEOREM 1. *The Λ-polynomials are dense in $C[0,1]$ if and only if* $\sum \lambda_i^{-1} = \infty$.

It is interesting to ask if the theorem remains true for integral Λ-polynomials, i.e. functions of the form (*) where the a_i's are integers. It is clear that the value of an integral Λ-polynomial at $x = 0$ or $x = 1$ is an integer, hence we can only hope to approximate continuous functions which take on integer values at $x = 0$ and $x = 1$. We don't know whether or not all such functions can be approximated on $[0,1]$ but if we add the assumption on Λ that

$$\lambda_n \longrightarrow \infty$$

then the following is true. For $0 < \alpha < 1$ let $C_0[0,\alpha]$ be the elements of $C[0,\alpha]$ which take on integer values at the origin.

THEOREM 2. For any $\alpha < 1$ the integral Λ-polynomials are dense in $C_0[0, \alpha]$ if and only if $\sum \lambda_i^{-1} = \infty$.

Proof. By means of the transformation $x \to \alpha x$ it is easy to see that the density of the Λ-polynomials in $C[0, \alpha]$ is equivalent to density in $C[0,1]$. If $\sum \lambda_i^{-1} < \infty$ then density fails for Λ-polynomials in $C[0, \alpha]$. Since the integral Λ-polynomials form a subset of the Λ-polynomials density fails for them also.

Conversely, assume $\sum \lambda_i^{-1} = \infty$. For each non-negative integer n let C_n be the number of λ_i's lying in the interval $[n,n+1)$. Then for $n > 0$

$$\frac{C_n}{n+1} \leq \sum_{n \leq \lambda_i \leq n+1} \lambda_i^{-1} < \frac{C_n}{n} . \tag{1}$$

Thus $\sum C_n/n = \infty$. At this point we assume that $C_n \leq n$, all n. In the lemma below we see that the sequence Λ can be replaced by a subsequence with divergent sum of reciprocals and $C_n \leq n$, hence there is no loss of generality in this assumption. Let f be in $C_0[0, \alpha]$ and $\varepsilon > 0$. Since $\alpha < 1$ the series $\sum n\alpha^n$ converges and there is a positive integer n_0 such that

$$\sum_{n \geq n_0} n . \alpha^n < \frac{\varepsilon}{3} . \tag{2}$$

Since $\sum_{n \geq n_0} \lambda_i^{-1} = \infty$ there is, by Theorem 1, a polynomial p of the form (*) with real coefficients but only involving powers x^{λ_i} with $\lambda_i \geq n_0$ such that

$$\| f - p \| < \varepsilon/3 . \tag{3}$$

Let $[p]$ represent p with each coefficient a_i replaced by a nearest integer $[a_i]$. By (3), since f(0) is an integer

$$|a_0 - [a_0]| < \varepsilon/3 . \tag{4}$$

In addition

$$\| p - [p] \| = \| (a_0 - [a_0]) + \sum_{i=1}^{n} (a_i - [a_i]) x^{\lambda_i} \|$$

$$\leq |a_0 - [a_0]| + \| \sum_{\lambda_i \geq n_0} x^{\lambda_i} \|$$

$$= \frac{\varepsilon}{3} + \sum_{\lambda_i \geq n_0} \| x \|^{\lambda_i}$$

$$\leq \frac{\varepsilon}{3} + \sum_{n \geq n_0} n \alpha^n < \frac{2\varepsilon}{3}$$

This last estimate together with (3) and the triangle inequality give

$$\| f - p \| < \varepsilon$$

which completes the proof.

It is clear from the proof that the following more general result holds. Let S be any subset of $\underset{\sim}{R}$ and S^- its closure. For $x \in \underset{\sim}{R}$ let dist $(x,S) = \inf \{ |x-s| : s \in S \}$. Suppose dist (x,S) is a bounded function on $\underset{\sim}{R}$. If Λ is a sequence as above then the closure in $C[0,\alpha]$, $\alpha < 1$, of the Λ-polynomials consists of those f in $C[0,\alpha]$ such that $f(0) \in S^-$.

LEMMA. *Let Λ and $\{C_n\}$ be defined as above. Then there is a subsequence of Λ , the sum of reciprocals of which diverges and whose corresponding sequence $\{C_n\}$ satisfies $C_n \leq n$ for all n .*

Proof. Let $\{C_n\}$ be defined from Λ as above. Then by (1) and the divergence of $\sum \lambda_i^{-1}$, $\sum \frac{C_n}{n}$ diverges. Let $C_n \wedge n = \min \{C_n, n\}$. Then $\sum (C_n \wedge n)/n$ diverges, as follows. Since $\sum C_n/n$ diverges it follows easily that $\sum (C_n/n)/(1 + C_n/n)$

diverges also.

Thus

$$\infty = \sum \frac{C_n/n}{1+C_n/n} = \sum \left(\frac{C_n n}{n+C_n}\right)\frac{1}{n}$$

$$\leq \sum \frac{C_n \wedge n}{n} .$$

From the sequence Λ we form a subsequence Λ' as follows. From each interval $[n,n+1)$ containing more than n λ_i's delete all but n of them. For this subsequence the number of λ_i's in $[n,n+1)$ is $C_n \wedge n$. We know that $\sum (C_n \wedge n)/n$ diverges hence so does $\sum (C_n \wedge n)/(n+1)$. By the left hand inequality in (1) the sum of the reciprocals of the sequence Λ' also diverges.

<u>Bibliography</u>.

1 Feller, William., On Müntz' theorem and completely monotone functions, Amer. Math. Monthly <u>75</u> (1968), 342-350.

INDUCTIVE LIMITS OF BANACH SPACES AND COMPLEX ANALYSIS

J.-P.Ferrier

1. The theory of Banach algebras is well-known and so are its applications to complex analysis. However, algebras of holomorphic functions which are not Banach algebras are also considered, in particular algebras of holomorphic functions with restricted growth. We shall give an example of such an algebra, which is sufficiently general so as to include most of the classical cases.

Denoting by $z = (z_1,\ldots,z_n)$ the identity mapping of C^n and setting

$$|z|^2 = |z_1|^2 + \ldots + |z_n|^2$$

we consider a Lipschitz non-negative function δ on C^n such that $\delta = O(1/|z|)$ at infinity. If Ω is the set where δ does not vanish, we introduce the algebra $\mathcal{O}(\delta)$ which is the union when N varies in the set of all positive integers, of the spaces ${}_N\mathcal{O}(\delta)$ of all holomorphic functions f on Ω such that $f\,\delta^N$ is bounded.

Each ${}_N\mathcal{O}(\delta)$, equipped with the norm $f \longrightarrow \|f\,\delta^N\|_\Omega$, is a Banach space and therefore $\mathcal{O}(\delta)$ is an inductive limit of such spaces. We shall not take on $\mathcal{O}(\delta)$ the direct limit locally convex topology because it is not easy to describe and not well adapted to our problems. We shall put the emphasis on bounded subsets instead of neighbourhoods of zero, and consider $\mathcal{O}(\delta)$ as a b-space in the sense of L.Waelbroeck(9); as the multiplication is bounded, $\mathcal{O}(\delta)$ is a b-algebra.

As an example of weight function, we may take the distance δ_Ω to the boundary of a bounded open subset Ω of C^n; in such a case, functions in $\mathcal{O}(\delta_\Omega)$ are called holomorphic functions with polynomial growth in Ω.

2. Every Banach algebra is a b-algebra and so is an inductive limit of Banach algebras; but b-algebras are even more general: for instance $\mathcal{O}(\delta)$ is not an inductive limit of Banach algebras. The

first idea, however, is to try to use the same tools; we shall see that this is not possible.

First a maximal ideal $\mathfrak{I}$ of a b-algebra A is not necessarily closed; examples of dense proper ideals of $\mathcal{O}(\delta_\Omega)$ exist when Ω is the unit disc of the complex plane ([7]).

When a maximal ideal $\mathfrak{I}$ happens to be closed, the quotient $A/\mathfrak{I}$ is both a b-algebra and a field. Unfortunately $A/\mathfrak{I}$ is not necessarily the field of complex numbers. For instance ([9]), the field of germs of meromorphic functions in the neighbourhood of 0 is the b-algebra

$$\varinjlim \mathcal{O}(\delta_{\Omega_n})$$

where Ω_n is the complement of 0 in the open disc with center at 0 and radius 1/n. In the case when A is the algebra $\mathcal{O}(\delta)$ itself, the existence of a closed maximal ideal $\mathfrak{I}$ such that $A/\mathfrak{I}$ is not the field of complex numbers is not known.

3. We indicate now a few results about algebras $\mathcal{O}(\delta)$. We suppose that $-\log \delta$ is plurisubharmonic in Ω; it can be proved this is not a restriction when Ω is pseudoconvex, or more generally when the envelope of holomorphy of Ω is simple.

Fixing $s \in \Omega$, obviously the mapping ${}^s\chi : f \longrightarrow f(s)$ is a bounded multiplicative linear form on $\mathcal{O}(\delta)$. Conversely,

PROPOSITION 1. *Every non trivial multiplicative linear form on $\mathcal{O}(\delta)$ is equal to some evaluation ${}^s\chi$, with $s \in \Omega$.*

It is an easy consequence of the following results, the proofs of which are based on the Hilbertian estimates of L. Hörmander ([4]) for the $\bar{\partial}$-operator.

THEOREM 1. *For each $s \notin \Omega$, there exist ${}^s u_1, \ldots, {}^s u_n$ in $\mathcal{O}(\delta)$ such that*

$$1 = (z_1 - s_1)\, {}^s u_1 + \dots + (z_n - s_n)\, {}^s u_n \ .$$

THEOREM 2. *If* $f \in \mathcal{O}(\delta)$ *vanishes at* $s \in \Omega$, *one can find* ${}^s u_1, \dots, {}^s u_n$ *in* $\mathcal{O}(\delta)$ *such that*

$$f = (z_1 - s_1)\, {}^s u_1 + \dots + (z_n - s_n)\, {}^s u_n \ .$$

Moreover ${}^s u_1, \dots, {}^s u_n$ *can be chosen in a bounded subset of* $\mathcal{O}(\delta)$ *when* s *varies in* $\complement\Omega$ *or when* f *varies in a bounded subset of* $\mathcal{O}(\delta)$ *and* s *in* Ω, *with* $f(s) = 0$.

The first statement is a particular case of results of I. Cnop (1) and L. Hörmander (5) and the second one can be deduced by means of spectral theory.

Now let χ be a non trivial multiplicative linear form on $\mathcal{O}(\delta)$ and set $s_i = \chi(z_i)$. From Theorem 1 it follows that s belongs to Ω and from Theorem 2 that $\chi(f) = f(s)$.

We note that every multiplicative linear form on $\mathcal{O}(\delta)$ is therefore bounded and its kernel is closed.

According to L. Waelbroeck, a b-ideal of a commutative b-algebra A is an ideal $\mathfrak{I}$, equipped with a structure of b-space such that the natural mappings $\mathfrak{I} \longrightarrow A$ and $A \times \mathfrak{I} \longrightarrow \mathfrak{I}$ are bounded. Using Theorems 1 and 2 and spectral theory, we obtain:

THEOREM 3. *Let* δ *be as above and let* $\mathfrak{I}$ *be a b-ideal of* $\mathcal{O}(\delta)$; *then* $\mathfrak{I} = \mathcal{O}(\delta)$ *if and only if there exists a bounded family* (h_α) *in* $\mathfrak{I}$ *such that* $\sup |h_\alpha| \geq \varepsilon\, \delta^N$ *on* Ω *for some* $\varepsilon > 0$ *and some positive integer* N.

Let us discuss a few particular cases.

1) An obvious necessary condition in order that $\mathfrak{I} = \mathcal{O}(\delta)$ is the existence of a bounded family (h_α) in $\mathfrak{I}$ such that $\sup |h_\alpha| \geq 1$; it is also sufficient as shown by Theorem 3.

2) Let $f_1,\ldots,f_m$ be functions of $\mathcal{O}(\delta)$; the ideal $f_1\mathcal{O}(\delta)+\ldots+f_m\mathcal{O}(\delta)$ generated by $f_1,\ldots,f_m$ is naturally equipped with a structure of b-ideal. In that case, the condition of Theorem 3 can be replaced by

$$|f_1| + \ldots + |f_m| \geqslant \varepsilon\, \delta^N$$

this is a famous result of L. Hörmander (9).

3) More generally, let (f_m) be a sequence of $\ell^p(\mathcal{O}(\delta))$ and $\mathfrak{I}$ the ideal of all $\sum f_m g_m$, where (g_m) is a sequence of $\ell^q(\mathcal{O}(\delta))$. Now the condition reduces to

$$\left(\sum |f_m|^p\right)^{1/p} \geqslant \varepsilon\, \delta^N$$

as the first member of the above inequality is the supremum of all $|\sum \lambda_m f_m|$, where λ_m is a sequence of the unit ball in $\ell^q(C)$. For $p = 2$, this result was obtained by H. Skoda (8).

4) Now let $\mathfrak{I}$ be an arbitrary ideal of $\mathcal{O}(\delta)$; we define $\overline{\mathfrak{I}}$ as the union, when N varies, of the closures of $\mathfrak{I} \cap {}_N\mathcal{O}(\delta)$ in ${}_N\mathcal{O}(\delta)$. Then $\overline{\mathfrak{I}} = \mathcal{O}(\delta)$ as soon as there exists a bounded family (h_α) in $\mathfrak{I}$ such that $\sup |h_\alpha| \geqslant \varepsilon\, \delta^N$ on Ω.

Using similar arguments, we show that a function g belongs to the radical of a b-ideal $\mathfrak{I}$ of $\mathcal{O}(\delta)$ if and only if there exists a bounded family (h_α) in $\mathfrak{I}$ such that $\sup|h_\alpha|$ is larger than some $\varepsilon\, \delta^N |g|$. In the case of example 2), this was obtained by I. Cnop (1) or J.J. Kelleher and B.A. Taylor(6).

4. We have considered b-spaces all through this paper. As a final remark, we mention that it is possible to avoid them if we assume the axiom of Solovay. A complex vector space E will be called a <u>Silva space</u> if it is the increasing union of a sequence (E_n) of Banach spaces such that each identity mapping $E_n \to E_{n+1}$ is compact. It is known (9) that a subset of E is bounded in

some E_n if and only if it is bounded in the locally convex direct limit $\mathcal{T}$ of the sequence (E_n). As $(E, \mathcal{T})$ is ultra bornological, the topology $\mathcal{T}$ is unique ([3]) and does not depend on the particular choice of the sequence (E_n). Therefore the concept of a bounded subset in a Silva space is well-defined.

A Silva algebra will be an algebra A which is a Silva space; clearly $A \times A$ is a Silva space and the multiplication of A is bounded. A Silva ideal of a commutative Silva algebra A will be an ideal $\mathfrak{I}$ of A which is a Silva space; both mappings $\mathfrak{I} \to A$ and $A \times \mathfrak{I} \to \mathfrak{I}$ are bounded.

It is easily shown that $\mathcal{O}(\mathcal{S})$ is a Silva algebra and that all the examples of b-ideals of $\mathcal{O}(\mathcal{S})$ are actually Silva ideals. We can therefore formulate our results in terms of Silva ideals of $\mathcal{O}(\mathcal{S})$.

Bibliography.

([1]) Cnop, I., Spectral study of holomorphic functions with bounded growth. Ann. Inst. Fourier.

([2]) Ferrier, J.P., Spectral theory and complex analysis. North - Holland mathematics studies 4, 1973.

([3]) Garnir, H.G., The axiom of Solovay and functional analysis. International conference on functional analysis and its applications, Madras, 1973.

([4]) Hörmander, L., L^2-estimates and existence theorems for the $\bar{\partial}$-operator. Acta Math. 113 (1965), 89-152.

([5]) Hörmander, L., Generators for some rings of analytic functions. Bull. Amer. Math. Soc. 73 (1967), 943-949.

([6]) Kelleher, J.J. and Taylor, B.A., Finitely generated ideals in rings of analytic functions. Math. Ann. 193 (1971), 225-237.

(7) Lavigne, J.P., Sur les idéaux de fonctions holomorphes. In preparation.

(8) Skoda, H., Système fini ou infini de générateurs dans un espace de fonctions holomorphes avec poids. C.R.Acad.Sci. Paris A 273 (1971), 389-392.

(9) Waelbroeck, L., Topological vector spaces and algebras. Springer Lecture Notes in Mathematics 230, 1971.

REPRESENTATION OF NONLINEAR OPERATORS WITH THE HAMMERSTEIN PROPERTY

Steven H. Fesmire

Let X be a compact Hausdorff topological space and let E be a Banach space. Denote by $C(X,E)$ the Banach space of continuous E-valued functions on X, equipped with the usual supremum norm. In [6], Bochner and Taylor obtained a concrete representation of the dual of $C([0,1], E)$, where $[0,1]$ is the closed unit interval. The recent advances in vector measures facilitated generalization of the Bochner-Taylor representation theorem to $C(X,E)^*$, the dual of $C(X,E)$ (See Singer [15]). More generally, the problem of representing continuous linear operators from $C(X,E)$ into F, where F is a Banach space, has been considered by Gelfand [11], Bartle, Dunford, and Schwartz [1], Dinculeanu [1,2] and Batt and Berg [5]. Subsequently the problem of representing a class of non-linear operators, known as Hammerstein operators, mapping $C(X,E)$ into F has been studied by Batt in [1,2]. The problem of representing non-linear operators from one function space into another, which are slightly different from the Hammerstein operators, has been extensively studied by Martin, Mizel and Sundaresan (see [12],[13], and [14]).

The purpose of this paper is to extend some of the results of Batt to the case when X is an arbitrary topological space.

Throughout this paper E and F will be used to denote real Banach spaces and X will denote a topological space. The ball of radius $\alpha > 0$ centered at 0 in E will be denoted by $U(E,\alpha)$. The field of subsets of X generated by the class of all zero sets in X will be denoted by $\mathfrak{F}$.

A finite collection $P = \{A_i \mid 1 \leq i \leq n\}$ of pairwise disjoint sets in $\mathfrak{F}$ is a $\mathfrak{F}$-partition of a set B in $\mathfrak{F}$ if

$$\bigcup_{i=1}^{n} A_i = B.$$

The integral needed for the representation theorems is defined by Batt in [2] . The definition is restated here for completness. The space $M(E,F)$ is the linear space of all mappings T from E into F with the following properties:

i) $T(0) = 0$;

ii) $\|T\|_\alpha = \sup\{\|Tx\| : x \in U(E,\alpha)\} < \infty$ for each $\alpha > 0$;

iii) $D_\delta T_\alpha = \sup\{\|T(x) - T(x')\| : x,x' \in U(E,\alpha)$ and $\|x - x'\| < \delta\}$ tends to 0 as δ tends to 0, for each $\alpha > 0$.

The space $M(E,F)$ is a locally convex space when equipped with the topology generated by the family of seminorms $\{\|\cdot\|_\alpha : \alpha > 0\}$. For each $\alpha > 0$, the space $M_\alpha(E,F) = \{T_\alpha \mid T_\alpha$ is the restriction of some T in $M(E,F)$ to $U(E,\alpha)\}$ is a normed linear space when equipped with the norm $\|\cdot\|_\alpha$.

A finitely additive set function μ from $\mathfrak{F}$ into $M(E,F)$ is a set function such that i) $\mu(\emptyset) = 0$, and ii) $\mu(B_1 \cup B_2) = \mu(B_1) + \mu(B_2)$ if $B_1, B_2 \in \mathfrak{F}$ and $B_1 \cap B_2 = \emptyset$. The set function μ is said to be of bounded semi-variation if for each $B \in \mathfrak{F}$

i) $SV(\mu_\alpha, B) = \sup\{\|\Sigma\mu(A_i)x_i\| : A_i \in P$, a $\mathfrak{F}$-partition of B, $x_i \in U(E,\alpha)$, $1 \leq i \leq n\} < \infty$ for each $\alpha > 0$;

ii) $SV_\delta(\mu_\alpha, B) = \sup\{\|\Sigma\mu(A_i)x_i - \mu(A_i)x_i'\| : A_i \in P$, a $\mathfrak{F}$-partition of B, $x_i, x_i' \in U(E,\alpha)$ and $\|x_i - x_i'\| \leq \delta$, $1 \leq i \leq n\}$ tends to 0 as δ tends to 0, for each $\alpha > 0$.

Let $\mathcal{E}(\mathfrak{F}, E)$ be the space of all E-valued $\mathfrak{F}$-simple functions on X equipped with the usual supremum norm. If

$$G = \sum_{i=1}^{n} \chi_{B_i} x_i$$

is a member of $\mathcal{E}(\mathfrak{F}, E)$ and if $\{B_i : 1 \leq i \leq n\}$ is a $\mathfrak{F}$-partition of X then <u>the integral of</u> g <u>with respect to the set function</u> μ is defined to be

$$\int_X g d\mu = \sum_{i=1}^{n} \mu(B_i) x_i .$$

Since μ is of bounded semi-variation this integral defines a continuous mapping from $\mathcal{E}(\mathfrak{F}, E)$ into F.

The integral is now extended to $\mathfrak{M}(\mathfrak{F}, E)$, the space of all $\mathfrak{F}$-totally measurable functions from X into E equipped with the supremum norm (see [3], p.83). If h is a member of $\mathfrak{M}(\mathfrak{F}, E)$ and if $\{g_n : n \geq 1\}$ is a sequence in $\mathcal{E}(\mathfrak{F}, E)$ which converges uniformly to h then <u>the integral of h with respect to the set function</u> μ is defined to be

$$\int_X h d\mu = \lim_{n \to \infty} \int_X g_n d\mu .$$

<u>The integral of</u> h <u>with respect to</u> μ <u>over a set</u> B <u>in</u> $\mathfrak{F}$ <u>is</u> defined to be

$$\int_B h d\mu = \int_X \chi_B h d\mu.$$

It follows from the definition that the integral is linear in μ but not linear in h. However it does have the following property: if h, h_1, h_2 are members of $\mathfrak{M}(\mathfrak{F}, E)$ such that

$\{t \in X : h_1(t) \neq 0\} \cap \{t \in X : h_2(t) \neq 0\} = \emptyset$ then

$$\int_X (h + h_1 + h_2) d\mu = \int_X (h + h_1) d\mu + \int_X (h + h_2) d\mu - \int_X h d\mu.$$

If $\alpha > 0$ is given then the <u>set function</u> μ_α mapping $\mathfrak{F}$

into the normed linear space $M_\alpha(E,F)$ is defined as $\mu_\alpha(B)$ being the restriction of $\mu(B)$ to $U(E,\alpha)$ for each $B \in \mathfrak{F}$. The set function μ_α is said to be regular if given a set $B \in \mathfrak{F}$ and an $\varepsilon > 0$ then there is a zero set Z and a cozero set G with $Z \subset B \subset G$ such that $\| \mu_\alpha(C) \|_\alpha = \| \mu(C) \|_\alpha < \varepsilon$ for all $C \in \mathfrak{F}$ with $C \subset G \sim Z$. (The symbol $\sim$ denotes set complimentation.)

Suppose that $F = R$, where R is the real numbers, and that μ maps $\mathfrak{F}$ into $M(E,R)$. For each $\alpha > 0$ and $\delta > 0$ define:

i) $\vee(\mu_\alpha,B) = \sup \{ \Sigma \| \mu(A_i) \|_\alpha : A_i \in P$ a $\mathfrak{F}$-partition of $B \}$;

ii) $\vee_\delta(\mu_\alpha,B) = \sup \{ \Sigma D_\delta \mu_\alpha(A_i) : A_i \in P$ a $\mathfrak{F}$-partition of $B \}$.

Condition i) defines the variation of the set function μ_α. It is readily verified that, for each $\alpha > 0$ and each $\delta > 0$

$$S \vee (\mu_\alpha,B) \leq \vee (\mu_\alpha,B) \leq 2S \vee (\mu_\alpha,B)$$

and

$$S \vee_\delta(\mu_\alpha,B) \leq \vee_\delta (\mu_\alpha,B) \leq 2S \vee_\delta (\mu_\alpha,B) \text{ for all } B \in \mathfrak{F}.$$

The collection $C_b(X,E) = \{ f : f$ maps X into E, f is continuous and $f(X)$ is bounded $\}$ is a Banach space when equipped with the norm, $\| f \| = \sup \{ \| f(t) \| : t \in X \}$. If $E = R$ then $C_b(X,R)$ will be written simply as $C_b(X)$. The subset of $C_b(X)$ of all functions φ such that $0 \leq \varphi(t) \leq 1$ for all $t \in X$ will be denoted by $C(X,I)$.

The collection $T(X,E) = \{ f : f \in C_b(X,E)$ and $f(X)$ is totally bounded $\}$ is a closed subspace of $C_b(X,E)$ when equipped with the norm of $C_b(X,E)$ and hence is a Banach space. It can be shown that the set of all finite sums

$$S(X,E) = \left\{ \sum_{i=1}^{n} \varphi_i x_i : \varphi_i \in C_b(X), \ x_i \in E, \ 1 \leq i \leq n \right\}$$

is dense in $T(X,E)$ (as shown in [10]).

For each set $B \in \mathfrak{J}$ define the following classes of sets:

i) $\pi(B) = \{ G : G$ is a cozero set in X and $B \subset G \}$;

ii) $\pi_o(B) = \{ Z : Z$ is a zero set in X and $Z \subset B \}$.

These classes of sets may be partially ordered if $G_1 \leq G_2$ is defined to mean $G_2 \subset G_1$ for $G_1, G_2 \in \pi(B)$ and if $Z_1 \leq Z_2$ is defined to mean $Z_1 \subset Z_2$ for $Z_1, Z_2 \in \pi_o(B)$.

If Z is a zero set in X and if G is a cozero set in $\pi(Z)$ then there is a function $\varphi_{Z,G}$ in $C(X,I)$ such that $\varphi_{Z,G}(t) = 1$ if $t \in Z$ and $\varphi_{Z,G}(t) = 0$ if $t \notin G$. If the zero set Z is specified and no confusion will arise, the function $\varphi_{Z,G}$ corresponding to a cozero set G in $\pi(Z)$ will be written simply as φ_G.

For a function $f \in T(X,E)$, <u>the support of</u> f is defined to be $\operatorname{supp}(f) = \{ t \in X : f(t) \neq 0 \}$. The closure of the support of f will be denoted by $S[f]$.

A function T in $M(T(X,E),F)$ is said to have the <u>Hammerstein property</u> if $T(f + f_1 + f_2) = T(f + f_1) + T(f + f_2) - T(f)$ for all $f, f_1, f_2 \in T(X,E)$ with $S[f_1] \cap S[f_2] = \emptyset$. (See Batt, [2] .) However, as the next lemma shows, the assumption that $S[f_1] \cap S[f_2] = \emptyset$ may be replaced with the assumption that $\operatorname{supp}(f_1) \cap \operatorname{supp}(f_2) = \emptyset$.

<u>LEMMA 1.</u> <u>Let</u> T <u>be a member of</u> $M(T(X,E),F)$ <u>which has the Hammerstein property. If</u> f_1, f_2 <u>are members of</u> $T(X,E)$ <u>such that</u>

$\text{supp}(f_1) \cap \text{supp}(f_2) = \emptyset$ then $T(f + f_1 + f_2) = T(f + f_1) + T(f + f_2) - T(f)$ for all $f \in T(X,E)$.

The proof of this lemma is omitted and may be found in [10].

This property has also been called strong additivity by some authors but the name Hammerstein property was chosen because, in particular, the class of HAMMERSTEIN operators satisfy this condition. Since an integral representation of non-linear functions T from $T(X,E)$ into F with respect to some finitely additive set function is desired, and since the integral is first defined on simple functions, it follows that some sort of assumption on T about additivity on functions with disjoint supports must be made. The assumption of the Hammerstein property is the correct assumption for functions T in $M(T(X,E),F)$.

In the following series of lemmas, A will denote a member of $M(T(X,E),R)$ which has the Hammerstein property. The proofs of these lemmas are modifications of the work of Batt found in [2] and may be found in [10].

LEMMA 2. Let B be a subset of X. If $g \in T(X,E)$ is such that $\text{supp}(g) \subset B$ and if $\varphi \in C_b(X)$ is such that $B \subset \{t : \varphi(t) = 1\}$ then $A(f-g) - A(f) = A(\varphi f - g) - A(\varphi f)$ for all $f \in T(X,E)$.

LEMMA 3. Let Z be a zero set in X and let G_1 be a cozero set containing Z. If $\varepsilon > 0$ is given and if $g \in T(X,E)$ with $\text{supp}(g) \subset X \sim Z$ then there is a cozero set $G \in \pi(Z)$ with $G \subset G_1$ and a $\varphi \in C(X,I)$ with $\text{supp}(\varphi) \subset X \sim G$ such that

$$|A(f + g) - A(f + \varphi g)| < \varepsilon \quad \text{for all} \quad f \in U(T(X,E), \|g\|).$$

LEMMA 4. If Z is a zero set in X and if $\alpha > 0$, $\varepsilon > 0$ are given then there is a cozero set $G \in \pi(Z)$ such that $|A(f + g) - A(f)| < \varepsilon$ for all $f, g \in U(T(X,E),\alpha)$ with $\text{supp}(g) \subset G \sim Z$. (Hence $|A(f_1) - A(f_2)| < \varepsilon$ for all

$f_1, f_2 \in U(T(X,E),\alpha)$ with $\{t : f_1(t) \neq f_2(t)\} \subset G \sim Z$.)

LEMMA 5. *Let Z be a zero set and let G_1 and G_2 be cozero sets such that $Z \subset G_1 \cup G_2$. If $\varepsilon > 0$ is given then there exists $\varphi_1, \varphi_2 \in C(X,I)$ with $\mathrm{supp}(\varphi_i) \subset G_i$ for $i = 1,2$ such that $|A(f) - (A(\varphi_1 f) + A(\varphi_2 f))| < \varepsilon$ for all $f \in U(T(X,E),\alpha)$ with $\{t : \|f(t)\| \geq \delta\} \subset Z$ and where $\alpha > 0$ and $\delta > 0$ satisfy the condition $D_\delta A_\alpha < \varepsilon/2$.*

Two set functions on the lattice of all cozero sets will now be defined.

I) Let $\alpha > 0$ be given. For a cozero set G define $\lambda_\alpha(G) = \sup \sum_{i=1}^{n} |A(f_i)|$, where the supremum is taken over $S(\alpha,G)$ of all finite sets of functions

$$\{f_i \in U(T(X,E),\alpha) : \mathrm{supp}(f_i) \subset G \text{ and } \|f_i(t)\| \, \|f_j(t)\| = 0 \text{ for all } t \in X \text{ and } i \neq j,\ 1 \leq i, j \leq n\}.$$

It can be verified that λ_α is a bounded, non-negative finitely additive set function on the lattice of all cozero sets in X.

II) Let $\alpha > 0$ and $\delta > 0$ be given. For a cozero set G define $\lambda_\delta^\alpha(G) = \sup \sum_{i=1}^{n} |A(f_i) - A(f_i')|$, where the supremum is taken over $S(\alpha,\Sigma;\delta)$ of all finite sets of functions

$$\{f_i, f_i' \in U(T(X,E),\alpha) : \{f_i : 1 \leq i \leq n\} \text{ and } \{f_i' : 1 \leq i \leq n\} \text{ belong to } S(\alpha,G) \text{ and } \|\sum_{i=1}^{n} f_i - \sum_{i=1}^{n} f_i'\| \leq \delta\}.$$

It can be verified that λ_δ^α is a bounded, non-negative finitely additive set function on the lattice of all cozero sets in X.

For each $\alpha > 0$ and $\delta > 0$, the set functions λ_α and λ_δ^α can now be extended uniquely to bounded, regular, non-negative, finitely additive set functions on $\mathfrak{B}$, which will be denoted by λ_α and λ_δ^α respectively. Furthermore, if $\delta \leq \xi$ then $\lambda_\delta^\alpha(G) \leq \lambda_\xi^\alpha(G) \leq 2\lambda_\alpha(G)$ for all cozero sets G. It follows from the regularity of these set functions that the inequalities $\lambda_\delta^\alpha(B) \leq \lambda_\xi^\alpha(B) \leq 2\lambda_\alpha(B)$ must hold for $B \in \mathfrak{B}$.

<u>LEMMA 6.</u> <u>Let</u> Z_1 <u>and</u> Z_2 <u>be disjoint zero sets in</u> X. <u>If</u> $\alpha > 0$ <u>and</u> $\delta > 0$ <u>are given then</u> $|A(f) - A(g)| \leq \lambda_\delta^\alpha((X \sim Z_1) \cap (X \sim Z_2))$ <u>for all</u> $f,g \in U(T(X,E),\alpha)$ <u>with</u> $Z_1 \cup Z_2 \subset \{t : f(t) = g(t)\}$ and $X \sim (Z_1 \cup Z_2) \subset \{t : \|f(t) - g(t)\| < \delta\}$.

<u>COROLLARY 7.</u> <u>Let</u> Z <u>be a zero set in</u> X <u>and let</u> $\alpha > 0$ <u>be given.</u> <u>If</u> G <u>is a cozero set in</u> $\pi(Z)$ <u>then</u> $|A(f) - A(g)| \leq 2\lambda_\alpha(G \sim Z)$ <u>for all</u> $f,g \in U(T(X,E),\alpha)$ <u>with</u> $\{t : f(t) \neq g(t)\} \subset G \sim Z$.

In order to state the representation theorem for $M_{HP}(T(X,E),R) = \{A \in M(T(X,E),R) : A$ has the Hammerstein property$\}$ the following space of additive set functions on $\mathfrak{B}$ is needed.

<u>The space</u> $N(\mathfrak{B}, M(E,R))$ is the set of all finitely additive set functions μ mapping $\mathfrak{B}$ into $M(E,R)$ with the properties:

i) for all $\alpha > 0$, $S\, V(\mu_\alpha, X) < \infty$ and $S\, V_\delta(\mu_\alpha, X)$ tends to 0 as δ tends to 0;

ii) for all $\alpha > 0$, μ_α mapping $\mathfrak{B}$ into $M_\alpha(E,R)$ is regular in the norm $\| \ \|_\alpha$. (Hence $V(\mu_\alpha)$ mapping $\mathfrak{B}$ into R is also regular.)

THEOREM 8. *Let* X *be a topological space and let* E *be a real Banach space. There is an isomorphism from* $M_{HP}(T(X,E),R)$ *onto the space* $N(\mathfrak{F}, M(E,R))$ *where corresponding elements* A *in* $M_{HP}(T(X,E),R)$ *and* μ *in* $N(\mathfrak{F}, M(E,R))$ *satisfy the following conditions*:

a) $A(f) = \int_X f d\mu$ *for all* $f \in T(X,E)$;

b) *for each zero set* Z *in* X, $\mu(Z)x = \lim \{A(\varphi_G x) : G \in \pi(Z)\}$ *uniformly for* x *in* $U(E,\alpha)$, *for each* $\alpha > 0$;

c) *for each* $\alpha > 0$ *and each* $\delta > 0$,

$$\|A\|_\alpha = S \vee (\mu_\alpha, X) \quad \text{and} \quad D_\delta A_\alpha = S \vee_\delta (\mu_\alpha, X).$$

Proof. Let $A \in M_{HP}(T(X,E),R)$ and let $\alpha > 0$ be given. For each zero set Z in X the limit

$$(*) \qquad \mu(Z)x = \lim \{A(\varphi_G x) : G \in \pi(Z)\}$$

exists uniformly for $x \in U(E,\alpha)$. In particular if G_0, G_1, G_2 are cozero sets in $\pi(Z)$ such that $G_i \subset G_0$ for $i = 1,2$ then by Corollary 7 it follows that $|A(\varphi_{G_1} x) - A(\varphi_{G_2} x)| \leq 2\lambda_\alpha(G_0 \sim Z)$ for all $x \in U(E,\alpha)$. Since λ_α is regular, $\lambda_\alpha(G_0 \sim Z)$ may be made as small as desired by choosing a suitable G_0.

Let $\alpha > 0$ be given, if Z is any fixed zero set in X then since the limit $(*)$ exists uniformly for $x \in U(E,\alpha)$ there is, given $\varepsilon > 0$, a cozero set $G \in \pi(Z)$ such that $|\mu(Z)x - A(\varphi_G x)| < \varepsilon/2$ for all $x \in U(E,\alpha)$. Thus

$$\begin{aligned} D_\delta \mu(Z)_\alpha &\leq \sup \{|A(\varphi_G x) - A(\varphi_G x')| : x, x' \in U(E,\alpha), \|x - x'\| \leq \delta\} + \\ &\quad + \sup \{|\mu(Z)x - A(\varphi_G x)| + |A(\varphi_G x') - \mu(Z)x'| : x, x' \in U(E,\alpha), \\ &\qquad \|x - x'\| \leq \delta\} \\ &\leq D_\delta A_\alpha + \varepsilon \quad \text{for all } \delta > 0. \end{aligned}$$

Since $\varepsilon > 0$ was arbitrary it follows that $D_\delta \mu(Z)_\alpha \leq D_\delta A_\alpha$ for all $\delta > 0$. Thus $D_\delta\, \mu(Z)_\alpha$ tends to 0 as δ tends to 0. Since A Has the Hammerstein property and since disjoint zero sets may be separated by disjoint cozero sets it follows that $\mu(Z_1)x + \mu(Z_2)x = \mu(Z_1 \cup Z_2)x$ for disjoint zero sets Z_1, Z_2 and for all $x \in E$. Thus it is verified that μ is an additive set function from the lattice of all zero sets into $M(E,R)$.

It is verified that for a zero set Z and a cozero set $G \in \pi(Z)$, $|\mu(Z)x - A(\varphi_G x)| \leq 2\lambda_\alpha(G \sim Z)$ for all $x \in U(E,\alpha)$. For a cozero set G in X define $\mu(G) = \mu(X) - \mu(X \sim G)$. For an arbitrary $B \in \mathfrak{F}$ define $\mu(B)x = \lim \{\mu(Z)x : Z \in \pi_o(B)\}$ ($= \lim \{\mu(G)x : G \in \pi B\}$) for each $x \in E$. These limits exist uniformly for $x \in U(E,\alpha)$ for each $\alpha > 0$. For if Z, Z_1, Z_2 are zero sets, if G is a cozero set such that $Z \subset Z_i \subset B \subset G$, for $i = 1,2$, and if $\varphi_G^i \in C(X,I)$, $i = 1,2$, are such that $\varphi_G^i(t) = 1$ if $t \in Z_i$ and $\varphi_G^i(t) = 0$ if $t \notin G$, then

$$|\mu(Z_1)x - \mu(Z_2)x| \leq |\mu(Z_1)x - A(\varphi_G^1 x)| + |A(\varphi_G^1 x) - A(\varphi_G^2 x)|$$
$$+ |A(\varphi_G^2 x) - \mu(Z_2)x| \leq 6\lambda_\alpha(G \sim Z).$$

If G_1, G_2 are cozero sets such that $Z \subset B \subset G_i \subset G$, $i = 1,2$, then

$$|\mu(G_1)x - \mu(G_2)x| = |\mu(X \sim G_1)x - \mu(X \sim G_2)x|$$
$$\leq 6\lambda_\alpha((X \sim Z) \sim (X \sim G)) = 6\lambda_\alpha(G \sim Z).$$

If Z_1 is a zero set and if G_1 is a cozero set such that $Z \subset Z_1 \subset B \subset G_1 \subset G$ then

$$|\mu(G_1)x - \mu(Z_1)x| = |A(\chi_X x) - \mu(Z_1 \cup X \sim G_1)x| \leq 2\lambda_\alpha(G_1 \sim Z_1)$$
$$\leq 2\lambda_\alpha(G \sim Z).$$

These inequalities hold for all $x \in U(E,\alpha)$. Since λ_α is regular it follows that the limits exist, are equal and exist uniformly for $x \in U(E,\alpha)$. Thus μ is a finitely additive set function from $\mathfrak{F}$ into

$M(E,R)$. By the above arguments it follows for $\alpha > 0$ and $B \in \mathcal{F}$ that $\| \mu(G) - \mu(B) \|_\alpha \leq 2\lambda_\alpha(G \sim B)$ for all $G \in \pi(B)$ and $\| \mu(B) - \mu(Z) \|_\alpha \leq 2\lambda_\alpha(B \sim Z)$ for all $Z \in \pi_0(B)$. Thus μ_α is a regular set function from $\mathcal{F}$ into $M(E,R)$. Thus it follows that $S \vee(\mu_\alpha, X) \leq \|A\|_\alpha$ and $S \vee_\delta(\mu_\alpha, X) \leq D_\delta A_\alpha$ for all $\alpha > 0$ and $\delta > 0$.

It will now be shown that $A(f) = \int_X f d\mu$ for all $f \in T(X,E)$. Let $f \in T(X,E)$ and let $\varepsilon > 0$ be given. Let $\alpha = \|f\|$ and choose $\delta > 0$ such that $D_\delta A_\alpha < \varepsilon/5$. If $g = \sum_{i=1}^{n} \chi_{B_i} x_i$ is a member of $U(\mathcal{E}(\mathcal{F},E),\alpha)$, where $\{B_i : 1 \leq i \leq n\}$ is a $\mathcal{F}$-partition of X, is such that $\|f-g\| \leq \delta$ then

$$\left| \int_X f d\mu - \int_X g d\mu \right| \leq S \quad V_\delta(\mu_\alpha, X) \leq D_\delta A_\alpha \leq \varepsilon/5.$$

Now there are zero sets Z_i in $\pi_0(B_i)$, $1 \leq i \leq n$ such that

$$\| \mu(B_i) - \mu(Z_i) \|_\alpha \leq 2\lambda_\alpha(B_i \sim Z_i) < \varepsilon/5n,$$

thus

$$\left| \int_X g d\mu - \sum_{i=1}^{n} \mu(Z_i) x \right| \leq \varepsilon/5.$$

Since the collection $\{Z_i : 1 \leq i \leq n\}$ is pairwise disjoint there is a pairwise disjoint collection $\{G_i : 1 \leq i \leq n\}$ of cozero sets with $G_i \in \pi(Z_i)$ and $\| f(t) - x_i \| < \delta$ if $t \in G_i$ for each $1 \leq i \leq n$. If $\varphi_i \in C(X,I)$, $1 \leq i \leq n$, is such that $\varphi_i(t) = 1$ if $t \in Z_i$ and $\varphi_i(t) = 0$ if $t \notin G_i$ then $\operatorname{supp}(\varphi_i) \cap \operatorname{supp}(\varphi_j) = \emptyset$ if $i \neq j$. It follows that

$$\left| \sum_{i=1}^{n} \mu(Z_i) x_i - \sum_{i=1}^{n} A(\varphi_i x_i) \right| \leq 2\lambda_\alpha\left(\bigcup_{i=1}^{n} (G_i \sim Z_i) \right)$$

$$\leq 2 \sum_{i=1}^{n} \lambda_\alpha(B_i \sim Z_i) < \varepsilon/5.$$

Also

$$\left| \sum_{i=1}^{n} A(\varphi_i x_i) - \sum_{i=1}^{n} A(\varphi_i f) \right| = \left| A\left(\sum_{i=1}^{n} \varphi_i x_i \right) - A\left(\sum_{i=1}^{n} \varphi_i f \right) \right|$$

$$\leq D_\delta A_\alpha < \varepsilon/5.$$

Finally since

$$\bigcup_{i=1}^{n} Z_i \subset \{t : f(t) = \sum_{i=1}^{n} \varphi_i(t)f(t)\}$$

it follows by Corollary 7 that

$$|\sum_{i=1}^{n} A(\varphi_i f) - A(f)| \leq 2\lambda_\alpha(X \sim \bigcup_{i=1}^{n} Z_i) = 2 \sum_{i=1}^{n} \lambda_\alpha(B_i \sim Z_i)$$

$$< \epsilon/5.$$

Thus

$$|A(f) - \int_X f d\mu| \leq |A(f) - \sum_{i=1}^{n} A(\varphi_i f)| + |\sum_{i=1}^{n} A(\varphi_i f) - \sum_{i=1}^{n} A(\varphi_i x_i)|$$

$$+ |\sum_{i=1}^{n} A(\varphi_i x_i) - \sum_{i=1}^{n} \mu(Z_i)x_i| + |\sum_{i=1}^{n} \mu(Z_i)x_i - \int_X g d\mu|$$

$$+ |\int_X g d\mu - \int_X f d\mu| < \epsilon.$$

Since $\epsilon > 0$ was arbitrary it follows that $A(f) = \int_X f d\mu$ for all $f \in T(X,E)$. From the definition of the integral it now follows that $\|A\|_\alpha \leq S\ V\ (\mu_\alpha, X)$ and $D_\delta A_\alpha \leq S\ V_\delta\ (\mu_\alpha, X)$ for all $\alpha > 0$ and $\delta > 0$. Hence $\|A\|_\alpha = S\ V\ (\mu_\alpha, X)$ and $D_\delta A_\alpha = S\ V_\delta\ (\mu_\alpha, X)$ for all $\alpha > 0$ and all $\delta > 0$.

Suppose that ν is an additive set function in $N(\mathfrak{F}, M(E,R))$. It is verified that the integral $\int_X f d\nu$ for $f \in T(X,E)$ defines a mapping in $M_{HP}(T(X,E),R)$. Suppose that $A(f) = \int_X f d\nu$ for all $f \in T(X,E)$. From the regularity of $V(\nu_\alpha)$ it follows that $\nu(Z)x = \lim \{A(\varphi_G x) : G \in \pi(Z)\}$ uniformly for $x \in U(E,\alpha)$ for each zero set Z in X. Hence $\nu(Z) = \mu(Z)$ for all zero sets Z which implies, by the regularity of ν_α for each $\alpha > 0$, that $\nu(B) = \mu(B)$ for all $B \in \mathfrak{F}$.

This completes the proof of Theorem 8.

It should be noted that the isomorphism of Theorem 8 is a topological isomorphism when $M_{HP}(T(X,E),R)$ is considered to have the topology generated by the family of semi-norms $\{\|\cdot\|_\alpha : \alpha > 0\}$ and the topology on $N(\mathfrak{I}, M(E,R))$ is defined by the family of semi-norms $\{p_\alpha : p_\alpha(\mu) = S \vee (\mu_\alpha, X)$ for $\mu \in N(\mathfrak{I}, M(E,R))$, $\alpha > 0\}$.

Before proving the representation theorem for $M_{HP}(T(X,E),F) = \{T \in M(T(X,E),F) \mid T$ has the Hammerstein property$\}$ it is necessary to introduce the following concept of adjoint which was first given by Bátt [4] .

Let S be a non-void set and let $\mathcal{A}$ be a collection of non-empty subsets of S with the property that if A_1 and A_2 are members of $\mathcal{A}$ then there is a member A of $\mathcal{A}$ such that $A_1 \cup A_2 \subset A$. Let S^β be a linear space of (not necessarily all) real valued functions on S with the property that the restriction of f to each set A in $\mathcal{A}$ is bounded.

For each $A \in \mathcal{A}$ the function $\|\cdot\|_A$ mapping S^β into R defined by $\|f\|_A = \sup\{|f(x)| : x \in A\}$ for $f \in S^\beta$, is a semi-norm on S^β. The topology defined by the famly $\{\|\cdot\|_A : A \in \mathcal{A}\}$ is a vector topology on S^β. It will be assumed that S^β and $\mathcal{A}$ are such that this topology is Hausdorff.

Let H be a locally convex topological vector space. A mapping T from S into F is said to be <u>bounded on</u> $\mathcal{A}$ if T maps sets in $\mathcal{A}$ into bounded subsets of H.

If T is a mapping from S into H which is bounded on $\mathcal{A}$ then T^*, the adjoint of T, from H^* into the space of all real valued functions on S which are bounded on $\mathcal{A}$ is defined by $T^* y^*(x) = y^*(T(x))$ for all $y^* \in H^*$ and all $x \in S$. In what follows, S^β will be chosen so that T^* maps H^* into S^β. It is vefified that T^* is a $\sigma(F^*,F^{**}) - \sigma(S^\beta,S^{\beta *})$ continuous linear mapping from H^* into S^β.

Now, let S be the set $T(X,E)$ and let $\mathcal{C} = \{U(T(X,E),\alpha): \alpha > 0\}$. Let the linear space S^β be $M_{HP}(T(X,E),R)$. The mappings T from $T(X,E)$ into F which are bounded on $\mathcal{C}$ to be considered are just those mappings T in $M_{HP}(T(X,E),F)$.

The following set of additive set functions on $\mathfrak{F}$ will be considered. <u>The space</u> $N(\mathfrak{F},M(E,F^{**}))$ is the space of all finitely additive set functions μ from $\mathfrak{F}$ into $M(E,F^{**})$ with the following properties:

i) $S\ V(\mu_\alpha,X) < \infty$ and $S\ V_\delta\ (\mu_\alpha,X)$ tends to 0 as δ tends to 0, for all $\alpha > 0$.

ii) For each $y^*\in F^*$, μ_{y^*} mapping $\mathfrak{F}$ into $M(E,R)$ is defined by $\mu_{y^*}(B)x = \mu(B)x(y^*)$ for all $B\in\mathfrak{F}$ and all $x\in E$. The set function $\mu_{y^*,\alpha}$ mapping $\mathfrak{F}$ into $M_\alpha(E,R)$ is regular in the norm $\|\cdot\|_\alpha$ for each $\alpha > 0$ and for all $y^*\in F^*$.

iii) If $\{y^*_\beta : \beta\in\Lambda\}$ is a net in F^* which converges to y^* in the weak * topology of F^* then the net

$$\{(\int_X f d\mu)(y^*_\beta) : \beta\in\Lambda\} \text{ converges to } (\int_X f d\mu)(y^*) \text{ for}$$

all $f\in T(X,E)$.

<u>THEOREM 9</u>. <u>Let</u> X <u>be a topological space and let</u> E <u>and</u> F <u>be real Banach spaces</u>. <u>There is an isomorphism from</u> $M_{HP}(T(X,E),F)$ <u>onto the space</u> $N(\mathfrak{F},M(E,F^{**}))$ <u>where corresponding elements</u> $T\in M_{HP}(T(X,E),F)$ <u>and</u> $\mu\in N(\mathfrak{F},M(E,F^{**}))$ <u>satisfy the following conditions</u>:

1) $T(f) = \int_X f d\mu$ <u>for all</u> $f\in T(X,E)$.

2) For each zero set Z in X and each $y^* \in F^*$,

$\mu(Z)\ x(y^*) = \lim \{ y^*(T(\varphi_G x)) : G \in \pi(Z) \}$ uniformly

for $x \in U(E,\alpha)$, for each $\alpha > 0$.

3) For all $\alpha > 0$ and all $\delta > 0$, $\|T\|_\alpha = S\ V(\mu_\alpha, X)$

and $D_\delta T_\alpha = S\ V_\delta(\mu_\alpha, X)$.

4) If T^* is the adjoint of T then $T^* y^* = \mu_{y^*}$ in the sense of the isomorphism of Theorem 8.

Proof. If $T \in M_{HP}(T(X,E),F)$ then T^*, the adjoint of T maps F^* into $M_{HP}(T(X,E),R)$. For each $y^* \in F^*$, let μ_{y^*} be the unique member of $N(\mathfrak{F}, M(E,R))$ guaranteed by Theorem 8 such that $T^* y^*(f) = \int_X f d\mu_{y^*}$ for all $f \in T(X,E)$. Define the set function μ mapping $\mathfrak{F}$ into $M(E,F^{**})$ by $\mu(B)\ x(y^*) = \mu_{y^*}(B)x$ for each $B \in \mathfrak{F}$, all $x \in E$ and all $y^* \in F^*$. From the regularity of $\mu_{y^*,\alpha}$ for all $y^* \in F^*$ and each $\alpha > 0$ it follows that $\mu(B)x$ is a linear map from F^* into R for each $B \in \mathfrak{F}$ and each $x \in E$. Since $\|T\|_\alpha < \infty$ for all $\alpha > 0$ it follows that $\mu(B)x \in F^{**}$ for each $B \in \mathfrak{F}$ and $x \in E$.

In addition if $\alpha > 0$ and $\delta > 0$ are given then the regularity of $\mu_{y^*,\alpha}$ for each $y^* \in F^*$ implies that $D_\delta \mu(B)_\alpha \leq D_\delta T_\alpha$ for all $B \in \mathfrak{F}$. Consequently $\mu(B)$ is a member of $M(E,F^{**})$ for each $B \in \mathfrak{F}$.

If B_1, B_2 are disjoint members of $\mathfrak{F}$ then

$$\begin{aligned}\mu(B_1 \cup B_2)\ x(y^*) &= \mu_{y^*}(B_1 \cup B_2)x = \mu_{y^*}(B_1)x + \mu_{y^*}(B_2)x \\ &= \mu(B_1)\ x(y^*) + \mu(B_2)\ x(y^*)\end{aligned}$$

for all $x \in E$ and all $y^* \in F^*$. Consequently μ is a finitely additive set function from $\mathfrak{F}$ into $M(E,F^{**})$.

Now, if $\alpha > 0$ is given then

$$
\begin{aligned}
S\vee(\mu_\alpha,X) &= \sup\{\|\Sigma\mu(B_i)x_i\| : \{B_1,\dots,B_n\} \text{ is a } \mathfrak{F}\text{-partition of } X,\ x_i \in U(E,\alpha),\ 1 \le i \le n\}\\
&= \sup\{S\vee(\mu_{y^*,\alpha},X) : y^* \in U(F^*,1)\}\\
&= \sup\{\|T^*y^*\|_\alpha : y^* \in U(F^*,1)\}\\
&= \sup\{\|T(f)\| : f \in U(T(X,E),\alpha)\} = \|T\|_\alpha .
\end{aligned}
$$

In addition if $\delta > 0$ then

$$
\begin{aligned}
S\vee_\delta(\mu_\alpha,X) &= \sup\{S\vee_\delta(\mu_{y^*,\alpha},X) : y^* \in U(F^*,1)\}\\
&= \sup\{D_\delta(T^*y^*)_\alpha : y^* \in U(F^*,1)\}\\
&= \sup\{\|T(f) - T(f')\| : f,f' \in U(T(X,E),\alpha),\ \|f-f'\| \le \delta\}\\
&= D_\delta T_\alpha .
\end{aligned}
$$

Let $f \in T(X,E)$ and let $\{g_n : n \ge 1\}$ be a sequence in $\mathcal{E}(\mathfrak{F},E)$, with $g_n = \sum_{i=1}^{r(n)} \chi_{B_i^n} x_i^n$, such that $\|g_n - f\|$ tends to 0 as n tends to ∞. Then

$$
\begin{aligned}
y^*(T(f)) &= T^*y^*(f) = \int_X f\,d\mu_{y^*}\\
&= \lim_{n\to\infty} \int_X g_n\,d\mu_{y^*} = \lim_{n\to\infty} \sum_{i=1}^{r(n)} \mu_{y^*}(B_i^n)x_i^n\\
&= \lim_{n\to\infty} \sum_{i=1}^{r(n)} \mu(B_i^n)\ x_i^n(y^*) = \left(\int_X f\,d\mu\right)(y^*)
\end{aligned}
$$

for all $y^* \in F^*$. Hence , $T(f) = \int_X f\,d\mu$ for all $f \in T(X,E)$. That μ has property iii) now follows from the fact that $\int_X f\,d\mu$ is in F for all $f \in T(X,E)$.

Let $\alpha > 0$ be given. Let Z be a zero set in X and let $y^* \in F^*$. Now by Theorem 8,

$$\mu(Z)\ (x)\ (y^*) = \mu_{y^*}(Z)(x) = \lim \{ T^* y^* (\varphi_G x) : G \in \pi(Z) \}$$

uniformly for $x \in U(E,\alpha)$. Since $T^* y^*(\varphi_G x) = y^*(T(\varphi_G x))$ it follows that $\mu(Z)\ x(y^*) = \lim \{ y^*(T(\varphi_G x)) : G \in \pi(Z) \}$ for $y^* \in F^*$, uniformly for $x \in U(E,\alpha)$.

Conversely, if ν is an additive set function in $N(\mathfrak{F}, M(E,F^{**}))$ then the integral $T_1(f) = \int_X f d\nu$ for $f \in T(X,E)$ defines an element in $M_{HP}(T(X,E),F^{**})$. Since property iii) implies that $T_1(f)$ is a weak $*$ continuous linear functional on F^* for each $f \in T(X,E)$, it follows that T_1 must take values in F. Property ii) implies that

$$y^*(\int_X f d\nu) = \int f d\nu_{y^*} \text{ for all } f \in T(X,E) \text{ and each } y^* \in F^*.$$

Hence $T_1^* y^* = \nu_{y^*}$ for each $y^* \in F^*$ in the sense of the isomorphism of Theorem 8, which, by the above arguments, implies that conditions 2) and 3) hold for T_1 and ν.

Finally, suppose for $T \in M_{HP}(T(X,E),F)$ there are $\mu, \nu \in N(\mathfrak{F}, M(E,F^{**}))$ such that $T(f) = \int_X f d\mu = \int_X f d\nu$ for all $f \in T(X,E)$. As seen above, this implies that $\mu_{y^*} = T^* y^* = \nu_{y^*}$ for each $y^* \in F^*$ in the sense of the isomrophism of Theorem 8. This implies that $\mu = \nu$.

This completes the proof the theorem 9.

Closing Remarks.

The last representation theorem yields a characterization of weakly and strongly compact mappings T in $M_{HP}(T(X,E),F))$ in terms of the corresponding set function μ_T in $N(\mathfrak{F},M(E,F^{**}))$ which is the same as that obtained by Batt in [2] for the case when X is a compact Hausdorff space. The proof, however, is quite distinct from Batt's proof as there is no hope for the countable additivity of any of the set functions (see [10]).

The material in this paper is a portion of a doctoral dissertation submitted to Carnegie-Mellon University. The author would like to thank Professor R.A.Alo and, especially, Professor K.Sundaresan for the guidance received from them in this endeavour.

References

1 Bartle, R.G., Dunford, N., and Schwartz, J., 'Weak compactness and vector measures', Cand.J.Math. 7 (1955), 289-305.

2 Batt,J., 'Non-linear integral operators on C(S,E)', Studia Mathematica, (To appear).

3 _______, 'Strongly additive transformations and integral representations with measures of non-linear operators', Bull. Amer. Math. Soc. 78 (1972), 474-478.

4 _______, 'Non-linear compact mappings and their adjoints', Math. Ann. 189 (1970), 5-25.

5 Batt, J., and Berg, E.J., 'Linear bounded transformations on the spaces of continuous functions', J. Functional Analysis 4 (1969), 215-239.

6 Bochner, S., and Taylor, A.E., 'Linear functionals on certain spaces of abstractly valued functions', Ann. of Math. 39(2) (1938), 913-944.

7 Dinculeanu, N., 'Measures vectorielles et operations lineaires', C.R.Acad. Sci. Paris 246 (1958), 2328-2331.

8 ______________, 'Sur la representation integrale de certaines operations lineaires III', Proc. Amer. Math. Soc. 10 (1959), 59-68.

9 ______________, 'Vector Measures', Pergamon Press, London,1967.

10 Fesmire, S., Ph.D. Thesis, Carnegie-Mellon University, 1973.

11 Gelfand, I., 'Abstrakte funktionen und lineare operatoren', Mat. Sbornik 4(46) (1938), 235-284.

12 Martin, A.D., and Mizel, V.J., 'A representation theorem for certain non-linear functionals', Arch. Rational Mech. Anal. 15 (1964), 353-367.

13 Mizel, V.J., 'Characterization of non-linear functionals possessing kernels', Canad. J. Math. 22 (1970), 449-471.

14 Mizel,V.J., and Sundaresan, K., 'Representation of vector valued non-linear functions', Trans. Amer. Math. Soc. 159 (1971), 111-127.

15 Singer, I., 'Linear functionals on the space of continuous mappings of a compact space into a Banach space', Revue Math. Pures et Appl. 2 (1957), 301-315.

ON POLYNOMIAL APPROXIMATION WITH RESPECT TO GENERAL WEIGHTS

Géza Freud

1. Introduction

In the present paper we are going to generalize some of our earlier results. Let $w(x)$ be a fixed nonnegative weight function on the real line, $\mathcal{P}_n$ be the set of polynomials of degree n at most. Let $1 \le p \le \infty$ and by $\|\cdot\|_p$ we denote the $\mathcal{L}_p(-\infty,\infty)$ norm. We define for an f satisfying $wf \in \mathcal{L}_p$ the measure of polynomial approximability

$$(1.1) \qquad E_n^{(p)}(w;f) = \inf_{q \in \mathcal{P}_n} \| w(f-q)\|_p .$$

We studied the order of $E_n^{(p)}(w;f)$ in our previous papers for the weights

$$(1.2) \qquad w_{\alpha,\beta}(x) = \exp\{-|x|^{\alpha}/2\} .$$

We introduced the generalized modulii of continuity

$$\omega(\mathcal{L}_p, w_{\alpha,\beta}; f, \delta) =$$

$$(1.3) \qquad = \sup_{0 \le t \le \delta} \| w_{\alpha,\beta}(x+t) f(x+t) - w_{\alpha,\beta}(x) f(x)\|_p + \\ + \| [\tau(\delta^{\frac{1}{\alpha-1}} x)]^{\alpha-1} f(x)\, w_{\alpha,\beta}(x)\|_p$$

where

$$(1.4) \qquad \tau(x) = \min(|x|, 1)$$

and proved that if f has an r-th order derivative satisfying $w_{\alpha,\beta} f^{(r)} \in \mathcal{L}_p$ then

$$(1.5)\quad \mathcal{E}_n^{(p)}(w;f) \le a_1 e^{a_2 r}\, \omega(\mathcal{L}_p, w_{\alpha,\beta}; f^{(r)}, n^{-1+1/\alpha}).$$

We proved (1.5) for $\alpha > 2$ and every real β respectively for $\alpha = 2$ and $\beta \geq 0$ (see [3] and [8]).

For $\alpha = 2$, $\beta \geq 0$ and for $\beta = 0$, $\alpha = 4, 6, 8, \ldots$ we proved the Bernstein-type converses of (1.5) in [4], [5]. Recently we proved also the converse theorem for every $\alpha > 2$ and every real β. As a special case of this result we have:

(*) Let $\alpha > 2$, β arbitrary real, r a natural number and $0 < \rho < 1$, then

$$\mathcal{E}_n^{(p)}(w;f) = O(n^{-(r+\rho)(1-1/\alpha)}) \quad \text{iff}$$

$$\omega(\mathcal{L}_p, w_{\alpha,\beta}; f^{(r)}, \delta) = O(\delta^\rho),$$ the same holds for $\alpha = 2$, $\beta \geq 0$.

It is our wellfounded conjecture that (1.4) as well as (*) holds also for $1 < \alpha < 2$ and arbitrary β but we have proofs only for particular cases.

In our present paper we generalize (1.4) to a more extended class of weights:

Let

$$(1.6)\qquad w_Q(x) = \exp\{-Q(x)\}$$

where $Q(t)$ $(-\infty < t < \infty)$ is even, positive and twice continuously differentiable in $(0,\infty)$. We assume further that

$$(1.7)\qquad 0 < Q''(t) < (1+C_1)\, Q''(x) \qquad (C_0 < t < x),$$

$$(1.8)\qquad Q''(2t) \geq (1+C_2)\, Q''(t) \qquad (t > C_0),$$

$$(1.9)\qquad t\,\frac{Q''(t)}{Q'(t)} \le C_3 \qquad (t > C_0).$$

Here and in what follows $C_0, C_1, C_2, \dots$ are positive constants depending only on the choice of $Q(t)$. By (1.7)

$$Q'(2^{r+1}C_0) \ge \int_{2^r C_0}^{2^{r+1}C_0} Q''(t)\,dt \ge (1+C_1)^{-1} C_0\, 2^r\, Q''(2^r C_0) \ge$$

$$\ge (1+C_1)^{-1} C_0\, 2^r (1+C_2)^r\, Q''(C_1)$$

so that

$$(1.10)\qquad Q'(t) \ge C_4\, t^{1+C_5} \qquad (t > 0).$$

All the more $t\,Q'(t)$ tends increasingly to ∞ for $t \to \infty$.

We define for every natural integer n the number q_n as the positive solution of the equation

$$(1.11)\qquad q_n\, Q'(q_n) = n \qquad (n = 1, 2, 3, \dots).$$

The sequence $\{q_n\}$ was introduced in our paper [7] . Clearly both sequences $\{q_n\}$ and $\{q_n^{-1}.n = Q'(q_n)\}$ are increasing. Let $s(t)$ be the inverse function of $Q'(x)$ $(x > 0)$ and let us define the generalized moduliiof continuity by

$$(1.12)\qquad \omega(\mathcal{L}_p, w_Q; f, h) = \sup_{0 < t \le h} \| f(x+t) w_Q(x+t) - f(x) w_Q(x) \|_p$$

$$+ h \,\| Q'\{s(h^{-1})\, \tau[x / s(h^{-1})]\}\, w_Q(x)\, f(x) \|_{\mathcal{L}_p}.$$

Note that $\omega(\mathcal{L}_p, w_Q; f, h)$ is a nondecreasing function of $h > 0$. The main result of the present paper is that

$$(1.13)\qquad E_n^{(p)}(w_Q; f) \le C_6 \left(C_7 \frac{q_n}{n}\right)^r \omega\left(\mathcal{L}_p, w_Q; f^{(r)}, \frac{q_n}{n}\right)$$

for every $1 \le p < \infty$ and $r = 0, 1, 2, \ldots$ [1]. In connection with (1.13) it is important to observe that

<u>LEMMA 1.1</u>. <u>We have for every</u> f <u>satisfying</u> $f\, w_Q \in \mathcal{L}_p$

$$(1.14) \qquad \lim_{h \to 0} \omega(\mathcal{L}_p, w_Q; f, h) = 0$$

<u>and (1.14) holds also for</u> $p = \infty$ <u>provided that</u> f <u>is continuous</u> and $\lim_{|x| \to \infty} w_Q(x)\, f(x) = 0$.

<u>Proof</u>. That the 'sup' term on the right of (1.12) tends to zero is clear. Since s is the inverse of Q' and $0 \le \tau(x) \le 1$ (see (1.4)) we have

$$h\; Q'\{s(h^{-1})\; \tau[x/s(h^{-1})]\} \le h\; Q'\{s(h^{-1})\} = 1$$

and for fixed value of x by $0 \le \tau(x) \le |x|$

$$h\; Q'\{s(h^{-1})\; \tau[x/s(h^{-1})]\} \le h\; Q'(|x|) \to 0 \text{ for } h \to 0 .$$

Thus for $1 \le p < \infty$ the last term of (1.12) tends to zero under our conditions by Lebesgue's dominated convergence theorem. For $p = \infty$ the proof is elementary.

The importance of Lemma 1.1 lies in the fact that (1.14) is valid for every f for which $\lim_{n \to \infty} \varepsilon_n^{(p)}(w_Q; f) = 0$. Consequently (1.13) gives an estimate for the speed of decrease of $\varepsilon_n^{(p)}(w_Q; f)$ whenever these expressions tend to zero at all for $n \to \infty$.

The essential ideas of the proof are the functional analytic considerations developed in our paper [1]. This must be combined with an investigation of certain orthonormal polynomials.

---

1. We apply the notation $f^{(0)} = f$.

In particular we are going to study the Christoffel functions

$$(1.15)\quad \lambda_n(w,x) = \left\{\sum_{\nu=0}^{n-1} p_\nu^2(w;x)\right\}^{-1} = \min_{q\in \mathcal{P}_n} q^{-2}(x)\int_{-\infty}^{\infty} q^2(x)\, w(x)\,dx$$

for the case $w = w_Q^2$. Here $\{p_\nu(w;x)\}$ is the sequence of orthonormal polynomials with respect to the weight w. (See e.g. [6]).

2. A polynomial inequality.

The purpose of the present chapter is to prove Theorem 2.6 (see below). Let us remark that in this chapter 2 the only assumptions needed concerning $Q(t)$ are that it is even, positive, differentiable and $Q'(t)$ is positive and increasing to ∞ for $0 < t < \infty$. It follows that the sequence $\{q_n\}$ defined by (1.11) satisfies $q_n \to \infty$ and $\frac{q_n}{n} \to 0$ and that we have

$$(2.1)\quad Q(x) - Q(q_n) = \int_{q_n}^{x} Q'(t)\,dt \geq n\int_{q_n}^{x}\frac{dt}{t} = n\log\frac{x}{q_n}, \quad (x \geq q_n).$$

Let

$$(2.2)\quad w_{Q_n}^2(x) = \begin{cases} e^{-2Q(x)} = w_Q^2(x) & (|x| \leq q_n) \\ 0 & (|x| > q_n) \end{cases}$$

and let

$$(2.3)\quad p_n(w_{Q_n}^2; x) = \gamma_n(w_{Q_n}^2)\,x^n + \dots \quad (n = 0, 1, 2, \dots)$$

be the orthonormal polynomials with respect to the weight $w_{Q_n}^2$.

LEMMA 2.1. If $w(x)$ is a nonnegative weight satisfying $\log w(\cos\theta) \in \mathcal{L}$ and the polynomial

$$p(x) = \eta_\nu x^\nu + q(x)\ , \ (q \in \mathcal{P}_{\nu-1})$$ satisfies

$$(2.4) \qquad \int_{-1}^{1} |p(x)|^2 w(x)\, dx \le 1$$

then

$$(2.5)\ |\eta_\nu| \le \pi^{-1/2}\, 2^\nu \exp\left\{-\frac{1}{2\pi}\int_0^\pi \log[w(\cos\theta)\sin\theta]\,d\theta\right\}.$$

Proof. By a well known Theorem of G.Szegö there exists a function $\mathcal{D}(z) \in H_2$ satisfying for a.e. $\theta \in [-\pi, \pi]$

$$(2.6) \qquad |\mathcal{D}(e^{i\theta})|^2 = w(\cos\theta)|\sin\theta|$$

and

$$(2.7) \qquad \mathcal{D}(0) = \exp\left\{\frac{1}{2\pi}\int_0^\pi \log[w(\cos\theta)\sin\theta]\,d\theta\right\}.$$

The function $\phi(z) = z^\nu p[\frac{1}{2}(z+z^{-1})]\,\mathcal{D}(z)$ belongs to H_2 and satisfies $\phi(0) = 2^{-\nu}\mathcal{D}(0)\,\eta_\nu$. Consequently by (2.5)

$$2^{-2\nu}|\mathcal{D}(0)|^2|\eta_\nu|^2 = |\phi(0)|^2 \le \frac{1}{2\pi}\int_{-\pi}^{\pi}|\phi(e^{i\theta})|^2\,d\theta$$

$$= \frac{1}{2\pi}\int_{-\pi}^{\pi}|p(\cos\theta)|^2\, w(\cos\theta)\,|\sin\theta|\,d\theta$$

$$= 2\int_{-1}^{1}|p(x)|^2\, w(x)\,dx.$$

Combining this with (2.4) and (2.7) we obtain (2.5).

LEMMA 2.2. We have

$$(2.8) \qquad \gamma_\nu(w^2_{Q_n}) \le c_8\, e^{c_9\nu + Q(q_n)}\, q_n^{-\nu-\frac{1}{2}}.$$

Proof. The polynomial $p(t) = q_n^{1/2} p_\nu(w^2_{Q_n}; q_n t) \in \mathcal{P}_\nu$ has the leading coefficient $\eta_\nu = q_n^{\nu+1/2} \gamma_\nu(w^2_{Q_n})$ and since $p_\nu \; w^2_{Q_n}$ is normed, we have

$$\int_{-1}^{1} [p(t)]^2 \, w^2_{Q_n}(q_n t)\, dt = \int_{-q_n}^{q_n} p_\nu^2(w^2_{Q_n}; t)\, w^2_{Q_n}(t)\, dt .$$

We apply Lemma 2.1. and exploit the fact that $Q(t)$ is increasing and consequently

$$-\log w_Q^2(q_n \cos\theta) = 2\, Q(q_n \cos\theta) \le 2\, Q(q_n) .$$

LEMMA 2.3. We have for every integer ν and n

$$(2.9) \quad |p_\nu(w^2_{Q_n}; x)| \le C_8\, e^{C_9 \nu}\, q_n^{-1/2} \left(\frac{q_n}{|x|}\right)^{n-\nu} e^{Q(x)}, \quad (|x| \ge q_n)$$

Proof. Since $w^2_{Q_n}$ is even, the zeros $x_{k\nu} = x_{k\nu}(w^2_{Q_n})$ of $p_\nu(w^2_{Q_n})$ are situated in $[-q_n, q_n]$ symmetrically around the origin. Consequently we have for $|x| \ge q_n$ applying (2.8)

$$|p_\nu(w^2_{Q_n}; x)| = \gamma_\nu(w^2_{Q_n})\, |x|^{\nu - 2[\nu/2]} \prod_{k=1}^{[\nu/2]} (x^2 - x_{k\nu}^2)| \le$$

(2.10)

$$\le \gamma_\nu(w^2_{Q_n})\, |x|^\nu \le C_8\, e^{C_9 \nu + Q(q_n)}\, q_n^{-\nu-1/2}\, |x|^\nu$$

$$(|x| \ge q_n) .$$

(2.9) is implied by (2.10) and (2.1).

LEMMA 2.4. We have

$$(2.11) \quad \lambda_n^{-1}(w_Q^2; x) \le C_{10}\, e^{C_{11} n}\, q_n^{-1} \left(\frac{q_{2n}}{|x|}\right)^{2n} e^{2Q(x)},$$

$$(|x| \ge q_{2n}) .$$

Proof. Since w_Q^2 is a majorant of $w_{Q_n}^2$ we have

$$(2.12) \qquad \lambda_n^{-1}(w_Q^2; x) \le \lambda_n^{-1}(w_{Q_n}^2; x)$$

(see e.g. [6], Theorem I.4.2).

Now for $|x| \ge q_n$ we have by (2.10)

$$(2.13) \quad \lambda_n^{-1}(w_{Q_n}^2; x) \le \sum_{\nu=0}^{n-1} c_8^2 e^{2c_9\nu + 2Q(q_n)} q_n^{-1} \left(\frac{|x|}{q_n}\right)^{2n}, \quad (|x| \ge q_n).$$

By (2.1) we have for $|x| \ge q_{2n} > q_n$

$$(2.14) \quad \begin{aligned} e^{Q(q_n)} &= e^{[Q(q_n) - Q(q_{2n})]} e^{[Q(q_{2n}) - Q(x)]} e^{Q(x)} \\ &= \left(\frac{q_n}{q_{2n}}\right)^n \left(\frac{q_{2n}}{|x|}\right)^{2n} e^{Q(x)}. \end{aligned}$$

We infer from (2.13) and (2.14) that

$$(2.15) \quad \lambda_n^{-1}(w_{Q_n}^2; x) \le c_{10} e^{c_{11}n} q_n^{-1} \left(\frac{q_{2n}}{|x|}\right)^{2n} e^{2Q(x)}, \quad (|x| \ge q_{2n}).$$

(2.11) follows from (2.12) and (2.15).

LEMMA 2.5. *We have for every* $p \in \mathcal{P}_n$

$$(2.16) \quad \begin{aligned} e^{-Q(x)} |p(x)| &\le \\ &\le c_{12} e^{c_{13}n} q_n^{-1/2} \left(\frac{q_{2n}}{|x|}\right)^n \left\{\int_{-q_n}^{q_n} p^2(t)\, w_Q^2(t)\, dt\right\}^{1/2}, \quad (p \in \mathcal{P}_n). \end{aligned}$$

Proof. (2.16) is a consequence of (1.15) and (2.15).

THEOREM 2.6. We have for every $p \in \mathcal{P}_n$ and sufficiently great c_{19}

(2.17) $$\int_{-\infty}^{\infty} p^2(x)\, w_Q^2(x)\, dx \le \left(1 + c_{17}\, e^{-c_{18} n}\right) \int_{-c_{19} q_{2n}}^{c_{19} q_{2n}} p^2(x)\, w_Q^2(x)\, dx .$$

Remark. For the weights $w_{\alpha,\beta}$ (see (1.2)) (2.15) was proved by my pupil G.P. Névai [10] . The present proof is new even for this case.

Proof. We have by (2.16)

$$\int_{c_{19} q_{2n}}^{\infty} p^2(x)\, w_Q^2(x)\, dx \le$$

$$\le c_{12}^2\, e^{2 c_{13} n}\, q_n^{-1}\, q_{2n}^{2n} \int_{c_{19} q_{2n}}^{\infty} \frac{dx}{x^{2n}} \int_{-q_n}^{q_n} p^2(t)\, w_Q^2(t)\, dt$$

$$= c_{19}\, c_{12}^2\, q_n^{-1}\, \frac{q_{2n}}{2n-1} \left(c_{19}^{-1}\, e^{c_{13}}\right)^{2n}$$

Since $q_n \longrightarrow \infty$ and $(q_n/n) \to 0$ (2.17) is valid for every $c_{19} > e^{c_{13}}$.

LEMMA 2.7. We have

(2.18) $$\frac{\gamma_{\nu-1}(w_Q^2)}{\gamma_\nu(w_Q^2)} \le c_{20}\, q_{2\nu} .$$

Proof. By the recursion formula

$$x\, p_{\nu-1}(w_Q^2; x) = \frac{\gamma_{\nu-1}(w_Q^2)}{\gamma_\nu(w_Q^2)}\, p_\nu(w_Q^2; x) + \frac{\gamma_{\nu-2}(w_Q^2)}{\gamma_{\nu-1}(w_Q^2)}\, p_{\nu-2}(w_Q^2; x),$$

so that

(2.19) $$\frac{\gamma_{\nu-1}^2(w_Q^2)}{\gamma_\nu^2(w_Q^2)} + \frac{\gamma_{\nu-2}^2(w_Q^2)}{\gamma_{\nu-1}^2(w_Q^2)} = \int_{-\infty}^{\infty} x^2\, p_{\nu-1}^2(w_Q^2; x)\, w_Q^2(x)\, dx .$$

(2.19) and (2.17) imply

$$\frac{\gamma_{\nu-1}^2(w_Q^2)}{\gamma_\nu^2(w_Q^2)} \leq c_{21}\int_{-c_{19}q_{2n}}^{c_{19}q_{2n}} x^2 p_{\nu-1}^2(w_Q^2;x)\, w_Q^2(x)\,dx \leq$$

$$\leq c_{21}c_{19}^2 q_{2n}^2 \int_{-\infty}^{\infty} p_{\nu-1}^2(w_Q^2;x)\, w_Q^2(x)\,dx \leq$$

$$\leq c_{21}\, c_{19}\, q_{2n}^2 \; .$$

3. Estimates for the Christoffel functions.

In the first part of the present chapter we give representations for w_Q and w_Q^{-1} as 'approximate polynomial envelopes' (see Lemma 3.2). In the remaining part of the chapter we apply these representations to estimate the Christoffel functions.

Let n be a natural integer, $\vartheta > 0$ and $|\mu| < [\frac{n}{4}]$. We consider (as functions of x) the two expressions

$$(3.1)\quad \mathcal{F}_{in}(\mu,\vartheta;x) = (\vartheta q_{2n} - x)^{[\frac{n}{4}]-\mu}(\vartheta q_{2n}+x)^{[\frac{n}{4}]+\mu}\exp\{(-1)^i Q(x)\}.$$

Denoting by „ ′ " differentiation with respect to x we get

$$(3.2)\quad \frac{\mathcal{F}_{in}'(\mu,\vartheta;x)}{\mathcal{F}_{in}(\mu,\vartheta;x)} = 2\,\frac{\mu\vartheta q_{2n} - [\frac{n}{4}]x}{\vartheta^2 q_{2n}^2 - x^2} + (-1)^i Q'(x)$$

and

$$(3.3)\quad \left[\frac{\mathcal{F}_{in}'(\mu,\vartheta;x)}{\mathcal{F}_{in}(\mu,\vartheta;x)}\right]' = -\frac{[\frac{n}{4}]-\mu}{(\vartheta q_{2n}-x)^2} - \frac{[\frac{n}{4}]+\mu}{(\vartheta q_{2n}+x)^2} + (-1)^i Q''(x) =$$

$$= \mathcal{G}_n(\mu,\vartheta;x) + (-1)^i Q''(x) \; .$$

LEMMA 3.1. a) For every $n \geq c_{22}$, $\vartheta > c_{23}$, $\mathcal{F}_{1n}$ has a single maximum ξ_{1n} in $[-\vartheta q_{2n}, \vartheta q_{2n}]$;
b) for every $n \geq c_{22}$ and $\vartheta \leq 2c_{23}$, also $\mathcal{F}_{2n}$ has a single maximum ξ_{2n} in $[-\vartheta q_{2n}, \vartheta q_{2n}]$;
c) considering ξ_{in} $(i=1,2)$ for fixed n and ϑ as functions of μ we have

$$(3.4)\quad c_{24}\frac{q_{2n}}{n} \leq -\frac{d\xi_{in}}{d\mu} \leq c_{25}\frac{q_{2n}}{n}, \quad (|\xi_{in}| \leq c_{26}\vartheta q_{2n}).$$

Proof. We have $\mathcal{F}_{in}(\mu, \vartheta; \pm\vartheta q_{2n}) = 0$, $i = 1, 2$ so that $\mathcal{F}_{1n}$ as well as $\mathcal{F}_{2n}$ have at least one local extremum in $(-\vartheta q_{2n}, \vartheta q_{2n})$.

By (1.7) and (1.9) we have for $|x| \leq q_{2n}$

$$(3.5)\quad \begin{aligned} Q''(x) &\leq (1+c_1)\,Q''(q_{2n}) \leq (1+c_1)\,c_3\,q_{2n}^{-1}\,Q'(q_{2n}) = \\ &= 2\,(1+c_1)\,c_3\,n\,q_n^{-2} \end{aligned}$$

and by (1.10)

$$(3.6)\quad 2n\,q_{2n}^{-2} = q_{2n}^{-1}\,Q'(q_{2n}) \geq c_4\,q_{2n}^{c_5} \longrightarrow \infty \quad \text{for } n \to \infty.$$

a) For $i = 1$, $-Q''(x) < 0$ for $|x| \geq c_0$. For every $x \in (-\vartheta q_{2n}, \vartheta q_{2n})$ we have

$$(3.7)\quad \mathcal{G}_n(\mu, \vartheta; x) \leq -\frac{(n/4)}{(2\vartheta q_{2n})^2}$$

so that for sufficiently large n by (3.7), (3.6) and $\vartheta > c_{23}$

$$(3.8a)\quad \left[\frac{\mathcal{F}_{1n}'(\mu, \vartheta; x)}{\mathcal{F}_{1n}(\mu, \vartheta; x)}\right]' \leq -c_{27}\,n\,q_{2n}^{-2} < 0 \qquad (|x| \leq \vartheta q_{2n}).$$

Thus $\mathcal{F}_{1n}'$ has not more than one zero in $(-\vartheta q_{2n}, \vartheta q_{2n})$; this proves assertion a).

b) We see from (3.5) and (3.7) that for $\vartheta \le c_{23}$ and sufficiently small $c_{23} < 1$

$$(3.8b)\quad \left(\frac{\mathcal{F}_{2n}'(\mu,\vartheta;x)}{\mathcal{F}_{2n}(\mu,\vartheta;x)}\right)' \le -\, c_{27}\, n\, q_{2n}^{-2} < 0$$

and this proves assertion b).

c) For $|x| \le \frac{1}{2}\vartheta q_{2n}$ we have

$$(3.7a)\quad \frac{n}{16\vartheta^2 q_{2n}^2} \le -\mathcal{G}_n(\mu,\vartheta;x) \le \frac{4n}{\vartheta^2 q_{2n}^2} \qquad \left(|x| \le \tfrac{1}{2}\vartheta q_{2n}\right)$$

so that by (3.3)

$$(3.8c)\quad c_{27}\, n\, q_{2n}^{-2} \le -\left(\frac{\mathcal{F}_{2n}'(\mu,\vartheta;x)}{\mathcal{F}_{2n}(\mu,\vartheta;x)}\right)' \le c_{28}\, n\, q_{2n}^{-2} \qquad \left(|x| \le \tfrac{1}{2}\vartheta q_{2n}\right).$$

Since

$$\frac{\partial}{\partial\mu}\left\{\frac{\mathcal{F}_{in}'(\mu,\vartheta;x)}{\mathcal{F}_{in}(\mu,\vartheta;x)}\right\} = \frac{2\vartheta q_{2n}}{\vartheta^2 q_{2n}^2 - x^2},$$

we have

$$(3.9)\quad \frac{2}{\vartheta q_{2n}} \le \frac{\partial}{\partial\mu}\left\{\frac{\mathcal{F}_{in}'(\mu,\vartheta;x)}{\mathcal{F}_{in}(\mu,\vartheta;x)}\right\} \le \frac{3}{\vartheta q_{2n}} \qquad \left(|x| \le \tfrac{1}{2}\vartheta q_{2n}\right).$$

(3.8c) and (3.9) imply (3.4).

We consider the polynomials

$$(3.10)\quad T_{in}(\vartheta,\xi;x) = \left(\frac{\vartheta q_{2n} - x}{\vartheta q_{2n} - \xi}\right)^{\left[\frac{n}{4}\right]-m_i} \left(\frac{\vartheta q_{2n} + x}{\vartheta q_{2n} + \xi}\right)^{\left[\frac{n}{4}\right]+m_i} \exp\{(-1)^i Q(\xi)\} \in \mathcal{P}_{n/2}$$

where $m_i = [\mu_i]$ and μ_i is the root with respect to μ of the equation $\mathcal{F}'_{in}(\mu, \vartheta; \xi) = 0$. This is uniquely defined by (3.4).

<u>LEMMA 3.2.</u> <u>We have for</u> $n \geq c_{29}$, $|\xi| \leq c_{30} \vartheta q_{2n}$ <u>and</u> $i = 1$, $\vartheta \geq c_{23}$ <u>resp.</u> $i = 2$ <u>and</u> $c_{23} \leq \vartheta \leq 2c_{23}$

$$(3.11) \qquad 0 \leq T_{in}(\vartheta, \xi; x) \leq c_{28} \exp\{(-1)^i Q(x)\}$$

<u>and</u>

$$(3.12) \qquad T_{in}(\vartheta, \xi; \xi) = \exp\{(-1)^i Q(\xi)\} .$$

<u>Proof.</u> (3.12) is trivial. Let $x = \xi_i$ be the unique zero situated in $(-\vartheta q_{2n}, \vartheta q_{2n})$ of the equation $\mathcal{F}'_{in}(m_i, \vartheta; x) = 0$. By Lemma 3.1., a,b

$$(3.13) \qquad \begin{aligned} &(\vartheta q_{2n} - x)^{[\frac{n}{4}] - m_i} (\vartheta q_{2n} + x)^{[\frac{n}{4}] + m_i} \exp\{(-1)^i Q(x)\} \leq \\ &\leq (\vartheta q_{2n} - \xi_i)^{[\frac{n}{4}] - m_i} (\vartheta q_{2n} + \xi_i)^{[\frac{n}{4}] + m_i} \exp\{(-1)^i Q(\xi)\} . \end{aligned}$$

From $0 \leq \mu_i - m_i < 1$ follows by (3.4)

$$(3.14) \qquad |\xi_i - \xi| \leq c_{31} \frac{q_{2n}}{n} ,$$

consequently

$$(3.15) \qquad \begin{aligned} &\left(\frac{\vartheta q_{2n} - \xi_i}{\vartheta q_{2n} - \xi}\right)^{[\frac{n}{4}] - m_i} \left(\frac{\vartheta q_{2n} + \xi_i}{\vartheta q_{2n} + \xi}\right)^{[\frac{n}{4}] + m_i} \leq \\ &\leq \left(1 + c_{32} \frac{|\xi_i - \xi|}{\vartheta q_{2n}}\right)^n \leq (1 + c_{32} c_{31} c_{23}^{-1} n^{-1})^n < e^{c_{33} n} \end{aligned}$$

and

$$(3.16)\quad |Q(\xi_i) - Q(\xi)| \le C_{31}\frac{q_{2n}}{n} Q'(q_{2n}) = 2C_{31}.$$

(3.11) is implied by (3.13), (3.15) and (3.16).

<u>THEOREM 3.3.</u> <u>We have for every natural</u> n <u>and</u> $|\xi| \le C_{35} q_{2n}$

$$(3.17)\quad \lambda_n(w_Q^2;\xi) \le C_{34}\frac{q_{2n}}{n} w_Q^2(\xi) \qquad (|\xi| \le C_{35} q_{2n}).$$

<u>Proof.</u> Let n be sufficiently large. We have by (1.15) and Theorem 2.6 for every $p \in \mathcal{P}_{n-1}$

$$(3.18)\quad \lambda_n(w_Q^2;\xi) \le C_{35}\, p^{-2}(\xi)\int_{-C_{19}q_{2n}}^{C_{19}q_{2n}} p^2(x) w_Q^2(x)\,dx .$$

We insert in (3.18) $p(x) = T_{1n}(\mathcal{D},\xi;x)q(x)$ where $q \in \mathcal{P}_{\frac{n}{2}-1}$ and $\mathcal{D} = \max(C_{19}, C_{23})$, $|\xi| \le C_{30}\mathcal{D} q_{2n} = C_{35} q_{2n}$.

By Lemma 3.2, setting $q(C_{19} q_{2n} x) = r(x)$,

$$\lambda_n(w_Q^2;\xi) \le C_{35} C_{28}^2 \min_{q \in \mathcal{P}_{\frac{n}{2}-1}} \int_{-C_{19}q_{2n}}^{C_{19}q_{2n}} q^2(x)\,dx =$$

$$(3.19)\quad = C_{35} C_{28}^2 C_{19} q_{2n} \min_{r \in \mathcal{P}_{\frac{n}{2}-1}} r^{-2}(\xi/C_{19}q_{2n}) \int_{-1}^{1} r^2(x)\,dx$$

$$= C_{35} C_{28}^2 C_{19} q_{2n}\, \lambda_{[\frac{n}{2}]}(w_1;\xi/C_{19}q_{2n})$$

where $w_1(x) = 1$ for $x \in [-1, 1]$ and $w_1(x) = 0$ for $x \notin [-1, 1]$. It is well known that

$$(3.20)\quad \frac{C_{36}}{\nu} \le \lambda_\nu(w_1;t) \le \frac{C_{37}}{\nu} \qquad (|t| \le \tfrac{1}{2}).$$

(see e.g. [6], Lemma III.3.1).

For sufficiently large n, (3.17) is a consequence of (3.19) and (3.20). By a proper choice of c_{34} it holds for every natural n.

THEOREM 3.4. We have for every natural n and every real ξ

$$\lambda_n(w_Q^2;\xi) \geq c_{38}\frac{q_{2n}}{n} w_Q^2(\xi). \tag{3.21}$$

Proof. We have by (1.15) and by (3.10) with $i = 2$

$$\lambda_n(w_Q^2;\xi) = \min_{p\in\mathcal{P}_{n-1}} p^{-2}(\xi)\int_{-\infty}^{\infty} p^2(x)\, w_Q^2(x)\,dx >$$

$$> \min_{p\in\mathcal{P}_{n-1}} p^{-2}(\xi)\int_{-c_{23}q_{2n}}^{c_{23}q_{2n}} p^2(x)\, w_Q^2(x)\,dx \geq$$

$$\geq c_{28}^{-2}\min_{p\in\mathcal{P}_{n-1}} \left[p(\xi)\,T_{2,n}(c_{23},\xi;\xi)\right]^{-2}\int_{-c_{23}q_{2n}}^{c_{23}q_{2n}}\left[p(x)\,T_{2,n}(c_{23},\xi;x)\right]^2 dx.$$

Setting $p(c_{23}q_{2n}x)\,T_{2,n}(c_{23},\xi;c_{23}q_{2n}x) = q(x)\in\mathcal{P}_{2n-1}$ we obtain by (3.20)

$$\lambda_n(w_Q^2;\xi) \geq c_{28}^{-2}\, c_{23}q_{2n}\min_{q\in\mathcal{P}_{2n-1}} q^{-2}\left(\frac{\xi}{c_{23}q_{2n}}\right)\int_{-1}^{1} q^2(x)\,dx =$$

$$= c_{28}^{-2}\, c_{23}q_{2n}\,\lambda_{2n}(w_1;\xi/c_{23}q_{2n}) \geq \tag{3.22}$$

$$\geq c_{39}\frac{q_{2n}}{n} w_Q^2(\xi), \quad (|\xi| \leq \tfrac{1}{2}c_{23}q_{2n}).$$

We replace in (3.22) n by $2^r n$ and make use of the fact that $\lambda_n(w_Q^2;\xi)$ is a decreasing function of n :

(3.23)
$$\lambda_n(w_Q^2;\xi) \ge \lambda_{2^r n}(w_Q^2;\xi) \ge$$
$$\ge c_{39}\frac{q_{2^{r+1}n}}{2^r n}\, w_Q^2(x) \ge 2^{-r} c_{39}\frac{q_{2n}}{n}\, w_Q^2(x)$$
$$\left(|\xi| \le \tfrac{1}{2} c_{23}\, q_{2^{r+1}n}\right).$$

By (1.11) and (1.9)

$$-\log\frac{q_{2\nu}}{q_\nu} + \log 2 = \log\frac{Q'(q_{2\nu})}{Q'(q_\nu)} = \int_{q_\nu}^{q_{2\nu}}\frac{Q''(t)}{Q'(t)}\,dt \le$$
$$\le c_3\int_{q_\nu}^{q_{2\nu}}\frac{dt}{t} = c_3\log\frac{q_{2\nu}}{q_\nu}$$

i.e.

(3.24)
$$\frac{q_{2\nu}}{q_\nu} > 2^{\frac{1}{1+c_3}}.$$

By (3.24) we can fix r so large that $\frac{1}{2}c_{23}\,q_{2^{r+1}n} > e^{c_{11}}q_{2n}$. Then for $|x| \ge e^{c_{11}}q_{2n}$ (3.21) is valid as a consequence of (2.15) and it is also valid for $|x| \le e^{c_{11}}q_{2n}$ by (3.23).

We observe that by (1.15), (3.21) is equivalent to

(3.25)
$$\lambda_n^{-1}(w_Q^2;x) = \sum_{\nu=0}^{n-1} p_\nu^2(w_Q^2;x) \le c_{38}^{-1}\, n\, q_{2n}^{-1}\, w_Q^{-2}(x)$$
$$(-\infty < x < \infty;\ n = 1,2,\ldots).$$

4. Study of the orthogonal polynomial series.

With the preceding chapter we ended the part of our investigation where new ideas were needed. From this point on we can apply the method developed by the author in [1] and [2].

Let

(4.1)
$$f(x) \sim \sum_{\nu=0}^{\infty} a_\nu(w_Q^2;f)\, p_\nu(w_Q^2;x)$$

be the orthogonal polynomial expansion of a measurable function f. Let us observe that (4.1) is meaningful if $w_Q f \in \mathcal{L}_p$ for some $1 \le p \le \infty$. We introduce the partial sums of (4.1)

$$\begin{aligned} s_m(w_Q^2; f; x) &= \sum_{\nu=0}^{m-1} a_\nu(w_Q^2; f)\, p_\nu(w_Q^2; x) \\ &= \int_{-\infty}^{\infty} K_m(w_Q^2; x, t)\, f(t)\, w_Q^2(t)\, dt . \end{aligned} \tag{4.2}$$

By the Christoffel-Darboux summation formula

$$\begin{aligned} K_m(w_Q^2; x, t) &= \sum_{\nu=0}^{m-1} p_\nu(w_Q^2; t)\, p_\nu(w_Q^2; x) = \\ &= \frac{\gamma_{m-1}(w_Q^2)}{\gamma_m(w_Q^2)} \cdot \frac{p_{m-1}(w_Q^2; t)p_m(w_Q^2; x) - p_m(w_Q^2; t)p_{m-1}(w_Q^2; x)}{x - t} \end{aligned} \tag{4.3}$$

<u>THEOREM 4.1</u> <u>We have for every</u> f <u>satisfying</u> $w_Q f \in \mathcal{L}_\infty$ <u>and every natural</u> n

$$w_Q(x)\, n^{-1} \sum_{m=1}^{n} |s_m(w_Q^2; f, x)| \le c_{40} \|w_Q f\|_\infty . \tag{4.4}$$

<u>Proof</u>. Let x be fixed, $f_n(t) = f(t)$ for $|t - x| \le \frac{q_{2n}}{n}$ and $f_n(x) = 0$ for $|t - x| > \frac{q_{2n}}{n}$ and $\mathcal{F}_n(t) = (x-t)^{-1}[f(t) - f_n(t)]$ so that by (4.3)

$$\begin{aligned} s_m(w_Q^2; f, x) &= s_m(w_Q^2; f_n, x) + \\ &+ \frac{\gamma_{m-1}(w_Q^2)}{\gamma_m(w_Q^2)} \left[p_m(w_Q^2; x)a_{m-1}(w_Q^2; \mathcal{F}_n) - p_{m-1}(w_Q^2; x)a_m(w_Q^2; \mathcal{F}_n)\right]. \end{aligned} \tag{4.5}$$

We have

$$|s_m(w_Q^2; f_n, x)| = \left| \int_{x-\frac{q_{2n}}{n}}^{x+\frac{q_{2n}}{n}} f(t)\, K_m(w_Q^2; x, t)\, w_Q^2(t)\, dt \right| \le$$

$$\le \left\{ \int_{x-\frac{q_{2n}}{n}}^{x+\frac{q_{2n}}{n}} f^2(t)\, w_Q^2(t)\, dt \right\}^{1/2} \left\{ \int_{-\infty}^{\infty} K_m^2(w_Q^2; x, t)\, w_Q^2(t)\, dt \right\}^{1/2} \le$$

$$(4.6) \qquad \leq \left\{ 2 \frac{q_{2n}}{n} \| f w_Q \|_\infty \right\}^{1/2} \lambda_m^{-\frac{1}{2}} (w_Q^2; x) .$$

By Schwarz's inequality, Bessel's inequality and (2.8)

$$\sum_{m=1}^{n} \frac{\gamma_{m-1}(w_Q^2)}{\gamma_m(w_Q^2)} [|p_m(w_Q^2;x)| . |a_{m-1}(w_Q^2; \mathcal{F}_n)| + |p_{m-1}(w_Q^2;x)| \, |a_m(w_Q^2;\mathcal{F}_n)|]$$

$$(4.7) \qquad \leq c_{20} q_{2n} \lambda_{n+1}^{-\frac{1}{2}}(w_Q^2;x) \| \mathcal{F}_n w_Q \|_2 \leq$$

$$\leq c_{20} q_{2n} \lambda_{n+1}^{-\frac{1}{2}} (w_Q^2; x) \| f w_Q \|_\infty \left(2 \frac{n}{q_n}\right)^{\frac{1}{2}} ;$$

(4.4) is implied by (4.5), (4.6), (4.7) and (3.25).

<u>THEOREM 4.2.</u> <u>The Féjer means</u>

$$(4.8) \qquad \sigma_n(w_Q^2; f; x) = \frac{1}{n} \sum_{m=1}^{n} s_m(w_Q^2; f; x)$$

<u>of the orthonormal expansion</u> (4.1) <u>satisfy for every</u> $1 \leq p \leq \infty$

$$(4.9) \qquad \| w_Q \sigma_n (w_Q^2; f) \|_p \leq c_{40} \| w_Q f \|_p .$$

<u>Proof.</u> For $p = \infty$ (4.9) is true by (4.4). For $p = 1$

$$\| w_Q \sigma_n (w_Q^2; f) \|_1 = \sup_{\| w_Q g \|_\infty \leq 1} \int_{-\infty}^{\infty} \sigma_n(w_Q^2; f; x) g(x) w_Q^2(x) \, dx =$$

$$= \sup_{\| w_Q g \|_\infty \leq 1} \int_{-\infty}^{\infty} \sigma_n(w_Q^2; g; x) f(x) w_Q^2(x) \, dx \leq$$

$$\leq \| f w_Q \|_1 \sup_{\| w_Q g \|_\infty \leq 1} \| \sigma_n(w_Q^2; g) w_Q \|_\infty \leq$$

$$\leq c_{40} \| w_Q f \|_1$$

i.e. (4.9) holds also for $p = 1$. Applying the Riesz-Thorin interpolation theorem to the linear transformations

$$A_n : h \longrightarrow w_Q \sigma_n (w_Q^2 ; w_Q^{-1} h)$$

we see that (4.9) holds for every $1 \le p \le \infty$.

We introduce the shifted arithmetic means (de la Vallée-Poussin means)

$$v_{2n}(w_Q^2 ; f ; x) = 2\sigma_{2n}(w_Q^2 ; f ; x) - \sigma_n(w_Q^2 ; f ; x) =$$

(4.10)

$$= \frac{1}{n} \sum_{m=n+1}^{2n} s_m(w_Q^2 ; f ; x) .$$

We will need the follwing properties of $v_{2n}(w_Q^2 ; f)$:

4(i). We have for every f satisfying $w_Q f \in \mathcal{L}_p$ for some $1 \le p \le \infty$ and every $q \in \mathcal{P}_n$

$$\int_{-\infty}^{\infty} [f(x) - v_{2n}(w_Q^2 ; f ; x)]\, q(x)\, w_Q^2(x)\, dx = 0 .$$

4(ii). We have for every $p \in \mathcal{P}_n$

$$v_{2n}(w_Q^2 ; p ; x) = p(x) .$$

4(iii). We have for every $1 \le p \le \infty$ and every natural n and every f satisfying $w_Q f \in \mathcal{L}_p$

$$\| w_Q(x)\, v_{2n}(w_Q^2 ; f ; x) \|_p \le 3 \cdot C_{40} \| w_Q(x)\, f(x) \|_p .$$

4(i) and 4(ii) follow trivially from the definition (4.10) of $v_{2n}(w_Q^2 ; f)$; 4(iii) is a consequence of (4.10) and (4.9).

5. Bohr-type inequalities and $\mathcal{L}_1$-approximation:

LEMMA 5.1. Let

(5.1)
$$\Gamma_\xi(x) = \begin{cases} 0 & (x \le \xi) \\ 1 & (x > \xi) \end{cases}$$

then

$$\mathcal{E}_n^{(1)}(w_Q;\Gamma_\xi) \le C_{41}\frac{q_{2n}}{n}w_Q(\xi). \tag{5.2}$$

Proof. Along with $Q(x)$ also $Q^*(x) = \frac{1}{2}Q(x)$ satisfies all our conditions we assumed about $Q(x)$ so that we can apply Theorem 3.3 to $w_{Q^*} = w_Q^{\frac{1}{2}}$. Evidently $q_n^* = q_{2n}$. By application of Theorem 3.3 we have

$$\Lambda_n(w_Q;\xi) \le C_{42}\frac{q_{4n}}{n}w_Q(\xi). \tag{5.3}$$

As a consequence of the Markov-Stieltjes construction (see e.g. [6], § I.5) for every weight w

$$\mathcal{E}_{2n-2}(w;\Gamma_\xi) \le \Lambda_n(w;\xi), \tag{5.4}$$

i.e.

$$\mathcal{E}_n(w;\Gamma_\xi) \le \Lambda_{[\frac{n}{2}]+2}(w;\xi). \tag{5.5}$$

Considering the facts that $\{q_\nu/\nu\}$ is decreasing and that $\{q_\nu\}$ is increasing, we infer from (5.3) and (5.5) that (5.2) is valid for $|\xi| \le C_{43}q_{2n}$. Approximating Γ_ξ for $\xi > 0$ by $p_0(x) \equiv 0$ resp. for $\xi < 0$ by $p_1(x) \equiv 1$ we see that

$$\begin{aligned} \mathcal{E}_n(w;\Gamma_\xi) &\le \int_{|\xi|}^{\infty} w_Q(x)\,dx = \int_{|\xi|}^{\infty} e^{-Q(x)}\,dx \le \\ &\le \frac{1}{Q'(|\xi|)}\int_{|\xi|}^{\infty} e^{-Q(x)}Q'(x)\,dx = \frac{1}{Q'(|\xi|)}w_Q(\xi) \le \\ &\le [Q'(C_{43}q_{2n})]^{-1}w_Q(\xi) \qquad (|\xi| > C_{43}q_{2n}). \end{aligned} \tag{5.6}$$

By (1.9)

$$\log \frac{Q'(q_{2n})}{Q'(c_{43}q_{2n})} = \int_{c_{43}q_{2n}}^{q_{2n}} \frac{Q''(t)}{Q'(t)}\,dt \le c_3 \int_{c_{43}q_{2n}}^{q_{2n}} \frac{dt}{t} = c_3 \log c_{43}^{-1}$$

i.e.

$$(5.7)\qquad \frac{1}{Q'(c_{43}q_{2n})} \le c_{43}^{-1} e^{c_3} \frac{1}{Q'(q_{2n})} = c_{43}^{-1} e^{c_3} \frac{q_{2n}}{2n}\,.$$

By (5.6) and (5.7), (5.5) is also valid for $|\xi| > c_{43}\, q_{2n}$.

<u>LEMMA 5.2.</u> (H.Bohr type inequality). <u>Let</u> $\| w_Q g \|_\infty \le 1$ <u>and for every</u> $p \in \mathcal{P}_n$ <u>let</u>

$$(5.8)\qquad \int_{-\infty}^{\infty} g(x)\, p(x)\, w_Q^2(x)\, dx = 0 \qquad (p \in \mathcal{P}_n)$$

<u>hold; then</u>

$$(5.9)\qquad |\mathcal{G}(\xi)| = \left| \int_{\xi}^{\infty} g(x)\, w_Q^2(x)\, dx \right| \le c_{41} \frac{q_{2n}}{n} w_Q(\xi)\,.$$

<u>Proof.</u> For an arbitrary $p \in \mathcal{P}_n$ we have

$$\int_{\xi}^{\infty} g(x)\, w_Q^2(x)\, dx = \int_{-\infty}^{\infty} g(x)\, \Gamma_\xi(x)\, w_Q^2(x)\, dx =$$

$$= \int_{-\infty}^{\infty} g(x) \left[\Gamma_\xi(x) - p(x)\right] w_Q^2(x)\, dx.$$

Consequently by (5.2)

$$\left| \int_{\xi}^{\infty} g(x)\, w_Q^2(x)\, dx \right| \le \| g\, w_Q \|_\infty \inf_{p \in \mathcal{P}_n} \| (\Gamma_\xi - p)\, w_Q \|_1$$

$$= \| g\, w_Q \|_\infty\, \mathcal{E}_n^{(1)}(w_Q; \Gamma_\xi) \le c_{41} \frac{q_{2n}}{n} w_Q(\xi)\,.$$

Let us denote by Δ_n the set of functions g satisfying (5.8) and $\| w_Q g \|_\infty \leq 1$. Then we have by Nikolski's duality principle of approximation [11] (see also e.g. [3]) for every $\mathcal{F}$ with $w_Q \mathcal{F} \in \mathcal{L}_1$

$$(5.10) \qquad \varepsilon_n^{(1)}(w_Q ; \mathcal{F}) = \sup_{g \in \Delta_n} \int_{-\infty}^{\infty} \mathcal{F}(x)\, g(x)\, w_Q^2(x)\, dx .$$

<u>THEOREM 5.3</u> <u>If</u> $\mathcal{F}(x)$ <u>has bounded variation in every finite interval then</u>

$$(5.11) \qquad \varepsilon_n^{(1)}(w_Q ; \mathcal{F}) \leq c_{41} \frac{q_{2n}}{n} \int_{-\infty}^{\infty} w_Q(\xi)\, |d\mathcal{F}(\xi)|$$

<u>Remark.</u> If $\mathcal{F}$ admits a derivative $\mathcal{F}'$ satisfying $w_Q \mathcal{F}' \in \mathcal{L}_1$ then follows

$$(5.12) \qquad \varepsilon_n^{(1)}(w_Q ; \mathcal{F}) \leq c_{41} \frac{q_{2n}}{n} \| w_Q \mathcal{F}' \|_1 .$$

<u>Proof.</u> Under the condition $\int w_Q |d\mathcal{F}| < \infty$ we have $\lim\limits_{|x| \to \infty} w_Q(x) \mathcal{F}(x) = 0$ so that by partial integration

$$(5.13) \qquad \int_{-\infty}^{\infty} \mathcal{F}(x)\, g(x)\, w_Q^2(x)\, dx = \int_{-\infty}^{\infty} \mathcal{G}(\xi)\, d\mathcal{F}(\xi)$$

where $\mathcal{G}(\xi) = \int_{\xi}^{\infty} g\, w_Q^2\, dx$.

(5.11) is a consequence of (5.10), (5.13) and Lemma 5.2.

<u>LEMMA 5.4</u> (Second H.Bohr type inequality). <u>Let</u> $g \in \Delta_n$, <u>i.e. let</u> $\| g\, w_Q \|_\infty \leq 1$ <u>and let (5.8) be satisfied, then</u>

$$(5.14) \qquad \left| \int_0^x g(t)\, dt \right| \leq c_{44} \frac{q_{2n}}{n} w_Q^{-1}(x) .$$

Proof. Let $x \geq 0$ and

$$\phi_x(t) = \begin{cases} w_Q^{-2}(t) & (t \in [0,x]) \\ 0 & (t \notin [0,x]) \end{cases}$$

Thus for every $p \in \mathcal{P}_n$ by (5.8)

$$\int_0^x g(t)\,dt = \int_{-\infty}^{\infty} g(t)\,[\phi_x(t) - p(t)]\,w_Q^2(t)\,dt$$

and by (5.11)

$$\left| \int_0^x g(t)\,dt \right| \leq \| g\, w_Q \|_\infty \, \varepsilon_n^{(1)}(w_Q; \phi_x) \leq$$

$$(5.15) \qquad \leq c_{41} \frac{q_{2n}}{n} \int_{-\infty}^{\infty} w_Q(\xi)\, |d\,\phi_x(\xi)| \; .$$

By direct calculation

$$(5.16) \qquad \int_{-\infty}^{\infty} w_Q(\xi)\, |d\,\phi_x(\xi)| \leq 3\, w_Q^{-1}(x)$$

(5.14) is implied in (5.15) and (5.16) if $x \geq 0$. For $x < 0$ it is also valid by symmetry.

6. The theorem on weighted polynomial approximation.

LEMMA 6.1 Let f be the integral of f' and $w_Q f \in \mathcal{L}_p$; then for $p = 1$ and $p = \infty$

$$(6.1) \qquad \varepsilon_n^{(p)}(w_Q; f) \leq c_{45} \frac{q_{2n}}{n} \| w_Q f' \|_p \qquad (n = 1, 2, \ldots).$$

Proof. For $p = 1$ (6.1) was proved; see the remark to Theorem 5.3. We turn to the proof of (6.1) for $p = \infty$. Let $\nu = [\frac{n}{2}] - 1$ so that $\Pi_n(x) = f(0) + \int_0^x v_{2\nu}(w_Q^2; f'; t)\,dt \in \mathcal{P}_n$.

By 4(i)

$$\rho = f' - v_{2\nu}(w_Q^2; f') \in \Delta_\nu$$

and by Lemma 5.4 applied to $\rho / \| w_Q \rho \|_\infty$ we obtain

$$\| w_Q(f - T_n) \|_\infty \le C_{44} \| w_Q[f' - v_{2\nu}(w_Q^2; f')] \|_\infty \frac{q_{2\nu}}{\nu} \le$$

(6.2)
$$\le C_{44} [\| w_Q f' \|_\infty + \| w_Q v_{2\nu}(w_Q^2, f') \|_\infty] \frac{q_{2n}}{[\frac{n}{2}] - 1} .$$

(6.1) for $p = \infty$ is a consequence of (6.2) and 4(iii).

LEMMA 6.2 <u>We have for every</u> $1 \le p \le \infty$ <u>and every</u> f <u>satisfying</u> $w_Q f \in \mathcal{L}_p$

(6.3) $$\| w_Q[f - v_{2n}(w_Q^2; f)] \|_p \le C_{46} \, \varepsilon_n^{(p)}(w_Q; f) .$$

Proof. We have by 4(ii) and 4(iii) for every $q \in \mathcal{P}_n$

$$\| w_Q[f - v_{2n}(w_Q^2; f)] \|_p \le$$

$$\le \| w_Q (f - q) \|_p + \| v_{2n}(w_Q^2; f - q) \|_p$$

$$\le (1 + 3 C_{40}) \| w_Q (f - q) \|_p ,$$

i.e.

$$\| w_Q[f - v_{2n}(w_Q^2; f)] \|_p \le$$

$$\le (1 + 3 C_{40}) \inf_{q \in \mathcal{P}_n} \| w_Q (f - q) \|_p \le$$

$$\le (1 + 3 C_{40}) \, \varepsilon_n^{(p)}(w_Q; f) .$$

LEMMA 6.3 Let $1 \le p \le \infty$, let f be the integral of f' and $f' w_Q \in \mathcal{L}_p$ then

$$(6.4) \qquad \varepsilon_n^{(p)}(w_Q; f) \le c_{46} \frac{q_{2n}}{n} \| w_Q f' \|_p \qquad (n = 1, 2, \dots) .$$

Proof. By Lemma 6.1 and (6.3) we have for $p = 1$ and $p = \infty$

$$(6.5) \quad \| w_Q [f - v_{2\nu}(w_Q^2; f)] \|_p \le c_{45} c_{46} \frac{q_{2\nu}}{\nu} \| w_Q f' \|_p .$$

Applying the Riesz-Thorin interpolation theorem to the uniquely defined linear operator

$$B_n(h) : h = w_Q f' \longrightarrow w_Q [f - v_{2n}(w_Q^2; f)]$$

we see that (6.4) holds also for every $1 \le p \le \infty$.

To prove (6.1) for $1 \le p \le \infty$ put in (6.4) $\nu = [\frac{n}{2}]$ and use $q_{2\nu} \le q_{2n}$.

Note. As a byproduct we proved also the fact (nontrivial in itself) that $w_Q f' \in \mathcal{L}_p$ implies $w_Q f \in \mathcal{L}_p$.

LEMMA 6.4 Let $1 \le p \le \infty$, let f be the integral of f' and $w_Q f' \in \mathcal{L}_p$ then

$$(6.5) \qquad \varepsilon_n^{(p)}(w_Q; f) \le 2 c_{46} \frac{q_{2n}}{n} \varepsilon_{n-1}^{(p)}(w_Q; f') \qquad (n = 1, 2, \dots).$$

Proof. For a properly chosen $q \in \mathcal{P}_n$ we have

$$\| w_Q (f' - q') \|_p \le 2 \varepsilon_{n-1}^{(p)}(w_Q; f') .$$

Applying Lemma 6.3 to $f - q$ in place of f we get

$$\varepsilon_n^{(p)}(w_Q; f) = \varepsilon_n^{(p)}(w_Q; f - q) \le c_{46} \frac{q_{2n}}{n} \| w_Q (f' - q') \|_p \le$$

$$\le 2 c_{46} \frac{q_{2n}}{n} \varepsilon_{n-1}^{(p)}(w_Q; f') .$$

We are now ready to prove the main approximation theorem announced in the Introduction. Let $w_Q f \in \mathcal{L}_p$ for some $1 \le p \le \infty$. We define $\delta_n = \frac{q_n}{n}$ and

$$(6.6) \qquad \varphi_n(x) = \begin{cases} w_Q(x)\, f(x) & (|x| \le q_n) \\ 0 & (|x| > q_n) \end{cases}$$

and

$$(6.7) \qquad f_n(x) = w_Q^{-1}(x)\, \delta_n^{-1} \int_x^{x+\delta_n} \varphi_n(t)\, dt \,.$$

<u>Lemma 6.5</u> <u>We have</u>

$$(6.8) \qquad \| w_Q (f - f_n) \|_p \le \omega(\mathcal{L}_p, w_Q ; f, \frac{q_n}{n})$$

<u>and</u>

$$(6.9) \qquad \| w_Q f_n' \|_p \le c_{47} \frac{n}{q_n} \omega(\mathcal{L}_p, w_Q ; f, \frac{q_n}{n}) \,.$$

<u>Proof</u>. Since $s(t)$ is the inverse of $Q'(x)$ and $Q'(q_n) = \frac{n}{q_n} = \delta_n^{-1}$ we have $s(n/q_n) = q_n$, consequently

$$(6.10) \quad Q'\{s(\delta_n^{-1})\ \tau[x/s(\delta_n^{-1})]\} = Q'\{q_n \tau[x/q_n]\} = \begin{cases} |Q'(x)|, & (|x| \le q_n) \\ Q'(q_n) = \frac{n}{q_n}, & (|x| \ge q_n). \end{cases}$$

We apply now several times Minkowski's inequality (all norms $\|\cdot\|_p$ refer to „ . " as a function of x):

$$\| w_Q(f - f_n) \|_p = \| \delta_n^{-1} \int_0^{\delta_n} [\varphi_n(x+t) - w_Q(x) f(x)] dt \|_p \leq$$

$$\leq \| \delta_n^{-1} \int_0^{\delta_n} [w_Q(x+t) f(x+t) - w_Q(x) f(x)] dt \|_p +$$

$$+ \| \delta_n^{-1} \int_0^{\delta_n} [\varphi_n(x+t) - w_Q(x+t) f(x+t)] dt \|_p \leq$$

$$\leq \sup_{0 \leq t \leq \delta_n} \| w_Q(x+t) f(x+t) - w_Q(x) f(x) \|_p + \tag{6.11}$$

$$+ \| w_Q f - \varphi_n \|_p \leq$$

$$\leq \sup_{0 \leq t \leq \delta_n} \| w_Q(x+t) f(x+t) - w_Q(x) f(x) \|_p +$$

$$+ \delta_n \| Q'\{q_n \tau(x/q_n)\} w_Q(x) f(x) \|_p .$$

By (1.12), (6.10) and (6.11) we proved that (6.8) holds. By (6.7)

$$\| w_Q f_n' \|_p \leq \| \delta_n^{-1} [\varphi_n(x+\delta_n) - \varphi_n(x)] \|_p + \| Q'(x) \delta_n^{-1} \int_x^{x+\delta_n} \varphi_n(t) dt \|_p \leq$$

$$\leq \delta_n^{-1} \| w_Q(x+\delta_n) f(x+\delta_n) - w_Q(x) f(x) \|_p +$$

$$+ 2 \delta_n^{-1} \| w_Q f - \varphi_n \|_p + \| Q' \varphi_n \|_p + \tag{6.12}$$

$$+ \| Q'(x) \delta_n^{-1} \int_0^{\delta_n} [\varphi_n(x+t) - \varphi_n(x)] dt \|_p \leq$$

$$\leq 2 \delta_n^{-1} \omega(\mathcal{L}_p, w_Q ; f, \delta_n) +$$

$$+ \| Q'(x) \delta_n^{-1} \int_0^{\delta_n} [\varphi_n(x+t) - \varphi_n(x)] dt \|_p .$$

In the last term the integral vanishes for

$|x| \geq q_n + \delta_n \geq q_n(1 + \frac{1}{n})$, and for $|x| \leq q_n + \delta_n$

we have by (1.6), (1.9) and (1.10) for $n \geq c_{48}$

$$|Q'(x)| \leq Q'(q_n + \delta_n) \leq Q'(2q_n) = Q'(q_n)\exp\left\{\int_{q_n}^{2q_n} \frac{Q''(t)}{Q'(t)}\,dt\right\} \leq$$

$$\leq Q'(q_n)\exp\left\{c_3 \int_{q_n}^{q_{2n}} \frac{dt}{t}\right\} \leq c_{49}\, Q'(q_n) = c_{49}\, \delta_n^{-1} .$$

Thus

$$\| Q'(x)\, \delta_n^{-1} \int_0^{\delta_n} [\varphi_n(x+t) - \varphi(x)]\,dt \|_p \leq$$

$$(6.13) \qquad \leq c_{49}\, \delta_n^{-1} \Big\{ \| \delta_n^{-1} \int_0^{\delta_n} [w_Q(x+t) f(x+t) - w(x) f(x)]\,dt \|_p + $$
$$+ 2 \| w_Q f - \varphi_n \|_p \Big\} \leq$$
$$\leq 3 c_{49}\, \delta_n^{-1}\, \omega(\mathcal{L}_p, w_Q; f, \delta_n) .$$

For $n \geq c_{48}$ (6.9) follows from (6.12) and (6.13). For $n < c_{48}$ (6.9) evidently holds provided that c_{47} is great enough.

<u>THEOREM 6.6</u> <u>If</u> f <u>has an</u> r-th <u>order derivative</u> $f^{(r)}$ <u>satisfying for some</u> $1 \leq p \leq \infty$ $w_Q f^{(r)} \in \mathcal{L}_p$ <u>then</u>[2]

$$(6.14) \qquad \varepsilon_n^{(p)}(w_Q; f) \leq c_{50}\, e^{c_{51} r} \left(\frac{q_n}{n}\right)^r \omega(\mathcal{L}_p, w_Q; f^{(r)}, \frac{q_n}{n}),$$
$$(n = 2r+1, 2r+2, \dots) .$$

<u>Proof</u>. We have by Lemma 6.4 for $n > 2r$

$$(6.15) \qquad \varepsilon_n^{(p)}(w_Q; f) \leq (2c_{46})^r \frac{q_{2n}\, q_{2(n-1)} \cdots q_{2(n-r+1)}}{n(n-1)\dots(n-r+1)} \times$$

$$\times\, \varepsilon_{n-r}^{(p)}(w_Q; f^{(r)}) \leq e^{c_{51} r} \left(\frac{q_n}{n}\right)^r \varepsilon_{n-r}^{(p)}(w_Q; f^{(r)}) .$$

2. (6.14) holds also for $r = 0$ if we define $f^{(0)} = f$.

Let $f_n^{(r)}$ be the function defined to $f^{(r)}$ by Lemma 6.5. Then by Lemma 6.5. and Lemma 6.3

$$\varepsilon_{n-r}^{(p)}(w_Q; f^{(r)}) \leq \| w_Q(f^{(r)} - f_n^{(r)}) \|_p + \varepsilon_{n-r}^{(p)}(w_Q; f_n^{(r)}) \leq$$

(6.16) $$\leq \omega(\mathscr{L}_p, w_Q; f^{(r)}, \frac{q_n}{n}) + C_{46} \frac{q_{2n-2r}}{n-r} \| w_Q (f_n^{(r)})' \|_p$$

$$\leq C_{50}\, \omega(\mathscr{L}_p, w_Q; f^{(r)}, \frac{q_n}{n}) .$$

(6.14) is implied by (6.15) and (6.16).

References[3].

1. FREUD, G., On approximation with weight by polynomials on the real line. Dokladii A.N. SSSR 191 (1970) 293-294. English translation: Soviet Math. Doklady 11 (1970), 370-371.

2. FREUD, G., On an inequality of Markov type. Dokladii A.N. SSSR 197 (1971) 790-793. English translation: Soviet Math. Dokl. 12 (1971) 570-573.

3. FREUD, G., A contribution to the problem of weighted polynomial approximation. ISNM vol.20 'Linear operators and approximation theory'. Conference report (Oberwolfach, 1970). Editors: P.J.Butzer, J.P.Kahane and B.Sz.-Nagy, Birkhäuser Verl. Basel 1972.

4. Freud, G., On direct and converse theorems in the theory of weighted polynomial approximation. Math. Zeitschr. 134 (1972) 123-134.

5. Freud, G., On converse theorems of weighted polynomial approximation. Acta Math. Acad. Sci. Hung., in print.

6. Freud, G., Orthogonale polynome, Birkhäuser Verl. Basel 1969. English translation by I.Földes, Pergamon Press 1971.

7. Freud, G., On weighted polynomial approximation on the whole real axis. Acta Math. Ac. Sci. Hung. 20 (1969) 223-225.

8. Freud, G., Investigations on weighted approximation by polynomials. Studia Sci. Math. Hung., in print.

9. Névai, G.P., Some properties of polynomials orthogonal to the weight $(1+x^{2k})^{\alpha} e^{-x^{2k}}$ and their application to approximation theory. (In Russian). Dokladii Ak. Nauk SSSR, in print.

3. A detailed bibliography of papers up to 1970 is to be found in [3]. In particular we refer the reader to the excellent paper of M.M.Džrbasian.

10. Névai, G.P., Polynomials orthogonal on the real axis to the weight $|x|^{\alpha} e^{-|x|^{\beta}}$. In Russian. Acta Math. Ac. Sci. Hung., in print.

11. Nikolski, S.M., Approximation of functions by trigonometric polynomials in the mean. (In Russian). Isvestia Ak. Nauk SSSR ser. matem. 10 (1946) 207-256.

DETERMINATION OF CONFORMAL MODULES OF RING DOMAINS AND QUADRILATERALS

Dieter Gaier

Introduction. Let G be a doubly connected region with $\infty \notin G$, bounded by two rectifiable Jordan curves C_1 (exterior boundary component) and C_2 (interior boundary component). It can be mapped conformally onto the concentric circular ring $R = \{w : 1 < |w| < M\}$, and $M = M(G)$ is called the conformal module of G. Let G be a Jordan region with four different points $P_i (i = 1,2,3,4)$ on $\partial G = C$ distinguished; we say a quadrilateral Q is given. Then G can be mapped conformally onto a rectangle $R: (0,a) \times (0,b)$ such that P_1, P_2, P_3, P_4 go into 0, a, $a+ib$, ib, and $m(Q) = \frac{a}{b}$ is called the conformal module of Q.

These domain functionals play an important role in many practical applications as well as in theoretical investigations. Some methods to determine $M(G)$ are described in the author's book ([3], chapter V), and it is the purpose of this lecture to survey the work that has been done since the publication of this book on the problem to determine $M(G)$ and $m(Q)$.

Part I: Modules of Ring Domains

1.1. Integral equation of Symm. Assume that 0 is interior to C_2, and let f be a conformal mapping of G onto $\{w: 1 < |w| < M\}$ such that C_2 is mapped onto $\{w : |w| = 1\}$. Then $\log \frac{f(z)}{z} = g(z) + ih(z)$ $(z \in G)$ is analytic in G and has a continuous extension to $\bar{G}$, so that

$$\begin{aligned} g(z) &= \log M - \log |z| \quad (z \in C_1), \\ g(z) &= -\log |z| \quad (z \in C_2). \end{aligned} \tag{1}$$

Since h is single valued in G we have furthermore $\int_{\gamma} \frac{\partial h}{\partial s} ds = 0$ and hence

$$\int_{\gamma} \frac{\partial g}{\partial n} ds = 0 \qquad (2)$$

for any $\gamma \subset G$ homotopic to C_1.

The idea of Symm [13] is to solve the boundary value problem for the harmonic function g satisfying (1) and (2) by means of a single layer potential; i.e. g is sought in the form

$$g(z) = \int_{\partial G} \log |z-t| \, \sigma(t) \, |dt| ,$$

where σ is an unknown density function. Once σ is found, we have

$$h(z) = \int_{\partial G} \arg (z-t) \, \sigma(t) \, |dt| ,$$

and f is determined. The conditions (1) and (2) mean for σ that

$$\int_{\partial G} \log |z-t| \, \sigma(t) \, |dt| = \begin{cases} \log M - \log |z| & (z \in C_1) \\ -\log |z| & (z \in C_2) \end{cases} \qquad (3)$$

plus

$$\int_{\partial G} \sigma(t) \, |dt| = 0. \qquad (4)$$

The mapping problem is therefore reduced to an integral equation of first kind. Although such equations are, in the absence of general existence theorems, not very popular, it can be shown that there is a unique σ satisfying (3) and (4) if ∂G satisfies additional smoothness conditions.

To solve (3), (4), Symm [13] approximates σ by a step function and performs the necessary integrations numerically. If N nodal points on ∂G are used, we obtain N+1 linear equations for the quantities $\sigma_1, \ldots, \sigma_N$, $\log M$, and so M is approximated as well as σ.

We have a check of accuracy of the method: If $\tilde{g}, \tilde{h}$ are approximations of g, h, the deviation of the image of G under $z \exp(\tilde{g}+i\tilde{h})$ from a concentric circular ring can serve as measure of accuracy. The method has been successfully tested by Symm.

1.2. Bergman's kernel function. Let $L^2(G)$ be the Hilbert space of functions f analytic and with single valued integral in G, for which $\iint_G |f|^2 \, db < \infty$, and assume that a complete orthonormal system $\{\phi_n\}$ in $L^2(G)$ has been constructed. Then the Bergman kernel $K(z,\bar{t}) = \sum_{n=1}^{\infty} \phi_n(z) \overline{\phi_n(t)}$, ($z,t$ in G) is a domain function independent of the system $\{\phi_n\}$, and so is

$$J_G(z) = \left[K(z,\bar{z}) \right]^{-1} \frac{\partial^2 \log K(z,\bar{z})}{\partial z \, \partial \bar{z}}, \quad (z \in G).$$

This function is positive in G, has boundary values 2π, and is conformally invariant: if f maps G onto R, $J_G(z) = J_R(f(z))$; see Bergman [1]. From this follows that J_G is constant on level curves in G, i.e.

$$J_G(z) = F(r) \quad \text{if} \quad |f(z)| = r \quad (1 < r < M).$$

Furthermore, Bergman showed that F is strictly increasing in $(1, \sqrt{M})$ and strictly decreasing in $(\sqrt{M}, M)$, so that $F(\sqrt{M}) = \max\{J_G(z) : z \in G\}$. This can be expressed explicitly

$$F(\sqrt{M}) = M + M \sum_{n=1}^{\infty} \frac{n^3 M^n}{M^{2n}-1} \Big/ \sum_{n=1}^{\infty} \frac{n M^n}{M^{2n}-1} . \tag{5}$$

Burbea [2] first used these ideas to determine M without constructing the mapping function. Of course the series for K has to be truncated giving K_n, and instead of J_G appears $J_G^{(n)}$. After the maximum of this function, as z ranges over G, has been obtained, inverse interpolation in (5) gives an approximation for the module $M(G)$.

In Burbea's experiments the functions z^k $(k = -n,\dots,+n, k \neq -1)$ are orthonormalized, n going up to 16. The inner products (z^k, z^j) are calculated explicitly (if ∂G is polygonal) or using high order Gaussian quadrature formulas, and the method was tested successfully in several cases.

1.3. Characterisation of M by Dirichlet integral. If u is the harmonic measure of C_2 in G, i.e. u is harmonic in G with boundary values $u = 0$ on C_1, $u = 1$ on C_2, the module $M(G)$ is connected with the Dirichlet integral

$$D_G[u] = \iint_G (u_x^2 + u_y^2)\,dxdy$$

by

$$M(G) = \exp\left\{2\pi / D_G[u]\right\}. \tag{6}$$

On the other hand, u has a minimum property. Let K denote the class of functions v continuous in $\bar{G}$ with $v = 0$ on C_1, $v = 1$ on C_2, for which v_x, v_y exist and are continuous in G with the exception of points on a finite number of rectifiable Jordan arcs. Then Dirichlet's principle asserts

$$D_G[u] \leq D_G[v] \qquad (v \in K). \tag{7}$$

Any function $v \in K$ gives an upper bound for $D_G[u]$ and so yields a lower bound for M(G), a very pleasant feature of this method.

In order to construct members of K, assume that G is composed of squares q. In each q we define a function of the form $F(x,y) = axy + bx + cy + d$ such that $F(x,y)=0$ if (x,y) is a corner of q on C_1, $F(x,y) = 1$ if (x,y) is a corner of q on C_2, and $F(x,y) = z_i$ arbitrary at a corner (x,y) of q in G. If there are N corners of squares in G, we obtain to each vector $z = (z_1, z_2, \ldots, z_N)$ a two dimensional spline function F_z defined on G and belonging to K, whence

$$D_G[F_z] \geq D_G[u] .$$

The Dirichlet integral for the spline function F_z can be evaluated and is a quadratic form in z. The minimum $\min_z D_G[F_z] = D_G[F_{z_0}]$ which gives a particularly good estimate of $D_G[u]$ is obtained if the components of z_0 define a discretely harmonic function on the interior mesh points of G, and so the computation of this optimal vector z_0 leads to the solution of a linear $N \times N$-system.

This method to approximate $M(G)$ from below is due to Opfer [8] who tested it with great success in numerous examples. Usually N is rather large (about 10,000) and the linear system is solved by iterative procedures which can be time-consuming. If G is not composed of squares, it can be approximated from within and without by regions of this special type, and estimates for $M(G)$ follow. However, the method can be recommended particularly for regions with non-smooth boundaries, where other (integral equation) methods fail or do not produce good approximations.

The theory of the method has been generalized to obtain the modules of regions of connectivity $n \geq 2$ (Mizumoto [7]) ; however, there are no experiments so far.

1.4. <u>New experiments to known methods</u>. Here we mention two papers. Reutter and Neukirchen [9] apply the function theoretic method of Komatu and its variations (see [3], chapter V, §4) to various cross sections including one in electronics. Very few iterations are sufficient to obtain the conformal mapping onto a concentric circular ring and $M(G)$ with an error of 1-2 $^o/_o$. If $M(G)$ is small, convergence is slower, as predicted by the theory.

Further experiments to be mentioned are by Richardson and Wilson [10] who use the integral equation of Gerschgorin-Krylow for ring domains as basis of their method; see Gaier [3] , p.191 - 192. After the boundary correspondence and M(G) are obtained, the authors find an explicit Laurent sum which approximately maps the circular ring onto G.

Part II . Modules of Quadrilaterals

2.1. Characterisation of m by Dirichlet integral. First of all, there is a method to obtain m(Q) without using the conformal mapping and similar to that described in 1.3. Let K denote the class of functions v continuous in $\bar{G}$ with $v = 0$ on P_1P_2, $v = 1$ on P_3P_4 for which v_x, v_y exist and are continuous in G with the exception of points on a finite number of rectifiable Jordan arcs. Then using Dirichlet's principle one obtains

$$m(Q) = \inf \left\{ D_G [v] \ : \ v \in K \right\} .$$

Any function $v \in K$ yields an upper bound for m(Q), and since replacing P_1P_2, P_3P_4 by P_2P_3, P_4P_1 gives the 'conjugate' quadrilateral Q' with $m(Q') = 1/m(Q)$, we can obtain a lower bound for m(Q) by considering Q'. This fact is a decisive advantage over other methods.

After 1.3 it is clear how the method works in practice assuming G composed of squares. Two dimensional bilinear spline functions F are constructed which are 0 on P_1P_2 and 1 on P_3P_4 and so belong to K :

$$m(Q) \leq D_G [F] .$$

An F with minimal Dirichlet integral is obtained if the restriction of F to the mesh points of G defines a discretely harmonic function, and if there are N such mesh points we have to solve a linear

$N \times N$ - system. In the author's experiments [4] N is about 12,000 and the solution of the linear system (by over relaxation) may be rather time-consuming. Special attention is given to the case where G is the L-shaped region composed of three squares and for different choices of the P_i, since in this example the modules can be calculated explicitly in terms of elliptic integrals.

The method admits a generalization when G is composed of rectangles and triangles. In the latter case we use spline functions which are of the form $ax + by + c$ over the triangular parts of G.

2.2. Determination of m by conformal mapping. Of course any method mapping G conformally onto the upper half plane or the unit disc can be used to determine m. If the P_i go into Q_i $(i=1,2,3,4)$ and if d is the cross ratio of the Q_i, we have

$$m(Q) = \frac{K'(k)}{K(k)} \quad \text{with } k = d^{-1/2}$$

(complete elliptic integrals; see [4] , p.190). In this connection we draw attention to Symm's integral equation method [11] in which the conformal mapping of G onto $\{w : |w| < 1\}$ is reduced to the Dirichlet problem for the region G and the boundary values $-\log |z|$ which in turn is solved via an integral equation of first kind similar to that discussed in 1.1. A slight modification of the method permits the mapping of infinite domains [12] , and a significant improvement in accuracy can be obtained by a more careful discretization of the integral equation; see the recent paper by Hayes, Kahaner and Kellner [5] .

Experiments for the L-shaped region mentioned in 2.2 have been carried out by Knierim [6] using the integral equation of Gerschgorin. As is well known ([3] , p.18), this equation can be modified so as to include regions with corners on ∂G. First results show that the computing time (in the order of seconds) is considerably shorter than for the method 2.1. In addition, this method works for infinite regions as well. However, if upper and lower bounds for m(Q) are desired, method 2.1 is preferable.

2.3. Analog method. If the region G is available on so called resistance paper, we can put an alloy on the sides P_2P_3 and P_4P_1 of the given quadrilateral. Then the resistance measured between P_2P_3 and P_4P_1 is proportional to m(Q), and comparison with the resistance of a square gives m(Q) itself. We have tested this simple method ([4] , p.193) and have obtained an accuracy of about 1 % .

References.

1 Bergman, S., The kernel function and conformal mapping. 2nd ed. Providence 1970. Zbl 208, 343 .

2 Burbea, J., A numerical determination of the modulus of doubly connected domains by using the Bergman curvature. Math. Comp. 25 (1971), 743-756, Zbl 234, 416 .

3 Gaier, D., Konstruktive Methoden der konformen Abbildung. Springer, Berlin 1964. MR 33, 1291; Zbl 132, 367.

4 ________, Ermittlung des konformen Moduls von Vierecken mit Differenzenmethoden. Numer.Math. 19 (1972), 179-194.

5 Hayes, J.K., Kahaner, D.K. and Kellner, R.G., An improved method for numerical conformal mapping. Math.Comp. 26 (1972), 327-334.

6 Knierim, H., Experimente zur konformen Abbildung des L-förmigen Gebiets. Studienarbeit. Giessen 1972.

7 Mizumoto, H., An application of Green's formula of a discrete function: Determination of periodicity moduli.I,II. Kōdai Math.Sem. Rep. 22 (1970), 231-243, 244-249. MR 42, 1429;Zbl 219, 437 .

8 Opfer, G., Zur Bestimmung des Moduls zweifach zusammenhängender Gebiete mit Hilfe von Differenzenverfahren. Arch.Rational Mech.Anal. 32 (1969), 281-297. MR 38, 830; Zbl 187, 403.

9 Reutter, F. and Neukirchen, H.J., Untersuchungen auf dem Gebiet der praktischen Mathematik: Vergleichende Untersuchung einiger numerischer Verfahren zur konformen Abbildung einfach und zweifach zusammenhängender Gebiete. Köln 1967. MR 36,467; Zbl 154, 412 .

10 Richardson, M.K. and Wilson, H.B., A numerical method for the conformal mapping of finite doubly connected regions. Developments in Theoretical and Applied Mechanics, Vol.3 (1967),305-321.

11 Symm, G.T., An integral equation method in conformal mapping. Numer. Math. 9 (1966), 250-258. MR 34,1296; Zbl 156, 169.

12 __________, Numerical mapping of exterior domains. Numer.Math. 10 (1967), 437-445. MR 36, 708; Zbl 155, 215.

13 __________, Conformal mapping of doubly-connected domains. Numer.Math. 13 (1969), 448-457. MR 40, 687; Zbl 174, 206.

SOLOVAY'S AXIOM AND FUNCTIONAL ANALYSIS

H.G. Garnir

1. Taking into account the recent progress in the study on the foundation of Mathematics, there are actually three ways to develop Functional Analysis which depend on the personal convictions of each mathematician.

a) Constructive Functional Analysis

This is the core of Functional Analysis we can get by using exclusively the traditional ways of reasoning universally admitted[1], including the axiom of countable choice and, more adequately, the axiom of inductive definition of sequences[2].

The contents of Constructive Functional Analysis is described, for instance, in my book with De Wilde and Schmets ''Analyse Fonctionnelle : théorie constructive des espaces à semi-normes'' [4]. With a minimum of reasonable separability assumptions, Constructive Functional Analysis contains all the essential facts of Functional Analysis, organized in an elegant theory quite efficient for the

---

(1) Of course, the admitted axioms may be listed : they have to secure a precise language, the rules of logical reasoning, the theory of sets, the integers and the axiom of inductive definition of sequences.
(2) This axiom states that there is a law which associates to each set of a sequence of sets an element of this set.
It has historically never been seriously contested by the majority of the mathematicians. Anyway, it cannot be dropped without losing the whole classical analysis.
It has some physical meaning which makes it easy to admit : for lack of an explicit choice function, it is possible to consider as such the listing of the successive choices ; this listing is in fact unlimited but may be effectively written as far as wanted.
The axiom of inductive definition of sequences is somewhat more powerful than the axiom of choice : here the choices are done in a sequence of sets, each of which being defined by the previous choices.

applications.

Constructive Functional Analysis is the firm basis from which new axioms may be added in a controlled way. This can be done at least in two directions and gives rise to the two following branches of Functional Analysis.

b) Functional Analysis with the additional axiom of non-countable choice : it is actually the Classical Functional Analysis.

c) Functional Analysis with the additional axiom of Solovay.

This seems to be a promising new branch of Functional Analysis: let us call it Solovayan Functional Analysis.

2. Let me first give some information about the axiom of non-countable choice and the new axiom of Solovay.

A. The classical exposition of Functional Analysis uses liberally the axiom of non-countable choice.

This axiom is used in the following three ways :

- directly : given a non-countable class of sets, there exists a law which associates an element of each set to each set of the class;
- under the equivalent form of Zorn : 'a set with a partial order relation has a maximal element if any ordered subset has a upper bound;
- indirectly by consequences : theorem of Tychonov, existence of a Hamel base in any linear space, existence of a ultrafilter finer than a given filter.

The quoted axiom is a very old and respectable axiom : it was born in 1904 under the form of Zermelo [9] and improved in power in 1935 in the form of Zorn [10] .

It has been often used - and sometimes abused - and its consequences are well-explored.

It secures proofs without separability restrictions of important propositions of Functional Analysis : theorems of Hahn-Banach, of Alaoglu, of compactness of precompact complete sets, of existence of multiplicative functionals (maximal ideals) in Banach Algebras, of Krein-Milman, ...

The major argument to use it freely is that, since a proof of Gödel [5] , the axiom of non-countable choice can be added to Constructive Analysis without leading to contradictions.

Nevertheless, many mathematicians (and some of them are famous) feel uncomfortable with this axiom. They are worried by its intuitively shocking consequences and use it with reluctance or by routine.

Indeed, this axiom is often used to prove the existence of mathematical objects which cannot be exhibited, which are not intuitively conceivable and for which we cannot even imagine a process of construction. For instance, except very special cases, nobody can show a Hamel base in the most usual spaces or exhibit a ultrafilter finer than a given filter.

Moreover, it leads to paradoxes of which one of the most striking is the paradox of Banach - Tarski [1] which says : in the three dimensional euclidian space it is possible to cut the unit ball in a finite number of pieces and, by using only translations and rotations, to reassemble them to form two unit balls.

B. The axiom of Solovay is bright new : it has been stated in 1971 [8] .

It refers to measure theory and states : any function defined in the euclidian space R^n is Lebesgue-measurable.

It would be fair to mention that this axiom has been used for a long time by the applied mathematicians to avoid measurability problems in the practical use of Lebesgue theory of integration. Their position was based on the fact that it seemed impossible to build any non-measurable functions in an honest way, that is without additional axioms.

The essential work done by Solovay was to prove for his new axiom the same result as Gödel : the axiom of Solovay may be added to Constructive Analysis without leading to contradictions.

The impact of the axiom of Solovay in Functional Analysis is not yet explored : this talk is an attempt in this direction.

There is a useful extension of the axiom of Solovay : any function in an open set Ω of R^n is μ-measurable for any measure μ on Ω.

By a measure on Ω, we mean a measure as defined in [4] II , that is a countably additive set function with finite variation on the semi-intervals with closure in Ω or, equivalently, a countably additive set function on the Borel sets with compact closure in Ω ([4] II p.126).

The proof is reduced to the case of a positive diffuse[1] measure, if we notice ([4] II p.202-208) that any measure is the sum of a series of point measures plus a diffuse measure, itself being a linear combination of four real positive diffuse measures.

For such a measure it is possible to build ([4] II § 41, p.221-3) two functions

$X(x)$, defined ℓ-pp in $]0, \mu(\Omega)[$, with values in Ω,

$x(X)$, defined μ-pp in Ω, with values in $]0, \mu(\Omega)[$,

ℓ denoting the Lebesgue measure, such that

$$\Omega = X(]0, \mu(\Omega)[), \ \mu \text{ - a.e.}$$

$$]0, \mu(\Omega)[= x(\Omega), \ \ell \text{ - a.e.}$$

and for any semi-interval i, with closure in $]0, \mu(\Omega)[$, or I, with closure in Ω,

$$\ell(i) = \mu(\{X : x(X) \in i\}),$$

$$\mu(I) = \ell(\{x : X(x) \in I\}).$$

---

(1) μ diffuse = μ continuous = $\mu(\{x\}) = 0, \ \forall x \in \Omega$.

So, by a well-known theorem for the change of variables ([4] II, § 23, p.177-180), f(X) is μ-measurable in Ω if $f[X(x)]$ is ℓ -measurable in $]0, \mu(\Omega)[$, which is always the case by the axiom of Solovay.

C. <u>It is essential to notice that the axiom of Solovay is contradictory to the axiom of Zorn</u>.

So, Constructive and Solovayan Functional Analysis are two different games : what is true in one may be false in the other.

This is strikingly pointed out by the problem of the existence of non-measurable functions :

- For a ''Zornian'' : any non-negligeable set contains a non-measurable subset, as it is shown in any text-book of measure theory.

- For a ''Solovayan'' : there are no non-measurable functions.

- For an ''agnostic'', that is a constructive analyst : an answer cannot be given to this question without additional axioms.

3. The aim of this talk is to develop a very important result of Functional Analysis which can be deduced from the axiom of Solovay and seems to be the point of departure of fruitful applications.

It concerns <u>exclusively</u> the general theory of <u>separated locally convex topological vector spaces</u> (LCTVS) or <u>spaces with s-n</u>(1).

In a linear space E we call <u>semi-norm</u> (abbreviated by s-n) a law which associates to every $f \in E$ a real number $\pi(f)$ such that

$$\pi(\sum_i c_i f_i) \leq \sum_i |c_i| \pi(f_i),$$

or, equivalently,

$$\pi(cf) = |c| \pi(f),$$
$$\pi(f+g) \leq \pi(f) + \pi(g).$$

A <u>space with s-n</u> is a linear space E whose topology is described by a system (= a separating and filtering set) of <u>natural</u> s-n,

(1) From here on, see [4] I for the definitions and notations.

$P = \{p\}$, defined up to an equivalence.

A s-n π is <u>continuous</u> in E if

$$\exists\ C > 0,\ p \in P : \pi(f) \leq C\ p(f),\ \forall\ f \in E .$$

The result we have in view states that <u>in a good space any s-n is continuous</u>.

Let us provisionaly define <u>good spaces</u> as the spaces where this statement is true : we will soon describe them.

The consequences of such a statement are very important.

a) A <u>linear functional</u> $\mathcal{T}$ defined in E with natural s-n $P = \{p\}$ is <u>continuous</u> if

$$\exists\ C > 0,\ p \in P : |\mathcal{T}(f)| \leq Cp(f),\ \forall\ f \in E.$$

Our statement implies that <u>in a good space any functional is continuous</u>.

This explains why it has always been impossible to build non-continuous functionals by honest means in a good space.

b) A <u>linear operator</u> T starting from a <u>departure space</u> E with the natural s-n $P = \{p\}$ to an <u>arrival space</u> F with the natural s-n $Q = \{q\}$ is <u>continuous</u> if

$$\forall\ q \in Q,\ \exists\ C > 0,\ p \in P : q(Tf) \leq Cp(f),\ \forall\ f \in E.$$

Our statement means here that <u>any linear operator starting from a good space is continuous</u>.

This supplies a perfect form of the general closed graph theorem [3a]: in fact, it suppresses any closed graph assumption and any restriction of the arrival space and keeps, as we will see, the full generality of the departure space.

4. It is time to give now some important examples of good spaces.

a) Any Fréchet space is a good space, which means that in a Fréchet space any s-n is continuous.

The following simple proof of this statement is based on a trick discovered by the Danish mathematician J.P.R. Christensen [2]. See also L.Schwartz [7].

Let F be a Fréchet space with the natural semi-norms p_k, chosen as to determine an increasing sequence.

First notice that if π is not continuous there is a sequence $f_m \in F$ such that

$$\sum_{m=1}^{\infty} p_m(f_m) < \infty \ , \ \pi(f_m) \rightarrow +\infty \ .$$

Indeed,

$$e_m = \left\{ f : p_m(f) \leq \frac{1}{2^m}, \ \pi(f) > m \right\} \neq \emptyset \ ,$$

unless π is such that

$$p_m(f) \leq \frac{1}{m2^m} \pi(f), \ \forall \ f \in E,$$

and continuous. We just have to choose f_m in e_m, $\forall$ m.

Our reasoning rests on the study of the Cantor set considered as a continuous group in R.

For the Cantor set in R we take

$$\mathcal{C} = \{\alpha = 0, \alpha_1 \alpha_2 \dots : \alpha_i = 0 \text{ or } 1\} \ .$$

We make $\mathcal{C}$ a commutative group by choosing a law of multiplication which consists of adding the figures modulo 2 : if

$$\gamma = \alpha . \beta \iff \gamma_i = \alpha_i + \beta_i \pmod 2 \ , \quad \forall i \ .$$

So,

0, 1 0 1 1 0 0 ...

times

$$0,\ 1\ 1\ 1\ 0\ 0\ 1\ldots$$

is

$$0,\ 0\ 1\ 0\ 1\ 0\ 1\ldots.$$

This multiplication is trivially commutative and associative. It has $0 = 0, 0\ 0 \ldots$ as unity.

Any element is the inverse of itself : $\alpha^{-1} = \alpha$.

The group $\mathcal{C}$ is continuous for the euclidian norm : we have

$$\left.\begin{array}{l}\alpha_m \longrightarrow \alpha \text{ in } \mathcal{C} \\ \beta_m \longrightarrow \beta \text{ in } \mathcal{C}\end{array}\right\} \Longrightarrow \alpha_m \cdot \beta_m \longrightarrow \alpha . \beta \text{ in } \mathcal{C}$$

because α_m and α, on one hand, and β_m and β, on the other, have more and more common figures and then so do $\alpha_m \cdot \beta_m$ and $\alpha . \beta$. Moreover,

$$\alpha_m^{-1} = \alpha_m \longrightarrow \alpha = \alpha^{-1}.$$

So, this group has a Haar measure μ, fulfilling the requirements of the last theorem of § 2, B. ([4] III, pp.166-177) such that

$$\mu \geq 0 \quad , \quad \mu(\mathcal{C}) = 1 \quad .$$

If we notice that for $\alpha \in \mathcal{C}$, the series

$$\sum_{j=1}^{\infty} \alpha_j f_j$$

is convergent in F, being Cauchy : if $\inf(r,s) \to \infty$, for any ℓ, we have, for r and s large enough,

$$p_\ell \left(\sum_{j=r}^{s} \alpha_j f_j \right) \leq \sum_{j=r}^{s} p_\ell (f_j) \leq \sum_{j=r}^{s} p_j(f_j) \longrightarrow 0 \ .$$

So, we may consider the sets

$$\mathcal{C}_m = \left\{ \alpha \in \mathcal{C} : \pi\left(\sum_{k=1}^{\infty} \alpha_k f_k \right) \leq m \right\} .$$

We have

$$\mathcal{C} = \bigcup_{m=1}^{\infty} \mathcal{C}_m .$$

As all the sets $\mathcal{C}_m$ are measurable by the axiom of Solovay, $\mu(\mathcal{C}) > 0$ implies $\mu(\mathcal{C}_{m_0}) > 0$ for at least one m_0.

Then, from a classical result about Haar measure ([6], corollary 20.17, p.296), we have

$$\mathcal{C}_{m_0}^{-1} . \mathcal{C}_{m_0} = \mathcal{C}_{m_0}^{2} = \{ \alpha . \beta : \alpha , \beta \in \mathcal{C}_{m_0} \} \supset \{ x : |x| \leq \varepsilon \} .$$

So,

$$e_k = 0, \underbrace{0 \; . \; . \; . \; 0 \; 1}_{k} \, 0 \; \ldots$$

belongs to $\mathcal{C}_{m_0}^2$ when k is large enough and there are $\alpha^{(k)}$, $\beta^{(k)} \in \mathcal{C}_{m_0}$ such that

$$e_k = \alpha^{(k)} \; \beta^{(k)} .$$

From the multiplication law of $\mathcal{C}$, $\alpha^{(k)}$ and $\beta^{(k)}$ differ only by the figure at the kth place.

This implies

$$\alpha^{(k)} - \beta^{(k)} = \pm e_k .$$

So,

$$\pm f_k = \sum_{j=1}^{\infty} \alpha_j^{(k)} f_j - \sum_{j=1}^{\infty} \beta_j^{(k)} f_j ,$$

and as

$$\alpha^{(k)}, \beta^{(k)} \in \mathcal{C}_{m_o} ,$$

$$\pi(f_k) \leq \pi(\sum_{j=1}^{\infty} \alpha_j^{(k)} f_j) + \pi(\sum_{j=1}^{\infty} \beta_j^{(k)} f_j) \leq 2m_o ,$$

which is a contradiction.

b) <u>More generally, any bornological sq-complete space</u> E <u>is a good space</u>.

If E is bornological to know that a s-n π defined in E is continuous in E, we only have to prove that

$$\pi(B) = \sup_{f \in B} \pi(f) < \infty ,$$

for any bounded set B in E.

We have

$$B \subset \langle \bar{B} \rangle \begin{cases} \text{absolutely convex,} \\ \text{bounded,} \\ \text{closed} \Rightarrow \text{sq-complete,} \end{cases}$$

and so

$$E_{\langle \bar{B} \rangle} = \{ f : \lambda > 0 : \lambda f \in \langle \bar{B} \rangle \}$$

is a Banach space for the gauge

$$p_{\langle \bar{B} \rangle}(f) = \inf_{f \in \lambda \langle \bar{B} \rangle} \lambda .$$

Then, π is continuous on E_B and

$$\pi \leq C p_{\langle \bar{B} \rangle} \Rightarrow \pi(B) \leq C p_{\langle \bar{B} \rangle}(B) \leq C \; ^{(1)}.$$

(1) We just prove that $\pi(B) < \infty$ for any B bounded absolutely convex such that E_B is a Banach space for its gauge. So, the proposition is true if this implies the continuity of π, in which case the space is called <u>ultrabornological</u>.

c) Let us mention some usual bornological sq-complete spaces :

- Fréchet spaces,
- Strict inductive limits of Fréchet spaces, (see also § 5, a) !),
- Strong duals of
  - a Schwartz space with countable s-n, ([4] I, § 21, p.256).
  - a hyperstrict inductive limit of such spaces, ([4] I, § 21, p.256)
  - $C_p(\Omega)$, Ω being an open set of R^n, ([4] III, ex 1, p.237).

5. The good spaces have some permanence properties.

a) Any countable inductive limit of good spaces is a good space.

Indeed, if $E = \mathcal{L} E_m$, a s-n π of E is a s-n in every E_k. So, π is continuous in every E_k, and then continuous in $\mathcal{L} E_m$.

b) Any finite product of good spaces is a good space.

The product

$$\underset{(P_1)}{E_1} \times \dots \times \underset{(P_N)}{E_N}$$

where the system of natural s-n of every space E_i is P_i, has the elements

$$\vec{f} = (f_1, \dots , f_N), \ f_k \in E_k, \ \forall \ k$$

and the natural s-n

$$\sup_{k=1,\dots,N} p_k(f_k), \ \forall \ N = 1,2,\dots, \forall \ p_k \in P_k.$$

So, we have

$$\pi [(f_1,\dots,f_N)] \leq \sum_k \pi [(\underbrace{0,\dots,f_k}_{k},\dots,0)]$$
$$\leq \sum_k C_k \, p_k \, (f_k) \leq (\sum_k C_k) \sup_k p_k \, (f_k),$$

because $\pi[(0,\ldots,f_k,\ldots 0)]$ is a s-n on the good space E_k.

c) Any countable product of good spaces is a good space.

Here, the countable product

$$\underset{(P_1)}{E_1} \times \underset{(P_2)}{E_2} \times \ldots$$

has as elements the vectors

$$\vec{f} = (f_1, f_2, \ldots), \; f_k \in E_k, \forall k$$

and its natural s-n are

$$\sup_{k=1,\ldots,N} p_k(f_k) \;\forall N = 1,2,\ldots, \;\forall p_k \in \mathcal{P}_k.$$

Our proof rests on a lemma of De Wilde [3b] : a s-n π defined in $E_1 \times E_2 \times \ldots$ is continuous if and only if

$\pi[(\underbrace{0,\ldots,0, f}_{k}, 0, \ldots)]$ is a continuous s-n of $f \in E_k$, $\forall k$,

$\pi[(z_1 f_1, z_2 f_2, \ldots)]$ is a continuous s-n of

$$\vec{z} \in C \times 0 \times C \times \ldots \text{ (0 if } f_k = 0, \text{ C otherwise).} \qquad (*)$$

From here, our theorem is trivial because the E_k and the Fréchet space $C \times 0 \times C \times \ldots$ are good spaces.

For the sake of completeness, let us prove the lemma.

Let us call $[\vec{f}]$, support of $\vec{f}$, the subset $\{i : f_i \neq 0\}$ of $N = \{1,2,\ldots\}$.

If $\varepsilon \subset N$, we denote by $\vec{f}\delta_\varepsilon$ the restriction of $\vec{f}$ in ε, that is the vector $\vec{f}$ where f_i is replaced by 0 if $i \notin \varepsilon$.

π satisfies $(*)$ for $\vec{f} \Rightarrow \exists\, e(\vec{f})$ finite in N:

$$\pi(\vec{f}\delta_\varepsilon) = 0, \qquad \forall \varepsilon \subset \complement e(\vec{f}).$$

We have just to take for $e(\vec{f})$, the finite set e such that

$$\pi\,[(z_1 f_1,\ z_2 f_2,\ \dots)] \leq C \sup_{i \in e} |z_i| \, .$$

As a consequence <u>for any</u> $\vec{f}$, $\vec{f}\delta_{e(\vec{f})}$ <u>has a finite support and is such that</u> $\pi(\vec{f}) = \pi(\vec{f}\delta_{e(\vec{f})})$.

In fact,

$$\pi(\vec{f}\delta_{\complement e(\vec{f})}) = 0 \Rightarrow \pi(\vec{f}) = \pi(\vec{f}\delta_{e(\vec{f})})$$

π <u>satisfies</u> (*) $\forall\, \vec{f} \Rightarrow \exists$ e <u>finite in</u> N : $\pi(\vec{f}) = 0$ if $[\vec{f}\,] \subset \complement e$.

If it is false then, for every $e \subset N$ there is an $\vec{f}$ such that $[\,f\,] \subset \complement e$ and $\pi(\vec{f}) \neq 0$. As we have seen, we may take $\vec{f}$ with a finite support in $\complement e$.

So, we construct by induction $\vec{f}^{(1)}, \vec{f}^{(2)}, \dots$ with finite disjoint supports and such that $\pi(\vec{f}^{(k)}) \neq 0$: if we know $\vec{f}^{(1)}, \dots, \vec{f}^{(k)}$, we take $\vec{f}^{(k+1)}$ with $\pi(\vec{f}^{(k+1)}) \neq 0$ and finite support in $\complement\{[\vec{f}^{(1)}] \cup \dots \cup [\vec{f}^{(k)}]\}$.

But, as seen before, for

$$\vec{f} = (\vec{f}^{(1)}, \vec{f}^{(2)}, \dots),$$

there is $e(\vec{f}) \subset N$ such that

$$\pi(\vec{f}\delta_{\varepsilon}) = 0, \ \forall\, \varepsilon \subset \complement e(\vec{f})$$

and then $\pi(\vec{f}^{(k)}) = 0$ if $[\vec{f}^{(k)}]$ is in $\complement e(\vec{f})$, which is a contradiction.

Therefrom

$$\pi(\vec{f}) = \pi(\vec{f}\delta_e), \ \forall\, \vec{f}.$$

This being settled, we have

$$\pi(\vec{f}) = \pi(\vec{f}\delta_e) = \sum_{i \in e} \pi(\vec{f}\delta_{\{i\}})$$

$$\leq \sum_{i \in e} C_i\, p_i(f_i) \leq \Big(\sum_{i \in e} C_i \Big) \sup_{i \in e} p_i(f_i) .$$

d) _If a good space_ E _admits a decomposition_

$$E = \sum_{i=1}^{N} E_i ,$$

$$E_i \cap \sum_{j \neq i} E_j = 0 \ , \ (i, j = 1, \ldots, N) \ ,$$

the projections P_i _from_ E _to_ E_i _are continuous,_

E_i _is closed,_

E_i _with the induced s-n of_ E _is a good space._

This result applies to algebraically complemented subspaces[(1)] and, of course, to subspaces with finite codimension.

Indeed, any $f \in E$ can be written $\sum_i f_i$, $f_i \in E_i$, in an unique way. So the projections P_i such that $f_i = P_i f$ are defined in E and continuous.

Then $E_i = \{ f \in E : (1-P_i)f = 0 \}$ is closed.

Moreover, let $P = \{p\}$ be the system of natural s-n of E and π an arbitrary s-n of E_i.

The s-n $\pi(P_i f)$ defined in the good space E is continuous and, if $f \in E_i$,

$$\exists\ C > 0,\ p \in P : \pi(f) = \pi(P_i f) \leq C\, p(f) .$$

---

(1) Any L with an algebraic complement ℓ such that $L \cup \ell = E$, $L \cap \ell = 0$ is closed. But we can no more secure the existence of an algebraic complement of L, by invoking the existence of a Hamel base.

Then π is continuous in E_i and E_i is a good space.

6. What about the open mapping theorem?

A linear operator T from E to F is <u>open</u> if

$$\forall\, p \in P, \exists C > 0,\; q \in Q : \quad \inf_{g=Tf} p(f) \leq C\, q(g),\; \forall\, g \in TE.$$

<u>If</u> T <u>is a linear operator from any space</u> E <u>to a good space</u> F,

$$TE = F \;(= T \text{ ''onto'' } F) \Rightarrow T \text{ open}\,.$$

Just notice that

$$\pi(g) = \inf_{g=Tf} p(f)$$

is a s-n on TE = F and then is continuous.

Indeed,

$$\pi(cg) = \inf_{cg=Tf} p(f) = |c| \inf_{g=T\frac{f}{c}} p\left(\frac{f}{c}\right) = |c|\, \pi(g),$$

$$\pi(f+g) = \inf_{f+g\,=\,Tf} p(f) \leq \inf_{\substack{f=f'+f''\\ f=Tf'\\ g=Tf''}} p(f) \leq \inf_{\substack{f=Tf'\\ g=Tf''}} p(f'+f'')$$

$$\leq \inf_{f=Tf'} p(f') + \inf_{g=Tf''} p(f'') = \pi(f) + \pi(g).$$

Bibliography

1 Banach, S. and Tarski, A., Sur la décomposition des ensembles de points en parties respectivement congruentes. Fund. Math. 6, 1924, pp.244-277.
(See also R.M.Robinson "On the decomposition of spheres", ibid 34, 1947, pp.246-260).

2 Christensen, J.P.R., Borel structure in groups and semi-groups. Math. Scand. 28, 1971, pp.124-128.

3 De Wilde, M., a) Réseaux dans les espaces linéaires à semi-normes. Mémoires Soc. Roy. Sc. de Liège, 18, 1969, no.2.
b) Vector topologies and linear maps on products of topological vector spaces.
Math. Ann. 196, 1972, pp.117-128.

4 Garnir, H.G., De Wilde, M. and Schmets, J., Analyse fonctionnelle : théorie constructive des espaces à semi-normes, Birkhäuser Verlag : I, General theory 1968, II, Measure theory 1970, III, Usual spaces 1973.

5 Gödel, K., The consistency of the axiom of choice and of the generalized continuum-hypothesis with the axioms of set theory. Ann. of Math. Studies no.3, Princeton Univ. Press, 1940.

6 Hewitt, E. and Ross, K., Abstract harmonic analysis I, Springer Verlag, 1963.

7 Schwartz, L., Sur le théorème du graphe fermé. C.R.Acad. Sc. Paris 263, 1966, pp.602-605.

8 Solovay, R., A model of set theory in which every set of reals is Lebesgue-measurable. Annals of Math. 92, 1970, pp.1-56.

9 Zermelo, E., Beweis dass jede menge wohlgeordnet werden kann. Math. Ann. 59, 1904, pp.514-516.

10 Zorn, M., A remark on method in transfinite algebra. Bull. Am. Soc. 41, 1935, pp. 667-670.

Q - UNIFORM ALGEBRAS AND OPERATOR THEORY

Bernard R. Gelbaum

1. The Gelfand theory of **commutative** Banach algebras leads to the spectral theorem for just those Hilbert space operators N, namely normal operators, for which such a theorem is possible:

$$N = \int \lambda \, d\, E(\lambda) ,$$

where $\{E(\lambda)\}$ is an (orthogonal) resolution of the identity, if and only if N is normal, i.e., $NN^* = N^*N$. The relevant commutative Banach algebra is the closure in some operator topology of the set of polynomials in I, N, N^*.

A commutative Banach algebra A enjoys, but is not characterized by, the property:

the algebra has only one nontrivial simple $\mathbb{C}$-eipmorphic image, namely $\mathbb{C}$ itself.

A Banach algebra A admitting, up to $\mathbb{C}$-isomorphisms, only one nontrivial simple $\mathbb{C}$-eipimorphic image, say Q, is called Q-uniform.

In earlier work [2,3,4], owing to the analogy with the situation where A is commutative, the preceding definition was supplemented by the restriction that A be a unitary Banach Q - bimodule. However, to close the circle of ideas, there is need for a fairly wide class of Hilbert space operators T for which the closure, in some operator topology, of the set of polynomials in I, T, T^* is Q-uniform for some Q. Although examples of such operators T exist [4], if there is insistence on the 'Banach Q-bimodule' restriction, the important class of shift operators is omitted.

Indeed, the closure $R_W(I, S, S^*)$, in the weak operator topology, of the set of polynomials in I, S, S^*, where S is the unilateral shift operator on separable Hilbert space H, is $\mathcal{B}(H)$, the set of all continuous endomorphisms of H. However, since the unique maximal ideal of $\mathcal{B}(H)$ is $\mathcal{K}(H)$, the set of compact operators, the only simple $\underset{\sim}{C}$-epimorphic image of $\mathcal{B}(H)$ is $Q \equiv \mathcal{B}(H)/\mathcal{K}(H)$. Furthermore, Catherine Olson in consultation with William Zame at the State University of New York at Buffalo has shown that $\mathcal{B}(H)$ cannot be a unitary Banach Q-bimodule. Consequently the present (and relaxed) form of the definition of a Q-uniform algebra is adopted.

[Note . Before the Olson - Zame result became known to the author, he attempted to show $\mathcal{B}(H)$ is a Banach Q-bimodule. In the course of the investigation, he explored to discover whether or not $\mathcal{K}(H)$ is complemented in $\mathcal{B}(H)$. A result of Thorp [7] shows that $\mathcal{K}(H)$ is not complemented. Since $\mathcal{K}(H)^{**} \cong \mathcal{B}(H)$ [5], there emerges another proof that $\mathcal{K}(H)$ is not a (Banach) conjugate space: if $\mathcal{K}(H) = E^*$, then $\mathcal{B}(H) = E^{***}$, whereas, by a result of A.E.Taylor [6], $E^* = \mathcal{K}(H)$ is always complemented in $E^{***} = \mathcal{B}(H)$.

It should be noted that the Bartle-Graves [1] section $\gamma: Q \to \mathcal{B}(H)$ cannot be differentiable at 0 since the complementedness of $\mathcal{K}(H)$ is equivalent to the existence of a differentiable section $\delta: Q \to \mathcal{B}(H)$. There appears to be a relationship between the Euclidean smoothness of the Hilbert space norm and the necessary existence of a differentiable section $\eta: F \to H$ for any quotient space of H.]

2. For an n-dimensional Hilbert space H_n and T in $\mathcal{B}(H_n)$, the Jordan normal form theorem for T may be stated as follows:

> There is in H_n an inner product [,], possibly different from that (,) originally given, so that with respect to [,] $H_n = \bigoplus_k M_k$, an orthogonal direct sum, $T = \bigoplus_k T_k$ where T_k is in $\mathcal{B}(M_k)$ and where $T_k - \lambda_k I_k$ is [,]-unitarily equivalent to the shift S_k on M_k for some λ_k in $\underset{\sim}{C}$.

Note that $R_w(I_k, T_k, T_k^*) = \mathcal{B}(M_k)$, a simple and hence Q-uniform algebra, and that $R_w(I,T,T^*) = \bigoplus_k \mathcal{B}(M_k)$, where adjoints are formed relative to [,] .

Clearly, if M_k is an infinite-dimensional separable Hilbert space, $k = 1,2,\ldots,m$, then $H = \bigoplus_k M_k$, the orthogonal direct sum, is a separable Hilbert space. If, in M_k, S_k is unitarily equivalent to the unilateral shift, and if λ_k is in $\mathbb{C}$, then $T \equiv \bigoplus_k(\lambda_k I_k + S_k)$ is such that $R_w(I,T,T^*) = \bigoplus_k \mathcal{B}(M_k)$ is Q-uniform where $Q = \mathcal{B}(H)/\mathcal{K}(H) = \mathcal{B}(M_k)/\mathcal{K}(M_k)$.

If $(X, \mathfrak{S}, \mu)$ is a measure space and if for each x in X, M_x is a separable and infinite-dimensional Hilbert space, there can be constructed first $H = \int_X M_x d\mu(x)$ and then an operator $T = \int_X(\lambda(x) I_x + S_x)d_\mu(x)$, where $\lambda(x)$ is in $\mathbb{C}$, S_x is unitarily equivalent to the unilateral shift in M_x and appropriate measurability, integrability conditions, etc., are imposed. The example of the preceding paragraph shows that this kind of construction can lead to an operator T for which $R_w(I,T,T^*)$ is Q-uniform. A problem of interest is the establishment of conditions on X, $\mathfrak{S}$, μ, $\lambda(x)$, S_x, etc., that imply $R_w(I,T,T^*)$ is Q-uniform. Problems of measurability, integrability, etc., aside, $R_W(I,T,T^*)$ will be of the form

$$\left\{ \int_X f(x)d\mu(x) : f(x) \text{ in } \mathcal{B}(M_x) \right\}$$

where $f(x)$ is appropriately restricted. The discussion will then center on the maximal ideal set structure of the set of functions $f(x)$.

Bibliography.

1 Bartle, R.G. and Graves, L.M., Mappings between function spaces. Trans.Amer.Math.Soc. (1952), 400-413

2 Gelbaum, B.R., Banach algebra bundles. Pac.J.Math.28 (1969), 337-349

3 ____________, Q-uniform Banach algebras. Proc.Amer.Math.Soc. 24 (1970), 344-353.

4 ____________, On a class of operators in Hilbert space. Stud. Math. XXXVIII(1970), 279-284.

5 Schatten, R., Norm ideals of completely continuous operators. Ergebnisse der Math. und ihrer Grenzgebiete, 27(1960),Berlin.

6 Taylor, A.E., The extension of linear functionals. Duke Math. J. 5 (1939), 538-547.

7 Thorp, E.O., Projections onto the subspace of compact operators. Pac. J. Math. 10 (1960), 693-696.

ON POLYNOMIALS WITH A PRESCRIBED ZERO

André Giroux

The importance in approximation theory of the following result is well known:

THEOREM (Bernstein). *If* $T_n(\theta)$ *is a trigonometric polynomial of order* n, *then:*

$$\max_{\theta} |T_n'(\theta)| \leq n \max_{\theta} |T_n(\theta)| ,$$

with equality if and only if $T_n(\theta) = c \sin n(\theta - \theta_o)$.

As a special case, if $p_n(z)$ is an algebraic polynomial of degree n, then $\max_{|z|=1} |p_n'(z)| \leq n \max_{|z|=1} |p_n(z)|$ with equality if and only if $p_n(z)$ has all its zeros at the origin. This remarkable fact has been investigated by many mathematicians who generalized the result or obtained sharper estimates for the derivative by restricting the polynomials considered to subclasses. In particular, we have:

THEOREM (Erdös-Lax). *If* $p_n(z)$ *is a polynomial of degree* n *not vanishing for* $|z| < 1$, *then:*

$$\max_{|z|=1} |p_n'(z)| \leq n/2 \max_{|z|=1} |p_n(z)|$$

with equality if and only if all the zeros of $p_n(z)$ *lie on* $|z|=1$

Here, we wish to consider the case where $p_n(z)$ has a prescribed zero on the unit circle. What we have been able to prove are the following results:

THEOREM A. *If* $p_n(z)$ *is a polynomial of degree* n *and* $p_n(1) = 0$, *then*

$$\max_{|z|=1} |p_n'(z)| \leq (n-A/n) \max_{|z|=1} |p_n(z)|$$

where A is an absolute positive constant.

THEOREM B. There exists a positive constant B and, for each n, a polynomial of degree $n, q_n(z)$ such that $q_n(1) = 0$ and

$$\max_{|z|=1} |q_n'(z)| \geq (n-B/n) \max_{|z|=1} |q_n(z)| .$$

Thus a single zero outside of $|z| < 1$ does not affect considerably the general result, in contrast to the Erdös-Lax Theorem. One can then ask how many zeros must be assumed in $|z| \geq 1$ in order to decrease the bound for the derivative appreciably. In this direction, we have:

THEOREM C. There exists a positive constant c and, for each n, a polynomial of degree $[n^{\frac{1}{2}}]^2$, $r_n(z)$ with $[n^{\frac{1}{2}}]$ zeros on $|z| = 1$ such that

$$\max_{|z|=1} |r_n'(z)| \geq ([n^{\frac{1}{2}}]^2 - c) \max_{|z|=1} |r_n(z)| .$$

Proof of Theorem A follows easily for the Riesz interpolation formula: for any trigonometric polynomial of order $n, T_n(\theta)$,

$$T_n'(\theta) = \frac{1}{2n} \sum_{k=1}^{2n} \frac{(-1)^k}{1-\cos\frac{2k+1}{2n}\pi} T_n(\theta + \frac{2k+1}{2n}\pi).$$

For proving Theorem B and C, we consider a polynomial of degree $n, F_n(z)$ such that $F_n(z) \neq 0$ if $|z| < 1$ and

$$|F_n(e^{i\theta})|^2 + \left(\frac{\sin(n+1)\theta/2}{(n+1)\sin\theta/2}\right)^2 = 1.$$

It can be shown that, when n is odd, one has

$$F_n'(-1) = F_n(-1) \cdot \frac{1}{2\pi} \int_{-\pi}^{\pi} \frac{e^{it}}{(e^{it}+1)^2} \log |F_n(e^{it})|^2 \, dt.$$

It follows from this that $|F_n'(-1)| = O(1/n)$. If one sets now $q_n(z) = z^n \overline{F_n(1/\bar{z})}$, one obtains Theorem B (in the case n odd). To get Theorem C, the polynomial $r_n(z) = q_m(z^m)$, where $m = [n^{\frac{1}{2}}]$ is to be considered.

This work was done with Professor Q.I.Rahman.

A PRIORI INEQUALITIES FOR SYSTEMS OF PARTIAL DIFFERENTIAL EQUATIONS

J.Gobert

1. Let Ω be a bounded open set of R^n with very regular boundary Γ. If $\vec{u}$ is a vector-valued function such that

1. $\vec{u} \in [L^2(\Omega)]^n$ (*)

2. $D_{x_i} u_j + D_{x_j} u_i \in L^2(\Omega)$, $i,j = 1,\dots,n$,

it is well-known that such $\vec{u} \in H^1(\Omega)$ and we have

$$\|\vec{u}\|^2_{H^1(\Omega)} \leq C\Big[\sum_{i,j} \|D_{x_i} u_j + D_{x_j} u_i\|^2_{L^2(\Omega)} + \|\vec{u}\|^2_{L^2(\Omega)}\Big],$$

where C does not depend on $\vec{u}$. It is the well known ([2],[4]) inequality of Korn, which is important in the theory of elasticity.

The operator of elasticity may be written in the matrix-form

$$\mathcal{A}(D) = \begin{pmatrix} D_{x_1}, & \dots, & D_{x_n} \\ \\ D_{x_i} & & D_{x_j} \\ (\text{col. } j^{th}) & & (\text{col. } i^{th}) \end{pmatrix}$$

If we replace that operator by

$$\mathcal{A}(D) = \begin{pmatrix} D_{x_1}, & \dots, & D_{x_n} \\ \\ D_{x_i} & & -D_{x_j} \\ (\text{col. } j^{th}) & & (\text{col. } i^{th}) \end{pmatrix}$$

---

* In the future, we'll no more write the exponent, its value coming right from the text.

for which

$$|\mathcal{A}(D)\vec{u}|^2 = \left|\sum_{i=1}^{n} D_{x_i} u_i\right|^2 + \sum_{\substack{i,j=1 \\ i<j}}^{n} |D_{x_j} u_i - D_{x_i} u_j|^2$$

that is, for $n = 3$,

$$|\mathcal{A}(D)\vec{u}|^2 = |\operatorname{div} \vec{u}|^2 + |\operatorname{curl} \vec{u}|^2 ,$$

that operator is elliptic too, but we have no more the mentioned-property.

2. Let us consider the space

$$V(\Omega) = \left\{ \vec{u} \in L^2(\Omega),\ \mathcal{A}(D)\vec{u} \in L^2(\Omega) \right\}.$$

For $\vec{u} \in V(\Omega)$, it is possible to define $\vec{n} \cdot \vec{u}|_\Gamma$ and $\vec{n} \wedge \vec{u}|_\Gamma$ which belong to $H^{-1/2}(\Gamma)$ and it follows too that $\vec{u}$ belongs to $H^{-1/2}(\Gamma)$.

We proved ([2] , [5] , [7]) that if $\vec{n}.\vec{u}|_\Gamma = 0$ or $\vec{n} \wedge \vec{u}|_\Gamma = 0$ then $\vec{u} \in H^1(\Omega)$ and consequently, for such $\vec{u}$, we have

$$\|\vec{u}\|^2_{H^1(\Omega)} \leq C\left[\|\operatorname{div} \vec{u}\|^2_{L^2(\Omega)} + \|\operatorname{curl} \vec{u}\|^2_{L^2(\Omega)} + \|\vec{u}\|^2_{L^2(\Omega)}\right].$$

It follows so that if $\vec{u} \in V(\Omega)$ is such that $\vec{n}.\vec{u}|_\Gamma \in H^{1/2}(\Gamma)$ or $\vec{n} \wedge \vec{u}|_\Gamma \in H^{1/2}(\Gamma)$ then $\vec{u} \in H^1(\Omega)$ and for such $\vec{u}$, we have

$$\|\vec{u}\|^2_{H^1(\Omega)} \leq C\left[\|\operatorname{div} \vec{u}\|^2_{L^2(\Omega)} + \|\operatorname{curl} \vec{u}\|^2_{L^2(\Omega)} + \|\vec{u}\|^2_{L^2(\Omega)} + \left\{\begin{matrix} \|\vec{n}.\vec{u}\|^2_{H^{1/2}(\Gamma)} \\ \text{or} \\ \|\vec{n} \wedge \vec{u}\|^2_{H^{1/2}(\Gamma)} \end{matrix}\right\}\right].$$

That problem reminds us the problem of a priori inequalities for system of partial differential equations.

3. In a previous work [4], we studied the conditions to have the inequality

$$\|\vec{u}\|^2_{H^m(\Omega)} \leq C\left[\|\mathcal{A}(D)\vec{u}\|^2_{L^2(\Omega)} + \|\vec{u}\|^2_{L^2(\Omega)}\right]$$

for $\vec{u} \in H^m(\Omega)$, if $\mathcal{A}$ is an operator of order m.

The conditions, which are necessary and sufficient, are

1) $\mathcal{A}(D)$ is elliptic

2) For each point $x_0 \in \Gamma$,

there is no function $\vec{u} \in H^m(R^1_+)$ other than 0 such that

$$\mathring{\mathcal{A}}\left(i \sum_{j=1}^{n-1} \xi_j \vec{\tau}_j + \vec{\eta}\, D_t\right)\vec{u}(t) = 0$$

where $\vec{\tau}_1, \ldots, \vec{\tau}_{n-1}$ are tangent vectors and $\vec{\eta}$ a normal vector to Γ at x_0, for each $\xi \in R^{n-1}$ such that $|\xi| = 1$.

Following the same idea, we tried to complete the question by adding to the second member, terms containing boundary norms.

If $B_1(D), \ldots, B_p(D)$ are boundary operators of order $m_1, \ldots, m_p$ such that $m_k \leq m-1$, then we have

$$\|\vec{u}\|^2_{H^m(\Omega)} \leq C\left[\|\mathcal{A}(D)\vec{u}\|^2_{L^2(\Omega)} + \|\vec{u}\|^2_{L^2(\Omega)} + \sum_{k=1}^{p}\|B_k(D)\vec{u}\|^2_{H^{m-m_k-1/2}(\Gamma)}\right]$$

for each $\vec{u} \in H^m(\Omega)$, if and only if

1) $\mathcal{A}(D)$ is elliptic

2) there is no function $\vec{u} \in H^m(R^1_+)$ other than 0, such that

$$\left.\begin{aligned} &\overset{\circ}{\mathcal{A}}\left(i \sum_{j=1}^{n-1} \xi_j \vec{\tau}_j + \vec{\eta}\, D_t\right) \vec{u}(t) = 0 \\ &\overset{\circ}{B}_k\left(i \sum_{j=1}^{n-1} \xi_j \vec{\tau}_j + \vec{\eta}\, D_t\right) \vec{u}(0) = 0 \end{aligned}\right\}$$

with the same notations as before.

The proof of that property, which is too long to be given here, is done by using " cartes locales " and in the case of R_+^n the conditions are the ellipticity of $\mathcal{A}$ and the following one : there is no function $\vec{u} \in H^m(\Omega)$ other than 0, such that

$$\left.\begin{aligned} &\overset{\circ}{\mathcal{A}}(i\,\xi_1, \ldots, i\,\xi_{n-1}, D_t)\, \vec{u}(t) = 0, \\ &\overset{\circ}{B}_k(i\,\xi_1, \ldots, i\,\xi_{n-1}, D_t)\, \vec{u}(0) = 0. \end{aligned}\right\}$$

I shall give some precisions for the passage from a neighbourhood of $x_0 \in \Gamma$ to a neighbourhood of 0 in R_+^n.

4. The first step consists to pass from a neighbourhood of x_0 in Ω to a neighbourhood of x_0 in the half space which is tangent to Γ at x_0. After that, we pass from that half-space to R_+^n; by a transformation we may write

$$x = x_0 + U x'$$

where $U = (\vec{\tau}_1, \ldots, \vec{\tau}_{n-1}, \vec{\eta})$ is an orthogonal matrix such that $\vec{\tau}_i$ is tangent and $\vec{\eta}$ normal to Γ at the point x_0.

By that transformation, the operators $\mathcal{A}(D)$ and $B_k(D)$ where D is in fact the operator gradient, become respectively $\mathcal{A}(U D')$ and $B_k(U D')$, from where follows the form given to the conditions in the general case.

But it often happens that $|\mathcal{A}(U D)\vec{u}| = |\mathcal{A}(D)\vec{u}|$ or that $|\mathcal{A}(U D)\vec{u}| = |\mathcal{A}(D)\, U^* \vec{u}|$.

We then may conclude the following results : if the inequality holds in R^n_+ for the operators $\mathcal{A}(D)$ and $B_k(D)$, then it is valid in a neighbourhood of x_o, in the first case for the operators $\mathcal{A}(D)$ and $B_k(U^*D)$ to which one arrives by considering the transformation $x' = U^*(x-x_o)$ and in the second case for the operators $\mathcal{A}(D)$ and $B_k(U^*D)U^*$ to which one arrives by writing the inequality in R^n_+ for $U^* \vec{u}$ and by applying, after that, the transformation $x' = U^*(x-x_o)$.

5. Let us consider some examples.

For the operator grad $= \begin{pmatrix} D_{x_1} \\ \vdots \\ D_{x_n} \end{pmatrix} = D$, we have

$$|U \ D\vec{u} \ | = |D\vec{u}|$$

but the operator div $= D^*$ becomes

$$(UD)^* = D^* U^* = \text{div } U^*.$$

From there it follows that the operator $\Delta = \text{div grad}$ becomes **div U* U** grad $= \Delta$ and is thus invariant.

In a similar manner, it is easy to verify that if we replace D by U D in the expression

$$\sum_{i,j} |D_{x_i} u_j \pm D_{x_j} u_i|^2 ,$$

it takes the same form where $\vec{u}$ must be replaced by $U^* \vec{u}$.

So, if we consider the operator $\mathcal{A}(D)$ given before and for which

$$|\mathcal{A}(D)\vec{u}|^2 = |\sum_{i=1}^{n} D_{x_i} u_i|^2 + \sum_{i,j} |D_{x_i} u_j - D_{x_j} u_i|^2 ,$$

in the open set R^n_+ , we see that the system

$$\overset{\circ}{\mathcal{A}}(i\,\xi_1, \ldots, i\,\xi_{n-1}, D_t)\,\vec{u}(t) = 0 \;, \quad |\xi| = 1 \;,$$

has solutions in $H^1(R^1_+)$ of the form

$$\vec{u}(t) = (-i\,\xi_1, \ldots, -i\,\xi_{n-1}, 1)\, C\, e^{-t} .$$

It follows from here that the operator B(D) of order 0, given by $B(D) = \vec{a}\,.$ ($\vec{a}$: constant vector) is such that

$$\overset{\circ}{B}(i\,\xi, D_t)\,\vec{u}(0) = 0$$

implies $C = 0$ if $a_n \neq 0$.

We thus have

$$\|\vec{u}\|^2_{H^1(R^n_+)} \leq C\Big[\|\operatorname{div}\vec{u}\|^2_{L^2(R^n_+)} + \sum_{i,j}\|D_{x_i}u_j - D_{x_j}u_i\|^2_{L^2(R^n_+)} + \\ + \|\vec{a}\,.\,\vec{u}\|^2_{H^{1/2}(R^{n-1})}\Big].$$

As we have an operator of the second type, this inequality is valid in Ω if we replace B(D) by $B(U^*D)U^*$ that is, if we replace $\vec{a}.\vec{u}$ by $\vec{a}.U^*\vec{u} = U\,\vec{a}.\vec{u} = \vec{b}.\vec{u}$, if

$$a_n = \vec{a}.\,\vec{e}_n = U\,\vec{a}\,.\,U\,\vec{e}_n = \vec{b}\,.\,\vec{\eta} \neq 0.$$

So, we have

$$\|\vec{u}\|^2_{H^1(\Omega)} \leq C\Big[\|\operatorname{div}\vec{u}\|^2_{L^2(\Omega)} + \sum_{i,j}\|D_{x_i}u_j - D_{x_j}u_i\|^2_{L^2(\Omega)} + \|\vec{b}\,.\,\vec{u}\|^2_{H^{1/2}(\Gamma)}\Big]$$

if $\vec{b}$ is a transverse vector-field, that is $\vec{b}.\vec{\eta} \neq 0$.

Instead of considering one vector which is transverse to Γ , we may consider a set of (n-1) tangent vectors linearly independent; in particular, it is allowed to consider the operator $\vec{a}\wedge$ if $\vec{a} \neq 0$.

Let us consider now the scalar operator Δ. In the half-space R^n_+, the differential equation $\mathring{A}(i\xi, D_t)\vec{u}(t) = 0$ writes $(D_t^2 - 1)\, u(t) = 0$ and $\vec{u}(t) = C\, e^{-t}$ is a solution belonging to $H^2(R^1_+)$.

The operator

$$B(D) = \sum_{j=1}^{n} a_j D_{x_j} + b \quad \text{where} \quad a_n \neq 0, \quad \text{is such}$$

that

$$\mathring{B}(i\xi, D_t)\, u(0) = \left[i \sum_{j=1}^{n-1} a_k \xi_k - a_n \right] C = 0$$

implies $C = 0$.

We thus have

$$\|u\|^2_{H^2(R^n_+)} \leq C \left[\|\Delta u\|^2_{L^2(R^n_+)} + \|u\|^2_{L^2(R^n_+)} + \left\| \frac{du}{d\vec{a}} + bu \right\|^2_{H^{1/2}(R^{n-1})} \right]$$

for each $u \in H^2(R^n_+)$, if $a_n \neq 0$.

As Δ is an operator of this first type, we have the same inequality in Ω if we replace $\vec{a}\,.\,\text{grad}\ u + b.u$ by $\vec{a}.U^*\ \text{grad}\ u + b.u = \vec{a}'.\ \text{grad}\ u + bu$ if $\vec{a}'.\vec{\eta} \neq 0$.

An other interesting example is the matrix operator ΔI_n completed by grad div, which is elliptic and for which it is possible to see that

$$\|\vec{u}\|^2_{H^2(\Omega)} \leq C \left[\|\Delta \vec{u}\|^2_{L^2(\Omega)} + \|\text{grad div}\ \vec{u}\|^2_{L^2(\Omega)} + \|\vec{u}\|^2_{L^2(\Omega)} + \left\{ \begin{matrix} \|A\vec{u}\|^2_{H^{3/2}(\Gamma)} \\ \\ \|\vec{a} \wedge \vec{u}\|^2_{H^{3/2}(\Gamma)} \end{matrix} \right\} \right]$$

if A is a non-singular-matrix or $\vec{a}$ a tranverse vector-field.

Bibliography

1 Agmon, S., The coerciveness Problem for integro-differential forms, J. D'An. Math., Vol. 6, 1958, pp.183-223.

2 Duvaut, G. and Lions, J., Les inéquations en mécanique et en physique, Dunod, 1972.

3 De Figueredo, D.G., The coerciveness problem for forms over vector valued functions, Comm. Pure and Appl., Math., 16 (1963), 63-94.

4 Gobert, J., Opérateurs matriciels de dérivation elliptique et problèmes aux limites, Mém. de la Soc. Roy. des Sc. de Liège, VI, 2, 1961.

5 Gobert, J., Sur une inégalité de coercivité, Journal of Math. Analysis and Appl., Vol. 36, n° 3, 1971.

6 Gobert, J., Inégalités à priori pour les systèmes d'équations aux dérivées partielles, Bull. Soc. Roy. des Sc. de Liège, 5-6, (1972), pp.261-267.

7 Goulaouic, C. and Hanouzet, B., Un résultat de régularité pour les solutions d'un système d'équations différentielles. (à paraître).

8 Lions, J.L. and Duvaut, G., cf [2] .

RECENT RESULTS ON SEGAL ALGEBRAS

Richard R. Goldberg

I wish in this paper to present some results on Segal algebras which have been obtained recently, principally those of two of my students from the University of Iowa - James T. Burnham and H.C. Wang. Although I include some historical material, this is not intended to be a comprehensive survey of the field. Those seeking a more complete introduction to Segal algebras should consult the book [8] and lecture notes [9] of Reiter.

Wiener's Space W.

In his famous paper, Tauberian Theorems [14], Wiener introduced continuous functions f on $R = (-\infty, \infty)$ such that

$$\sum_{n=-\infty}^{\infty} \max_{I_n} |f(x)| < \infty . \tag{1}$$

Here, $I_n = [n, n+1]$. He also made use of functions α on R which are locally of bounded variation and satisfy

$$\operatorname*{var}_{I_n} \alpha = \int_{I_n} |d\alpha(t)| \leq M \quad (n=0, \pm 1, \pm 2, \ldots) \tag{2}$$

for some $M > 0$.

Although the Tauberian theorem involving L^1 has become more well known, the Tauberian theorem involving functions which satisfy (1) has led to some interesting developments as we shall see. It goes as follows.

THEOREM. *Suppose* α *satisfies* (2). *Let* f_1 *be a continuous function satisfying* (1) *whose Fourier transform never vanishes. If*

$$\lim_{x \to \infty} \int_{-\infty}^{\infty} f_1(x-t)\, d\alpha(t) = 0$$

then

$$\lim_{x \to \infty} \int_{-\infty}^{\infty} f(x-t)\, d\alpha(t) = 0$$

for any continuous f satisfying (1).

Indeed, it was this Tauberian theorem that has the Prime Number Theorem as a corollary. Wiener was interested in the functions satisfying (1) or (2) themselves, rather than in the spaces of these functions. However, interest in the space of functions (1) developed. Accordingly, we make the

Definition. Let W denote the linear space of all continuous f on R satisfying (1). We norm W by

$$\| f \|_W = \sum_n \max_{I_n} |f(x)| \qquad (f \in W).$$

It is not difficult to verify that, with this norm, W is a Banach space. Moreover,

$$\| f * g \|_W \leq 2 \| f \|_W \| g \|_W \qquad (f, g \in W)$$

so that W (with norm $2 \| \|_W$) is a Banach algebra. Details may be found in [6].

Many results about W have been established. Edwards [5] showed that its maximal ideal space is identical to that of L^1. Goldberg [6] showed that every bounded linear functional on W is given by

$$f \longrightarrow \int_{-\infty}^{\infty} f d\alpha \qquad (f \in W)$$

for some α satisfying (2). Using this result, Unni and Murthy [12] have shown that the multipliers for W are all given by convolutions with these same functions α.

Definition of Segal Algebra and Examples.

The space W is a typical example of a Segal algebra.

Definition. Let G denote a locally compact abelian group. A Segal algebra S is a proper dense subalgebra of $L^1(G)$ such that S is itself a Banach algebra with respect to a norm $\| \|_S$. It is also assumed that if $f \in S$ then every translate f_a of f is also in S and, finally, that

S1. $$\| f_a \|_S \leq A \| f \|_S \qquad (f \in S;\ a \in G)$$

where A is independent of f, and that

S2. The map $a \longrightarrow f_a$ is continuous from G to $(S, \| \ \|_S)$.

Two important consequences of this definition are:

S3. $$\| f \|_{L^1} \leq M \| f \|_S \qquad (f \in S).$$

S4. The algebra S is a dense ideal in L^1 and

$$\| f * g \|_S \leq N \| f \|_S \| g \|_{L^1} \text{ for all } f \in S,\ g \in L^1.$$

The inequalities in S3 and S4 are the chief computational tools in dealing with Segal algebras.

The name Segal algebra was coined by Reiter [7] with reference to Segal's paper [11] in which the properties S1 and S2 were isolated as crucial to the structure of many algebras.

We would now like to outline some recent developments in this area. First we shall list several examples. Indeed, it is the richness of examples that is one of the most interesting features of the subject. We denote the reals by R, the circle group by T, an arbitrary l.c.a.g. by G and its character group by Γ.

E1. The algebras $C^{(k)}(T)$ consisting of all f on T with continuous derivatives. Use the norm

$$\| f \| = \sum_{j=0}^{k} \| f^{(j)} \|_\infty .$$

Multiplication for $C^{(k)}$ is ordinary pointwise multiplication of functions.

In examples E2 through E7, multiplication is convolution.

E2. The algebras $L^p(T)$ for $1 < p < \infty$ with the usual norm. (The space $L^\infty(T)$ is not a Segal algebra since S2 of the definition does not hold.)

E3. The algebras $L^1 \cap L^p(G)$ for $1 < p < \infty$, G non-compact with norm $\| f \| = \| f \|_{L^1} + \| f \|_{L^p}$.

E4. The algebras $A^p(G)$ for $1 \leq p < \infty$ consisting of all $f \in L^1(G)$ such that $\hat{f} \in L^p(\Gamma)$, with $\| f \| = \| f \|_{L^1} + \| \hat{f} \|_{L^p}$.

E5. The algebra $L^1 \cap C_0(G)$ for G non-compact, with $\| f \| = \| f \|_{L^1} + \| f \|_{C_0}$.

E6. The algebra W.

E7. The algebra $L_A(R)$ consisting of all $f \in L^1(R)$ such that f is absolutely continuous and $f' \in L^1(R)$. Use the norm $\| f \| = \| f \|_{L^1} + \| f' \|_{L^1}$.

Here are some recent results.

Ideal Theory.

In [8] Reiter shows that the ideal theory for any Segal algebra $S \subset L^1(G)$ is precisely the same as that of $L^1(G)$. More precisely:

THEOREM: 1. *If I is a closed ideal in L^1 then $I \cap S$ is a closed ideal in S.*

2. *If J is any closed ideal in S then the closure J^{L_1} of J in L^1 is a closed ideal in L^1 and*

$$J = J^{L_1} \cap S.$$

(The inclusion $J^{L_1} \cap S \subset J$ is the significant content of the theorem.) In particular, the maximal ideal space for S is the same as that of L^1. We have already mentioned that Edwards [5] proved this for $S = W$.

Burnham [1] has handled the ideal theory for Segal algebras on non-abelian groups in an interesting manner. First note that the definition of Segal algebra makes sense for non-abelian G if translation is taken as left (or right) translation. The property S3 must be taken as an axiom in the non-commutative case. If ideal means right (or left) ideal, the property S4 will hold. Burnham uses the following definition to generalize Segal algebras.

Definition. Let $(A, \| \|_A)$ be a (not necessarily commutative) Banach algebra. The subalgebra B of A is called an A-Segal algebra if

A1. B is a dense left ideal of A.

A2. B is a Banach algebra with respect to a norm $\| \|_B$.

A3. There exists $M > 0$ such that

$$\|f\|_A \leq M \|f\|_B \qquad (f \in B).$$

A4. There exists $N > 0$ such that

$$\|fg\|_B \leq N \|f\|_B \|g\|_A \qquad (f \in B, g \in A).$$

Burnham proves the precise analog of Reiter's theorem.

THEOREM. _Let_ B _be an_ A-_Segal algebra_. _If_ B _has a right approximate identity then_

1. _If_ I _is a closed left ideal in_ $(A, \| \|_A)$ _then_ $I \cap B$ _is a closed ideal in_ B.

2. _If_ J _is a closed left ideal in_ $(B, \| \|_B)$ _then the closure_ J^A _of_ J _in_ $(A, \| \|_A)$ _is a closed left ideal in_ $(A, \| \|_A)$ _and_

$$J = J^A \cap B.$$

(The inclusion $J^A \cap B \subset J$ is the significant part of the theorem.)

Now any (ordinary) Segal algebra S on a non-abelian group will have a right approximate identity provided only that the properties S1 and S2 hold for both right and left translations (see [9, p. 43]).

Hence Burnham's theorem (with $A = L^1$, $B = S$) settles the question of closed ideals in such algebras. Examples of such algebras include $L^1 \cap L^p$ and $L^1 \cap C_0$ on a non-abelian, non-compact group.

Multipliers.

By a multiplier of a Segal algebra S we mean a bounded linear operator $T : S \longrightarrow S$ that commutes with translations. Unlike the ideal theory, the multiplier theory for Segal algebras is not uniform. The following examples illustrate this.

1. Every multiplier for $L^1(T)$ is given by

$$Tf = f * \mu \qquad (f \in L^1(T))$$

where μ is a measure on T.

2. Every multiplier for $L^2(T)$ is given by

$$Tf = f * \varphi \qquad (f \in L^2(T))$$

where φ is a pseudo-measure on T.

3. If $p \neq 1,2$ the multipliers for $L^p(T)$ cannot be neatly characterized.

4. On R, however, the multipliers for L^1 as well as for $L^1 \cap L^p$ $(1 < p < \infty)$ are all the same — namely, convolution with measures.

We wish to call attention here to a recent result of Unni and Murthy [12]

Every multiplier for W *is given by*

$$Tf = f * \alpha = \int f(x-t)\,d\alpha(t)$$

where α *satisfies* (2).

Non-Factorization.

Salem (see [16; p.378]) showed that every $f \in L^1(T)$ could be factored $f = g * h$ for some $g,h \in L^1(T)$. (We will refer to this, somewhat loosely, as '$L^1(T)$ factors.') In [10] Rudin proved a similar result for $L^1(R^n)$. Finally, Cohen [4] showed that any Banach algebra factors provided that it contains a left approximate identity bounded in norm. In particular, if G is any locally compact group then $L^1(G)$ factors.

Precisely the opposite seems to be true for Segal algebras. Several special results along this line have accumulated. For example, it is easy to see that $L^2(T)$ does not factor. For if every $f \in L^2$ could be expressed $f = g * h$ for some $g, h \in L^2$ then $\hat{f} = \hat{g}.\hat{h} \in \ell^1$ by the Schwarz inequality. This would imply $\ell^2 = \widehat{L^2} \subset \ell^1$ which is false.

Yap [15] proved that $L^1 \cap L^p(G)$ (example E3) does not factor. His proof is complicated and makes use of Hardy's rearrangement of functions and properties of $L(p,q)$ spaces.

In [13] H.C.Wang gives what is by far the most complete treatment of non-factorization. He shows that all of the Segal algebras E1 - E7, and many others as well, fail to factor. His approach is beautifully simple and can be easily described:

Suppose $A \subset L^1(G)$ is an algebra such that

1. A factors,
2. $\hat{A} \subset L^{p_0}(\Gamma)$ for some p_0, $1 < p_0 < \infty$.

If $f \in A$ then, by 1, $f = g * h$ so that $\hat{f} = \hat{g}.\hat{h}$. But, by 2, $\hat{g}, \hat{h} \in L^{p_0}(\Gamma)$. Hence, by the Schwarz inequality, $\hat{f} \in L^{p_0/2}(\Gamma)$. Proceeding in the same fashion we find that $\hat{f} \in L^{p_0/2^n}(\Gamma)$ for $n = 0, 1, 2, \ldots$. This implies $\hat{f} \in L^p(\Gamma)$ for $0 < p \leq p_0$. On the other hand, $\hat{f} \in L^\infty(\Gamma)$ since $f \in L^1(G)$. So $\hat{f} \in L^p(\Gamma)$ for all $p > 0$. Thus, if A factors and $\hat{A} \subset L^{p_0}(\Gamma)$ for some p_0 then $\hat{A}$ must be contained in $L^p(\Gamma)$ for <u>all</u> $p > 0$. Therefore, to show that an algebra $A \subset L^1(G)$ does not factor it suffices to show that $\hat{A}$ is contained in some $L^{p_0}(\Gamma)$ but not in all $L^p(\Gamma)$.

Now it is easy to show that if A is any of the algebras E1-E7 then $\hat{A}$ is contained in some L^{p_o}. For A^p this is true by definition. Since $W \subset L^1 \cap L^\infty$ we have $W \subset L^2$ so that $\hat{W} \subset L^2$. If $p \leq 2$ then $\widehat{L^1 \cap L^p(G)} \subset L^{p'}(\Gamma)$ by Hausdorff-Young. If $p > 2$, then $L^1 \cap L^p \subset L^2$ so that $\widehat{L^1 \cap L^p} \subset L^2$. The other algebras are also easily handled. Thus, to prove that any A among E1-E7 fails to factor, it is enough to show that $\hat{A}$ contains a function which fails to be in some $L^p(\Gamma)$. For E2 - E6 this follows from the following general theorem of Wang:

THEOREM. *Let* $S^{(G)}$ *be a Segal algebra with the property that*

$$\|\chi f\|_S = \|f\|_S \quad \text{for all} \quad \chi \in \Gamma. \tag{3}$$

Then there exists $f \in S$ *such that* $\hat{f} \in L^1(\Gamma)$ *but* $\hat{f} \notin L^{1/3}(\Gamma)$.

If $G = R$, the property (3) says simply that

$$\|e^{iat}f(t)\|_S = \|f\|_S \qquad (a \in R).$$

This clearly holds for $L^1 \cap L^p$, A^p, and W, but does not hold for L_A which (together with $C^{(k)}$) Wang handles separately. The theorem of course takes care of a wide class of examples other than E2 - E6.

Wang's theory will not handle a Segal algebra A with $\hat{A}$ not contained in any L^p. An example of such an algebra is the set of all $f \in L^1(T)$ such that

$$\lim_{n\to\infty} \hat{f}(n) \log n = 0,$$

with $\|f\| = \|f\|_{L^1} + \sup_{n \geq 1} |\hat{f}(n)| \log n$, and with convolution as multiplication.

Dependence of Basic Properties.

Cigler [3] has raised the following question: If $(B, \| \|_B)$ is a Banach algebra which is a dense ideal in $L^1(G)$, can

$$\| f \|_{L^1} \leq M \| f \|_B \qquad (f \in B)$$

be deduced from

$$\| f * g \|_B \leq N \| f \|_B \| g \|_{L^1} \qquad (f \in B, g \in L^1)?$$

In a recent paper [2] Burnham and I have supplied the following answer in a more general context.

THEOREM. _Let_ $(B, \| \|_B)$ _be a Banach algebra which is a dense left ideal in the Banach algebra_ $(A, \| \|_A)$. _Suppose that_ A _contains a right approximate identity. If, in addition,_

$$\| fg \|_B \leq N \| f \|_B \| g \|_A \qquad (f \in B, g \in A) \quad (4)$$

then

$$\| f \|_A \leq M \| f \|_B \qquad (f \in B). \quad (5)$$

In the other direction we prove

THEOREM. _Let_ $(B, \| \|_B)$ _be a Banach algebra which is a left ideal in the Banach algebra_ $(A, \| \|_A)$. _Then if_ (5) _holds, so does_ (4).

Bibliography.

1 Burnham, J.T., Closed ideals in subalgebras of Banach algebras I. Proc.Amer.Math.Soc. 32(1972), 551-555.

2 Burnham, J.T. and Goldberg, R.R., Basic properties of Segal algebras. J.Math.Anal.Appl. (to appear).

3 Cigler, J., Normed ideals in $L^1(G)$. Nederl.Akad.Wetensch. Proc. Ser. A, 72 (1969), 273-282.

4 Cohen, P.J., Factorization in group algebras. Duke Math. J. 26 (1959), 199-206.

5 Edwards, R.E., Comments on Wiener's Tauberian theorems. J. London Math. Soc. 33(1958), 462-466.

6 Goldberg, R.R., On a space of functions of Wiener. Duke Math. J. 34 (1967), 683-691.

7 Reiter, H., Subalgebras of $L^1(G)$. Nederl.Akad.Wetensch. Ser. A, 68(1965), 691 - 696.

8 ----------, Classical harmonic analysis and locally compact groups, Oxford, 1968.

9 ----------, L^1-algebras and Segal algebras. Lecture Notes in Mathematics 231, Springer-Verlag, Berlin, 1971.

10 Rudin, W., Representation of functions by convolutions. J.Math. Mech. 7(1958), 103-115.

11 Segal, I.E., The group algebra of a locally compact group. Trans.Amer.Math.Soc. 61 (1947), 69-105.

12 Unni, K.R. and Keshavamurthy, G.N., Multipliers on a space of Wiener, Nanta Math. (to appear)

13 Wang, H.C., Nonfactorization in group algebras. Studia Math. 42(1972), 231-241.

14 Wiener, N., Tauberian theorems. Ann.of Math. 33(1932), 1-100.

15 Yap, L.Y.H., Ideals in subalgebras of the group algebra. Studia Math. 35(1970), 165-175.

16 Zygmund, A., Trigonometrical series, v.I. Cambridge University Press, New York, 1959.

MEASURABILITY OF LATTICE OPERATIONS IN A CONE

Kohur Gowrisankaran

Let X be a locally convex Hausdorff topological vector space over the real numbers. Let C be a closed proper convex cone with vertex 0 and let C generate X. Let further C be a lattice in its own order. There are well known results asserting the continuity of the mappings $(x,y) \mapsto \sup(x,y)$ and $(x,y) \mapsto \inf(x,y)$ of $C \times C \to C$ under suitable restrictions on the cone C [4, Ch. V], [2, Appendix]. In this note we shall give conditions under which the lattice operations are Borel mappings of $C \times C \to C$. This Borel measurability was found to be very useful in our recent work in potential theory [1]. The following is the result.

THEOREM. *Let X be a Hausdorff locally convex real topological vector space. Let C be a closed proper convex cone with vertex at the origin, generating X and such that C is a lattice in its own order. Let B be a compact metrisable base for C and let further the continuous positive linear functionals on X separate the points of C.*

Then, the mappings $C \times C \to C$ *given by* $(x,y) \mapsto \sup(x,y)$ *and* $(x,y) \mapsto \inf(x,y)$ *are Borel, viz., the inverse image of any Borel set of C under each of these mappings is a Borel set of* $C \times C$.

Proof. *Step (1)*. Let us denote by K, the set of all positive continuous linear functionals on X and Y = K-K the vector space generated by this cone. Let τ be the given topology on X. We note that (C, τ) is locally compact, metrisable and separable and hence it is a polish space. Hence, for any Hausdorff topology τ' weaker than τ on X, the (C, τ') Borel sets and (C, τ) Borel sets are identical [5]. Also, the $\tau \times \tau$ -Borel sets and $\tau' \times \tau'$-Borel sets are the same on $C \times C$; this σ-algebra is nothing but the product of the Borel σ-algebra of C with itself. Hence, to prove the theorem, we may work with any weaker topology on X. We shall indeed work with a weaker topology τ' which is the $\sigma(X,Y)$ topology on X. Since the elements of Y (in fact K) separate the points of C (and therefore X) this topology τ' is Hausdorff. Clearly, the induced topology on C is the coarsest one such that functions $x \mapsto L(x)$ are continuous for every L in K. It is in fact fairly easy to see that τ and τ' induce the same topology on C, though we shall not make use of this property.

Step (2). Let L_0 be the positive continuous linear functional on X defining the base B, i.e. $B = \{x \in C : L_0(x) = 1\}$ [3]. Let x and y be any two elements of C. Let $L_0(x+y) = a$. Then the set $\{z \in C : L_0(z) \leq a\}$ is τ-compact. Consider the set $\{z_i\}_{i \in I}$ of all elements z_i in C majorising both x and y and such that $z_i \leq x+y$. Clearly, this set is non-void and is decreasingly directed since C is a lattice. Also sup(x,y) belongs to this set. Hence for every L in K, $\{L(z_i)\}_{i \in I}$ is a decreasingly directed system of positive real numbers and converges to the lower bound $L[\sup(x,y)]$. Consider now the filter of sections $\mathfrak{F}$. Suppose z is any τ-adherent point of this filter. Then, τ' being weaker than τ, we get that $L(z) = L[\sup(x,y)]$ for every L in K. Hence we conclude that $z = \sup(x,y)$ and that $\mathfrak{F}$ converges to sup(x,y) in τ. However, due to the metrisability of (C,τ), we may get a sequence of elements z'_n such that $z'_n \leq x+y$ and z'_n converges in τ to sup(x,y). Now, let $z_1 = z'_1$, $z_2 = \inf(z'_2, z_1)$, ..., $z_n = \inf(z'_n, z_{n-1})$ ..... Then $\{z_n\}$ is a decreasing sequence in C, $z_1 \leq x+y$ and it is straightforward to deduce that $\{z_n\}$ converges to sup(x,y).

Step (3). Let us fix a positive continuous linear functional L on X. We shall show that $(x,y) \mapsto L[\sup(x,y)]$ (resp. $(x,y) \mapsto L[\inf(x,y)]$) is a lower (resp. upper) semi-continuous function on C x C. For this, suppose (x_n, y_n) in C x C converges to (x_o, y_o) in C x C. We may assume that x_n, y_n, for all n and x_o, y_o belong to a fixed compact set A of **C**, A of the form $\{z : L_o(z) \leq b\}$. For every n, let $z_{n,k}$ be elements of C such that (1) $z_{n,k} \leq x_n + y_n$ for each k (in particular, $z_{n,k}$ belongs to **A+A**) and (2) $z_{n,k}$ decreases with k and converges to $\sup(x_n, y_n)$ as k tends to infinity.

Given $\epsilon > 0$, choose for each n an integer k_n such that

$$L[\sup(x_n, y_n)] > L(z_{n,k}) - \epsilon/2 \qquad \forall k \geq k_n .$$

Let us denote $s_n = z_{n,k_n}$, for each n. Then, for each n,

$$L[\sup(x_n, y_n)] > L(s_n) - \epsilon/2 . \tag{1}$$

Now, consider the sequence of elements $\{s_n\}$ in C . This sequence is contained in A+A which is τ -compact. Hence, given any subsequence, $\{n \in M' \subseteq N\}$, we may choose a further subsequence, say $\{n \in M \subset M' \subseteq N\}$, such that as n in M tends to infinity, $\{s_n\}$ converges to an element z in C . Hence, for all sufficiently large n in M, $L(s_n) > L(z) - \epsilon/2$. It follows from (1) that

$$L[\sup(x_n, y_n)] > L(z) - \epsilon . \tag{2}$$

We now claim that z is a majorant of $\sup(x_o, y_o)$. It is enough to show that $z \geq x_o$ and $z \geq y_o$. Since s_n is a majorant of both x_n and y_n there are elements t_n and u_n in C such that $s_n = x_n + t_n$ and $s_n = y_n + u_n$ for every n. But $\{x_n\}$, $\{y_n\}$ and $\{s_n\}$ converge in τ to x_o, y_o and z respectively as n in M tends to $+\infty$, and hence $\{t_n\}$ and $\{u_n\}$ converge respectively to t_o and u_o as n in M tends to infinity. Since C is closed, t_o and u_o are both elements of C, and we have $z = x_o + t_o$ and $z = y_o + u_o$. It follows that $z \geq \sup(x_o, y_o)$. Hence $L(z) \geq L[\sup(x_o, y_o)]$ and we get from (2) that

$$L[\sup(x_n, y_n)] > L[\sup(x_o, y_o)] - \epsilon$$

for all sufficiently large n belonging to M. Hence, we have shown that for arbitrary $\epsilon > 0$, and any infinite subset $M' \subseteq N$, there is a further subset $M \subseteq M' \subseteq N$ such that

$$\liminf_{n \in M,\ n \to \infty} L[\sup(x_n, y_n)] \geq L[\sup(x_o, y_o)] - \epsilon.$$

It follows that,

$$\liminf_{n \to \infty} L[\sup(x_n, y_n)] \geq L[\sup(x_o, y_o)] .$$

This proves the lower semi-continuity of the function $(x,y) \longmapsto L[\sup(x,y)]$. The upper semi-continuity of the function $(x,y) \longmapsto L[\inf(x,y)]$ is deduced from (1) $(x,y) \mapsto x+y$ is continuous and (2) $x+y = \inf(x,y) + \sup(x,y)$.

Step(4). Define for any L in K and non-negative real numbers a and b

$$W(L,a,b) = \{x \in C;\ a < L(x) < b\}.$$

The sets of the form W(L,a,b) form a subbasis for the open sets of (C, τ'). However, (C, τ') (being Lusin) is a strongly Lindelöf space, i.e., every open subspace is Lindelöf. Hence, each τ'open set in C is a countable union of finite intersections of sets of the form W(L,a,b). Hence the sets of the form W(L,a,b) generate the Borel σ-algebra of (C, τ'). But

$$\{(x,y) \in C \times C : \sup(x,y) \in W(L,a,b)\} = \{(x,y) \in C \times C : a < L[\sup(x,y)] < b\}$$

and

$$\{(x,y) \in C \times C : \inf(x,y) \in W(L,a,b)\} = \{(x,y) \in C \times C : a < L[\inf(x,y)] < b\}$$

are Borel subsets of C x C. The theorem is proved.

COROLLARY. For every Radon measure on C x C, the above lattice operation mappings are Lusin measurable.

The corollary follows immediately by observing that (C, τ) is a polish space [5].

Bibliography

1 Gowrisankaran, K., Integral representation for a class of multiply superharmonic functions, Annales Inst. Fourier (to appear).

2 Kelley, J.L., Namioka, I., et al., Linear topological spaces, Van Nostrand, Princeton 1963.

3 Peressini, A.L., Ordered topological vector spaces, Harper and Row, New York 1967.

4 Schaefer, H.H., Topological vector spaces, The MacMillan Co. New York 1964.

5 Schwartz, L., Radon measures on General Topological Spaces, Tata Institute of Fundamental Research Monographs (to appear).

ON SOME NONLINEAR ELLIPTIC BOUNDARY VALUE PROBLEMS

Peter Hess

Introduction.

The intensive development of the theory of monotone operators in reflexive Banach spaces started about a decade ago when it was realized that these mappings form a very powerful tool in discussing variational boundary value problems for nonlinear equations of the form

$$(\mathcal{A}u)(x) \equiv \sum_{|\alpha|\leq m}(-1)^{|\alpha|} D^{\alpha} A_{\alpha}(x,u(x),\ldots,D^{m}u(x))=f(x) \quad (x \in \Omega \subset R^{N}),$$

provided the functions A_{α} satisfy a polynomial growth condition in u and its partial derivatives of order $\leq m$. In connection with the decomposition of $\mathcal{A}$ into its top order part ($|\alpha| = m$) and its lower order terms, the concept of monotonicity was weakened, and various classes of nonlinear operators which we summarize as 'mappings of monotone type' were introduced.

If the A_{α}'s do no longer satisfy a polynomial growth condition, one works with operators of monotone type in Orlicz-Sobolev spaces which may not be reflexive. The study of questions arising in that context is in full progress at the moment; for a review of the present state of the theory cf. Gossez [5].

In this paper we propose to investigate nonlinear elliptic problems which lie somewhat between the two extrema mentioned. We first discuss the solvability of boundary value problems for equations of the form

$$(\mathcal{A}u)(x) \equiv (\mathcal{A}_0 u)(x) + p(x,u(x)) = f(x) \qquad (x \in \Omega),$$

where $\mathcal{A}_o$ is a second order linear elliptic differential operator, while $p : \Omega \times R \longrightarrow R$ is a function on which no growth restriction is imposed. The study of those equations was initiated by Browder [3];

his results were subsequently sharpened by the writer in [7] by the introduction of 'operators of monotone type with respect to two Banach spaces'. Without proof we mention the basic facts of [7] in Sections 1 and 2, but add also various new results. Note that the restriction to equations of second order is mostly pure convenience.

By reduction to the same abstract class of mappings we prove in Section 3 the solvability of some linear elliptic equations subject to nonlinear boundary conditions.

Section 1. An abstract existence theorem.

Let W,V be real reflexive separable Banach spaces with norms $\|\cdot\|_W$ and $\|\cdot\|_V$, and assume $W \subset V$, with a continuous injection mapping of W into V. Let W^*, V^* be the conjugate spaces of W,V, respectively. Then $V^* \subset W^*$ in the sense that if $f/_W$ denotes the restriction of the functional $f \in V^*$ to W, then $f/_W \in W^*$. By (w,u) we denote the duality pairing, either between $w \in V^*$, $u \in V$, or between $w \in W^*$, $u \in W$. The symbols '$\longrightarrow$' and '$\rightharpoonup$' denote strong and weak convergence, respectively.

Definition 1.1. Let A be a mapping with domain $D(A) : W \subset D(A) \subset V$, and range contained in W^*. We say that A is

of type (M) with respect to V,W, provided

(i) A is continuous from finite-dimensional subspaces of W to the weak topology on W^*;

(ii) if $\{v_n\}$ is a sequence in W and $u \in V$, $g \in V^*$ elements such that $v_n \rightharpoonup u$ in V, $Av_n \rightharpoonup g/_W$ in W^*, and $\limsup(Av_n, v_n) \leq (g,u)$, then $u \in D(A)$ and $Au = g/_W$.

quasi-bounded if for any sequence $\{v_n\}$ in W which is bounded in the V-norm, and for which $(Av_n, v_n) \leq$ const.$\times \|v_n\|_V$, the boundedness in W^* of the sequence $\{Av_n\}$ follows.

THEOREM 1.2. *Let the mapping* A *be quasi-bounded and of type (M) with respect to W, V, and suppose*

$$(Aw, w) \, \|w\|_V^{-1} \longrightarrow +\infty \quad (\|w\|_V \to \infty, \; w \in W).$$

Then the equation $Au = f/_W$ *is solvable for each* $f \in V^*$.

Mappings of type (M) with respect to two Banach spaces were introduced by the writer in [7], where also Theorem 1.2 is proved.

Section 2. Strongly nonlinear elliptic boundary value problems.

Let $\Omega \subset R^N (N \geq 1)$ be an open bounded subset with smooth boundary Γ. We apply Theorem 1.2 in the discussion of variational boundary value problems for an equation of the form

$$(Au)(x) \equiv -\sum_{i,j=1}^{N} \frac{\partial}{\partial x_i}\left(a_{ij}(x) \frac{\partial u}{\partial x_j}(x)\right) + \sum_{i=1}^{N} b_i(x) \frac{\partial u}{\partial x_i}(x) + c(x)u(x) + p(x, u(x)) = f(x), \quad (x \in \Omega). \quad (I)$$

The following is assumed:

a_{ij}, b_i, c *are* $L^\infty(\Omega)$ - *functions*; (2.1)

the function $p(x,\eta): \Omega \times R \longrightarrow R$ *is measurable in* x, *continuous in* η, *essentially bounded for* η *bounded, and satisfies the sign condition* $p(x,\eta)\eta \geq 0$ *for a.a.* $x \in \Omega$ *and* $\eta \in R$. (2.2)

Note that no growth restriction is imposed on p.

Let V be a given closed subspace : $H_0^1(\Omega) \subset V \subset H^1(\Omega)$, which determines the boundary conditions, and set $\|\cdot\|_V = \|\cdot\|_1$. By assumption (2.1), the bilinear form

$$a(u,v) = \sum_{i,j=1}^{N} \int_\Omega a_{ij} \frac{\partial u}{\partial x_j} \frac{\partial v}{\partial x_i} dx + \sum_{i=1}^{N} \int_\Omega b_i \frac{\partial u}{\partial x_i} v \, dx + \int_\Omega c u v \, dx \quad (2.3)$$

is defined for all u, v in V and is bounded.

Let $W = H^m(\Omega) \cap V$, $m > 1 + [\frac{N}{2}]$ (so that $W \subset C(\bar{\Omega})$ by Sobolev's imbedding theorem), and provide W with the norm $\|\cdot\|_W = \|\cdot\|_m$. Let further V_1 be the set defined by

$$V_1 = \{u \in V : p(.,u) \in L^1(\Omega), p(.,u)u \in L^1(\Omega)\} .$$

Clearly $W \subset V_1 \subset V$. Let the (semilinear) form b be given by

$$b(u,w) = \int_\Omega p(.,u)w\,dx \qquad (u \in V_1, w \in W).$$

Definition 2.1. For given $f \in W^*$, a function u is said to be a weak solution (variational solution) of equation (I) with respect to the boundary conditions imposed by V, provided

(i) $u \in V_1$,

(ii) $a(u,w) + b(u,w) = (f,w) \quad \forall\, w \in W$.

THEOREM 2.2. *Let the previously introduced notations and hypotheses hold, and assume*

$$\|w\|_V^{-1} \{a(w,w) + b(w,w)\} \longrightarrow +\infty \quad (\|w\|_V \to \infty, w \in W). \tag{2.4}$$

Then, for each $f \in V^*$ there exists a weak solution of the considered boundary value problem.

Let $A_0 : V \longrightarrow V^*$ be the bounded linear operator induced by

$$(A_0 u, v) = a(u,v) \qquad (u,v \in V),$$

and let $\tilde{A}_0 : V \longrightarrow W^*$ be given by $\tilde{A}_0 u = A_0 u/_W$. For fixed $u \in V_1$ there exists $u^* \in W^*$ such that

$$(u^*, w) = b(u,w) \qquad \forall\, w \in W.$$

The correspondence $u \rightsquigarrow u^*$ determines a (nonlinear) mapping $A_1 : V_1 \longrightarrow W^*$ by $A_1 u = u^*$.

LEMMA 2.3. *The mapping $A = \tilde{A}_0 + A_1 : V_1 \to W^*$ is quasi-bounded and of type (M) with respect to the spaces W, V.*

A proof of Lemma 2.3 may be found in Hess [7] . The assertion of Theorem 2.2 now follows immediately from Theorem 1.2 and Lemma 2.3.

The coerciveness assumption (2.4) is of course satisfied if a is coercive on V: $a(v,v) \geq \alpha \|v\|_1^2 \quad \forall v \in V \ (\alpha > 0)$. A different sufficient condition is given in

PROPOSITION 2.4. <u>Suppose the form (2.3) is uniformly elliptic:</u>

$$\sum_{i,j=1}^{N} a_{ij}(x)\,\xi_i\,\xi_j \geq \alpha |\xi|^2 \quad \underline{\text{a.e. on}}\ \Omega,$$

$$\forall \xi \in R^N,\ (\alpha > 0), \tag{2.5}$$

<u>and suppose there exists a continuous function</u> $\psi : \Omega \times R \to R^+$, <u>with</u> $\lim_{|\eta| \to \infty} \psi(x,\eta) = +\infty$ <u>for</u> $x \in \Omega$, <u>such that</u> $|p(x,\eta)| \geq \psi(x,\eta)\,|\eta|$ <u>for a.a.</u> $x \in \Omega$, $\forall \eta$. <u>Then (2.4) holds.</u>

<u>Proof</u>. Assume to the contrary that there exists a constant $e > 0$ and a sequence $\{w_n\}$ in W with $\|w_n\|_V \to \infty$ $(n \to \infty)$, such that

$$a(w_n,w_n) + b(w_n,w_n) \leq e\,\|w_n\|_V \tag{2.6}$$

for all n. Let $z_n = \|w_n\|_V^{-1}\, w_n$, and let $A_0' : V \to V^*$ be the bounded linear operator induced by the principal part of a:

$$(A_0' u, v) = \sum_{i,j=1}^{N} \int_\Omega a_{ij} \frac{\partial u}{\partial x_j} \frac{\partial v}{\partial x_i}\, dx \qquad \forall u, v \in V.$$

Further let $A_0'' = A_0 - A_0'$. From (2.6) we deduce

$$(A_0' z_n, z_n) + \int_\Omega \psi(\cdot,w_n)\, z_n^2\, dx \leq e\|w_n\|_V^{-1} - (A_0'' z_n, z_n). \tag{2.7}$$

By reflexivity of V and compactness of the imbedding $V \subset L^2(\Omega)$ we may assume (after passage to subsequences) that

$$\begin{cases} z_n \rightharpoonup z & \text{in } V, \\ z_n \to z & \text{in } L^2(\Omega), \\ z_n \to z & \text{a.e. in } \Omega. \end{cases}$$

Let $\Omega_0 = \{x \in \Omega : z(x) \neq 0,\ z_n(x) \to z(x)\}$. It follows from (2.7) that

$$\limsup \int_{\Omega_0} \psi(\cdot, w_n) z_n^2 \, dx \leq -(A_0'' z, z) < \infty .$$

But

$$\psi(x, w_n(x)) (z_n(x))^2 = \psi(x, \|w_n\|_V \, z_n(x)) (z_n(x))^2 \longrightarrow +\infty$$

for $x \in \Omega_0$. Thus meas $(\Omega_0) = 0$ by Fatou's lemma, and consequently $z(x) = 0$ a.e. on Ω. Since $z_n \longrightarrow z = 0$ in $L^2(\Omega)$ and $\|z_n\|_V \equiv 1$, we infer that $\int_\Omega |\text{grad } z_n|^2 \, dx \longrightarrow 1$ as $n \longrightarrow \infty$.

On the other hand, (2.7) also implies that

$$\alpha \int_\Omega |\text{grad } z_n|^2 \, dx \leq e \|w_n\|_V^{-1} - (A_0'' z_n, z_n) \longrightarrow 0 \ (n \longrightarrow \infty)$$

and hence $\int_\Omega |\text{grad } z_n|^2 \, dx \longrightarrow 0$. We arrive at a contradiction, q.e.d.

We now assume that the function $p(x,\eta)$ is odd in η and prove the solvability of the Dirichlet problem for equation (I) under a condition 'weaker than coerciveness'.

THEOREM 2.5. *Let* $V = H_0^1(\Omega)$, *and let the bilinear form (2.3) be uniformly elliptic. Let further the function* $p(x,\eta)$ *be odd in* η. *Suppose that for given* $f \in V^*$ *the following a priori estimate holds:*

If v *denotes any solution of the Dirichlet problem*

$$(DP)_t \quad \begin{cases} v \in V_1, \\ a(v,w) + b(v,w) = (1-t)(f,w) \quad \forall \, w \in W \end{cases}$$

for some $t \in [0,1]$, *then* $\|v\|_V < M < \infty$.

Then the problem $(DP)_0$ *admits a solution.*

Remark 2.6. Let $V = H_0^1(\Omega)$, and suppose the coercivity assumption (2.4) holds in the slightly stronger form

$$\|w\|_V^{-1} \left\{ a(w,w) + \int_\Omega p(\cdot, w) w \, dx \right\} \longrightarrow +\infty \qquad (2.4')$$
$$(\|w\|_V \longrightarrow \infty, \ w \in V_1) .$$

Then for each $f \in V^*$ the solutions of $(DP)_t$ for $0 \leq t \leq 1$ are uniformly bounded in the V-norm.

Indeed, let v be any solution of $(DP)_t$ for some $t \in [0,1]$:

$$a(v,w) + \int_\Omega p(.,v)w\,dx = (1-t)\,(f,w) \quad \forall\, w \in W. \qquad (2.8)$$

For (fixed) $K > 0$ let $v^{(K)} \in V$ be the function derived from v by truncation:

$$v^{(K)}(x) = \begin{cases} v(x) & \text{if } |v(x)| \leq K \\ K & \text{if } v(x) \geq K \\ -K & \text{if } v(x) \leq -K, \end{cases}$$

and let $\{\varphi_n\} \subset C_0^\infty(\Omega)$ be a sequence of functions with $|\varphi_n(x)| \leq K$ $\forall$ n and $x \in \Omega$, such that $\varphi_n \longrightarrow v^{(K)}$ in V as $n \longrightarrow \infty$ (for the proof of existence of such a sequence cf. [9]). Setting $w = \varphi_n$ in (2.8) and passing to the limit $n \longrightarrow \infty$ we infer that

$$a(v,v^{(K)}) + \int_\Omega p(.,v)v^{(K)}\,dx = (1-t)\,(f,v^{(K)}). \qquad (2.9)$$

Now let $K \longrightarrow +\infty$. Then $v^{(K)} \longrightarrow v$ in V, and it follows from (2.9) that

$$a(v,v) + \int_\Omega p(.,v)v\,dx = (1-t)(f,v) \leq \|f\|_{V*}\,\|v\|_V .$$

Thus $\|v\|_V < M$ by (2.4'), for some M.

<u>Proof of Theorem 2.5</u>. For each n let $p_n(x,\eta)$ denote the truncated function p:

$$p_n(x,\eta) = \begin{cases} p(x,\eta) & \text{if } |p(x,\eta)| \leq n \\ n & \text{if } p(x,\eta) \geq n \\ -n & \text{if } p(x,\eta) \leq -n, \end{cases}$$

and consider the associated problem

$$(DP)_{t,n} \quad \begin{cases} v \in V, \\ a(v,w) + \int_\Omega p_n(.,v)w\,dx = (1-t)(f,w) \quad \forall\, w \in V. \end{cases}$$

(i) We claim that there exists n_0 with the following property:

If $v \in V$ is any solution of $(DP)_{t,n}$ for some $t \in [0,1]$ and $n > n_0$, then $\| v \|_V \neq M$.

Suppose to the contrary that to each n we find elements $v_n \in V$ with $\| v_n \|_V = M$ and $t_n \in [0,1]$ such that

$$a(v_n,w) + \int_\Omega p_n(\cdot,v_n) w \, dx = (1 - t_n)\,(f,w) \quad \forall\, w \in V. \qquad (2.10)$$

We may assume that $v_n \rightharpoonup v$ in V and $t_n \longrightarrow t$ as $n \longrightarrow \infty$. Setting $w = v_n$ in (2.10) we conclude that

$$\int_\Omega p_n(\cdot,v_n) v_n \, dx \leq \text{const.} \quad \forall\, n.$$

It follows from a result by Strauss [11] that $v \in V_1$ and $p_n(\cdot,v_n) \longrightarrow p(\cdot,v)$ in $L^1(\Omega)$ for some subsequence. By passing to the limit $n \longrightarrow \infty$ in (2.10) for $w \in W$ we obtain

$$a(v,w) + \int_\Omega p(\cdot,v) w \, dx = (1 - t)(f,w) \qquad (w \in W).$$

In order to derive a contradiction to the a priori assumption in Theorem 2.5 we now show that $\| v \|_V = M$. For fixed $K > 0$ we set $w = v_n - v^{(K)}$ in (2.10) ($v^{(K)}$ is the truncated function v) and get

$$a(v_n,v_n - v^{(K)}) = - \int_\Omega p_n(\cdot,v_n)(v_n - v^{(K)}) \, dx + \\ + (1 - t_n)\,(f,v_n - v^{(K)}).$$

Hence

$$\limsup a(v_n,v_n - v) = \limsup a(v_n,v_n - v^{(K)}) + \\ + a(v,v^{(K)} - v) \leq - \int_\Omega p(\cdot,v)\,(v - v^{(K)}) + \\ + (1-t)\,(f,v - v^{(K)}) + a(v,v^{(K)} - v). \qquad (2.11)$$

Since the left side of (2.11) is independent of $K > 0$, we may pass to the limit $K \longrightarrow +\infty$ on the right side of (2.11) and obtain

$$\limsup a(v_n - v,\, v_n - v) = \limsup a(v_n,v_n - v) \leq 0.$$

The Gårding inequality

$$a(v_n - v,v_n - v) \geq c_1 \| v_n - v \|_V^2 - c_2 \| v_n - v \|_{L^2}^2, \; (c_1 > 0)$$

implies that $v_n \longrightarrow v$ in V. Thus $\|v\|_V = M$.

(ii) For fixed $n > n_0$ we consider the problems $(DP)_{t,n}$ $(0 \leq t \leq 1)$. We write them in the abstract form

$$A_0 v + B_n v = (1 - t)\, f, \tag{2.12}$$

where $B_n : V \longrightarrow V^*$ is the (nonlinear) compact operator defined by $(B_n u, v) = \int_\Omega p_n(\cdot, u) v \, dx \quad \forall\, u, v \in V$. For each $t \in [0,1]$ the mapping $v \leadsto (A_0 + B_n) v - (1-t) f$ is of so-called monotone type $(S)^+$ (cf. [2]). It is further odd for $t = 1$. Since $\|v\|_V \neq M$ for all possible solutions v of (2.12) with $0 \leq t \leq 1$, the existence of a solution $v_n \in V$ with $\|v_n\|_V < M$ of the equation

$$A_0 v_n + B_n v_n = f$$

follows by a well-known homotopy result on mappings of monotone type (cf. Hess [6], Theorem 1). We have proved that $(DP)_{0,n}$ admits a solution for each $n > n_0$.

(iii) The existence of a solution of $(DP)_0$ now follows by a limiting argument $n \longrightarrow \infty$ similar to that employed in (i), q.e.d.

Remark 2.7. The assertion of Theorem 2.5 also holds for von Neumann boundary condition: $V = H^1(\Omega)$. Theorem 2.5 generalizes a related result by Browder [3] who imposes a rather restrictive additional condition on the function p.

Remark 2.8. With a slightly more sophisticated method one can prove Theorem 2.2 also for equations with nonlinear highest order part (cf. Hess [9]). The same is true for Theorem 2.5.

Remark 2.9. If p is monotone, more generally if p is replaced by a maximal monotone graph β in R x R, some of the previous results can be sharpened. See [1,10].

Section 3. Nonlinear Boundary Conditions

Let again Ω denote an open bounded subset of R^N with smooth boundary Γ . In this section we investigate the solvability of the following boundary value problem :

$$\left.\begin{aligned}
(\mathcal{A} u)(x) &\equiv - \sum_{i,j=1}^{N} \frac{\partial}{\partial x_i}\left(a_{ij}(x) \frac{\partial u}{\partial x_j}(x)\right) + \sum_{i=1}^{N} b_i(x) \frac{\partial u}{\partial x_i}(x) + \\
&\qquad + c(x)\, u(x) = f(x) \qquad x \in \Omega \\
\left(- \frac{\partial u}{\partial n_a}\right)(x) &\equiv - \sum_{i,j=1}^{N} a_{ij}(x) \frac{\partial u}{\partial x_j}(x) \cos(n(x), x_i) = \\
&= p(x, u(x)) \qquad x \in \Gamma
\end{aligned}\right\} \text{(II)}$$

Here $n(x)$ denotes the outward normal to Γ at $x \in \Gamma$, (2.1) is assumed, and $p = p(x,\eta)$: $\Gamma \times R \longrightarrow R$ is a function satisfying (2.2) (with Ω replaced by Γ). To $\mathcal{A}$ we associate the bilinear form(2.3), which is bounded and defined on $V \times V$, where $V = H^1(\Omega)$. Let $u/_\Gamma$ denote the restriction of $u \in V$ to the boundary. Further let $W = H^m(\Omega)$, $m > 1 + [\frac{N}{2}]$, and define the set V_1 by

$$V_1 = \{u \in V : p(\cdot, u/_\Gamma) \in L^1(\Gamma),\ p(\cdot, u/_\Gamma)\, u/_\Gamma \in L^1(\Gamma)\} .$$

Clearly $W \subset V_1 \subset V$. Finally let the (semilinear) form b be given by

$$b(u,w) = \int_\Gamma p(.,u/_\Gamma)\, w/_\Gamma \, d\Gamma \qquad (u \in V_1, w \in W).$$

<u>Definition 3.1</u>. Given $f \in W^*$, a function u is weak solution of problem (II) if

(i) $u \in V_1$,

(ii) $a(u,w) + b(u,w) = (f,w) \quad \forall w \in W$.

THEOREM 3.2. <u>Under the assumptions stated above, problem (II) admits a weak solution for each</u> $f \in V^*$ <u>provided</u>

$$\|w\|_V^{-1} \{a(w,w) + b(w,w)\} \longrightarrow +\infty \qquad (\|w\|_V \longrightarrow \infty,\ w \in W). \tag{3.1}$$

The proof of Theorem 3.2 runs parallel to that of Theorem 2.2. Condition (3.1) is trivially satisfied if a is coercive on V. Here another sufficient condition:

PROPOSITION 3.3. Let the form a:

$$a(u,v) = \sum_{i,j=1}^{N} \int_{\Omega} a_{ij} \frac{\partial u}{\partial x_j} \frac{\partial v}{\partial x_i} dx \qquad (u,v \in V) \qquad (3.2)$$

be uniformly elliptic, and suppose there exists a subset $\Gamma_0 \subset \Gamma$ with positive measure and a continuous function $\psi(x,\eta)$: $\Gamma_0 \times R \longrightarrow R^+$, with $\lim_{|\eta| \to \infty} \psi(x,\eta) = d(x) > 0$ for $x \in \Gamma_0$, such that $|p(x,\eta)| \geq \psi(x,\eta)|\eta|$ for a.a $x \in \Gamma_0$ and $\eta \in R$. Then (3.1) holds.

Proof. Suppose $\{w_n\}$ is a sequence in W with $\|w_n\|_V \longrightarrow \infty$ and $e > 0$ a constant such that

$$a(w_n, w_n) + b(w_n, w_n) \leq e\, \|w_n\|_V$$

for all n, and let $z_n = \|w_n\|_V^{-1} w_n$. Then

$$a(z_n, z_n) + \int_{\Gamma_0} \psi(\cdot, w_n/_{\Gamma})(z_n/_{\Gamma})^2 d\Gamma \leq e \|w_n\|_V^{-1}. \qquad (3.3)$$

By passage to subsequences we may assume that

$$\begin{cases} z_n \rightharpoonup z & \text{in } V, \\ z_n \longrightarrow z & \text{in } L^2(\Omega), \\ z_n/_{\Gamma} \longrightarrow z/_{\Gamma} & \text{in } L^2(\Gamma), \\ z_n(x) \longrightarrow z(x) & \text{for a.a. } x \in \Gamma. \end{cases}$$

(3.3) implies that

$$\alpha \int_{\Omega} |\operatorname{grad} z_n|^2 dx \leq e \|w_n\|_V^{-1} \longrightarrow 0 \qquad (n \to \infty).$$

Thus

$$1 \geq \|z\|_V \geq \|z\|_{L^2(\Omega)} = \lim \|z_n\|_{L^2(\Omega)} = 1.$$

Therefore $z(x) = \text{const.} \neq 0$ $(x \in \bar{\Omega})$.

Also by (3.3),

$$\int_{\Gamma_0} \psi(\cdot, w_n/_\Gamma)(z_n/_\Gamma)^2 d\Gamma \leq e\|w_n\|_V^{-1} \to 0 \quad (n \to \infty) \tag{3.4}$$

But $\psi(x,w_n(x))\ z_n(x)^2 \longrightarrow$ const. $d(x) > 0$ for a.a. $x \in \Gamma_0$, which, together with (3.4), gives a contradiction to the Fatou lemma, q.e.d.

If we replace the function p by a maximal monotone graph $\beta \subset R \times R$ (i.e. a mapping of R to 2^R such that (i) $\forall\ t_1 \in \beta(s_1)$, $t_2 \in \beta(s_2)$, the inequality $(t_1 - t_2)(s_1 - s_2) \geq 0$ holds, and (ii) there exists no monotone graph in $R \times R$ extending β properly), we can sharpen the preceding results.

THEOREM 3.4. Let a denote the uniformly elliptic bilinear form (3.2) defined on $V \times V$ ($V = H^1(\Omega)$), and suppose in addition that $a_{ij} \in C^1(\bar{\Omega})$ $(i,j = 1, \ldots, N)$. Let further β be a maximal monotone graph in $R \times R$ with $0 \in \beta(0)$, and let β° be defined by

$$|\beta^\circ(s)| = \begin{cases} \mathrm{Min}_{t \in \beta(s)} |t| & \text{if } \beta(s) \neq \phi \\ +\infty & \text{if } \beta(s) = \phi \end{cases}$$

Suppose $|\beta^0(s)| \longrightarrow \infty$ $(|s| \longrightarrow \infty)$. Then to each $f \in L^2(\Omega)$ there exist $u \in V$ and $b \in L^2(\Gamma)$ with $b(x) \in \beta(u(x))$ for a.a. $x \in \Gamma$, such that

$$a(u,w) + \int_\Gamma b.w/_\Gamma \, d\Gamma = \int_\Omega f\, w\, dx \quad \forall\ w \in V.$$

Employing a result of Brézis [1] we may conclude that even $u \in H^2(\Omega)$.

Bibliography.

1 Brézis, H., Problèmes unilatéraux, J.Math.Pures Appl.

2 Browder, F.E., Existence theorems for nonlinear partial differential equations. Proc.Sympos.Pure Math.,Vol.16,Amer.Math.Soc., Providence, R.I.,1970.

3 ---- Existence theory for boundary value problems for quasilinear elliptic systems with strongly nonlinear lower order terms. Proc.Sympos.Pure Math. (to appear).

4 Browder,F.E. and Hess,P., Nonlinear mappings of monotone type in Banach spaces. J.Functional Analysis 11 (1972),251-294.

5 Gossez, J.P., Nonlinear elliptic boundary value problems for equations with rapidly increasing coefficients. Trans.Amer. Math. Soc. (to appear).

6 Hess, P., On nonlinear mappings of monotone type homotopic to odd operators. J.Functional Analysis 11 (1972), 138-167.

7 --------, On nonlinear mappings of monotone type with respect to two Banach spaces. J.Math.Pures Appl. (to appear).

8 --------,Variational inequalities for strongly nonlinear elliptic operators. J.Math.Pures Appl. (to appear).

9 ---------,A strongly nonlinear elliptic boundary value problem. J.Math.Anal.Appl. (to appear).

10 ---------,On a unilateral problem associated with elliptic operations. Proc. Amer. Math.Soc. (to appear).

11 Strauss, W.A., On weak solutions of semi-linear hyperbolic equations. An.Acad.Brasil. Cienc. 42 (1970), 645-651.

QUASICOMPLEMENTED BANACH ALGEBRAS

T. Husain

I. Introduction

Let A be a complex topological algebra. Let $\mathcal{L}_r$ denote the set of all closed right ideals in A. A is said to be a right quasi-complemented algebra if there exists a mapping $q: R \to R^q$ of $\mathcal{L}_r$ into itself such that:

(1) $R \cap R^q = (0) \quad \forall R \in \mathcal{L}_r$

(2) $(R^q)^q = R$

(3) $R_1 \subset R_2 \Rightarrow R_2^q \subset R_1^q, \quad (R_1, R_2 \in \mathcal{L}_r)$

The mapping q is called the right quasicomplementor and R^q, the right quasicomplement of R. If in addition, a right quasi-complementor satisfies the following property:

(4) $R + R^q = A \quad \forall R \in \mathcal{L}_r$

then q is called the right complementor and A a right complemented algebra. Similarly one can define left quasicomplemented algebras by replacing 'right' by 'left' in the above definitions. Since the results for the 'left' case are analogous to that of the 'right' case, we study only the right quasicomplemented algebras.

The class of right quasicomplemented algebras introduced herein generalizes the class of right complemented Banach algebras studied earlier by several authors, e.g., Alexander, [1,2], Alexander and Tomiuk [3], Tomiuk [11] and Wong [12].

The notion of right quasicomplemented algebras was first introduced in [9]. Some of the results cited below will appear in [8] and [9]. Standard definitions have been taken from [10].

First we observe that each right complemented algebra is a right quasicomplemented algebra but not conversely. An example for the converse can be constructed as in [1] and [3]. However, under certain conditions the converse holds. For example, each finite-dimensional right quasicomplemented normed algebra is a right complemented algebra. Further, if A is a simple B^*-algebra, then each right quasicomplementor on A is a right complementor [9].

For the study of right quasicomplemented algebras, the following types of problems arise:

(1) What is the structure of such algebras?

(2) What are its representations?

(3) When is a quasicomplementor a complementor?

(4) Does there exist a right quasicomplementor (resp. complementor) on a topological algebra?

(5) Generalize right quasicomplemented algebras by weakening the axioms:

We give the following results in answer to the above questions for right quasicomplemented algebras.

II. Structure Theorem.

Before we give the structure theorem, we observe that in a right quasicomplemented algebra for each $R \in \mathcal{L}_r$, $R + R^q$ is dense in A and thus the socle of a semisimple quasicomplemented algebra in which every maximal modular right ideal is closed, is dense in A [9]. Further in a semisimple dual algebra in which every maximal modular right ideal is closed, the set of all minimal idempotents, $\{e_\alpha\}$, is nonempty and each nonzero closed right ideal R of A can be written as:

$$R = \overline{\sum_\alpha e_\alpha A}$$

(see [5]). Using these results we have:

THEOREM 1. *Let A be a semisimple quasicomplemented algebra in which every maximal modular right ideal is closed and $x \in \overline{xA}$ for each $x \in A$. Then $A = \Sigma_\alpha \oplus A_\alpha$ (topological direct sum), where each A_α is a minimal closed two-sided ideal. Further, each A_α is a simple quasicomplemented algebra.*

Proof. See [9].

III. Representation Theorem.

Before we give the main theorem, we first observe that in a semisimple Banach algebra A, each minimal left ideal I can be written as I = Ae, where e is a minimal idempotent of A. (See [13]). Hence for each closed right ideal R in A, $RI = R \cap I$. Now let A be a primitive quasicomplemented Banach algebra and I a minimal ideal of A. Then the left regular representation: $a \to T_a$ of A is a faithful, continuous and strictly dense representation of A on I, where I can be given a Hilbert space structure with inner product (x,y) and the norm induced by the inner product is equivalent with that induced from A.

THEOREM 2. *Let A be a primitive right quasicomplemented Banach algebra in which every maximal closed right ideal is modular and $x \in \overline{xA}$ for all $x \in A$. Then A is continuously isomorphic with an algebra A_c of completely continuous operators on a Hilbert space.*

Proof. See [9].

By using Theorem 2, in answer to problem (3), we have:

THEOREM 3. *Let A be a primitive right quasicomplemented Banach algebra in which every maximal closed right ideal is modular and $x \in \overline{xA}$ for all $x \in A$. Let A_c be the representation set of A as given in Theorem 2. Assume A_c is a two-sided ideal of B(I), the set of all bounded linear operators on I as indicated above. Then each right* quasicomplementor q on A is a right complementor.

Proof. See [9].

COROLLARY 1. Let A be a primitive quasicomplemented commutative Banach algebra with identity. Then each quasicomplementor is a complementor.

Proof. In each commutative Banach algebra with identity, each maximal ideal is closed and modular. Also for each $x \in A$, $x \in xA$ because the identity $e \in A$. Hence Theorem 3 applies.

IV. Existence of Quasicomplementors:

Let A be a dual semisimple Banach algebra with norm $\|\cdot\|$ which is a dense subalgebra of a semisimple Banach algebra B with norm $|\cdot|$ satisfying the properties:

(i) There exists a constant $M > 0 \ni |x| \leq M \|x\|$.

(ii) Every proper closed left (right) ideal in A is the intersection of maximal modular left (right) ideals in B. Let p be a quasicomplementor on B. With these conditions, we have:

THEOREM 3. The mapping $q : R \to (\bar{R})^p \cap A$ on the closed right ideals R of A is a quasicomplementor.

COROLLARY 2. Let A be a dual A^*-algebra. Then for each right ideal R in A, the mapping $q : R \to [\ell(R)]^*$ defines a quasicomplementor in A, where $\ell(R) = \{y \in A: yx = 0 \ \forall x \in R\}$.

COROLLARY 3. Let A be a commutative dual A^*-algebra. Then there is only one quasicomplementor q on A.

V. Generalizations:

There are several ways of generalizing the class of right quasicomplemented algebras. It is clear from the above discussion that semisimplicity of an algebra in the above results plays an important role. Thus we define the following:

Definition. Let A be a topological algebra, $\mathcal{R}$ the set of all nonzero closed right ideals and $\mathcal{M}$ the set of all closed modular maximal right ideals of A. A mapping $p: \mathcal{M} \to \mathcal{R}$ is said to be a generalized right modular complementor if

(i) for each $M \in \mathcal{M}$, $M \cap M^p = \{0\}$

(ii) $\cap \{M : M \in \mathcal{M}\} = \{0\}$.

(See [8]) A with p is called a generalized right modular complemented algebra.

We prove a similar structure theorem for such algebras and other results (See [8]).

References

1. Alexander, F.E., Representation theorems for complemented algebras, Trans. Amer. Math. Soc. 148 (1970), 385-397.
2. Alexander, F.E., On complemented and annihilator algebras, Glasgow, J. Math. 10 (1969) 38-45.
3. Alexander, F.E. and Tomiuk B.J., Complemented B*-algebras, Trans. Amer. Math. Soc. 137 (1969), 459-480.
4. Bachelis, G.F., Homomorphisms of annihilator Banach algebras, Pacific J. Math. 25 (1968), 229-247.
5. Barnes, B.A., Modular annihilator algebras, Canad. J. Math.18 (1966), 566-578.
6. Dixmier, J., Les C*-algèbres et leurs reprèsentations, Cahiers Scientifiques fasc. 29, Gauthier-Villars, Paris 1964.
7. Husain, T. and Wong, P.K. On modular complemented and annihilator algebras, Proc. Amer. Math. Soc. 34 (1972), 457-462.
8. Husain, T. and Wong, P.K., On generalized right modular complemented algebras, Studia Math. (to appear).
9. Husain, T. and Wong, P.K., Quasicomplemented Algebras, Trans. Amer. Math. Soc. (to appear).
10. Richart, C.E., General theory of Banach algebras, University series in higher Math., Van Nostrand, Princeton, N.J. 1960.
11. Tomiuk, B.J., Structure theory of complemented Banach algebras, Canad. J. Math. 14 (1962), 651-659.
12. Wong, P.K., Continuous complementors on B*-algebras, Pacific J. Math. 33 (1970), 255-260.
13. Yood, B., Ideals in topological rings, Canad. J. Math. 16 (1964), 28-45.

ON THE (L^p, L^p) MULTIPLIERS

Satoru Igari

1. Introduction.

Let G be a locally compact abelian group. The translation operator τ_y, $y \in G$, is defined by

$$\tau_y \, f(x) = f(x-y)$$

for a function f on G. For $p,q \geq 1$ let $L_p^q(G)$ be the set of all bounded linear operators of $L^p(G)$ to $L^q(G)$ which commute with translation operators. The norm in $L_p^q(G)$ is defined by the operator norm.

Operators in $L_p^q(G)$ appear often in Fourier analysis and it is interesting to study the characterizations of elements in $L_p^q(G)$, sufficient conditions for T to belong to $L_p^q(G)$, structure of the space, and so on.

Our aim is to study some results on the third problem when $p=q$. For T in $L_p^p(G)$ there corresponds uniquely a bounded function ϕ on $\hat{G}$, the dual to G such that

$$(Tf)^{\wedge}(\gamma) = \phi(\gamma) \, \hat{f}(\gamma), \quad f \in L^2 \cap L^p(G),$$

so that we denote T by T_ϕ. Put $M_p(\hat{G}) = \{ \phi : T_\phi \in L_p^p(G) \}$ and define the norm of ϕ by the norm of T_ϕ. Since $T_\phi T_\psi = T_{\phi\psi}$, $M_p(\hat{G})$ is a commutative Banach algebra with pointwise multiplication. We have $M_p(\hat{G}) = M_{p'}(\hat{G})$, $1/p + 1/p' = 1$, and $B(\hat{G}) = M_1(\hat{G})$ $M_p(\hat{G})$ $M_2(\hat{G}) = L^\infty(\hat{G})$ where $B(\hat{G})$ is the set of Fourier-Stieltjes transforms of bounded regular measures on G.

2. Wiener-Pitt phenomenon of L^p-multipliers.

The following theorem gives an answer proposed by L. Hörmander [3].

THEOREM 1. Let G be a non-compact, locally compact abelian group and $p \neq 2$. Suppose Φ is a function on the interval $[-1,1]$. Then $\Phi(\phi) \in M_p(\hat{G})$ for every $\phi \in M_1(\hat{G})$ whose range is contained in $[-1,1]$ if and only if Φ can be extended to an entire function.

When $p = 1$, this theorem reduces to a theorem of H.Helson, J.-P.Kahane, Y.Katznelson and W.Rudin [2] and for general p, a proof is given in S.Igari [4] .

COROLLARY 1. Assume the conditions of Theorem 1. For every complex number z there exists a real valued ϕ in $M_1(\hat{G})$ and a homomorphism h such that $h(\phi) = z$.

COROLLARY 2. Assume the conditions of Theorem 1. Then the Banach algebra $M_p(\hat{G})$ is not regular and not symmetric.

COROLLARY 3. Assume the conditions of Theorem 1. Then there exists ϕ in $M_1(\hat{G})$ such that $\phi \geq 1$ but $1/\phi \notin M_p(\hat{G})$.

The important subalgebras in which the last two corollaries hold positively are given in [1] and [3] .

3. In the approximation theory it is important to know whether $[1 - \phi(\xi)] / |\xi|^{\alpha}$ is a multiplier or not under the condition that it is continuous and ϕ is a multiplier. The case of Fourier - Stieltjes transform is proved negatively by P.Malliavin (see [5] for the proof and its application).

As an application of Theorem 1 we have the following :

Let $0 < \alpha$ and $p \neq 2$. Then there exists $\phi \in M_1(\mathbb{R})$ such that $[1 - \phi(\xi)] / |\xi|^{\alpha}$ is continuous but does not belong to $M_p(\mathbb{R})$.

Proof. Let X and Y be the Banach spaces of continuous functions on $\mathbb{R}$ with norms $\|\phi\|_X = \|\phi\|_{\infty} + \| |\xi|^{\alpha} \phi \|_{M_1}$ and $\|\phi\|_Y = \|\phi\|_{M_p} + \| |\xi|^{\alpha} \phi \|_{M_1}$. We shall prove that Y is a proper subspace of X. Otherwise by the open mapping theorem there is a constant $c > 0$ such that

$$\| \phi \|_{M_p} \leq c \, \| \phi \|_X .$$

Let ψ be an element in M_1 such that $\psi \geq 1$ and $\frac{1}{\psi} \notin M_p$. Let u be a C^∞ - function with compact support such that $u = 1$ for $|\xi| < 1$, and $|u| \leq 1$. Put $u_N(\xi) = u(\xi/N)$ and choose N so that $\| u_N^2 / \psi \|_{M_p} > c$. Since the multiplier norm of $\phi_a(\xi) = u_N^2(a\xi) / \psi(a\xi)$ is independent on a,

$$c < \| \phi_a \|_{M_p} \quad \text{and} \quad \| \phi_a \|_\infty \leq 1 .$$

On the other hand

$$\| |\xi|^\alpha \phi_a \|_{M_1} \leq \| |\xi|^\alpha u_N(a\xi) \|_{M_1} \, \| u_N(a\xi) / \psi(a\xi) \|_{M_1}$$

$$\leq \| |\xi|^\alpha u_N(a\xi) \|_{M_1} \, \| u_N / \psi \|_{M_1} .$$

Put

$$f_a(x) = \int_{-\infty}^{\infty} |\xi|^\alpha u_N(a\xi) \, e^{i\xi x} \, d\xi .$$

Then $|f_a(x)| \leq \text{const } a^{-\alpha - 1}$. Since

$$f_a(x) = i x^{-1} \int_{-\infty}^{\infty} \frac{\partial}{\partial \xi} |\xi|^\alpha u_N(a\xi) \, e^{i\xi x} \, d\xi ,$$

by Hölder inequality and Hausdorff-Young inequality

$$\int_{|x|>1} |f_a(x)| \, dx \leq \text{const.} \left(a^{(1-\alpha)q - 1} \int_{-\infty}^{\infty} \left| \frac{\partial}{\partial \xi} |\xi|^\alpha u_N(a\xi) \right|^q d\xi \right)^{\frac{1}{q}}$$

$$\leq \text{const. } a^{1 - \alpha - 1/q} ,$$

where q is a number such that $2 > q > 1$ and $\alpha-1 + 1/q > 0$. Thus $\|f_a\|_{L^1} = \| |\xi|^{\alpha} u_N(a\xi)\|_{M_1} \to 0$ as $a \to \infty$, which is impossible.

Now let $\phi \in X \setminus Y$. Then $\phi(\xi) = [1 - (1 - |\xi|^{\alpha}\phi(\xi))]/|\xi|^{\alpha}$ is the desired function.

<u>References</u>.

1 Calderón, A.P. and Zygmund, A., Algebras of certain singular operators. Amer. J. Math., 78(1956), 310-320

2 Helson, H., Kahane, J.-P., Katznelson, Y. and Rudin, W., The functions which operate on Fourier transforms. Acta Math.,102 (1959), 135-157.

3 Hörmander,L., Estimates for translation invariant operators in L^p spaces. Acta Math., 104 (1960), 93-140.

4 Igari, S., Functions of L^p-multipliers. Tôhoku Math.J., 21 (1969), 304-320.

5 Sunouchi,G., Direct theorems in the theory of approximation. Acta Math.Acad.Sci. Hungaricae, 20(1969), 409-420.

HEREDITY IN METRIC PROJECTIONS

J.-P.Kahane

I. The theory of metric projection appeared in 1853, when Čebyšev introduced the best approximation of a given continuous function on an interval by polynomials of order $\leq$ n. This was made for a practical purpose, that is, improving the construction of the mechanisms used in steam-engines. Between 1853 and 1892, Čebyšev developed the theory, including best approximation by rational functions, in a highly refined way [13] .

It may be pointed out that only in 1885 Weierstrass gave his celebrated theorem, that every continuous function on an interval can be approximated uniformly by polynomials. In this sense, the theory of best approximation happens to be older than the usual approximation theory.

The general frame for metric projection is as follows. A metric space E is given, together with a subset F of E. For each $x \in E$ we consider the subset of F which consists of points y such that $d(x,y) = d(x,F)$, we denote it by $\mathcal{P}x$, and we study the mapping $\mathcal{P}$ from x to (F) (set of all parts of F). Some authors name $\mathcal{P}$ metric projection. In this paper we are mainly interested in the case where $\mathcal{P}x$ consists in one point for each x; we denote this point by Px, and we consider the mapping P from E to F. That is what we call metric projection now. To avoid trivialities, we suppose $F \neq \emptyset$ and $F \neq E$.

The classical Čebyšev case is $E = C(I)$, the Banach space of all continuous functions on E, and F consists of all polynomials of degree $\leq$ n.

The easiest and most important case is when E is a Hilbert space and F a closed subspace of E. Then P is the ordinary projection of F.

When E is a Banach space and F is a closed subspace of E, the metric projection may or may not exist. It exists, for example when E is a uniformly convex Banach space, and then it is a continuous mapping (in general, of course, non-linear) from E to F. For instance, $L^p(X)$ is uniformly convex when $1 < p < \infty$. But $L^1(X)$ is not uniformly convex (except for a trivial X) and C(I) is not uniformly convex either. It appears that the Čebyšev theory was made in the difficult case. Information on the general topic can be found in [10] .

The L^1-theory was not so much developed as the C-theory, though it already appears (under another name of course) in the works of Čebyšev ([13] , II, pp.189-215). In the sequel of this paper we give some recent results in case $E = L^1(T)$ ($T = \mathbb{R}/\mathbb{Z}$) and F is a subspace of E, closed and translation-invariant. In particular we define the subspaces F for which the metric projection exists. A similar question was studied by Y. Domar in case $E = L^1(\mathbb{R})$; then the only closed and translation-invariant subspaces for which P exists consist in functions f whose Fourier transform $\hat{f}$ vanish on a given half-line [2] . The L^1-theory exhibits already interesting features if we consider a finite set X with the same mass at every point; we shall see in a moment how that is related to the case $E = L^1(T)$.

Apart from the general theory (does P exist, is it continuous, is it uniformly continuous on bounded sets, and so on), an interesting question was raised by H.S.Shapiro. Suppose that P exists for a given space E of functions and for a subset F of E (generally, a closed subspace). Which are the hereditary properties for P, that is the properties enjoyed by Pf as soon as they are enjoyed by f ?

Even in the cases where the abstract theory is completely known, this concrete problem can be quite attractive ("abstract" means that E consists of points, and "concrete", here, that E consists of functions). If E is an abstract Hilbert space and F a closed linear subspace, there is nothing to say, except that P is the ordinary projection. But if $E = L^2(T)$ and $F = L^2_\Lambda(T)$, where Λ is a subset of $\mathbb{Z}$ ($\Lambda \neq \emptyset$ and $\Lambda \neq \mathbb{Z}$), and $L^2_\Lambda(T)$ is the closed subspace of $L^2(T)$ generated by the exponentials $e_\lambda(t) = \exp(2\pi i\lambda t)$, $\lambda \in \Lambda$, the study of hereditary properties can be quite deep. For example, consider the class $L^p(T)$ ($p > 2$); for which Λ's is it true that $f \in L^p \Longrightarrow Pf \in L^p$? This is true (and not obvious at all) when $\Lambda = \mathbb{N} = \{0, 1, 2, \ldots\}$: it is a classical theorem of Marcel Riesz. On the other hand, if we choose a random Λ (in the most natural sense: for each $n \in \mathbb{Z}$ the event $n \in \Lambda$ has probability $\frac{1}{2}$ and these events are mutually independent), this is not true at all : for there exists $f \in \bigcap_{\infty > p > 2} L^p(T)$ such that $Pf \notin \bigcup_{p > 2} L^p(T)$ almost surely. This is easy, and we sketch the proof in Appendix 1. For the same reason, the L^p-heredity does not hold when Λ is chosen at random in a given set Λ_o, as soon as $L^2_{\Lambda_o}(T)$ contains a function which is not in $L^p(T)$ (which means that Λ_o is not too sparse). On the other hand, the L^p-heredity holds for lacunary Λ's, because then $L^2_\Lambda(T) \subset \bigcap_{\infty > p > 2} L^p(T)$ ([14], I chap. V).

Again with $E = L^2(T)$ and $F = L^2_\Lambda(T)$, let us consider the continuity property. Here the Λ's that provide heredity can be characterized: they are finite unions of translates of one subgroup of $\mathbb{Z}$, upto a finite set. For our Λ's are exactly the supports of Fourier transforms of idempotent measures (in other words, the function 1_Λ which is 1 on Λ and 0 on $\mathbb{Z}-\Lambda$ is a multiplier of $\mathfrak{F}C(T)$, the space of Fourier transforms of continuous functions on T) and the characterization was made by H.Helson [4]. All properties like analyticity, C^∞ and others which depend only on the size of Fourier coefficients are hereditary for all Λ's. For the class Lip α, there is no obvious characterization of good Λ's, though the class of multipliers of $\mathfrak{F}$ Lip α is known [15] .

Let us consider now the case $E = L^p(T)$ $(1 < p < \infty)$ and $F = L^p_\Lambda(T)$ ($f \in F$ if its Fourier series is $\sum_{\lambda \in \Lambda} \hat{f}(\lambda) e_\lambda$). It has been studied recently by H.S.Shapiro [11] , and strange things can happen. Suppose $\mathbb{Z} - \Lambda = \{\alpha, \beta\}$, consider $g = (e_\alpha - e_\beta)^{-1} | e_\alpha - e_\beta |^{p'}$, where $p' = \frac{p-1}{p}$, and $f = \hat{g}(\alpha) e_\alpha + \hat{g}(\beta) e_\beta$. Then $Pf = f - g$ [11]. Therefore f has all possible good properties, and in case $p > 2$ Pf is not in C^1, and in case $p < 2$ Pf is not even bounded !. The particular case $\Lambda = \mathbb{N}$ ($L^p_{\mathbb{N}}(T)$ is the Hardy class H^p) was studied before [7] , [9],[12] ; nevertheless not much was known about hereditary properties even in this case before 1973.

When $E = L^\infty(T)$ and $F = L^\infty_{\mathbb{N}}(T) = H^\infty$, $\mathcal{P}(f)$ consists of one point Pf whenever f is continuous. In this case, hereditary properties are studied in a recent paper of L.Carleson and S.Jacobs [1] ; analyticity (this being already established in [12]), C^∞, Lip α are hereditary properties. Moreover, a sharp condition is given on the modulus of continuity of f in order that Pf is continuous.

II. We consider now the case $E = L^1(T)$, $F = L^1_\Lambda(T)$. The references are [5] and [6]; we shall give an improved version of [5] . Some proofs are given in the Appendix.

For any Λ, define $\Gamma = \Gamma_\Lambda = \{g \in L^1(T), \|g\| = d(g, L^1_\Lambda(T))\}$. From Hahn-Banach theorem $g \in \Gamma$ if and only if there exists $u \in L^\infty(T)$, $\|u\|_\infty \leq 1$, $u \perp L^1_\Lambda(T)$ and $\int_T ug = \int_T |g|$. Writing $\Lambda' = \mathbf{Z} \setminus \Lambda$ and $\tilde{\Lambda}' = -\Lambda'$, we have

(1) $g \in \Gamma \iff \exists u \in L^\infty_{\tilde{\Lambda}'}(T)$, $\|u\|_\infty \leq 1$ $\quad ug = |g|$ a.e.

Using (1) the following can be proved [6] .

<u>The metric projection from</u> $L^1(T)$ <u>to</u> $L^1_\Lambda(T)$ <u>exists if and only if</u> Λ <u>is an arthmetical progression with an odd ratio. Moreover it is continuous whenever it exists.</u>

There are essentially three cases : 1) $\Lambda = \mathbf{N}^+ = \{1,2,\ldots\}$ 2) $\Lambda = (2\nu+1)\mathbf{Z}$ (ν positive integer) 3) $\Lambda = (2\nu+1)\mathbf{N}^+$.

Case 1) was considered by J.L. Doob, who proved the existence (that is, existence and uniqueness) of the metric projection P [3] . Already in 1920 F.Riesz proved a theorem which can be stated as follows (the theorem of F.Riesz is actually stronger): if $\Lambda = N^+$ and if f is a trigonometric polynomial of degree $\leq N$, Pf is also a trigonometric polynomial of degree $\leq N$ [8]. In his thesis [12] , H.S.Shapiro proved that analyticity is hereditary. Easy proofs are given below.

Case 2) is quite elementary, though not trivial, and case 3) is the most difficult.

Before stating more refined results, let us summarize some of them; we answer (when we can) the question " is the property hereditary ?".

	$L^p (1 \leq p \leq \infty)$	L^∞, C	Lip α	C^∞, analyticity
case 1) $\Lambda = \mathbb{N}^+$	yes	no	yes	yes
2) $\Lambda = (2\nu+1)\mathbb{Z}$	yes	yes	$\nu=1$ yes; $\nu\neq 1$ no	no
3) $\Lambda = (2\nu+1)\mathbb{N}^+$	yes	no	?	?

In all cases let us write $h = Pf$.

<u>Case 1</u>. We have essentially three positive results, whose proofs are given in Appendix 2. First we have the inequality

$$(2) \qquad |\hat{h}(n) - \hat{f}(n)| \leq \sum_{m=0}^{\infty} |\hat{f}(-m-n)| \qquad (n \geq 1).$$

In case $\hat{f}(m) = 0$ for $|m| > N$, it gives F.Riesz's theorem. In case $|\hat{f}(m)| \leq Ae^{-\varepsilon |m|}$ it gives Shapiro's result. If $\hat{f}(m) = O(|m|^{-A})$ $(|m| \to \infty)$ for each A we have preservation of C^∞. In the same way, Gevrey classes and the so-called "derivable" classes of infinitely differentiable functions are preserved by P.

Secondly <u>all linear classes invariant under multiplication by bounded functions, complex conjugacy and Hilbert conjugacy</u> (in the sense of Hilbert transforms) <u>are preserved</u>. The simplest example is L^p $(1 < p < \infty)$.

Finally let us consider the L^p-best approximation of a function $\varphi \in L^p$ by trigonometric polynomials of order $\leq n$, that is

$$E_n^p(\varphi) = \inf_{\alpha_j} \|\varphi - \sum_{-n}^{n} \alpha_j e_j\|_p \quad (\text{here } 1 \leq p \leq \infty).$$

Given a sequence $\omega_n \nearrow \infty$, let us write $\mathcal{C}^p_\omega$ for the class of all functions $\varphi \in L^p$ such that $E^p_n(\varphi) = O(\omega_n)$ $(n \nearrow \infty)$. <u>If the class $\mathcal{C}^p_\omega$ is closed under Hilbert-conjugacy and if moreover</u>

$$\omega_{2n} = O(\omega_n) \quad \underline{\text{and}} \quad \sum_{j=k}^{\infty} \omega^{-1}_{2^j} = O(\omega^{-1}_{2^k}),$$

$\mathcal{C}^p_\omega$ <u>is preserved</u>. The most interesting example is $p = \infty$ and $\omega_n = n^\alpha$. Then $\mathcal{C}^\infty_\omega$ is Lip α when $0 < \alpha < 1$, $\mathcal{C}^\infty_\omega$ is the class Λ^* of Zygmund when $\alpha = 1$, and $\mathcal{C}^\infty_\omega$ consists of all functions f having continuous derivatives $f^1, \dots, f^{(\nu)}$, and $f^{(\nu)} \in \text{Lip}\,\alpha'$ (or Λ^* if $\alpha' = 1$) if $\alpha = \nu + \alpha'$ with $0 < \alpha' \leq 1$. This improves a statement made in [5].

The negative result is that $f \in C$ exists such that $h \notin L^\infty$. This is almost obvious by choosing $F(z)$ holomorphic in $|z| < 1$ with $\text{Re}F(z)$ continuous in $|z| \leq 1$ and $\text{Im}\,F(z) \geq 0$ and unbounded. Writing $g(t) = -i \lim_{r \nearrow 1} \text{Im}\,F(re^{2\pi it})$ we have $g \in \Gamma$ by (1). Choosing $f(t) = \text{Re}F(e^{2\pi it}) - F(0)$ we have $f = g + h$, $g \in F$ and $h \in H^1_0$, $f \in C$ and $h \notin L^\infty$.

A stronger version, obtained in the same way, is this: given any modulus of continuity $\varpi(t)$ such that $\int_0^1 \varpi(t) \frac{dt}{t} = \infty$, there exists f in the corresponding lip_ϖ class such that $h \notin L^\infty$.

<u>Case 2</u>. Given f, we write $f_j(t) = f(t + \frac{j}{2\nu+1})$. Then h is defined by the condition

$$\sum_{1}^{2\nu+1} | f_j(t) - h(t) | \quad \text{minimum}$$

for almost every t. We are led to the following question. Given $z = (z_1, z_2, \dots, z_{2\nu+1})$, we define ζ to be <u>the</u> complex number such that $\sum_{1}^{2\nu+1} | z_j - \zeta |$ is minimum and we write $\zeta = Q(z)$.

Then Q is a well defined mapping from $\mathbb{C}^{2\nu+1}$ to $\mathbb{C}$ (such a mapping does not exist when $\mathbb{C}^{2\nu+1}$ is replaced by $\mathbb{C}^{2\mu}$!). If $\nu = 1$, ζ is the Steiner point of the triangle z_1, z_2, z_3, and elementary geometry shows that Q is Lipschitzian. If we restrict Q to $\mathbb{R}^{2\nu+1}$, Q is Lipschitzian again, and ζ is the medial point among $z_1, z_2, \dots z_{2\nu+1}$. In the general case, Q satisfies a Lipschitz-Hölder condition of order $\frac{1}{2}$ [6] , but not order 1; in other words, when $\nu \neq 1$, Q is not uniformly continuous. We give the proof in Appendix 3.

Choosing now $f(t) = \sin t$, it is easy to check that $h(t)$ is not continuously differentiable.

Since $\zeta = Q(z)$ belongs to the convex hull of $z_1, z_2, \dots z_{2\nu+1}$, any condition of the type $\int \varphi(|f|) < \infty$ (φ being an increasing function) implies $\int \varphi(|h|) < \infty$; in particular L^p defines a hereditary property. The fact that Lip α is hereditary for $\nu = 1$ and not hereditary for $\nu > 1$ depends on the Lipschitzian character of Q. In any case $f \in \text{Lip}\,\alpha \implies h \in \text{Lip}\,\frac{\alpha}{2}$ (here $0 < \alpha \leq 1$). Continuity is hereditary, and also real-Lip α.

<u>Case 3</u>. We write f_j as above, $u_j(t) = u(t + \frac{j}{2\nu+1})$, and $v = \sum_1^{2\nu+1} u_j$. Then $v \in L^\infty_{(2\nu+1)\mathbb{N}^+}$, therefore $hv \in H^1_0$. We have $uf = uh + |g|$, i.e. $u_i f_i = u_i h + |g|$, hence $\text{Im}\, u_i f_i = \text{Im}\, u_i h$ and adding $|\text{Im}\, vh| \leq \sum_1^{2\nu+1} |f_j|$. Suppose $f \in L^p$. Then $\text{Im}\, vh \in L^p$, therefore (Marcel Riesz) $vh \in L^p$. Write $E = \left\{ t \;\middle|\; |v(t)| < \frac{1}{2} \right\}$; then

$hl_{T \setminus E} \in L^p$. Divide now E into two parts : if $t \in E_1$, at least one $u_j(t)$ has modulus strictly < 1; on E_2, all $|u_j| = 1$. If $t \in E_1$, at least one $g_j(t) = 0$, therefore $h(t) = f_j(t)$ for the corresponding j ; therefore $hl_{E_1} \in L^p$. On E_2 we have $|u_j| = 1$ and $|\sum_1^{2\nu+1} u_j| < \frac{1}{2}$; let us denote by $\alpha = \alpha(\nu)$ the infimum of $\sup_j |\mathrm{Im}\, u_j|$ under these conditions ($\alpha > 0$ because $2\nu+1$ is odd); then $\sup_j |\mathrm{Im}\, u_j h| \geq \alpha |h|$, therefore $hl_{E_2} \in L^p$. This proves the hereditary property for L^p.

The negative result derives from the negative result in case 1.

III. Here are some comments about the results described in section II.

a) Case 2 deals with a particular case of the following problem. Give a finite measure space (E,μ), and study the metric projection from $L^1(E,\mu)$ on the one dimensional subspace which consist of constants. In other words, study the mapping $Q: z = (z_1, \ldots, z_n) \to \zeta$ defined by $\sum_1^n \mu_j |\zeta - z_j|$ minimum, when it exists (here $E = \{1,2, \ldots, n\}$ and $\mu(j) = \mu_j$). It is easy to see that Q exists if and only if E cannot be split into two subsets of equal mass; in case all μ_j are 1, this means that n is odd. Though this question is very simple, Appendix 3 shows that the properties of Q can be rather tricky. It is also possible to study the case of a 'vector valued' $L^1(E,\mu)$ in a normed space X, i.e. the mapping Q from X^n to X (when it exists) which maps $(x_1, x_2, \ldots, x_n)$ on ζ such that $\sum_1^n \mu_j \|\zeta - x_j\|$ is minimum. May be some examples in metric projections can be found either with $L^1(E,\mu)$, or $L^p(E,\mu)$, or some other natural spaces, and looking at the metric projections on constants.

b) Instead of T (in the problem $L^1(T) \longrightarrow L^1_\Lambda(T)$), we could have considered another compact abelian group. For example, in case of $D = \{-1, 1\}^{\mathbb{N}}$, it is not difficult to prove that there exists no Λ for which the metric projection exists. In case of T^2, Λ must be, roughly speaking, either a subgroup of $\mathbb{Z}^2$ of odd index, or the intersection of such a subgroup with a open half plane, plus something at the boundary of the half plane (the investigation has not yet been carried out).

c) Quite recently L. Carleson proved some hereditary properties in the metric projection from L^p to H^p $(1 < p < \infty)$ in particular L^q is mapped into H^q when $p \leq q < \infty$. (See C.R. Acad. Sc. Paris, 1973).

d) Though the metric projection from $L^1(T)$ to $L^1_{m\mathbb{N}}(T)$ does not exist in the strict sense when m is even, it is possible to assert and to prove that L^p is hereditary in this case too. The assertion means the following: given $f \in L^1(T)$, there exists at least one $h \in L^1_{m\mathbb{N}}(T)$ such that $\| f-h \|$ is minimum; whenever $f \in L^p$, then $h \in L^p$. The proof is a slight refinement of what is given in case 3; here it is necessary to look at $\operatorname{Re} u_j h$ and split E into three parts : E_1 as before, E_2 defined in the complement of E_1 by $\sup |\operatorname{Im} u_j f| \geq \alpha |h|$ and E_3 by $\sup |\operatorname{Im} u_j h| < \alpha |h|$; on E_3 the $u_j h$ are almost real, one of them must be nearly positive and therefore satisfy $|u_j h| < 2 |u_j f_j| = 2 \cdot |f_j|$.

Appendix 1.

Choose $\varphi \in L^2(T)$, $\varphi \notin \bigcup_{p>2} L^p(T)$. The function f obtained from φ in changing signs of Fourier coefficients randomly belongs to $\bigcap_{\infty > p > 2} L^p(T)$ a.s. ([14], I, chap. V.). If Λ is the set of integers n where $\hat{f}(n) = \hat{\varphi}(n)$, we have $Pf = \frac{1}{2}(f + \varphi)$, hence

$Pf \notin \bigcup_{p>2} L^p(T)$.

Appendix 2.

When $\Lambda = \mathbb{N}^+$, the usual notation for L^1_Λ is H^1_0 and $L^\infty_{\tilde{\Lambda}'}$ is H^∞. Therefore (1) is written as

$$\text{(i)} \qquad g \in \Gamma \iff \exists\, u \in H^\infty, \quad \|u\|_\infty \leq 1, \quad ug = |g| \quad \text{a.s.}$$

For $f \in L^1(T)$ let us write $f = g + h$, $g \in \Gamma$, $h \in H^1_0$, therefore $h = Pf$. Choose u as in (i). We shall derive properties of g from the equalities $\hat{g}(n) = \hat{f}(n)$ $(n \leq 0)$ and $\bar{g} = bg$ with $b = u^2$ (note that $b \in H^\infty$). Formally we have

$$\text{(ii)} \qquad \sum_{n=1}^{\infty} \hat{g}(n) e_n = \sum_{m=0}^{\infty} \overline{\hat{b}(m)}\, e_{-m} \sum_{-\infty}^{\infty} \overline{\hat{f}(p)}\, e_{-p} + \sum_{-\infty}^{0} x_n e_n$$

that is
$$\hat{g}(n) = \sum_{m=0}^{\infty} \overline{\hat{b}(m)}\; \overline{\hat{f}(-m-n)} \quad \text{for} \quad n \geq 1.$$

From this we have inequality (2).

From (ii) we have the second result of case 1.

Finally let us fix $b \in H^\infty$ and study the linear map $\varphi \to \psi$ defined by

$$\psi = \sum_{n=1}^{\infty} \hat{\psi}(n) e_n = \bar{b}\,\bar{\varphi} + \sum_{-\infty}^{0} x_n e_n \quad \text{for some } x_n.$$

Let X_j be the trigonometric polynomial of order $\leq 2^j$ which approximates φ in L^p, and V_j the de la Vallée Poussin kernel defined by

$$\hat{V}_j(n) = 1 \text{ for } |n| \leq 2^j, \quad \hat{V}_j(n) = 2 - |n|\, 2^{-j} \text{ for } 2^j \leq |n| < 2^{j+1},$$
$$\hat{V}_j(n) = 0 \text{ for } |n| \geq 2^{j+1}.$$

We defined

$$Y_{j+1} = (\bar{b}.\overline{X_{j+1} - X_j}) * V_{j+1}, \quad Y_0 = (\bar{b}\,\bar{X}_0) * V_0.$$

Since $\| b \|_{\infty} \leq 1$, $\| X_{j+1} - X_j \|_p < 2\, \omega^{-1}_{2^j}$ and $\| V_{j+1} \|_1 \leq 3$, we have $\| Y_{j+1} \|_p \leq 6\, \omega^{-1}_{2^j}$. Defining $\psi_o = \sum_{o}^{\infty} Y_j$ we obtain

$$E^p_{2^{j+1}}(\psi_o) \leq 6 \sum_{i=j}^{\infty} \omega^{-1}_{2^i} = O(\omega^{-1}_{2^{j+1}})$$

(we use the assumptions made on ω). Therefore $\psi_o \in \mathcal{C}^p_\omega$. Since $\psi = \sum_{n=1}^{\infty} \hat{\psi}_o(n) e_n$ and $\mathcal{C}^p_\omega$ is closed under conjugacy, $\psi \in \mathcal{C}^p_\omega$.

<u>Appendix 3</u>.

The fact that Q is Lip $\frac{1}{2}$ is in [6], formula (13).

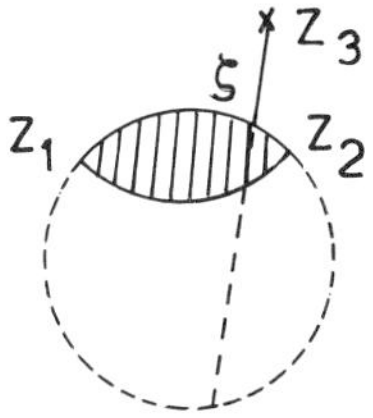

If $\nu = 1$, the figure shows how ζ depends on z_3 when z_1 and z_2 are given. If z_3 is inside the loop, $\zeta = z_3$. The loop contains all points from which the half-lines through z_1 and z_2 make an angle $\geq \frac{2\pi}{3}$.

If $\nu > 1$, ζ depends on $z_{2\nu+1}$ in analogous way. But we shall show that the angle between the line L from $z_{2\nu+1}$ to ζ and the curve Γ (see figure) is arbitrarily small; therefore, moving $z_{2\nu+1}$ by a small ε perpendicular to L produces a big move of ζ (compared to ε), that is, Q is not Lip 1.

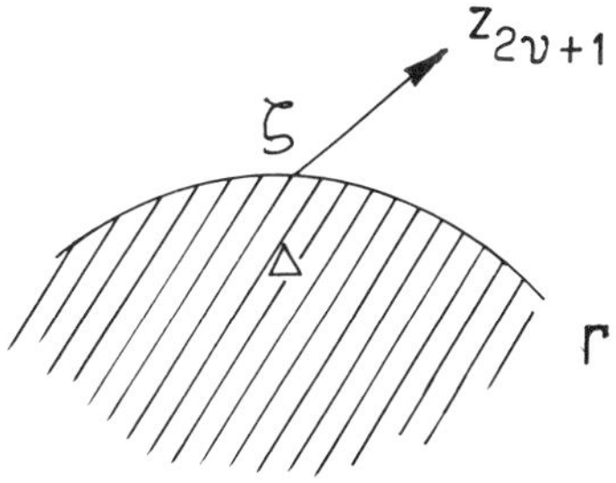

We write $z_j - \zeta = u_j \rho_j$, $|u_j| = 1$ and $\rho_j > 0$, or $\rho_j = 0$. Given $z_1, z_2, \ldots, z_{2\nu}$, all of them different, $u_j = u_j(\zeta)$

for $j = 1,2, \ldots 2\nu$, and $\zeta \neq z_j$ for any $j \leq 2\nu$. Let Γ be the curve $\{\zeta \mid \sum_1^{2\nu} |u_j(\zeta)|=1\}$ and Δ the domain $\{\zeta \mid \sum_1^{2\nu} |u_j(\zeta)| \leq 1\}$. Suppose $0 \in \Gamma$ and $-\sum_1^{2\nu} u_j(0) = 1$; this is possible by a translation and a rotation. We write as usual $z_j = x_j + iy_j$, $\zeta = \xi + i\eta$, and we consider $f(\xi,\eta) = \sum_1^{2\nu} |u_j(\zeta)|^2$ and $g = \frac{\partial f}{\partial \xi}(0,0) + i\frac{\partial f}{\partial \eta}(0,0)$. A simple computation gives

$$g = \sum_1^{2\nu} \frac{y_j^2}{\rho_j^3} - i \sum_1^{2\nu} \frac{x_j y_j}{\rho_j^3} = g_x + ig_y.$$

We want to prove that g_x can be small compared to g_y (for here $\zeta = 0$ and $z_{2\nu+1}$ belongs to the positive real axis). We choose all z_j out of the region $y^2 > (x^2+y^2)^{3/2}$, which is bounded by a closed curve γ , tangent to $0x$ at 0 (see figure). Now we choose z_1 on γ very near 0, so that $\frac{x_1 y_1}{\rho_1^3} = \frac{x_1}{y_1}$ is very large, and the $2\nu-1$ remaining z_j far away, so that

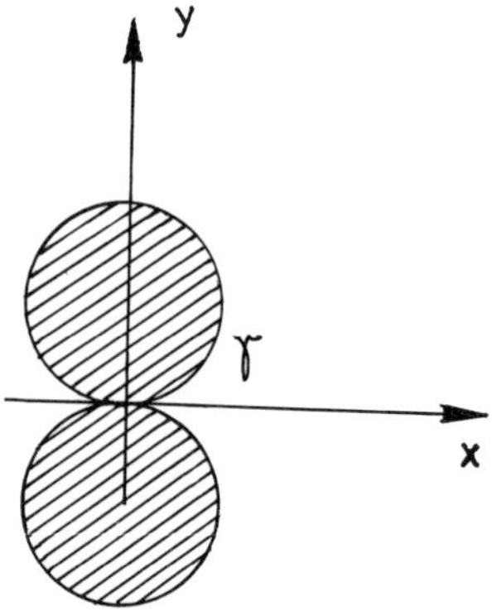

$\sum_2^{2\nu} \frac{x_j y_j}{\rho_j^3}$ is not large, and moreover near the summits of a regular polygon centered at 0, in such a way that $\sum_1^{2\nu} \frac{x_j}{\rho_j} = -1$ and $\sum_1^{2\nu} \frac{y_j}{\rho_j} = 0$ (the simplest way to do it is simply to move a regular polygon in the right position). Here we are done : L can make an angle with Γ as small as we want.

References

1 Carleson, L., and Jacobs, S., Best Uniform Approximation by Analytic Functions, Arkiv för Matematik 10(1972),219-229.

2 Domar, Y., On the uniqueness of minimal extrapolation, Arkiv för Matematik 4 (1960), 19-29.

3 Doob, J.L., A minimum problem in the theory of analytic functions, Duke Math. J. 8 (1941), 413-424.

4 Helson, H., Note on harmonic functions, Proceedings Amer. Math. Soc. 4 (1953), 686-691.

5 Kahane, J.-P, Projections métriques dans $L^1(T)$, C.R.Acad. Sc. Paris, série A, 276 (1973), 621-623.

6 Kahane, J.-P, Projection métrique de $L^1(T)$ sur des sous-espaces fermés invariants par translation, in Proceedings of the Conference on Approximation (Oberwolfach 1971), INSM 1972.

7 Macintyre, A.S., and Rogosinski, W.W., Extremum problems in the theory of analytic functions, Acta Mathematica 82 (1950), 275-325.

8 Riesz, F., Über Potenzreihen mit vorgeschriebenen Anfangsgliedern, Acta Mathematica 42 (1920), 145-171.

9 Rogosinski, W.W., and Shapiro, H.S., On certain extremum problems for analytic functions, Acta Mathematica 90 (1953), 287-318.

10 Singer, I., Best Approximation in normed linear spaces by elements of linear subspaces, Grundlehren der mathematischen Wissenschaften 171 (1970).

11 Shapiro, H.S., Regularity properties of the element of closest approximation, To appear.

12 Shapiro, H.S., Extremal problems for polynomials and power series, Dissertation M.I.T. 1952.

13 Tchebychef, P.L., Oeuvres, tomes I et II, St Petersbourg 1899 et 1907.

14 Zygmund, A., Trigonometric series, vol I, Cambridge.

15 Zygmund, A., On the preservation of classes of functions, Journal of Math. and Mech. 8 (1959), 889-896.

MULTIPLIERS ON WEIGHTED SPACES

G.N. Keshava Murthy and K.R. Unni

1. Introduction.

This paper is concerned with the characterisation of the space of bounded linear operators which commute with translations on weighted spaces defined on a locally compact abelian group. These characterization theorems are motivated by those of Figa-Talamanca [2,3], Figa-Talamanca and Gaudry [4] and Rieffel [9] wherein the concept of tensor product has been used as a basic tool for obtaining such characterizations.

2. Weighted spaces and some of their properties.

Let G be a locally compact abelian group with Haar measure dx. Let Ω be the set of all functions ω satisfying the two conditions

(i) ω is a measurable function on G, positive a.e. for the Haar measure dx.

(ii) for each $p \in [1,\infty)$, both ω^p and ω^{-p} are locally integrable.

The elements of Ω are called weights (see P.Kree [7]).

If $\omega \in \Omega$, let $L^{p,\omega}(G)$ denote the space of equivalence classes of complex valued measurable functions f on G such that $|f|\omega$ has its pth power summable and has norm

$$\|f\|_{p,\omega} = \left(\int_G |f(x)\,\omega(x)|^p dx \right)^{1/p} .$$

Then $L^{p,\omega}(G)$ is a Banach space and its conjugate space is $L^{p',\omega^{-1}}(G)$ where $\frac{1}{p} + \frac{1}{p'} = 1$. Moreover, if $1 < p < \infty$, then $L^{p,\omega}(G)$ is a reflexive Banach space.

Suppose now that ω is a non-negative function on G satisfying

$$\omega(x+y) \leq \omega(x)\,\omega(y) \qquad x,y \in G \qquad (1)$$

If $\frac{1}{\omega}$ is bounded away from zero, then ω being locally bounded, both ω^p and ω^{-p} are locally integrable. Hence $\omega \in \Omega$. Hereafter we shall assume that all our weight functions ω will satisfy (1).

Let $y \in G$. Then the translation operator τ_y is defined by

$$\tau_y f = f(y) \quad y \in G .$$

If $f \in L^{p,\omega}(G)$, then it is easy to verify that

$$\| \tau_y f \|_{p,\omega} \leq \omega(y) \| f \|_{p,\omega} . \tag{2}$$

Thus τ_y is a bounded linear operator on $L^{p,\omega}(G)$ with norm bounded by $\omega(y)$.

LEMMA 1. *If* $\mathcal{K}(G)$ *denotes the space of continuous functions on* G *with compact support and* $1 \leq p < \infty$, *then*

$$\mathcal{K}(G) \subset L^{p,\omega}(G) \subset L^p(G) .$$

Proof. Trivial.

LEMMA 2. $\mathcal{K}(G)$ *is dense in* $L^{1,\omega}(G)$

Proof. See Gaudry [6, Lemma 3] and also Edwards [1] .

LEMMA 3. $L^{1,\omega}(G)$ *has approximate identities, that is there exists* ϕ_α *with the following properties*

(i) $\phi_\alpha \in \mathcal{K}(G)$, $\phi_\alpha \geq 0$, $\|\phi_\alpha\|_1 = 1$,

(ii) $\phi_\alpha * f \longrightarrow f$ *in* $L^{1,\omega}(G)$ *for each* $f \in L^{1,\omega}(G)$,

(iii) (ϕ_α) *is bounded in* $L^{1,\omega}(G)$.

Proof. See Gaudry [6, Lemma 2] .

LEMMA 4. *Every weight* ω *is equivalent to a continuous weight*.

Proof. See R.Spector [10, Prop. III. 1-3] .

LEMMA 5. $\mathcal{K}(G)$ <u>is dense in</u> $L^{p,\omega}(G)$.

<u>Proof</u>. By Lemma 4, we may assume that ω is continuous. Let $f \in L^{p,\omega}(G)$. Then $f\omega \in L^p(G)$. Since $\mathcal{K}(G)$ is dense in $L^p(G)$, given $\varepsilon > 0$, there exists $f_c \in \mathcal{K}(G)$ such that

$$\| f_c - f\omega \|_p < \varepsilon .$$

Set $g_c = \frac{f_c}{\omega}$. Then $g_c \in \mathcal{K}(G)$ and $\| g_c - f \|_{p,\omega} < \varepsilon$. This completes the proof.

LEMMA 6. <u>Let</u> ϕ_α <u>be as in Lemma 3. Then for each</u> $f \in L^{p,\omega}(G)$ <u>we have</u>

$$\phi_\alpha * f \longrightarrow f \quad \underline{\text{in}} \quad L^{p,\omega}(G) .$$

<u>Proof</u>. First we prove that $\phi_\alpha * g \longrightarrow g$ in $L^{p,\omega}(G)$ for each $g \in \mathcal{K}(G)$. Now

$$\phi_\alpha * g(x) - g(x) = \int_G (g(x-y) - g(x))\, \phi_\alpha(y)\, dy ,$$

so that

$$\| \phi_\alpha * g - g \|_{p,\omega} \le \left(\int_G \omega^p(x)\, dx \, \left| \int_G (g(x-y) - g(x))\, \phi_\alpha(y)\, dy \right|^p \right)^{1/p} .$$

By Minkowsky's inequality

$$\| \phi_\alpha * g - g \|_{p,\omega} \le \int_G \phi_\alpha(y)\, dy \left(\int_G \omega^p(x)\, | g(x-y) - g(x) |^p\, dx \right)^{1/p} .$$

This can be made arbitrarily small because g is uniformly continuous on supp g + supp ϕ_α and ω^p is locally integrable.

Now let $f \in L^{p,\omega}(G)$. By Lemma 5, there exists $f_c \in \mathcal{K}(G)$ such that

$$\| f - f_c \|_{p,\omega} < \varepsilon .$$

Now

$$\|\phi_\alpha * f - f\|_{p,\omega} \le \|\phi_\alpha * (f - f_c)\|_{p,\omega} + \|\phi_\alpha * f_c - f_c\|_{p,\omega} + \|f_c - f\|_{p,\omega}$$

$$\le \|\phi_\alpha\|_{1,\omega} \|f - f_c\|_{p,\omega} + \|\phi_\alpha * f_c - f_c\|_{p,\omega} + \|f_c - f\|_{p,\omega}$$

$$= (\|\phi_\alpha\|_{1,\omega} + 1)\|f - f_c\|_{p,\omega} + \|\phi_\alpha * f_c - f_c\|_{p,\omega} .$$

Since $\|\phi_\alpha\|_{1,\omega}$ is bounded, the right hand side can be made arbitrarily small first by choosing f_c and then the functions ϕ_α. This completes the proof of Lemma 6.

3. Multipliers from $L^{p,\omega}(G)$ to $L^{q,\omega}(G)$.

Let Ω_o denote the subset of Ω consisting of those even continuous functions ω satisfying (1). It then follows that

$$1 \le \omega(o) \le \omega(x)$$

for all $x \in G$. Moreover

$$\frac{1}{\omega(x)} \le \frac{\omega(y)}{\omega(x-y)}$$

for all $x, y \in G$. Suppose that $1 < p \le q < \infty$. A multiplier from $L^{p,\omega}(G)$ to $L^{q,\omega}(G)$ is a bounded linear operator from $L^{p,\omega}(G)$ to $L^{q,\omega}(G)$ which commutes with translations. Let $M(L^{p,\omega}(G), L^{q,\omega}(G))$ denote the space of all multipliers from $L^{p,\omega}(G)$ to $L^{q,\omega}(G)$. For $p \ne q$ let $\mathfrak{A}(p,q,\omega)$ be the space of all those functions u which can be represented as

$$u = \sum_{i=1}^{\infty} f_i * g_i \quad a.e. , \tag{3}$$

where $f_i \in \mathcal{K}(G)$ and $g_i \in L^{q',\omega^{-1}}(G)$ and such that $\sum_{i=1}^{\infty} \|f_i\|_{p,\omega} \|g_i\|_{q',\omega^{-1}} < \infty$. We take a norm on $\mathfrak{A}(p,q,\omega)$ by

$$\|u\| = \inf\left\{\sum_{i=1}^{\infty} \|f_i\|_{p,\omega} \|g_i\|_{q',\omega^{-1}}\right\}$$

where the infimum is taken over all such representations of u. Then $\mathfrak{A}(p,q,\omega)$ is a Banach space in this norm. Since

$$\| f * g \|_{r,\omega^{-1}} \leq \| f \|_{p,\omega} \| g \|_{q',\omega^{-1}}$$

for $f \in \mathcal{K}(G)$ and $g \in L^{q',\omega^{-1}}(G)$ where $\frac{1}{r} = \frac{1}{p} - \frac{1}{q}$, it follows that $\mathfrak{A}(p,q,\omega) \subset L^{r,\omega^{-1}}(G)$.

When $p = q$, $\mathfrak{A}(p,q,\omega)$ is defined as before where (3) is assumed to hold everywhere. For each ω, let $C_\omega(G)$ denote the class of all functions h such that $h\omega \in C_0(G)$ the space of all continuous functions on G which vanish at infinity. Then it is easy to verify that $\mathfrak{A}(p,p,\omega)$ is a subspace of $C_{\frac{1}{\omega}}(G)$ where the norm in $C_{\frac{1}{\omega}}(G)$ is given by

$$\| h \|_{\infty,\omega^{-1}} = \sup_{x \in G} \frac{|h(x)|}{\omega(x)} .$$

THEOREM 1. <u>Let</u> G <u>be a locally compact abelian group and</u> $\omega \in \Omega_0$ **with** $\omega(0)=1$. <u>If</u> $1 < p \leq q < \infty$ <u>then the space of multipliers</u> $M(L^{p,\omega}(G), L^{q,\omega}(G))$ <u>is isometrically isomorphic to the dual</u> $\mathfrak{A}(p,q,\omega)^*$ <u>of</u> $\mathfrak{A}(p,q,\omega)$.

<u>Proof</u>. Let $T \in M(L^{p,\omega}(G), L^{q,\omega}(G))$ and define the linear functional t on $\mathfrak{A}(p,q,\omega)$ by

$$t(u) = \sum_{i=1}^{\infty} T f_i * g_i(0)$$

for $u = \sum_{i=1}^{\infty} f_i * g_i$ in $\mathfrak{A}(p,q,\omega)$. We claim now that t(u) is unambiguously defined. To show this it is enough to show that if

$$u = \sum_{i=1}^{\infty} f_i * g_i = 0 \quad \text{in} \quad \mathfrak{A}(p,q,\omega) \quad \text{and}$$

$$\sum_{i=1}^{\infty} \| f_i \|_{p,\omega} \| g_i \|_{q',\omega^{-1}} < \infty$$

then

$$\sum_{i=1}^{\infty} Tf_i * g_i(0) = 0 .$$

We first notice that if $\phi \in \mathcal{K}(G)$ and T_ϕ is defined by

$$T_\phi(f) = \phi * f \qquad f \in L^{p,\omega}(G)$$

then $T_\phi \in M(L^{p,\omega}, L^{q,\omega})$. To see this, let $\phi \in \mathcal{K}(G)$ and $f \in L^{p,\omega}(G)$. From the relation

$$|(\phi * f)(x)\,\omega(x)| \leq \int_G |f(x-y)\,\omega(x-y)|\,|\phi(y)\omega(y)|\,dy$$

we obtain

$$\|(\phi * f)\,\omega\|_\infty \leq \|f\,\omega\|_p \|\phi\,\omega\|_{p'}$$

and

$$\|(\phi * f)\,\omega\|_p \leq \|f\,\omega\|_p \|\phi\,\omega\|_1 .$$

Then

$$\begin{aligned}\|(\phi * f)\omega\|_q^q &= \|(\phi * f)\,\omega\|_\infty^{q-p} \|(\phi * f)\,\omega\|_p^p \\ &\leq \|\phi\,\omega\|_{p'}^{q-p} \|\phi\,\omega\|_1^p \|f\,\omega\|_p^q\end{aligned}$$

so that

$$\|(\phi * f)\,\omega\|_q \leq \left(\|\phi\,\omega\|_{p'}^{q-p} \|\phi\,\omega\|_1^p\right)^{1/q} \|f\,\omega\|_p .$$

Moreover $\tau_y(\phi * f) = \phi * \tau_y f$ for each $f \in L^{p,\omega}(G)$ which implies that $\tau_y T_\phi = T_\phi \tau_y$. Thus $T_\phi \in M(L^{p,\omega}, L^{q,\omega})$ and

$$\|T_\phi\| \leq \|\phi\,\omega\|_1^{p/q} \|\phi\,\omega\|_{p'}^{1-p/q} .$$

We next show that every element of $M(L^{p,\omega}, L^{q,\omega})$ can be approximated boundedly in the strong operator topology by operators of the form T_ϕ, $\phi \in \mathcal{K}(G)$. We show that if $T \in M(L^{p,\omega}, L^{q,\omega})$, then there exists a net (ϕ_α) in $\mathcal{K}(G)$ such that $\lim_\alpha \phi_\alpha * f = Tf$ in the norm of $L^{q,\omega}(G)$ for every $f \in L^{p,\omega}(G)$ and there exist constants $K_\alpha(\omega)$ which depend on ω such that

$$\| \phi_\alpha * f \|_{q,\omega} \leq K_\alpha(\omega) \| f \|_{p,\omega} \|T\|$$

where $\lim_\alpha K_\alpha(\omega) = 1$, and $\{K_\alpha(\omega)\}$ is bounded. It is sufficient to show that $\lim_\alpha \phi_\alpha * f = Tf$ weakly in $L^{q,\omega}(G)$ and then a net of convex combinations of the ϕ_α's will satisfy our requirements. Let $\{h_\beta\}$ be an approximate identity in $L^{1,\omega}(G)$ with $h_\beta \in \mathcal{K}(G) * \mathcal{K}(G)$, $\|h_\beta\|_1 = 1$ and h_β vanishes outside some fixed compact set for all β. Let $\{k_\delta\}$ be an approximate identity in $L^1(\Gamma)$ such that $\hat{k}_\delta \in \mathcal{K}(G)$, $\|k_\delta\|_1 = 1$. Since it is easy to prove that T commutes with convolutions by functions in $\mathcal{K}(G)$, it then follows that Th_β is continuous for all β. Now we set $\phi_\alpha = \phi_{(\beta,\delta)} = \hat{k}_\delta T h_\beta$ and give $\alpha = (\beta,\delta)$ the usual product ordering. Then $\phi_\alpha \in \mathcal{K}(G)$ for each α. If $f, g \in \mathcal{K}(G)$, we have

$$\hat{k}_\delta T h_\beta * f * g(0) = \int_G (\hat{k}_\delta T h_\beta)(-y) \, f * g(y)\, dy$$

$$= \int_G \int_G \int_\Gamma k_\delta(\gamma)\, T h_\beta(-y)\, \gamma(-y)\, f(y-t) g(t)\, d\gamma\, dt\, dy .$$

Since $\bar{\gamma}(y) = \gamma(-y)$, we have by Fubini's theorem

$$|\phi_\alpha * f * g(0)| \leq \int_\Gamma |k_\delta(\gamma)| \left| \int_G \int_G T h_\beta(-y) f(y-t) g(t) \bar{\gamma}(y)\, dt\, dy \right| d\gamma$$

$$\leq \|k_\delta\|_1 \cdot \sup_{\gamma \in \Gamma} \left| \int_G T h_\beta(-y)(\bar{\gamma} f * \bar{\gamma} g)(y)\, dy \right|$$

$$= \sup_{\gamma \in \Gamma} | T h_\beta * (\bar{\gamma} f * \bar{\gamma} g)(0) |$$

$$= \sup_{\gamma \in \Gamma} | T(h_\beta * \bar{\gamma} f) * \bar{\gamma} g(0) |$$

$$\leq \| T \| \sup_{\gamma \in \Gamma} \| h_\beta * \bar{\gamma} f \|_{p,\omega} \| \bar{\gamma} g \|_{q',\omega^{-1}}$$

$$\leq \| T \| \| h_\beta \|_{1,\omega} \| f \|_{p,\omega} \| g \|_{q',\omega^{-1}} .$$

Using the relation

$$\| h_\beta \|_{1,\omega} \leq K_\beta(\omega) \| h_\beta \|_1 = K_\beta(\omega)$$

where $K_\beta(\omega) = \max \{ \omega(x) : x \in \operatorname{supp} h_\beta \}$, it follows that

$$| \phi_\alpha * f * g(0) | \leq \| T \| K_\beta(\omega) \| f \|_{p,\omega} \| g \|_{q',\omega^{-1}}$$

so that

$$\| \phi_\alpha * f \|_{q,\omega} \leq \| T \| K_\alpha(\omega) \| f \|_{p,\omega}$$

where $K_\alpha(\omega) = K_\beta(\omega)$. It is clear that $\{ K_\alpha(\omega) \}$ is bounded and $\lim_\alpha K_\alpha(\omega) = 1$ since $\omega(0) = 1$. The operators T_{ϕ_α} satisfy

$$\| T_{\phi_\alpha} \| \leq K_\alpha(\omega) \| T \| \leq K(\omega) \| T \| .$$

Since each closed ball of $M(L^{p,\omega}, L^{q,\omega})$ is compact in the weak operator topology, the net $\{ T_{\phi_\alpha} \}$ has a limit point $U \in M(L^{p,\omega}, L^{q,\omega})$ (for this same topology) with $\| U \| \leq \| T \|$. We suppose that $\lim_\alpha T_{\phi_\alpha} = U$ in the weak operator topology. Then we have

$$\lim_{\beta} \lim_{\delta} (\hat{k}_{\delta} T h_{\beta}) * f * g(0) = Tf * g(0)$$

for $f, g \in \mathcal{K}(G)$ since $\hat{k}_{\delta} \to 1$ locally uniformly, $\{h_{\beta}\}$ is an approximate identity and T commutes with convolutions by functions from $\mathcal{K}(G)$. Hence $T = \mathcal{U}$ and our assertion is proved.

Now suppose that $\sum_{i=1}^{\infty} f_i * g_i$ is a representation of 0 as an element of $\mathfrak{A}(p, q, \omega)$ and consider the net (ϕ_{α}) given in the preceding paragraph. Since the series $\sum_{i=1}^{\infty} \phi_{\alpha} * f_i * g_i(0)$ converges uniformly with respect to α and $\phi_{\alpha} * f_i \longrightarrow Tf_i$ in $L^{q,\omega}(G)$ for each i we have

$$\sum_{i=1}^{\infty} Tf_i * g_i(0) = \lim_{\alpha} \sum_{i=1}^{\infty} \phi_{\alpha} * f_i * g_i(0).$$

Now for each α,

$$\begin{aligned} \sum_{i=1}^{\infty} \phi_{\alpha} * f_i * g_i(0) &= \sum_{i=1}^{\infty} \int \phi_{\alpha}(-y)\,(f_i * g_i)(y)\,dy \\ &= \int \phi_{\alpha}(-y) \sum_{i=1}^{\infty} (f_i * g_i)(y)\,dy \\ &= 0 \end{aligned}$$

since $\phi_{\alpha} \in \mathcal{K}(G)$ and hence can be viewed as an element of $L^{r',\omega}(G)$ and $f_i * g_i \in L^{r,\omega^{-1}}(G)$ where $\frac{1}{r} = \frac{1}{p} - \frac{1}{q}$. This proves that t is well defined.

It is obvious that the mapping $T \longrightarrow t$ is linear. We now show that it is an isometry. In fact

$$|t(u)| \leq \sum_{i=1}^{\infty} |Tf_i * g_i(0)| \leq \sum_{i=1}^{\infty} \|Tf_i\|_{p,\omega} \|g_i\|_{q',\omega^{-1}}$$

$$\leq \|T\| \sum_{i=1}^{\infty} \|f_i\|_{p,\omega} \|g_i\|_{q',\omega^{-1}}$$

implies that

$$|t(u)| \leq \|T\| \, \|u\| .$$

Hence $\|t\| \leq \|T\|$. On the other hand

$$\|T\| = \sup\{|Tf * g(0)| : \|f\|_{p,\omega} \leq 1, \|g\|_{q',\omega^{-1}} \leq 1\}$$

$$= \sup\{|t(f*g)| : \|f\|_{p,\omega} \leq 1, \|g\|_{q',\omega^{-1}} \leq 1\}$$

$$\leq \|t\| .$$

Therefore $\|t\| = \|T\|$.

Finally we show the mapping $T \longrightarrow t$ is onto. Suppose $t \in \mathfrak{A}(p,q,\omega)^*$ and $f \in \mathcal{K}(G)$. Define

$$g \longrightarrow t(f*g)$$

on $L^{q',\omega^{-1}}(G)$. This is a bounded linear functional on

$L^{q',\omega^{-1}}(G)$, since

$$|t(f*g)| \leq \|t\| \, \|f\|_{p,\omega} \|g\|_{q',\omega^{-1}} .$$

Now $L^{q,\omega}(G)$ is the conjugate space of $L^{q',\omega^{-1}}(G)$. Hence there exists a unique $Tf \in L^{q,\omega}(G)$ such that

$$Tf * g(o) = t(f*g) \text{ for all } g \in L^{q',\omega^{-1}}(G)$$

and $\|Tf\|_{q,\omega} \leq \|t\| \, \|f\|_{p,\omega}$. Thus we have a continuous linear operator T defined on the dense subset $\mathcal{K}(G)$ of $L^{p,\omega}(G)$. We extend T continuously and linearly to the whole of $L^{p,\omega}(G)$ without changing the norm. We claim that this T belongs to $M(L^{p,\omega}(G), L^{q,\omega}(G))$. Let $y \in G$. If $f \in \mathcal{K}(G) \subset L^{p,\omega}(G)$ and $g \in L^{q',\omega^{-1}}(G)$ we have

$$T(\tau_y f) * g(o) = t(\tau_y f * g) = t(f * \tau_y g) = Tf * \tau_y g(o)$$

$$= \tau_y Tf * g(o) .$$

Hence $T\tau_y f = \tau_y Tf$ for all $f \in \mathcal{K}(G)$ and hence the same holds for all $f \in L^{p,\omega}(G)$. Thus T belongs to $M(L^{p,\omega}(G), L^{q,\omega}(G))$ and our assertion is proved.

4. <u>Multipliers on</u> $L^{p_1,\omega_1}(G) \cap L^{p_2,\omega_2}(G)$.

Let $\omega_1, \omega_2 \in \Omega$ and $p_1, p_2 \in (1, \infty)$. Let $S(p_1,\omega_1,p_2,\omega_2)$ be the set of all complex valued functions g which can be written as

$$g = g_1 + g_2 \text{ with } (g_1, g_2) \in L^{p_1,\omega_1}(G) \times L^{p_2,\omega_2}(G)$$

We define a norm on $S(p_1,\omega_1,p_2,\omega_2)$ by

$$\| g \|_S = \inf \{ \| g_1 \|_{p_1,\omega_1} + \| g_2 \|_{p_2,\omega_2} \}$$

where the infimum is taken over all such decompsotions of g. It is not hard to see that $S(p_1,\omega_1,p_2,\omega_2)$ is a Banach space under this norm.

Similarly if $D(p_1,\omega_1,p_2,\omega_2)$ denotes the set of all complex valued functions defined on G which are in $L^{p_1,\omega_1}(G) \cap L^{p_2,\omega_2}(G)$ we introduce a norm by

$$\| f \|_D = \max \left(\| f \|_{p_1,\omega_1} , \| f \|_{p_2,\omega_2} \right) .$$

Then $D(p_1,\omega_1,p_2,\omega_2)$ is also a Banach space with norm $\| \cdot \|_D$. Following Liu and Wang [8], it is possible to show that $D(p_1,\omega_1,p_2,\omega_2)$ and $S(p_1,\omega_1,p_2,\omega_2)$ are reflexive Banach spaces and the following duality relations hold

$$D(p_1,\omega_1,p_2,\omega_2)^* \cong S(p_1', \omega_1^{-1}, p_2', \omega_2^{-1}),$$

$$D(p_1,\omega_1^{-1},p_2,\omega_2^{-1}) \cong S(p_1', \omega_1, p_2', \omega_2).$$

Let us now consider the case when one of the p_i's is 1. It turns out that

$$D(1,\omega_1,p,\omega_2) = L^{1,\omega_1}(G) \cap L^{p,\omega_2}(G)$$

with the norm

$$\| f \|_D = \max \left(\| f \|_{1,\omega_1} , \| f \|_{p,\omega_2} \right)$$

is a Banach space and is the dual of $S(\infty, \omega_1^{-1}, p', \omega_2^{-1})$ where

$$S(\infty, \omega_1, p, \omega_2) = \{ g = g_1 + g_2 : (g_1, g_2) \in C_{\omega_1}(G) \times L^{p,\omega_2}(G) \}$$

with the norm

$$\|g\|_S = \inf\{\|g_1\|_{\infty,\omega_1} + \|g_2\|_{p,\omega_2}\}$$

where the infimum is taken over all the decompositions of g.

To obtain the space of multipliers on $D(p_1,\omega_1,p_2,\omega_2)$ as a certain dual, we define the space $\mathfrak{A}(p_1,\omega_1,p_2,\omega_2)$ to be set of all functions u which can be written in the form

$$u = \sum_{j=1}^{\infty} f_j * g_j$$

where $f_j \in \mathcal{K}(G) \subset D(p_1,\omega_1,p_2,\omega_2)$ and

$g_j \in S(p_1{}',\omega_1^{-1},p_2{}',\omega_2^{-1})$ with $\sum_{j=1}^{\infty} \|f_j\|_D \|g_j\|_S < \infty$.

Notice that $\mathcal{K}(G)$ is dense in $D(p_1,\omega_1,p_2,\omega_2)$, by Lemma 5. Define a norm $u \longrightarrow \|u\|$ by

$$\|u\| = \inf\left\{\sum_{j=1}^{\infty} \|f_j\|_D \|g_j\|_S\right\},$$

where the infimum is taken over all such representations of u. It is then easy to verify that $\|\cdot\|$ defines a norm on $\mathfrak{A}(p_1,\omega_1,p_2,\omega_2)$ and that a Banach space results.

Let $C_{\omega_1,\omega_2}(G)$ denote the space of all functions h which can be written as

$$h = h_1 + h_2, \quad (h_1, h_2) \in C_{\omega_1}(G) \times C_{\omega_2}(G),$$

with a definition of norm given by

$$|||h||| = \inf\left\{\|h_1\|_{\infty,\omega_1} + \|h_2\|_{\infty,\omega_2}\right\}$$

where the infimum is taken over all such decompositions of h, it follows that $\mathcal{C}_{\omega_1, \omega_2}(G)$ is a Banach space under $|||\cdot|||$.

Now let $f \in \mathcal{K}(G)$ and $g \in S(p_1', \omega_1^{-1}, p_2', \omega_2^{-1})$. Let $g = g_1 + g_2$ be a decomposition of g. Since $f \in L^{p_i, \omega_i}(G), i=1,2$, it follows that $f * g_1 \in \mathcal{C}_{\omega_1^{-1}}(G)$ and $f * g_2 \in \mathcal{C}_{\omega_2^{-1}}(G)$. Furthermore

$$\|f * g_1\|_{\infty, \omega_1^{-1}} + \|f * g_2\|_{\infty, \omega_2^{-1}} \leq$$

$$\leq \|f\|_{p_1, \omega_1} \|g_1\|_{p_1', \omega_2^{-1}} + \|f\|_{p_2, \omega_2} \|g_2\|_{p_2', \omega_2^{-1}}$$

$$\leq \|f\|_D \left(\|g_1\|_{p_1', \omega_2^{-1}} + \|g_2\|_{p_2', \omega_2^{-1}} \right)$$

so that $f * g \in \mathcal{C}_{\omega_1^{-1}, \omega_2^{-1}}(G)$ and

$$|||f * g||| \leq \|f\|_D \|g\|_S .$$

From this it is clear that $\mathfrak{A}(p_1, \omega_1, p_2, \omega_2) \subset \mathcal{C}_{\omega_1^{-1}, \omega_2^{-1}}(G)$ and that the topology on $\mathfrak{A}(p_1, \omega_1, p_2, \omega_2)$ is not weaker than the topology induced from $\mathcal{C}_{\omega_1^{-1}, \omega_2^{-1}}(G)$.

We say that T is a multiplier on $D(p_1, \omega_1, p_2, \omega_2)$ if T is a bounded linear operator on $D(p_1, \omega_1, p_2, \omega_2)$ which commutes with translation. The space of all multipliers on $D(p_1, \omega_1, p_2, \omega_2)$ is denoted by $M(D(p_1, \omega_1, p_2, \omega_2))$

THEOREM 2. <u>Let</u> G <u>be a locally compact abelian group and let</u> $1 < p_1, p_2 < \infty$. <u>If</u> $\omega_1, \omega_2 \in \Omega_0$, <u>then the space of multipliers</u> $M(D(p_1, \omega_1, p_2, \omega_2))$ <u>is isometrically isomorphic to</u> $\mathfrak{A}(p_1, \omega_1, p_2, \omega_2)^*$, <u>the conjugate space of</u> $\mathfrak{A}(p_1, \omega_1, p_2, \omega_2)$.

<u>Proof</u>. For any $T \in M(D(p_1, \omega_1, p_2, \omega_2))$ define

$$t(u) = \sum_{j=1}^{\infty} Tf_j * g_j(0)$$

for $u = \sum_{j=1}^{\infty} f_j * g_j$ in $\mathfrak{A}(p_1,\omega_1,p_2,\omega_2)$. First we show that t is well defined. To this end it is sufficient to show that if $u = \sum_{j=1}^{\infty} f_j * g_j = 0$ in $\mathfrak{A}(p_1,\omega_1,p_2,\omega_2)$ and $\sum_{j=1}^{\infty} \|f_j\|_D \|g_j\|_S < \infty$ then $\sum_{j=1}^{\infty} Tf_j * g_j(0) = 0$.

Let $\{\xi_\alpha\}$ be an approximate identity for $L^{1,\omega_1}(G)$ with $\|\xi_\alpha\|_1 = 1$ and $\{\eta_\beta\}$ an approximate identity for $L^{1,\omega_2}(G)$ with $\|\eta_\beta\|_1 = 1$. Let $\phi_\gamma = \phi_{(\alpha,\beta)} = \xi_\alpha + \eta_\beta - \xi_\alpha * \eta_\beta$ and give $\gamma = (\alpha,\beta)$ the usual product ordering. Now let $f \in \mathcal{K}(G)$ and consider $\phi_\gamma * f - f$. From

$$\phi_\gamma * f = \xi_\alpha * f + \eta_\beta * f - \xi_\alpha * \eta_\beta * f$$

it follows that

$$\|\phi_\gamma * f - f\|_{p_1,\omega_1} \le \|\xi_\alpha * f - f\|_{p_1,\omega_1} + \|(\eta_\beta * f) - \xi_\alpha * (\eta_\beta * f)\|_{p_1,\omega_1}$$

and

$$\|\phi_\gamma * f - f\|_{p_2,\omega_2} \le \|\eta_\beta * f - f\|_{p_2,\omega_2} + \|\xi_\alpha * f - \eta_\beta * (\xi_\alpha * f)\|_{p_2,\omega_2}.$$

Hence

$$\|\phi_\gamma * f - f\|_D \longrightarrow 0,$$

taking the limit over the index γ . Then

$$|T(\phi_\gamma * f_j) * g_j(0) - Tf_j * g_j(0)| \leq$$

$$\leq \|T\| \, \|\phi_\gamma * f_j - f_j\|_D \, \|g_j\|_S \longrightarrow 0,$$

so that

$$\lim_\gamma \; T(\phi_\gamma * f_j) * g_j(0) = Tf_j * g_j(0) .$$

Since $u = \sum_{j=1}^{\infty} f_j * g_j = 0$ and the series $\sum_{j=1}^{\infty} f_j * g_j$ converges uniformly we get as in Theorem 1

$$\sum_{j=1}^{\infty} T(\phi_\gamma * f_j) * g_j(0) = 0 .$$

We shall now show that $\sum_{j=1}^{\infty} T(\phi_\gamma * f_j) * g_j(0)$ converges uniformly with respect to γ .

We shall suppose that support of ξ_α is contained in a fixed compact set K_1 for all α and the support of η_β is contained in a fixed compact set K_2 for all β. Since (ξ_α) and (η_β) are bounded respectively in $L^{1,\omega_1}(G)$ and $L^{1,\omega_2}(G)$ there exist M_1, M_2 such that $\|\xi_\alpha\|_{1,\omega_1} \leq M_1$ and $\|\eta_\beta\|_{1,\omega_2} \leq M_2$. Set $M = M_1 + M_2$. If $y \in G$, then from the relations

$$\|\tau_y f\|_{p_1,\omega_1} \leq \omega_1(y) \; \|f\|_{p_1,\omega_1}$$

and

$$\|\tau_y f\|_{p_2,\omega_2} \leq \omega_2(y) \; \|f\|_{p_2,\omega_2}$$

it follows that the translation operator τ_y on $D(p_1,\omega_1,p_2,\omega_2)$ has a norm bounded by $\max\{\omega_1(y), \omega_2(y)\}$. Let m_1 and m_2 be the maxima of the continuous functions ω_1,ω_2 respectively on the compact set $K = K_1 + K_2$. Set $m = \max(m_1,m_2)$. Then m and M are independent of α and β and hence of γ . Now if $f \in \mathcal{K}(G)$

$$\|\phi_\gamma * f\|_{p_1,\omega_1} \leq \|\xi_\alpha * f\|_{p_1,\omega_1} + \|\eta_\beta * f\|_{p_1,\omega_1} + \|\xi_\alpha * \eta_\beta * f\|_{p_1,\omega_1}$$

$$\leq \|\xi_\alpha\|_{1,\omega_1}\|f\|_{p_1,\omega_1} + \|\eta_\beta * f\|_{p_1,\omega_1} + \|\xi_\alpha\|_{1,\omega_1}\|\eta_\beta * f\|_{p_1,\omega_1}.$$

We shall now calculate $\|\eta_\beta * f\|_{p_1,\omega_1}$. Now using the definition of $\eta_\beta * f$ and Minkowsky's inequality, we get

$$\|\eta_\beta * f\|_{p_1,\omega_1} \leq \int \eta_\beta(t)\,dt \left(\int |f(x-t)\,\omega_1(x)|^{p_1}\,dx\right)^{1/p_1}$$

$$\leq \sup_{t \in \operatorname{supp}\eta_\beta} \|\tau_t f\|_{p_1,\omega_1}$$

$$\leq \sup_{t \in \operatorname{supp}\eta_\beta} \|\tau_t\|\,\|f\|_{p_1,\omega_1} \leq m\|f\|_{p_1,\omega_1}.$$

Then

$$\|\phi_\gamma * f\|_{p_1,\omega_1} \leq \|\xi_\alpha\|_{1,\omega_1}\|f\|_{p_1,\omega_1} + \left[1 + \|\xi_\alpha\|_{1,\omega_1}\right] m\,\|f\|_{p_1,\omega_1}$$

$$\leq [M + m(1+M)]\,\|f\|_{p_1,\omega_1}.$$

Similarly for $f \in \mathcal{K}(G)$

$$\|\phi_\gamma * f\|_{p_2,\omega_2} \leq [M + m(1+M)]\,\|f\|_{p_2,\omega_2}.$$

Hence

$$\| \phi_\gamma * f \|_D \leq (M + m + mM) \|f\|_D .$$

Then

$$\left| \sum_{j=1}^{\infty} T(\phi_\gamma * f_j) * g_j(0) \right| \leq \sum_{j=1}^{\infty} \|T(\phi_\gamma * f_j)\|_D \|g_j\|_S$$

$$\leq \|T\| \sum_{j=1}^{\infty} \|\phi_\gamma * f_j\|_D \|g_j\|_S$$

$$\leq \|T\| (M + m + mM) \sum_{j=1}^{\infty} \|f_j\|_D \|g_j\|_S$$

and the convergence of $\sum_{j=1}^{\infty} T(\phi_\gamma * f_j) * g_j(0)$ is uniform with respect to γ . Hence

$$\sum_{j=1}^{\infty} Tf_j * g_j(0) = \lim_{\gamma} \sum_{j=1}^{\infty} T(\phi_\gamma * f_j) * g_j(0) = 0$$

since $T(\phi_\gamma * f_j) * g_j(0) \longrightarrow Tf_j * g_j(0)$ for each j. Thus t is well defined. It is clearly linear. It is easy to check as in Theorem 1 that the mapping $T \longrightarrow t$ is an isometry.

To see that $T \longrightarrow t$ is onto, we proceed as follows. Let $t \in \mathfrak{A}(p_1,\omega_1,p_2,\omega_2)^*$ and let $f \in \mathcal{K}(G) \subset D(p_1,\omega_1,p_2,\omega_2)$. Define

$$g \longrightarrow t(f * g) \qquad g \in S(p_1', \omega_1^{-1}, p_2', \omega_2^{-1}).$$

Then $|t(f * g)| \leq \|t\| \|f\|_D \|g\|_S$ implies that this mapping gives a bounded linear functional on $S(p_1', \omega_1^{-1}, p_2', \omega_2^{-1})$. Hence there exists a unique element, denoted by Tf , in $D(p_1,\omega_1,p_2,\omega_2)$ such that

$$Tf * g(0) = t(f * g) \qquad g \in S(p_1', \omega_1^{-1}, p_2', \omega_2^{-1})$$

and $\| Tf \|_D \leq \| t \| \ \| f \|_D$. Hence T is a bounded operator from $\mathcal{K}(G)$ into $D(p_1,\omega_1,p_2,\omega_2)$. It is clearly linear. Since $\mathcal{K}(G)$ is dense in $D(p_1,\omega_1,p_2,\omega_2)$, it can be extended uniquely as a bounded linear operator on $D(p_1,\omega_1,p_2,\omega_2)$. We have to prove that this extended T is a multiplier. Let $y \in G$ and $f \in \mathcal{K}(G)$. If $g \in S(p_1', \omega_1^{-1}, p_2', \omega_2^{-1})$, then

$$\tau_y Tf * g(0) = Tf * \tau_y g(0) = t(f * \tau_y g)$$

$$= t(\tau_y f * g) = T\tau_y f * g(0)$$

holds for all functions g in $S(p_1', \omega_1^{-1}, p_2', \omega_2^{-1})$. Hence

$$\tau_y T f = T \tau_y f \qquad f \in \mathcal{K}(G)$$

from which the result follows and $T \in M(D(p_1,\omega_1,p_2,\omega_2))$. This completes the proof.

For the case when one of p_i's is 1, we define the space $\mathfrak{A}(1,\omega_1,p,\omega_2)$ to be the set of all those functions u which can be expressed as

$$u = \sum_{i=1}^{\infty} f_i * g_i \qquad f_i \in \mathcal{K}(G),\ g_i \in S(\infty, \omega_1^{-1}, p', \omega_2^{-1})$$

with $\sum_{i=1}^{\infty} \| f_i \|_D \| g_i \|_S < \infty$, with the norm defined by

$$\| u \| = \inf \left\{ \sum_{i=1}^{\infty} \| f_i \|_D \| g_i \|_S \right\}$$

where the infimum is taken over all such representations of u, $\mathfrak{A}(1,\omega_1,p,\omega_2)$ becomes a Banach space.

Quite analogously, we can prove

THEOREM 3. <u>Let</u> G <u>be a locally compact abelian group. Suppose</u> $1 < p < \infty$ <u>and</u> $\omega_1;\omega_2 \in \Omega_0$. <u>Then the multiplier space</u> $M(D(1,\omega_1,p,\omega_2))$ <u>is isometrically isomorphic to the dual</u> $\mathfrak{A}(1,\omega_1,p,\omega_2)^*$ <u>of</u> $\mathfrak{A}(1,\omega_1,p,\omega_2)$.

References.

1 Edwards, R.E., The stability of weighted Lebesgue spaces, Trans.Amer.Math.Soc. 93 (1959) 369-394.

2 Figa-Talamanca, A., Multipliers of p-integrable functions, Bull.Amer.Math.Soc. 70 (1964) 666-669.

3 ------------ , Translation invariant operators in L^p . Duke Math. J. 32 (1965) 495-502

4 Figa-Talamanca,A. and Gaudry, G.I., Density and representation theorems for multipliers of type (p,q), J.Austr.Math. Soc. 7 (1967) 1-6.

5 Gaudry, G.I., Quasi-measures and operators commuting with convolutions, Pac.J.Math. 18 (1966) 505-514.

6 ------------, Multipliers of weighted Lebesgue and measure spaces, Proc.Lond.Math.Soc. (3) 19 (1969) 327-340.

7 Kree, P., Sur les multiplicateurs dans FL^p avec poids, Ann. Inst. Fourier 2 (1966) 91-121.

8 Lui, T.S. and Wang, J.K., Sums and intersections of Lebesgue spaces, Math. Scand. 23 (1968) 241-251.

9 Rieffel, M.A., Multipliers and Tensor products of L^p-spaces of locally compact groups, Stud. Math. 33 (1969) 71-88.

10 Spector, R., Sur la structure locale des groupes abeliens localement compacts, Bull.Soc. Math. France, Memoire 24 (1970)

ON PROPERTIES OF TRACES OF FUNCTIONS BELONGING TO WEIGHT SPACES

L.D. Kudryavtsev

We shall consider weight spaces of functions defined on some regions of Euclidian space. A functional space is called a weight space if it consists of functions having finite norms or, more generally, semi-norms with some weight, i.e. with some multiplier.

1. An example

Before proceeding to the investigation of some properties of functions belonging to weight spaces let us consider an example (which will be of use later on) indicating advisability of introducing the weight spaces.

Let us take the following theorem by S.V.Uspenskiĭ [1] : let function u(x), $x=(x_1,\dots,x_n)$, have derivatives on the whole Euclidian space E^n and let absolute values of these derivatives be integrable to the power p, $n>p$:

$$\mathcal{D}_p(u) = \int \sum_{i=1}^{n} \left| \frac{\partial u}{\partial x_i} \right|^p dE^n < +\infty \tag{1}$$

Then the function u(x) can be modified on a set of zero measure so that after this modification the function u(x) tends to a constant, when its argument tends to infinity along a radius, and this constant does not depend on a choice of any direction of the radius. This theorem is an analog of the Liouville theorem for analytic functions. In fact, the essence of Uspenskiĭ's theorem consists in the following: out of the finiteness of a semi-norm of the function it follows that this function is "almost constant" at infinity.

According to this theorem, in particular, each function with the finite Dirichlet integral

$$\mathcal{D}_2(u) = \int \sum_{i=1}^{n} \left(\frac{\partial u}{\partial x_i}\right)^2 dE^n, \qquad n \geq 3 \tag{2}$$

tends to the same constant at the infinity in the sense indicated above.

The statement of the Uspenskiĭ's theorem is connected with the following: If one changes variables $x_1, \ldots, x_n$ to the spherical variables ρ, $\varphi_1, \ldots, \varphi_{n-1}$ (ρ is radius, $\varphi_1, \ldots, \varphi_{n-1}$ are corresponding angles) then according to the rule for effecting a change of variables and to the rule for differentiating of composite functions some multipliers will appear at the derivatives. Consequently, the derivatives $\frac{\partial u}{\partial \rho}$, $\frac{\partial u}{\partial \varphi_i}$, $i=1,2,\ldots,n-1$, will be integrable with some weights. One can prove that the weight at the derivative $\frac{\partial u}{\partial \rho}$ ensures limiting values of the function $u(x)$ at infinity and the weights at the derivatives $\frac{\partial u}{\partial \varphi_i}$, $i=1,2,\ldots,n-1$, ensure the independence of these limiting values from a choice of some direction.

Thus in studying the behaviour of the functions, which have the finite integral (1), at infinity, weight spaces emerge in a natural way. It concerns, in particular, functions with the finite integral (2) and therefore the Laplace equation

$$\Delta u = 0$$

(see [2]) because it is the Euler equation for the functional (2). It is necessary to explain what is the meaning of limiting (or boundary) values for functions belonging to functional spaces, which are given by finiteness of some integrals, in particular, by finiteness of the integral (1).

For the sake of simplicity we shall give a needed definition for the case of (n - 1)-dimensional hyperplane.

Let $u=u(x)$, $x=(x_1,\dots,x_n)$, be a function defined on the half-space

$$\overset{+}{E}{}^n = \{ x : x_n > 0 \} \tag{3}$$

A function $\varphi = \varphi(x^{(n-1)})$, $x^{(n-1)}=(x_1,\dots,x_{n-1})$, defined on the hyperplane

$$E^{n-1} = \{ x : x_n = 0 \} \quad , \tag{4}$$

is called <u>the trace</u> (or the limiting values or boundary values) of the function $u=u(x)$ on the hyperplane E^{n-1} in the sense of the space L_p, if after a possible modification of the function $u(x)$ on a set of zero measure, the condition

$$\lim_{x_n \to +0} \int |f(x^{(n-1)}, x_n) - \varphi(x^{(n-1)})|^p \, dE^{n-1} = 0$$

is valid.

It can be proved that the defined trace of the function does not depend on a choice of an indicated modification of the given function on a set of zero measure. It means that the trace of a function does not change if one substitutes an equivalent function for the given function. In this sense the traces of functions are stable.

It is evident that in the same way one can easily define a trace of function on a hyperplane of any dimension m, $m = 0,1,2,\dots, n-1$ and at infinity.

The operator which assigns to the function its trace is a linear operator.

In the cases to be considered in the present paper the indicated notion of the trace is equivalent to that of the trace in the sense of convergence almost everywhere.

It is interesting to give conditions, when there exists a trace of function, and to describe its local and global properties. It is convenient to do in terms of imbedding theorems. Therefore let us make it clear how we understand the term ''imbedding''.

The norm of element x in a Banach space B will be denoted by $|x, B|$.

If two Banach spaces B_1 and B_2 are given, then a bounded operator A mapping the space B_1 into the space B_2, i.e. such an operator for which there exists a constant $c > 0$ that the inequality

$$|Ax, B_2| \leq c\, |x, B_1|, \quad \forall\, x \in B_1$$

is valid, is called <u>the imbedding operator</u> of the space B_1 into the space B_2. If such an operator exists then one says that the space B_1 is imbedded into the space B_2 and one writes

$$B_1 \longrightarrow B_2.$$

Let us introduce base functional weight spaces, which we need for our aims.

2. Definitions and notations

We shall confine ourselves to the consideration of only the case of functions defined on the whole space E^n or on its half-space $\overset{+}{E}{}^n$.

Let $\varphi = \varphi(x)$, $x = (x_1,\dots,x_n)$, be a non-negative measurable function defined on the half-space $\overset{+}{E}{}^n$ (see (3)). For each integer τ_i (i is fixed, $1 \le i \le n$) and for each p, $1 \le p \le +\infty$, we shall denote by $L^{\tau_i}_{p,\varphi} = L^{\tau}_{p,\varphi}(\overset{+}{E}{}^n)$ the linear semi-normed space of functions f, which have generalized derivatives of the order τ_i with respect to the variable x_i belonging to the space $L_p(\overset{+}{E}{}^n)$ with weight φ :

$$\left| f, L^{\tau_i}_{p,\varphi}(\overset{+}{E}{}^n)\right| = \left\{ \int\!\!\int \left|\varphi(x)\, \frac{\partial^{\tau_i} f}{\partial x_i^{\tau_i}}\right|^p d\overset{+}{E}{}^n \right\}^{1/p} < +\infty . \tag{5}$$

The function φ is called <u>weight</u> or <u>weight function</u> in this case.

Let us introduce the following notation

$$L^{\tau_{i_1}\dots\tau_{i_m}}_{p,\varphi_{i_1}\dots\varphi_{i_m}} = \bigcap_{j=1}^{m} L^{\tau_{i_j}}_{p,\varphi_{i_j}} \tag{6}$$

$$\left| f, L^{\tau_{i_1}\dots\tau_{i_m}}_{p,\varphi_{i_1}\dots\varphi_{i_m}} \right| = \sum_{j=1}^{m} \left| f, L^{\tau_{i_j}}_{p,\varphi_{i_j}} \right|$$

where admissable values for integer m are $1,2,\dots,n$. In the case when $m = n$, $\tau_1 = \dots = \tau_n = \tau$ and $\varphi_1(x) = \dots = \varphi_n(x) = \varphi(x)$ instead of $L^{\tau_1\dots\tau_n}_{p,\varphi_1\dots\varphi_n}(\overset{+}{E}{}^n)$ we will merely write $L^{(\tau)}_{p,\varphi}(\overset{+}{E}{}^n)$.

In a similar way one can define the spaces $L^{\tau_{i_1}\dots\tau_{i_m}}_{p,\varphi_{i_1}\dots\varphi_{i_m}}(E^n)$, i.e., consider the case, when functions are defined on the whole space E^n, but not on the half-space $\overset{+}{E}{}^n$.

The functional weight spaces of the type $L^{r_{i_1} \dots r_{i_m}}_{p, \varphi_{i_1} \dots \varphi_{i_m}}$ i.e. the functional spaces of functions norms or semi-norms of which are defined by some weight and which have generally speaking, different properties in the directions of different coordinate axes, are called the <u>anisotropic weight spaces</u>.

In the case when the weights $\varphi_i(x)$ are equal to unity, the space $L^{r_{i_1} \dots r_{i_m}}_{p, \varphi_{i_1} \dots \varphi_{i_m}}$ transforms into the ordinary Sobolev space $L^{r_{i_1} \dots r_{i_m}}_{p}$.

Let us first of all indicate a condition of existence of function trace in terms of the properties of the weight.

We shall say that a function $\varphi(x)$ is the weakly degenerating function at the hyperplane E^{n-1}, if for a certain b , $0 < b \leq +\infty$, and for each finite a the inequality

$$\int_{-a}^{a} \dots \int_{-a}^{a} dx_1 \dots dx_{n-1} \int_0^b \varphi^q(x)\, dx < +\infty \qquad (7)$$

is valid, where $\frac{1}{p} + \frac{1}{q} = 1$, $1 < p < +\infty$.

If

$$f \in L^{1_n}_{p,\varphi}(\overset{+}{E}{}^n)$$

and the weight function φ is a weakly degenerating function at the hyperplane E^{n-1}, then there exists a trace of function f on this hyperplane E^{n-1}. If the weight function φ is not a weakly degenerating one, then there exist functions $f \in L^{1_n}_{p,\varphi}(\overset{+}{E}{}^n)$ which have no traces on the hyperplane E^{n-1}.

For each function $g(x)$ of one variable x, $x \in E^1$, let us denote by $\Delta^k(h)\ g(x)$ the k-th order difference of function $g(x)$ with the step h:

$$\Delta^k(h)\ g(x) = \sum_{j=0}^{k} (-1)^{j+k} \binom{k}{j}\ g(x+jh)$$

If function $f = f(x)$ is a function defined on the whole space E^n then by $\Delta_i^k (h)\ f(x)$ we shall denote the k-th order difference of the function $f(x)$ with respect to the variable x_i with the step h $(i = 1,2,\dots,n)$.

To describe properties of function traces we shall need to consider the spaces of functions, which have partial derivatives satisfying the integral Hölder's condition. Let us define these spaces more explicitly.

Let $1 \leq p \leq +\infty$, $\tau > 0$, $\tau = \bar{\tau} + \alpha$, where $\bar{\tau}$ is a non-negative integer and $0 < \alpha \leq 1$.

We shall say that function $u = u(x)$, $x \in E^n$, belongs to space $H^{\tau}_{px_i}$ (i is fixed, $i = 1,2,\dots,n$) if it has generalized unmixed partial derivatives $\frac{\partial^k u}{\partial x_i^k}$ ($k = 0,1,\dots,\bar{\tau}$) which are p-th power integrable on E^n, and, in addition, the higher derivative satisfies the following condition (integral Hölder condition of the order α):

$$\left| \Delta_i^2(h) \frac{\partial^{\bar{\tau}} u(x)}{\partial x_i^{\bar{\tau}}} , L_p(E^n) \right| \leq M |h|^{\alpha} \tag{8}$$

(M is a constant depending on the function $u = u(x)$).

In the case $0 < \alpha < 1$ we shall have an equivalent definition, if we substitute the first difference for the second one in condition (8).

It is possible to prove (see [3]) that the function $u(x)$ belongs to the space $H^{\tau}_{px_i}$ if and only if for all k and s such that $k > \tau - s > 0$ the following condition

$$\left| \Delta_i^k(h) \frac{\partial^s u(x)}{\partial x_i^s}, L_p(E^n) \right| \leq M |h|^{\tau - s}$$

is valid (here M is, generally speaking, some other constant, than the constant M in inequality (8)).

Further, by definition (see [3]):

$$H_p^{\tau_1 \dots \tau_n} = \bigcap_{i=1}^{n} H^{\tau_i}_{px_i} ,$$

$$H_p^{(\tau)} = H_p^{\tau \dots \tau}$$

3. Properties of function traces on a hyperplane

Let us consider functions defined on the half-space $\overset{+}{E}{}^n$ (see (3)) and belonging to some weight spaces. We shall study characteristics of their traces on the hyperplane E^{n-1} (see (4)).

We shall assume here that weight functions (probably with some indexes) depend only on one variable x_n and that all weight functions satisfy the condition of weak degeneration, i.e. the condition (7), which in this case is equivalent to the following one : there exists b , $0 < b < +\infty$ such that

$$\int_0^b \varphi^{-q}(x_n)\, dx_n < +\infty , \qquad \frac{1}{p} + \frac{1}{q} = 1 . \tag{9}$$

For the description of the characteristics of function traces it is convenient to consider functions

$$G_i(t) = \frac{1}{t}\left[\int_0^t \varphi^{-q}(x_n)\, dx_n \right]^{1/q} \tag{10}$$

($i = 1,2,\dots,n$) besides weight function $\varphi_i(x_n)$. In terms of these functions it is possible to describe both differential properties of function traces belonging to weight spaces and the rate of the convergence of a function to its trace.

The following theorem can be proved (see [2] , [3] , [5]):

<u>THEOREM 1</u>. <u>Let</u> $\tau_j \geq 1$ (j <u>be fixed</u>, $j = 1,2,\dots,n-1$), $\tau_n \geq 1$, $0 < b \leq +\infty$

$$\int_0^b \varphi_j^{-q}(t)\,dt < +\infty, \qquad \int_0^b \varphi_n^{-q}(t)\,dt < +\infty \tag{11}$$

<u>If</u> $f \in L^{\tau_j \tau_n}_{p,\varphi_j\varphi_n}(\overset{+}{E}{}^n)$ <u>then for each</u> $t > 0$ <u>and</u> $k > \tau_j$ <u>the following inequality is valid</u>:

$$\begin{aligned} &\left| \Delta_j^k(h) f,\ L_p(E^{n-1}) \right| \leq \\ &\leq c \left| f,\ L^{\tau_j \tau_n}_{p,\varphi_j\varphi_n} \right| \left[|h|^{\tau_j} G_j(t) + t^{\tau_n} G_n(t) \right] \end{aligned} \tag{12}$$

Let us note that under available assumptions

$$\lim_{t \to +0} t^{\tau_n} G_n(t) = 0 \tag{13}$$

It follows from (11) and the condition $\tau_n \geq 1$.

Therefore if we suppose in addition to the conditions of Theorem 1, that

$$\lim_{t \to +0} G_j(t) = 0$$

then the right-hand side of inequality (12) tends to zero, when its argument t tends to zero as well. Consequently, the left-hand side of this inequality, which is independent of the variable t, is equal to zero. Hence it follows, that for all h and for almost all points $x^{(n-1)} = (x_1, \dots, x_{n-1})$ on the hyperplane E^{n-1} the equality

$$\Delta_j^k (h) \; f(x^{(n-1)}, 0) = 0$$

is valid. From this condition (at any rate for locally integrable functions) it follows that the function f is a polynomial with respect to the variable x_j and that the order of this polynomial is not higher than $\tau_j - 1$.

Thus we obtain

COROLLARY 1. *If in addition to the conditions of Theorem 1 the condition* (13) *is fulfilled, then the trace* $f(x^{(n-1)}, 0)$ *of the function* f *is a polynomial with respect to the variable* x_j *and the order of this polynomial is not higher than* $\tau_j - 1$.

In case, when the weight functions $\varphi_j(x)$ and $\varphi_n(x)$ are power functions

$$\varphi_j (t) = t^{\alpha_j} , \qquad \varphi_n (t) = t^{\alpha_n} \tag{14}$$

then the condition (9) of weak degeneration of weight functions gives the inequalities

$$\alpha_j < \frac{1}{q} , \quad \alpha_n < \frac{1}{q} , \quad \frac{1}{p} + \frac{1}{q} = 1$$

and the condition (13) gives the inequality

$$\alpha_j < - \frac{1}{p} .$$

The parametric inequality (12) gives us a chance to estimate an integral continuity modulus of traces of functions belonging to the space $L^{\tau_j \, \tau_n}_{p, \varphi_j \varphi_n} (\overset{+}{E}{}^n)$. Namely, let us suppose in addition to the given assumptions, that the function $G_j(t)$ is a strictly decreasing one and denote by t(h) the function inverse to the function

$$h = t^{\frac{\tau_n}{\tau_j}} G_n^{\frac{1}{\tau_j}}(t)\, G_j^{-\frac{1}{\tau_j}}(t) . \tag{15}$$

This has sense, as the function

$$t^{\frac{\tau_n}{\tau_j}} G_n^{\frac{1}{\tau_n}}(t)$$

is a strictly increasing one under our assumption and, consequently, the function in the right-hand side of the formula (15) also strictly increases in virtue of the strict decrease of the function $G_j(t)$.

Taking $t = t(|h|)$ we have

$$|h|^{\tau_j} G_j(t) = t^{\tau_n} G_n(t) ,$$

then putting $t = t(|h|)$ in (12) we obtain

$$|\Delta_j^k (h) f, L_p(E^{n-1})| \leq$$

$$\leq c \, |f, L^{\tau_j \tau_n}_{p, \varphi_j \varphi_n}| \, G_j[t(|h|)] \, |h|^{\tau_j} \tag{16}$$

where the constant c does not depend on the function f.

Thus we have

<u>COROLLARY 2</u>. <u>If in addition to the assumption of Theorem 1 function</u> $G_j(t)$ <u>decreases strictly then the inequality</u> (16) <u>is valid</u>.

This inequality is a generalization of well-known properties of function traces studied before in the more simple cases : the case when weight is absent, i.e. when $\varphi_j(x) = \varphi_n(x) = 1$, was considered in a monograph by S.L.Sobolev [6] and the case of power weights (14) was considered in a paper by S.V.Uspenskiǐ (see [7]). The relations between properties of weights and estimates of differences of function trace are of rather complicated nature, which one can see well in the case of power weights (14). Here

$$G_j(t) = O(t^{-\alpha_j - \frac{1}{p}}) \qquad t \to +0,$$

$$G_n(t) = O(t^{-\alpha_n - \frac{1}{p}}) \qquad t \to +0;$$

therefore (see (15))

$$t(h) = O\left(h^{\frac{\tau_j}{\tau_n + \alpha_j - \alpha_n}}\right), \qquad t \to +0.$$

Consequently, if $\tau_n + \alpha_j - \alpha_n \neq 0$, then from (15) we have

$$|\Delta_j^k(h) f, L_p(E^{n-1})| \leqslant c\, |f, L^{\tau_j\ \tau_n}_{p,\varphi_j\ \varphi_n}|\, |h|^{\mu}$$

where

$$\mu = \frac{\tau_j\,(\tau_n - \alpha_n - 1/p)}{\tau_n + \alpha_j - \alpha_n}.$$

In other words:

$$L^{\tau_j\ \tau_n}_{p,x_n^{\alpha_j}\ x_n^{\alpha_n}}(\overset{+}{E}{}^n) \longrightarrow H_p^{\mu}(E^{n-1})$$

(see § 2 and [7]).

If $\tau_j = \tau$ and $\alpha_j = \alpha$, $j=1,2,\dots,n$ then $\mu = \tau - \alpha - \frac{1}{p}$

i.e. $L^{\tau\dots\tau}_{p,x_n^{\alpha}\dots x_n^{\alpha}}(\overset{+}{E}{}^n) \to H_p^{\tau-\alpha-\frac{1}{p}}(E^{n-1})$(see [10]).

One should yet note the following : if we are to consider weight functions similar, in a certain sense, to the power ones, then in this case it is possible (see [2], [4]) to obtain results analogous to Theorem 1 and its Corollaries in some more accurate terms like so-called B-spaces (see [3]), in which the semi-norm is defined by an ''integrated Hölder condition''.

It was mentioned above that the functions $G_n(t)$ (see (10)) give a possibility to estimate the rate of the converging of a function to its trace. In fact, an estimate is valid : if

$$f \in L^{1_n}_{p,\varphi_n}(\overset{+}{E}{}^n)$$

and the weight function φ_n depends on one variable x_n and satisfies the condition of weak degeneration (9), then the function f can be modified on a set of zero measure so that after this modification the following estimate holds true :

$$\left[\int | f(x^{(n-1)}, h) - f(x^{(n-1)}, 0) |^p \, dE^{n-1} \right]^{\frac{1}{p}} =$$

$$= o(hG_n(h)), \; h \longrightarrow +0.$$

4. Degeneracy of a function into a trigonometrical polynomial at infinity

S.L.Sobolev (see [8]) proved that if

$$u(x) \in L_p^{\tau}(E^n)$$

τ is integer, and $n > \tau p$, then there exists a polynomial $P(x)$ the degree of which is not higher than $\tau - 1$ and such a modification of $u(x)$ on a set of zero measure, that after this modification we have

$$\lim_{|x| \to +\infty} [u(x) - P(x)] = 0, \quad |x| = \sqrt{x_1^2 + \dots + x_n^2}$$

where x tends to infinity along radii.

The result by S.Uspenskiĭ mentioned above in §1 is a particular case of this theorem. Generalization of Sobolev's theorem was given by V.N.Sedov (see [9]).

From Sobolev's theorem one can easily obtain the following result.

THEOREM 2. *Let the function* $u = u(x)$ *have all generalized partial derivatives to the order* τ *inclusive*, (τ *is an integer*) *and let*

$$\mathcal{D}^*_\tau(u) = \int\limits_{|x|\geq 1} \sum_{m=0}^{\tau} \sum_{|\mu|=m} \left| x^{m-1} \mathcal{D}^\mu u(x)\right|^p dx < +\infty$$

$$\mu = (\mu_1, \dots, \mu_n), \quad |\mu| = \mu_1 + \mu_2 + \dots + \mu_n$$

$$\mathcal{D}^\mu = \frac{\partial^{|\mu|}}{\partial x_1^{\mu_1} \dots \partial x_n^{\mu_n}}, \qquad n > \tau p.$$

Then there exists a trigonometrical polynomial $T(\varphi_1, \dots, \varphi_{n-1})$ ($\varphi_1, \dots, \varphi_{n-1}$ *are the angles, which the radius vector of a point forms with the coordinate axes) the degree of which is not higher than* $p - 1$ *and a modification of the function* $u(x)$ *on a set of zero measure such that after this modification*

$$\lim_{|x| \to +\infty} u(x) = T(\varphi_1, \dots, \varphi_{n-1})$$

where x *tends to the infinity along radii*.

Proof. Let us set $v(x) = |x|^{\tau-1} u(x)$ and

$$\mathcal{D}_\tau(v) \equiv \int\limits_{|x|\geq 1} \sum_{|\mu|=\tau} \left| \mathcal{D}^\mu u(x)\right|^p dx$$

Since

$$\mathcal{D}^{\tau} v(x) = \sum_{m=0}^{\tau} \sum_{|\mu|=m} O(|x|^{m-1})\, \mathcal{D}^{\mu} u(x), \qquad |x| \to +\infty$$

there exists a constant $c > 0$ such that

$$\mathcal{D}_{\tau}(v) \le c\, \overset{*}{\mathcal{D}}_{\tau}(u)$$

According to Sobolev's theorem there exists such a polynomial $P(x)$, the degree of which is not higher than $\tau - 1$, that after a corresponding modification of the function $v = v(x)$ we have

$$\lim_{|x| \to +\infty} [v(x) - P(x)] = 0$$

where x tends to the infinity along radii. If we now change the variable $x_1, \ldots, x_n$ to the spherical variables $\rho, \varphi_1, \ldots, \varphi_{n-1}$ (ρ is a radius and $\varphi_1, \ldots, \varphi_{n-1}$ are angles) in polynomial $P(x)$ and put

$$T(\varphi_1, \ldots, \varphi_{n-1}) = \frac{P(x)}{\rho^{\tau-1}}$$

then $T(\varphi_1, \ldots, \varphi_{n-1})$ is a desirable trigonometrical polynomial and moreover

$$u(x) - T(\varphi_1, \ldots, \varphi_{n-1}) = o\left(\frac{1}{\rho^{\tau-1}}\right), \quad \rho \to +\infty .$$

Theorem 2 is proved.

It is possible to give conditions from which it follows, that a polynomial, into which a function degenerates at infinity, does not depend on the angles $\varphi_1, \ldots, \varphi_{n-1}$. Let us formulate this theorem in a model form for two variables.

THEOREM 3. *Let the function* $u = u(\rho, \varphi)$ *satisfy the condition*

$$\int_0^{2\pi} d\varphi \int_1^{+\infty} \left[\left| \frac{\partial^\tau u}{\partial \rho^\tau} \right|^p + \sum_{k=1}^{\tau} \frac{1}{\rho^{(\tau-k+1)p}} \left| \frac{\partial^k u}{\partial \rho^{k-1} \partial \varphi} \right|^p \right] \rho^{n-1} d\rho < +\infty$$

and let $n > \tau p$, then there exists such a polynomial

$$P(\rho) = a_0 + a_1\rho + \dots + a_{\tau-1}\rho^{\tau-1} \tag{17}$$

that after possible modification of the function $u(\rho, \varphi)$ on a set of zero measure one has:

$$\lim_{\rho \to +\infty} \left[u(\rho, \varphi) - P(\rho) \right] = 0. \tag{18}$$

The proof of this theorem is based on the following three lemmas:

LEMMA 1. If

$$\int_1^{\infty} \rho^{\alpha} \left| \frac{\partial f(\rho)}{\partial \rho} \right|^p d\rho < +\infty$$

and $\alpha > p - 1$, then there exists a limit

$$\lim_{\rho \to +\infty} f(\rho).$$

The conditions of this lemma means that the condition of weak degeneration at infinity is fulfilled, and therefore there exists the trace of the function $f(\rho)$ at infinity.

LEMMA 2. If

$$\int_1^{+\infty} \rho^{\beta} \left| \frac{\partial f(\rho, \varphi)}{\partial \varphi} \right|^p d\rho < +\infty , \quad \beta > -1$$

and if for each $\varphi \in [0, 2\pi]$ there exists a limit

$$\lim_{\rho \longrightarrow +\infty} f(\rho, \varphi)$$

then this limit does not depend on φ.

Proof of Lemma 2. Let us consider the difference

$$f(\rho, \varphi'') - f(\rho, \varphi') = \int_{\varphi'}^{\varphi''} \frac{\partial f(\rho, \varphi)}{\partial \varphi} d\varphi$$

where φ'' and φ' are arbitrary but fixed. If one integrates this equality with respect to the variables φ' and φ'' over the interval $[0, 2\pi]$, takes the lower bound of the left-hand side for $a \le \rho \le 2a$ where a is arbitrary positive number, and applies the Hölder inequality then one obtains

$$\inf_{a \le \rho \le 2a} \int_0^{2\pi} d\varphi' \int_0^{2\pi} |f(\rho, \varphi'') - f(\rho, \varphi')| \, d\varphi'' \le$$

$$\le \frac{4\pi^2}{a} \left(\int_a^{2a} d\rho \int_0^{2\pi} \rho^{\beta} \left| \frac{\partial f}{\partial \varphi} \right|^p d\varphi \right)^{\frac{1}{p}} \left(\int_0^{2\pi} d\varphi \int_a^{2a} \rho^{-\frac{\beta}{p-1}} d\rho \right)^{\frac{1}{q}} \le$$

$$\le 2^{3+\frac{1}{q}} \pi^{2+\frac{1}{q}} \left(\int_0^{2\pi} d\varphi \int_a^{2a} \rho^{\beta} \left| \frac{\partial f}{\partial \varphi} \right|^p d\varphi \right)^{\frac{1}{p}} \left(\int_a^{2a} \rho^{-\frac{\beta+p}{p-1}} d\rho \right)^{\frac{1}{q}} \tag{19}$$

If $a \longrightarrow +\infty$, then under the condition of the lemma the right-hand side of the inequality tends to zero. Let us choose such a sequence a_k, $k = 1,2,\ldots$, that the right-hand side of (19) be less than $\frac{1}{k}$ for $a = a_k$, then

$$\inf_{a_k \le \rho \le 2a_k} \int_0^{2\pi} d\varphi' \int_0^{2\pi} |f(\rho, \varphi'') - f(\rho, \varphi')| \, d\varphi'' < \frac{1}{k}$$

Therefore there exists a sequence ρ_k, $k=1,2,\ldots$, such that

$$\lim_{k \to +\infty} \rho_k = +\infty \quad \text{and}$$

$$\lim_{k \to +\infty} \int_0^{2\pi} d\varphi' \int_0^{2\pi} |f(\rho_k, \varphi'') - f(\rho_k, \varphi')|^p d\varphi'' = 0 . \qquad (20)$$

Since the limit

$$\lim_{k \to +\infty} \left[f(\rho_k, \varphi'') - f(\rho_k, \varphi') \right]$$

exists as well, it follows from (20) that this limit is equal to zero, i.e.

$$\lim_{k \to +\infty} f(\rho_k, \varphi'') = \lim_{k \to +\infty} f(\rho_k, \varphi')$$

This means that the limit $\lim_{\rho \to +\infty} f(\rho, \varphi)$ does not depend on φ .

Lemma 2 is proved.

LEMMA 3. *If the limit*

$$\lim_{\rho \to +\infty} f(\rho) = A$$

exists,

$$\int_1^{+\infty} \rho^\lambda \left| \frac{df}{d\rho} \right|^p d\rho < +\infty$$

and $\lambda > p - 1$, *then*

$$\int_1^{+\infty} \rho^{\lambda - p} |f(\rho) - A|^p d\rho \le \left(\frac{p}{p-1} \right)^p \int_1^{+\infty} \rho^\lambda \left| \frac{df(\rho)}{d\rho} \right|^p d\rho$$

Lemma 3 directly follows from the Hardy inequality:

$$\int_1^{\infty} \rho^{\lambda-p} |f(\rho) - A|^p d\rho = \int_1^{+\infty} \rho^{\lambda-p} \left| \int_\rho^{+\infty} \frac{df(t)}{dt} dt \right|^p d\rho \leq$$

$$\leq \left(\frac{p}{p-1}\right)^p \int_1^{+\infty} \rho^{\lambda} \left| \frac{df(\rho)}{d\rho} \right|^p d\rho$$

Theorem 3 can be proved by sequential application of three lemmas. Really, in virtue of the conditions of the Theorem 3

$$\int_1^{+\infty} \rho^{n-1} \left| \frac{\partial}{\partial \rho} \left(\frac{\partial^{\tau-1} u}{\partial \rho^{\tau-1}} \right) \right|^p d\rho < +\infty$$

and $n - 1 > p - 1$, therefore according to Lemma 1 there exists a limit

$$\lim_{\rho \to +\infty} \frac{\partial^{\tau-1} u}{\partial \rho^{\tau-1}} = A$$

Further, in virtue of the same conditions

$$\int_1^{+\infty} \rho^{n-1-p} \left| \frac{\partial}{\partial \varphi} \left(\frac{\partial^{\tau-1} u}{\partial \rho^{\tau-1}} \right) \right|^p d\rho < +\infty$$

and $n-1-p > -1$, therefore according to Lemma 2 the limit A does not depend on φ. Lastly, according to Lemma 3

$$\int_1^{+\infty} \rho^{n-1-p} \left| \frac{\partial^{\tau-1}[u - (\tau-1)!\, A \rho^{\tau-1}]}{\partial \rho^{\tau-1}} \right|^p d\rho =$$

$$= \int_1^{+\infty} \rho^{n-1-p} \left| \frac{\partial^{\tau-1} u}{\partial \rho^{\tau-1}} - A \right|^p d\rho$$

$$\leq \left(\frac{p}{p-1}\right)^p \int_1^{+\infty} \rho^{n-1} \left| \frac{\partial^{\tau} u}{\partial \rho^{\tau}} \right|^p d\rho < +\infty .$$

Repeating the same argument with respect to the function $u - (\tau-1)!\, A\rho^{\tau-1}$ and keeping on one obtains a polynomial (17) for which the property (18) is valid.

Theorem 3 is proved.

References

1 Uspenskiĭ, S.V., ''On imbedding theorems for weight classes''. Trudy Steklov Math. Institute, 1961, vol. 60, 282-303.

2 Kudryavtsev, L.D. ''On traces of functions of many variables and fields of directions regular at infinity for Laplace operator''. Trudy Steklov Math. Institute, 1971, vol.112, 256-270.

3 Nikol'skiĭ, S.M. ''Approximation of functions of many variables and imbedding theorems''. 1969. Moscow.

4 Kudryavtsev, L.D. ''On polynomial traces and on modules of smoothness of functions of many variables''. Trudy Steklov Math. Inst. 1972, vol. 117, 180-211.

5 Kudryavtsev, L.D. ''On parametric inequalities for functions of several variables''. Doklady Akademii Nauk SSSR. 1972, vol.202, No.6, 1261-1264.
English transl. Soviet Math. Dokl. vol. 13 (1972), No.1, 276-280.

6 Sobolev, S.L. ''Some applications of functional analysis to mathematical physics''. 1950, Leningrad.

7 Uspenskiĭ, S.V. ''A theorem of imbedding and extension for a class of functions, II''. Siberian Mathem. Journal. 1966, vol.8, No.2, 409-418.

8 Sobolev, S.V. ''Density of finite functions in the space $L_p^{(m)}(E_n)$''. Siberian Mathem. Journal. 1963, vol.4, No.3, 673-682.

9 Sedov, V.N. ''On functions degenerating into polynomial at infinity''. In the book ''Imbedding theorems and their applications''. Moscow. 1970, 204-211.

10 Kudryavtsev, L.D. ''Direct and inverse imbedding theorems. Applications to the solution of elliptic equations by variational methods''. Trudy Steklov Mathem. Inst., 1959, vol.55.

FUNDAMENTAL SOLUTIONS OF HYPERBOLIC DIFFERENTIAL EQUATIONS

P.Leonard

The purpose of this lecture is to present an approach, using Fourier transform methods, to the study of gaps in the support of fundamental solutions of first order hyperbolic systems of partial differential equations with constant coefficients.

1. Let P(D) be a partial differential operator of the form

$$\sum_{|\mu| \leq m} a_\mu D^\mu$$

where μ is a multi-index, $|\mu| = \sum_{k=1}^{n} \mu_k$, $D^\mu = D_{x_1}^{\mu_1} \cdots D_{x_n}^{\mu_n}$ and the coefficients a_μ are complex numbers.

The polynomial associated to P(D) is P(ix), $\overset{\circ}{P}(ix)$ is the principal part of P(ix).

An operator P(D) is called <u>hyperbolic with respect to</u> $N \in \mathbb{R}^n \setminus 0$ if $\overset{\circ}{P}(N) \neq 0$ and there exists a real number τ_0 such that $P(ix - \tau N) \neq 0$ if $x \in \mathbb{R}^n$ and $\tau < \tau_0$, [3,5].

The following results are proved in [5] .

The set

$$\Gamma(P,N) = \{ x \in \mathbb{R}^n : \overset{\circ}{P}(x + \tau N) \text{ has only negative zeros } \tau \}$$

is a convex cone in $\mathbb{R}^n$ and is the connected component containing N of the set $\{ x : \overset{\circ}{P}(x) \neq 0 \}$.

If P is hyperbolic with respect to N, it is hyperbolic with respect to any $\theta \in \Gamma(P,N)$ and the inequality

$$| P(ix - \tau\theta) | \geq | \overset{\circ}{P}(\theta) | \; | \tau - \tau_0 |^m$$

holds.

A fundamental solution of $P(D)$ is given by

$$\mathcal{E}(\varphi) = (2\pi)^{-n} \int_{\mathbb{R}^n} \frac{1}{P(iy - \tau\theta)}\, dy \int_{\mathbb{R}^n} e^{ix.y}\, e^{-\tau x.\theta}\, \varphi(x)\, dx \,, \qquad \varphi \in \mathcal{D}(\mathbb{R}^n) .$$

It is independent of $\tau < \tau_0$ and of $\theta \in \Gamma(P,N)$.

Its support is contained in the convex cone

$$\Gamma^*(P,N) = \bigcap_{\theta \in \Gamma(P,N)} \{x : x.\theta \geq 0\}$$

and in no smaller convex cone.

2. An $M \times M$ matrix valued partial differential operator is hyperbolic with respect to $N \in \mathbb{R}^n \setminus 0$, if its formal determinant is hyperbolic with respect to N.

If $\mathcal{E}$ is a fundamental solution of $\operatorname{dtm} P(D)$ and if the matrix operator $P'(D)$ is the formal transposed cofactor matrix associated to $P(D)$, then the distribution defined by

$$P'(D)\,\mathcal{E} \,, \quad [P'(D)\,\mathcal{E}]_{ij} = [P'(D)]_{ij}\,\mathcal{E}$$

is obviously a fundamental solution of $P(D)$.

The support of $P'(D)\,\mathcal{E}$ is contained in the support of $\mathcal{E}$ and these two sets do not generally coincide. For instance, the operator $D_t^2 - D_x^2$, x and $t \in \mathbb{R}$, has a fundamental solution supported by the cone

$$\{(x,t) : |x| \leq t\}$$

but the corresponding one for the operator

$$\begin{pmatrix} D_t & D_x \\ D_x & D_t \end{pmatrix}$$

is supported by $\{(x,t) : |x| = t\}$.

3. First order $M \times M$ hermitian systems of the form

$$D_t + \sum_{k=1}^{n} A_k D_{x_k}$$

are obviously hyperbolic with respect to $(0,1) \in \mathbb{R}^n \times \mathbb{R}$ and a fundamental solution with support in $\{(x,t) : t \geq 0\}$ is given by

$$Q(\vec{\varphi}) = (2\pi)^{-n} \int_0^{+\infty} dt \int_{\mathbb{R}^n} e^{-it A\cdot y} dy \int_{\mathbb{R}^n} e^{ix.y} \vec{\varphi}(x,t)\, dx$$

where we write $A \cdot y$ for $\sum_k A_k y_k$, $\vec{\varphi} \in [\mathcal{D}(\mathbb{R}^{n+1})]^M$.

Its support is in the cone

$$\Gamma^*(\operatorname{dtm} L, (0,1))$$

for Q coincides with the distribution

$$\mathcal{T}(\vec{\varphi}) = (2\pi)^{-(n+1)} \int_{\mathbb{R}^{n+1}} [is - \tau + i A\cdot y]^{-1} ds\, dy \int_{\mathbb{R}^{n+1}} e^{i(x.y+ts)} e^{-\tau t} \vec{\varphi}(x,t)\, dx\, dt$$

obtained from the construction given in the preceding sections.

One obtains Q from $\mathcal{T}$ by performing the Fourier transform with respect to s in $\mathcal{T}(\vec{\varphi})$.

The support of Q is obviously contained in the cone in $\mathbb{R}^{n+1}$ with vertex at 0 and through the support of the distribution $\mathcal{R}$ defined by

$$\mathcal{R}(\vec{\varphi}) = \int_{\mathbb{R}^n} e^{-i A\cdot y} dy \int_{\mathbb{R}^n} e^{ix\cdot y} \vec{\varphi}(x)\, dx$$

translated in the hyperplane $t = 1$.

Generalising a result of Anderson [1] , Nelson has proved that the support of $\mathcal{R}$ is contained in the numerical range of the n-tuple $(A_1,\ldots,A_n)$, i.e.

$$\bigcup_{|\vec{u}|=1} (A_1\vec{u}.\vec{u}\,,\ \ldots\ ,\ A_n\vec{u}.\vec{u})$$

and is exactly this set if the identity matrix I and the considered n-tuple span $\mathcal{L}(\mathbb{R}^n)$ [7] .

Applied to the preceding example, this method gives the interval [-1,1] .

4. The polynomial $dtm(\lambda I - A.y)$ in $(y,\lambda) \in \mathbb{R}^{n+1}$ has a unique factorization

$$Q_1^{m_1}(y,\lambda)\ldots\ Q_p^{m_p}(y,\lambda)$$

where Q_k is a homogeneous irreducible polynomial in (y,λ).

The product $Q_1(y,\lambda)\ldots Q_p(y,\lambda)$ is a.e. the minimal polynomial of A.y (see [9]). This implies that the zeros λ of the polynomials $Q_1(y,\lambda)\ \ldots\ Q_p(y,\lambda)$ are different a.e.. They are also positively homogeneous of degree one.

If $\lambda_1(y),\ldots,\lambda_r(y)$ are the a.e. different eigenvalues of A.y, then

$$A.y = \sum_{k=1}^{r} P_k(y)\,\lambda_k(y) \qquad \text{a.e.}$$

and

$$e^{-i\,A.y} = \sum_{k=1}^{r} P_k(y)\,e^{-i\lambda_k(y)} \qquad \text{a.e.}\,.$$

The hermitian projectors $P_1(y),\ldots,P_r(y)$ are homogeneous functions of y of degree zero and rational functions of y and $\lambda_k(y)$. Since the eigenvalues are continuous functions of y , the P_k's are measurable functions of y .

5. In the expression of $\mathcal{R}(\vec{\varphi})$, let us introduce polar coordinates $y = \rho\omega$, $|\omega| = 1$, and coordinates in the plane $x.\omega = s$. We then get

$$(2\pi)^n \mathcal{R}(\vec{\varphi}) = \int_{|\omega|=1} d\omega \int_0^{+\infty} e^{-i\rho A\cdot\omega} \rho^{n-1} d\rho \int_{-\infty}^{\infty} e^{i\rho s} ds \int_{x.\omega=s} \vec{\varphi}(x)\, dS .$$

Since the inner integral is integrated over the sphere $|\omega| = 1$, it may be replaced by its even part in ω and, after a simple computation, one obtains if n is odd

$$2(2\pi)^n \mathcal{R}(\vec{\varphi}) = \int_{|\omega|=1} d\omega \int_{-\infty}^{\infty} e^{-i\rho A\cdot\omega} d\rho \int_{-\infty}^{\infty} e^{i\rho s} ds \int_{x.\omega=s} (-\Delta)^{(n-1)/2} \vec{\varphi}(x)\, dS$$

and, if n is even

$$2(2\pi)^n \mathcal{R}(\vec{\varphi}) = \int_{|\omega|=1} d\omega \int_{-\infty}^{\infty} e^{-i\rho A\cdot\omega} |\rho|\, d\rho \int_{-\infty}^{\infty} e^{i\rho s} ds \int_{x\cdot\omega=s} (-\Delta)^{(n-2)/2} \vec{\varphi}(x)\, dS .$$

Because of the presence of the absolute value of ρ in the last integral, the case of even n is not simple and we shall only consider the case of odd n.

Since $\int_{x\cdot\omega=s} (-\Delta)^{(n-1)/2} \vec{\varphi}\, dS$ belongs to $\mathcal{D}(\mathbb{R})$ as a function of s, introducing the value of $e^{-i\rho A\cdot\omega}$ given at the end of the preceding section, one gets

$$(2\pi)^n \mathcal{R}(\vec{\varphi}) = \sum_{k=1}^{r} \int_{|\omega|=1} P_k(\omega)\, d\omega \int_{x\cdot\omega=\lambda_k(\omega)} (-\Delta)^{(n-1)/2} \vec{\varphi}(x)\, dS .$$

6. If no hyperplane $x.\omega = \lambda_k(\omega)$ meets the support of $\vec{\varphi}$, the corresponding term vanishes and we shall say that, if it is not empty, the interior of the convex set of $\mathbb{R}^n$ supported by the

hyperplanes $x.\omega = \lambda_k(\omega)$, $|\omega| = 1$, is a <u>gap for</u> $\lambda_k(\omega)$.

The intersection of all the gaps is a gap for the fundamental solution.

For instance, if every $\lambda_k(\omega)$ is bounded away from zero, the classical result that the inner core of the ray cone is a gap follows.

7. What happens outside the convex hull of the envelope of the hyperplanes $x.\omega = \lambda_k(\omega)$, $|\omega| = 1$, depends on the value of $P_k(\omega)$. In the commutative case for instance (n odd or even), all these projectors are constant and the considered envelopes are points which constitute exactly the support of $\mathcal{R}$.

Finally, according to the values of $P_k(\omega)$, there may be no corresponding gap for the fundamental solution but one may exist for a solution of a particular problem.

Consider for instance, Maxwell's equations in a homogeneous anisotropic medium.

$$D_t\, Su + S^{-1}\begin{pmatrix} \cdot & -\mathrm{curl} \\ \mathrm{curl} & \cdot \end{pmatrix} S^{-1}\, Su = S^{-1} f$$

with $S = \mathrm{diag}\,(\sqrt{\varepsilon_1}\,, \sqrt{\varepsilon_2}\,, \sqrt{\varepsilon_3}\,, 1, 1, 1\,)$ and $u = (E,H)$ (μ is assumed to be 1).

The projector $P(\omega)$ corresponding to the double zero eigenvalue of $A.y$ is of the form

$$\begin{pmatrix} P_1 & 0 \\ 0 & P_2 \end{pmatrix}$$

where P_1 and P_2 are 3×3 matrices defined by

$$(P_1)_{ij} = \sqrt{\varepsilon_i \varepsilon_j}\; \omega_i \omega_j\,, \quad (P_2)_{ij} = \omega_i \omega_j \;.$$

It follows that

$$P(\omega)\int_{x\cdot\omega=0} \Delta\vec{\varphi}\,dS = \begin{pmatrix} \dfrac{1}{\sum_1^3 \varepsilon_k \omega_k^2} \begin{pmatrix} \sqrt{\varepsilon_1}\ \omega_1 \\ \sqrt{\varepsilon_2}\ \omega_2 \\ \sqrt{\varepsilon_3}\ \omega_3 \end{pmatrix} \displaystyle\int_{x\cdot\omega\leq 0} \Delta\, \mathrm{div}(\sqrt{\varepsilon_1}\varphi_1, \sqrt{\varepsilon_2}\varphi_2, \sqrt{\varepsilon_3}\varphi_3)\,dx \\ \\ \begin{pmatrix} \omega_1 \\ \omega_2 \\ \omega_3 \end{pmatrix} \displaystyle\int_{x\cdot\omega\leq 0} \Delta\, \mathrm{div}(\varphi_4, \varphi_5, \varphi_6)\,dx \end{pmatrix}$$

and therefore there is a gap provided that the data of a Cauchy problem satisfy the divergence conditions, (see Courant-Hilbert [2] John [6] , Petrowsky [8] , Garding [4] for other methods).

Bibliography

1 Anderson, R.F., The Weyl Functional Calculus, J. Functional Analysis, 4, 240-267, 1969.

2 Courant, R. and Hilbert, D., Methods of Mathematical Physics, Vol.2, New York 1962.

3 Gårding,L., Linear Hyperbolic Partial Differential Equations with constant coefficients, Acta Math., 85, 1-62, 1950.

4 Gårding, L., The Solution of Cauchy's problem for two totally hyperbolic linear differential equations by means of Riesz integrals, Ann. of Math. 48, 785 - 826, 1947. ibid. 52, 506-507, 1950.

5 Hörmander, L., Linear Partial Differential Operators, Berlin 1963.

6 John, F., Plane Waves and Spherical Means Applied to Partial Differential Equations, New York 1955.

7 Nelson, E. Operants : A Functional Calculus for Non-Commuting Operators, Funct. Analysis and Related Fields, 172 - 187,1970.

8 Petrowsky, I.G., On the diffusion of waves and the lacunas for hyperbolic equations, Mat. Sb. 17 (59), 289 - 370, 1945.

9 Wilcox, C.H., The Singularities of the Green's Matrix in Anisotropic Wave Motion, Indiana Univ. Math. J., 20, N° 12, 1093 - 1117, 1971.

NON-SELF-CONJUGATE DIFFERENTIAL DIRAC OPERATORS EXPANSION IN EIGENFUNCTIONS THROUGH THE WHOLE AXIS

F.G.Maksudov and S.G.Veliev

Introduction

In the work of V.A.Marchenko [1] the equation

$$-y''(x) + q(x)y(x) = \lambda^2 y(x) \quad (1)$$

was considered on a semiaxis $[0,\infty)$ with a boundary condition

$$y'(0) - hy(0) = 0 \quad (2)$$

where $g(x)$ is an arbitrary complex-valued function summable in every finite interval of the axis, and h is an arbitrary complex number. The following result has been obtained.

THEOREM. Let us denote linear topological space of even entire functions of exponential type summable on a real axis by Z, and the space conjugate to it by $T(Z)$. Then to every boundary value problem (1)-(2) there corresponds a generalized function $R(\lambda) \in T(Z)$ such that

$$\int_0^\infty f(x)g(x)dx = \left(R(\lambda),\ E_f(\lambda)\ E_g(\lambda) \right)$$

where $f(x)$ and $g(x)$ are arbitrary finite functions belonging to $L^2[0,\infty)$,

$$E_f(\lambda) = \int_0^\infty f(x)\ \omega^{(1)}(\lambda, x)dx$$

($\omega^1(x,\lambda)$ is the solution of equation (1) satisfying the condition (2)).

In the present paper the results got by V.A.Marchenko are transformed to Dirac equation given through the whole interval $(-\infty,\infty)$. An analogous problem for Sturm-Liuvill equation is solved in the work [2] .

1. Some information about generalized functions.

We denote the set of all finite vector functions with a summable square by K^2, and the set of all vector functions of K^2 equal to zero out of the interval $[-\sigma,\sigma]$ by K^2_σ.

The entire function $F(\lambda)$ is called the function of exponential type if the inequality

$$F(\lambda) \leq C(\sigma)\, e^{\sigma|\lambda|} \tag{1.1}$$

where σ and $C(\sigma)$ are some constants takes place for all complex λ. The greatest lower boundary of those σ which inequalities (1.1) are carried out is called the degree of function $F(\lambda)$.

Definition. Linear topological space Z is a set of all entire exponential type functions $F(\lambda)$ summable on the real axis $-\infty < \lambda < \infty$ with ordinary operations of addition and multiplication by complex numbers. The sequence $F_n(\lambda) \in Z$ converges to $F(\lambda)$ if

$$\lim_{n\to\infty} \int_{-\infty}^{\infty} |F(\lambda) - F_n(\lambda)|\, d\lambda = 0$$

and degrees σ_n of functions $F_n(\lambda)$ are bounded in the set $\sigma_n \leq \sigma < \infty$.

Taking Z for the main space we shall call linear continuous functionals $R[F(\lambda)]$ defined on Z generalized functions setting

$$R[F(\lambda)] = (R, F(\lambda)).$$

The set of all generalized functions defined in this way we shall denote by $T(Z)$.

The sequence $R_n \in T(Z)$ converges to $R \in T(Z)$ if

$$\lim_{n\to\infty} (R_n, F(\lambda)) = (R, F(\lambda))$$

in all main functions $F(\lambda) \in Z$.

Generalized function $R \in T(Z)$ is called regular if it is given by formula

$$(R,F(\lambda)) = \int_{-\infty}^{\infty} R(\lambda)\, F(\lambda)\, d\lambda$$

where $R(\lambda)$ is an arbitrary function bounded on the axis $-\infty < \lambda < \infty$.

2. Generalized spectral matrix function.

Let us consider differential equation

$$By'(x) + Q(x)\, y(x) = \lambda y(x)$$

$$B = \begin{pmatrix} 0 & 1 \\ -1 & 0 \end{pmatrix}, \quad Q(x) = \begin{pmatrix} p(x) & q(x) \\ q(x) & -p(x) \end{pmatrix}, \quad y(x) = \begin{pmatrix} y_1(x) \\ y_2(x) \end{pmatrix} \qquad (2.1)$$

where $p(x)$, $q(x)$ are arbitrary complex-valued continuous functions on $(-\infty, \infty)$.

We denote the solution of differential equation (2.1) with initial conditions

$$\omega_{11}(0,\lambda) = 1, \quad \omega_{12}(0,\lambda) = 0,$$
$$\omega_{21}(0,\lambda) = 0, \quad \omega_{22}(0,\lambda) = 1 \qquad (2.2)$$

by

$$\omega_1(x,\lambda) = \begin{pmatrix} \omega_{11}(x,\lambda) \\ \omega_{12}(x,\lambda) \end{pmatrix}, \qquad \omega_2(x,\lambda) = \begin{pmatrix} \omega_{21}(x,\lambda) \\ \omega_{22}(x,\lambda) \end{pmatrix}.$$

Equation (2.1) on semiaxis $[0,\infty)$ with boundary condition $y_1(0) = 0$ has been considered in [3] .

The following two theorems will play further an essential role.

THEOREM 2.1 Solutions $\omega_1(x,\lambda)$, $\omega_2(x,\lambda)$ of equation (2.1) may be expressed by $\begin{pmatrix} \cos \lambda x \\ \sin \lambda x \end{pmatrix}$ and $\begin{pmatrix} -\sin \lambda x \\ \cos \lambda x \end{pmatrix}$ correspondingly by means of transformation operators in the form:

$$\begin{pmatrix} \omega_{11}(x,\lambda) \\ \omega_{12}(x,\lambda) \end{pmatrix} = \begin{pmatrix} \cos \lambda x \\ \sin \lambda x \end{pmatrix} + \int_{-x}^{x} K(x,t) \begin{pmatrix} \cos \lambda t \\ \sin \lambda t \end{pmatrix} dt, \tag{2.3}$$

$$\begin{pmatrix} \omega_{21}(x,\lambda) \\ \omega_{22}(x,\lambda) \end{pmatrix} = \begin{pmatrix} -\sin \lambda x \\ \cos \lambda x \end{pmatrix} + \int_{-x}^{x} K(x,t) \begin{pmatrix} -\sin \lambda t \\ \cos \lambda t \end{pmatrix} dt,$$

$$\begin{pmatrix} \cos \lambda x \\ \sin \lambda x \end{pmatrix} = \omega_1(x,\lambda) - \int_{-x}^{x} H(x,t)\, \omega_1(t,\lambda)\, dt, \tag{2.4}$$

$$\begin{pmatrix} -\sin \lambda x \\ \cos \lambda x \end{pmatrix} = \omega_2(x,\lambda) - \int_{-x}^{x} H(x,t)\, \omega_2(t,\lambda)\, dt$$

where kernels $K(x,t)$ and $H(x,t)$ are absolutely continuous through both variables and satisfy the following boundary conditions:

$$BK(x,-x) + K(x,-x)\,B = 0, \quad BH(x,-x) + H(x,-x)B = 0$$

$$K(x,x)\,B - BK(x,x) = Q(x), \quad H(x,x)B - BH(x,x) = Q(x) \tag{2.5}$$

$$K(x,t) = 0, \quad H(x,t) = 0$$

for $|t| > |x|$.

For all this if matrix function $Q(x)$ has $n \geq 0$ absolutely continuous derivatives then $K(x,t)$ and $H(x,t)$ have n continuous derivatives through both variables and satisfy the following partial equations.

$$B \frac{\partial}{\partial x} K(x,t) + Q(x)\, K(x,t) + \frac{\partial}{\partial t} K(x,t)\, B = 0 \tag{2.6}$$

$$B \frac{\partial}{\partial x} H(x,t) - Q(t)\, H(x,t) + \frac{\partial}{\partial t} H(x,t)\, B = 0 \tag{2.7}$$

And vice versa, if continuously differentiable matrix function $K(x,t)$ satisfies equation (2.6) and boundary conditions (2.5) then vector functions $\omega_i(x,\lambda)$ $(i=1,2)$ defined by formula (2.3) satisfy equation (2.1) and initial conditions (2.2).

This theorem is well-known from [4].

THEOREM 2.2 Let

$$u(x,y) = \int_{-\infty}^{\infty} \left(\omega_1(x,\lambda),\, \omega_2(x,\lambda)\right) T(\lambda) \begin{pmatrix} \tilde{\omega}_1(x,\lambda) \\ \tilde{\omega}_2(x,\lambda) \end{pmatrix} d\lambda \tag{2.8}$$

$$\varphi(x) E = \mathcal{U}(x,0) = \int_{-\infty}^{\infty} (\omega_1(x,\lambda), \omega_2(x,\lambda)) T(\lambda)\, d\lambda \qquad (2.9)$$

where $T(\lambda)$ is an arbitrary matrix function summable on the axis $-\infty < \lambda < \infty$; E is a unit matrix, and the sign $\sim$ denotes transposition. Then

$$\mathcal{U}(x,y) = \varphi(x-y) E + \int_{x-y}^{x+y} W(x,y,t) \varphi(t) dt \qquad (2.10)$$

for $|y| \leq |x|$

where $W(x,y,t)$ is some matrix function bounded in every finite domain.

Proof. If $T(\lambda)$ is finite, and $Q(x)$ is continuously differentiable then direct verification shows that matrix function $\mathcal{U}(x,y)$ satisfies equation

$$B \frac{\partial \mathcal{U}(x,y)}{\partial x} + Q(x) \mathcal{U}(x,y) = - \frac{\partial \mathcal{U}(x,y)}{\partial y} B + \mathcal{U}(x,y) Q(y) \qquad (2.11)$$

and boundary conditions

$$\mathcal{U}(x,0) = \varphi(x) E,$$

$$\frac{\partial}{\partial y} \mathcal{U}(x,0) = - \varphi'(x) E + \varphi(x) [Q(x) - Q(0)].$$

We can show that equation (2.11) is equivalent to the following equation

$$\frac{\partial^2}{\partial x^2} \mathcal{U}(x,y) - \frac{\partial^2}{\partial y^2} \mathcal{U}(x,y) = [Q^2(x) + BQ^1(x)] \mathcal{U}(x,y) - \mathcal{U}(x,y) [Q^2(y) - Q^1(y) B]. \qquad (2.12)$$

Solving equation (2.12) by Riemann's method we come directly to the formula (2.10).

In the general case we need to approximate matrix function $T(\lambda)$ by finite matrix functions and $Q(x)$ by continuously differentiable matrix functions, and perform then the limit transition.

We define Fourier ω -transformation of vector function f(x) of K^2 by equalities

$$E_f^i(\lambda) = \int_{-\infty}^{\infty} \tilde{f}(x)\, \omega_i(x,\lambda)\,dx, \qquad (i=1,2) .$$

It is evident from formulas (2.3) that for any f(x) and g(x) of K^2 functions $E_f^i(\lambda)$, $E_g^i(\lambda) \in Z$.

The following theorem is valid.

THEOREM 2.3 *To every differential equation* (2.1) *with solutions satisfying condition* (2.2) *there corresponds a generalized spectral matrix function of the second order* $R(\lambda) = \| R_{ik}(\lambda) \|$, $R_{ik}(\lambda) \in T(Z)$ *such that*

$$\int_{-\infty}^{\infty} [f_1(x)\; g_1(x) + f_2(x)\; g_2(x)]\; dx = \sum_{i,k=1}^{2} \left(R_{ik}, E_f^i(\lambda)\; E_g^k(\lambda) \right) ,$$

where $f(x), g(x) \in K^2$ *and* $E_f^i(\lambda)$, $E_g^i(\lambda)$ *are their Fourier* ω *-transformations*.

Spectral matrix function $R(\lambda)$ is connected with kernel $H(x,t)$ of transformation operator (2.4) by the formula

$$R(\lambda) = \frac{1}{2\pi} E - \frac{1}{2\pi} \int_{-\infty}^{\infty} \begin{pmatrix} \cos\lambda x & \sin\lambda x \\ -\sin\lambda x & \cos\lambda x \end{pmatrix} H(x,0)\, dx .$$

Proof. Let degrees of functions $E_{f,g}^i(\lambda)$, $(i=1,2)$ not exceed σ, and hence vector functions f(x), g(x) belong to K_σ^2. Suppose that we managed to construct matrix functions of the second order $K_n^\sigma(\lambda) = \| R_{ik,n}^\sigma(\lambda) \|_{i,k>1}^2$ summable on the axis $-\infty < \lambda < \infty$ such that the sequence

$$u_n(x,y) = \int_{-\infty}^{\infty} (\omega_1(x,\lambda), \omega_2(x,\lambda)) \begin{pmatrix} R_{11,n}^\sigma(\lambda) & R_{12,n}^\sigma(\lambda) \\ R_{21,n}^\sigma(\lambda) & R_{22,n}^\sigma(\lambda) \end{pmatrix} \begin{pmatrix} \tilde{\omega}_1(y,\lambda) \\ \tilde{\omega}_2(y,\lambda) \end{pmatrix} \qquad (2.13)$$

converges as $n \to \infty$ in the domain $|x| < \sigma$, $|y| < \sigma$ at $\delta(x-y)E$.

Multiplying both sides of equality (2.13) by

$$\begin{pmatrix} f_1(x)\ g_1(y) & f_2(x)\ g_1(y) \\ f_1(x)\ g_2(y) & f_2(x)\ g_2(y) \end{pmatrix}$$

we denote the matrices obtained after multiplication by $C_n = \| C_{ij,n} \|$. After integration of the elements $C_{11,n}$ and of matrix $C_{22,n}$ by both variables and addition we have

$$\int_{-\sigma}^{\sigma}\int_{-\sigma}^{\sigma}\left[\mathcal{U}_{n,n}(x,y)f_1(x)g_1(y) + \mathcal{U}_{12,n}(x,y)f_1(x)g_2(y)\right] dxdy + \int_{-\sigma}^{\sigma}\int_{-\sigma}^{\sigma}\left[\mathcal{U}_{21,n}(x,y)\times\right.$$

$$\left.\times f_2(x)g_1(y) + \mathcal{U}_{22,n}(x,y)f_2(x)g_2(y)\right] dxdy = \sum_{i,k=1}^{n}\int_{-\infty}^{\infty}\left[R^{\sigma}_{ik,n}(\lambda)E^i_f(\lambda)\times\right.$$

$$\left.\times E^k_g(\lambda)\right]d\lambda$$

from which as $n\to\infty$ we obtain

$$\int_{-\infty}^{\infty}\left[f_1(x)\,g_1(x) + f_2(x)g_2(x)\right]dx = \lim_{n\to\infty}\left[\sum_{i,k=1}^{2}\int_{-\infty}^{\infty} R_{ik,n}(\lambda)E^i_f(\lambda)E^k_g(\lambda)d\lambda\right].$$

Functions $R^{\sigma}_{ik,n}(\lambda)$ are regular generalized functions of $T(Z)$, and if $R^{\sigma}_{ik,n}(\lambda)\to R^{\sigma}_{ik}(\lambda)$ as $n\to\infty$ (in the sense of convergence of generalized functions) then according to the last equality

$$\int_{-\infty}^{\infty}\left[f_1(x)g_1(x) + f_2(x)g_2(x)\right]\ dx = \sum_{i,k=1}^{n}\left(R^{\sigma}_{ik,n},\ E^i_f(\lambda)\ E^k_g(\lambda)\right)$$

whatever vector functions $f(x)$ and $g(x)$ of K^2_{σ} might be. If at last $R^{\sigma}_{ik} \longrightarrow R_{ik}$ as $\sigma\to\infty$ then

$$\int_{-\infty}^{\infty}\left[f_1(x)g_1(x) + f_2(x)g_2(x)\right]\ dx = \sum_{i,k=1}^{n}\left(R_{ik}(\lambda),\ E^i_f(\lambda)\ E^k_g(\lambda)\right)$$

for all vector functions $f(x)$ and $g(x)$ of K^2, and that is what we were going to prove.

Now we shall construct necessary sequences $R^{\sigma}_n(\lambda) = \| R^{\sigma}_{ik,n}(\lambda) \|$. First of all we must provide the equality

$$\mathcal{U}_n(x,0) = \int_{-\infty}^{\infty} (\omega_1(x,\lambda),\ \omega_2(x,\lambda)) R_n^{\sigma}(\lambda)\, d\lambda = \delta_n(x)E \rightarrow \delta(x)E,\ |x|<\delta.$$

Applying transformation (2.4) of the theorem 2.1 to this equality we get

$$\int_{-\infty}^{\infty} \begin{pmatrix} \cos\lambda x & -\sin\lambda x \\ \sin\lambda x & \cos\lambda x \end{pmatrix} \begin{pmatrix} R_{11,n}^{\sigma}(\lambda) & R_{12,n}^{\sigma}(\lambda) \\ R_{21,n}^{\sigma}(\lambda) & R_{22,n}^{\sigma}(\lambda) \end{pmatrix} d\lambda = \delta_n(x)E - \int_{-x}^{x} H(x,t)\, \delta_n(t)\, dt$$

from which it follows that cosine-sine of Fourier transformation of the searched matrix functions $R_n^{\sigma}(\lambda)$ at least in the interval $|x| < \delta$ must be set equal to

$$\delta_n(x)E - \int_{-x}^{x} H(x,t)\, \delta_n(t)\, dt$$

where $\delta_n(x) \longrightarrow \delta(x)$ as $n \rightarrow \infty$. Therefore we shall take the same way as in [1] : to choose two smooth functions $\delta_n(x)$ and $\gamma_\sigma(x)$ satisfying conditions

$$\int_{-\infty}^{\infty} \delta_n(x)\, dx = 1$$

$$\delta_n(x) = 0 \quad \text{for} \quad x \geq \frac{1}{n}, \qquad \delta_n(x) > 0 \quad \text{for} \quad |x| < \frac{1}{n}$$

$$\gamma_\sigma(x) = 1 \quad \text{for} \ |x| \leq 2\sigma, \quad \gamma_\sigma(x) = 0 \quad \text{for} \quad |x| > 2\sigma + 1$$

and to set

$$R_n^{\sigma}(\lambda) = \frac{1}{2\pi} \int_{-\infty}^{\infty} \begin{pmatrix} \cos\lambda x & \sin\lambda x \\ -\sin\lambda x & \cos\lambda x \end{pmatrix} \Big[\delta_n(x)E - \gamma_\sigma(x) \int_{-x}^{x} H(x,t)\delta_n(t)\, dt \Big] dx. \tag{2.14}$$

Since matrix function

$$\delta_n(x)E - \gamma_\sigma(x) \int_{-x}^{x} H(x,t)\, \delta_n(t)\, dt$$

belongs to K_σ^2 the integral

$$\int_{-\infty}^{\infty} \begin{pmatrix} \cos\lambda x & -\sin\lambda x \\ \sin\lambda x & \cos\lambda x \end{pmatrix} R_n^{\sigma}(\lambda)\, d\lambda = \delta_n(x)E - \gamma_\sigma(x) \int_{-x}^{x} H(x,t)\, \delta_n(t)\, dt$$

converges absolutely.

Applying transformation (2.3) inverse to (2.4) (theorem 2.1) to both sides of this equality and noticing that $\gamma_\sigma(x) = 1$ for $|x| < 2\sigma$ we get

$$\int_{-\infty}^{\infty} (\omega_1(x,\lambda), \omega_2(x,\lambda)) R_n^\sigma(\lambda)\, d\lambda = \delta_n(x) E, \quad |x| \le 2\sigma .$$

Now let

$$\mathcal{U}_n(x,y) = \int_{-\infty}^{\infty} (\omega_1(x,\lambda), \omega_2(x,\lambda)) R_n^\sigma(\lambda) \begin{pmatrix} \tilde{\omega}_1(y,\lambda) \\ \tilde{\omega}_2(y,\lambda) \end{pmatrix} d\lambda .$$

Since

$$\mathcal{U}_n(x,0) = \int_{-\infty}^{\infty} (\omega_1(x,\lambda), \omega_2(x,\lambda)) R_n^\sigma(\lambda)\, d\lambda = \delta_n(x) E, \quad |x| \le 2\sigma$$

then according to theorem 2.2

$$\mathcal{U}_n(x,y) = \delta_n(x-y) E + \int_{x-y}^{x+y} W(x,y,t)\, \delta_n(t)\, dt$$

in the domain $|y| \le |x| \le 2\sigma$. From the definition (2.13) it follows that $\mathcal{U}_n(x,y) = \mathcal{U}_n(y,x)$. Therefore in the whole domain $|x| \le \sigma$, $|y| \le \sigma$

$$\mathcal{U}_n(x,y) = \delta_n(x-y)\, E + \theta_n(x,y) \tag{2.15}$$

where matrix function $\theta_n(x,y)$ symmetrical in regard to x,y is defined for $|y| \le |x| \le \sigma$ by the equality

$$\theta_n(x,y) = \int_{x-y}^{x+y} W(x,y,t)\; \delta_n(t) dt . \tag{2.16}$$

Matrix function $W(x,y,t)$ is bounded in every finite domain (theorem 2.2). Hence there exists a constant $C(\sigma)$ dependent on only σ such that

$$\| W(x,y,t) \| \le C(\sigma), \qquad (|y| \le |x| \le \sigma)$$

and according to (2.16) it gives us

$$\| Q_n(x,y) \| \le C(\sigma) \int_{-\infty}^{\infty} \delta_n(t)\, dt = C(\sigma), \quad |y| \le |x| \le \sigma .$$

Moreover since $\delta_n(t) = 0$ for $|t| > \frac{1}{n}$, then $\Theta_n(x,y) = 0$ for $|x-y| > \frac{1}{n}$.

From these estimations and formula (2.15) it follows that if f(x) and g(x) belong to K_σ^2 then

$$\int_{-\infty}^{\infty}\int_{-\infty}^{\infty} U_n(x,y) \begin{pmatrix} f_1(x) g_1(y) & f_2(x) g_1(y) \\ f_1(x) g_2(y) & f_2(x) g_2(y) \end{pmatrix} dx\, dy =$$

$$= \int_{-\infty}^{\infty}\int_{-\infty}^{\infty} [\delta_n(x-y) E + \Theta_n(x,y)] \begin{pmatrix} f_1(x) g_1(y) & f_2(x) g_1(y) \\ f_1(x) g_2(y) & f_2(x) g_2(y) \end{pmatrix} dx\, dy$$

and

$$\left\| \int_{-\infty}^{\infty}\int_{-\infty}^{\infty} \Theta_n(x,y) \begin{pmatrix} f_1(x) g_1(y) & f_2(x) g_1(y) \\ f_1(x) g_2(y) & f_2(x) g_2(y) \end{pmatrix} dx\, dy \right\| \le C(\sigma) \iint_{(\mathcal{D}_n)} \begin{pmatrix} f_1(x) g_1(y) & f_2(x) g_1(y) \\ f_1(x) g_2(y) & f_2(x) g_2(y) \end{pmatrix} dx\, dy$$

where the domain $\mathcal{D}_n$ is defined by inequalities

$$|x-y| < \frac{1}{n}, \quad |x| \le \sigma, \quad |y| \le \sigma .$$

Since measure of the domain $\mathcal{D}_n$ tends to zero as $n \to \infty$ and function

$$\left\| \begin{pmatrix} f_1(x) g_1(y) & f_2(x) g_1(y) \\ f_1(x) g_2(y) & f_2(x) g_2(y) \end{pmatrix} \right\|$$

is summable then

$$\lim_{n\to\infty} \left\| \int_{-\infty}^{\infty}\int_{-\infty}^{\infty} \Theta_n(x,y) \begin{pmatrix} f_1(x) g_1(y) & f_2(x) g_1(y) \\ f_1(x) g_2(y) & f_2(x) g_2(y) \end{pmatrix} dx\, dy \right\| = 0$$

and hence

$$\int_{-\infty}^{\infty} [f_1(x) g_1(x) + f_2(x) g_2(x)]\, dx = \lim_{n\to\infty} \sum_{i,k=1}^{2} \int_{-\infty}^{\sigma} R_{ik,n}(\lambda)\, E_f^i(\lambda)\, E_g^k(\lambda)\, d\lambda \tag{2.17}$$

whatever vector functions f(x) and g(x) of K^2_σ might be.

It follows from definition (2.14) of matrix functions $R^\sigma_n(\lambda)$ that

$$\lim_{n\to\infty} R^\sigma_n(\lambda) = \frac{1}{2\pi} E - \frac{1}{2\pi}\int_{-\infty}^{\infty}\begin{pmatrix}\cos\lambda x & \sin\lambda x\\ -\sin\lambda x & \cos\lambda x\end{pmatrix}\gamma_\sigma(x)\, H(x,o)\,dx = R^\sigma(\lambda).$$

Since $\gamma_\sigma(x)\ H(x,o)$ as $\sigma \to \infty$ tends to H(x,o) uniformly in every finite interval,

$$\lim_{\sigma\to\infty}\left\{\lim_{n\to\infty} R^\sigma_n(\lambda)\right\} = \frac{1}{2\pi} E - \frac{1}{2\pi}\int_{-\infty}^{\infty}\begin{pmatrix}\cos\lambda x & \sin\lambda x\\ -\sin\lambda x & \cos\lambda x\end{pmatrix} H(x,o)\,dx = R(\lambda)$$

where both limits exist in the sense of convergence of generalized functions. According to formula (2.17) it follows that

$$\int_{-\infty}^{\infty}\left[f_1(x)\, g_1(x) + f_2(x)\, g_2(x)\right]dx = \sum_{i,k=1}^{2}\left(R_{ik}(\lambda),\ E^i_f(\lambda)\ E^k_g(\lambda)\right).$$

This proves the theorem.

<u>COROLLARY.</u> <u>If vector function</u> $f(x) \in K^2$ <u>then the formula</u>

$$f(x) = \sum_{i,k=1}^{2}\left(R_{ik}(\lambda),\ E^i_f(\lambda)\ \omega_k(x,\lambda)\right)$$

<u>holds</u>

3. <u>Analytical expression for a spectral matrix function.</u>

As it is well-known we must investigate resolvent behaviour of the corresponding boundary value problem for finding a spectral function in the self-conjugate case. It is convenient to construct a resolvent using H.Weil's theorem connected with the notion of a limit point and a limit circle. In the works of V.B.Lidskyi [6] and V.A.Marchenko [1] it is proved that this theorem and its proof are transposed without any modifications to some types of non-self-conjugate boundary value problems. It will allow us to find the analytical expression of a spectral matrix function of Dirac differential operators.

We consider the differential expression

$$\ell\, y = By'(x) + Q(x)\, y(x)$$

where p(x), q(x) are arbitrary complex-valued continuous functions on $(-\infty, \infty)$, and

$$-\infty < \tau_0 \leq p_2(x) = \operatorname{Im} p(x) \leq \tau_1 < \infty$$
$$-\infty < \tau_0 \leq q_2(x) = \operatorname{Im} q(x) \leq \tau_1 < \infty \tag{3.1}$$

It is possible to construct a differential operator in the space of vector functions $L^2(-\infty, \infty)$ by this differential expression in the following way. Let us denote by the set of all vector functions $y(x) = \begin{pmatrix} y_1(x) \\ y_2(x) \end{pmatrix} \in L^2(-\infty, \infty)$ such that

1^0 Vector function y(x) is absolutely continuous in every finite interval;

2^0 $\ell\, y \in L^2(-\infty, \infty)$.

We define the operator L as follows : the domain of definition of L is $\mathcal{D}(L)$ and

$$Ly = \ell\, y$$

for $y(x) \in L(\lambda)$.

We shall call operator L a differential operator generated by a differential expression ℓ.

According to Green's formula the identity

$$\left[y_2(x)\overline{y_1(x)} - y_1(x)\overline{y_2(x)} \right]\Big|_0^b - 2i\int_0^b \Big\{ [p_2(x) - v]\, |y_2(x)|^2 - [p_2(x) + v]\, |y_1(x)|^2 - q_2(x)\left[y_2(x)\overline{y_1(x)} + y_1(x)\overline{y_2(x)} \right] \Big\}\, dx = 0 \tag{3.2}$$

where $v = \operatorname{Im}\lambda$ is valid for any solution of equation

$$\ell\, y = \lambda y\ . \tag{3.3}$$

Preserving notations of the preceding sections we shall denote by $\Psi_2(x,\lambda)=\begin{pmatrix}\psi_{21}(x,\lambda)\\ \psi_{22}(x,\lambda)\end{pmatrix}$ the solution of equation (2.14) of the type

$$\Psi_2(x,\lambda)+\omega_1(x,\lambda)+\ell_1\,\omega_2(x,\lambda) \tag{3.4}$$

where ℓ_1 yet is an arbitrary parameter.

Applying identity (2.15) to the solution $\Psi_2(x,\lambda)$ we obtain

$$\psi_{22}(\lambda,b)\,\overline{\psi_{21}(\lambda,b)}-\psi_{21}(\lambda,b)\,\overline{\psi_{22}(\lambda,b)}-(\ell_1-\overline{\ell}_1)-$$

$$-2i\int_0^b\{[p_2(x)-v]\,|\psi_{12}|^2-[p_2(x)+v]\,|\psi_{21}|^2-q_2(x)[\psi_{22}\overline{\psi}_{21}+\overline{\psi}_{22}\psi_{21}]\}\,dx=0$$

or

$$\int_0^b\{[v-p_2(x)]\,|\psi_{22}|^2+[v+p_2(x)]\,|\psi_{21}|^2+q_2[\psi_{22}\overline{\psi}_{21}+\psi_{21}\overline{\psi}_{22}]\}\,dx=$$

$$=\operatorname{Im}\ell_1-\psi_{21}(\lambda,b)\,\overline{\psi_{21}(\lambda,b)}\,\operatorname{Im}W \tag{3.5}$$

where

$$W=\frac{\psi_{22}(\lambda,b)}{\psi_{21}(\lambda,b)}\ . \tag{3.6}$$

Substituting for functions $\psi_{22}(x,\lambda)$, $\psi_{21}(x,\lambda)$ their expressions (3.4) into (3.6), and solving the obtained equation in regard to ℓ_1 we find that

$$\ell_1=\ell_1(\lambda,b,W)=-\frac{\omega_{12}(\lambda,b)-W\,\omega_{11}(\lambda,b)}{\omega_{22}(\lambda,b)-W\,\omega_{21}(\lambda,b)}\ . \tag{3.7}$$

Applying identity (3.2) to the solution $\omega_2(x,\lambda)$ we obtain

$$\int_0^b\{[v-p_2(x)]\,|\omega_{22}|^2+[v+p_2(x)]\,|\omega_{21}|^2+q_2(x)[\omega_{22}\overline{\omega}_{12}+$$

$$+\omega_{12}\overline{\omega}_{22}]\}\,dx=\omega_{21}(\lambda,b)\,\overline{\omega_{21}(\lambda,b)}\,\operatorname{Im}\frac{\omega_{22}(\lambda,b)}{\omega_{21}(\lambda,b)}\ . \tag{3.8}$$

Taking in account inequality (3.1) it follows that

$$\operatorname{Im} \frac{\omega_{22}(\lambda,b)}{\omega_{21}(\lambda,b)} < 0 \qquad \text{if} \quad \operatorname{Im} \lambda = \nu > \tau_1 \tag{3.9}$$

$$\operatorname{Im} \frac{\omega_{22}(\lambda,b)}{\omega_{21}(\lambda,b)} > 0 \qquad \text{if} \quad \operatorname{Im} \lambda = \nu < \tau_0 . \tag{3.10}$$

Let $\operatorname{Im}\lambda > \tau_1$. From formula (3.7) and inequality (3.9) it follows that ℓ_1 is a fractionally-linear function of W, its pole lying in the lower half-plane. Fractionally-linear function (3.7) transforms therefore half-plane $\operatorname{Im} W \geq 0$ to some circle $C(\lambda, b)$ of ℓ_1-plane, such that $\ell_1 \leq C(\lambda, b)$ if and only if $\operatorname{Im} W \geq 0$. Hence, according to formulas (3.1) and (3.5) it follows that inequality

$$\int_0^b \{[\nu - p_2(x)]|\psi_{22}|^2 + [\nu + p_2(x)]|\psi_{21}|^2 + q_2(x)[\psi_{22}\bar{\psi}_{21} + \psi_{21}\bar{\psi}_{22}]\}dx \leq \operatorname{Im}\ell_1$$

is the condition for ℓ_1 to belong to the circle $C(\lambda, b)$, this inequality transitting to equality if and only if ℓ_1 lies on the boundary of the circle $C(\lambda,b)$ (i.e. $\operatorname{Im} W = 0$).

Now let $0 < a < b < \infty$. Therefore if $\ell_1 \in C(\lambda,b)$ then

$$\int_0^a \{[\nu - p_2(x)]|\psi_{22}|^2 + [\nu + p_2(x)]|\psi_{21}|^2 + q_2(x)[\psi_{22}\bar{\psi}_{21} + \psi_{21}\bar{\psi}_{22}]\}dx <$$

$$< \int_0^b \{[\nu - p_2(x)]|\psi_{22}|^2 + [\nu + p_2(x)]|\psi_{21}|^2 + q_2(x)[\psi_{22}\bar{\psi}_{21} + \psi_{21}\bar{\psi}_{22}]\}dx \leq \operatorname{Im}\ell_1$$

and hence $\ell_1 \in C(\lambda, a)$. So the circle $C(\lambda, a)$ includes the circle $C(\lambda, b)$ wherefrom it follows that for every fixed value of λ ($\operatorname{Im}\lambda < \tau_1$) circles $C(\lambda,b)$ converge to some limit circle or point $C(\lambda,\infty)$ as $b \to \infty$, and if $\ell_1 \in C(\lambda,\infty)$ then

$$\int_0^\infty \{[\nu - p_2(x)]|\psi_{22}|^2 - [\nu + p_2(x)]|\psi_{21}|^2 + q_2(x)[\psi_{22}\bar{\psi}_{21} + \psi_{21}\bar{\psi}_{22}]\}dx \leq -\operatorname{Im}\ell_1 .$$

Let $\operatorname{Im}\lambda < \tau_0$. Reasoning the same way we can also obtain an inequality

$$\int_a^b \left\{ [p_2(x) - \nu] |\varphi_{22}|^2 - [\nu + p_2(x)] |\varphi_{21}|^2 + q_2[\varphi_{22}\overline{\varphi}_{21} + \varphi_{21}\overline{\varphi}_{22}] \right\} dx \le -\operatorname{Im} \ell_1 .$$

Now we take some real value of W (for example $W = \infty$) and consider ℓ_1 as a function of λ. It follows from formula (3.7) that for each value in $\ell = \ell_1(\lambda, b, \infty)$ there is a meromorphic function of λ, its poles lying inside the strip

$$\tau_0 \le \operatorname{Im} \lambda \le \tau_1 . \tag{3.11}$$

Preceding considerations show that $\ell_1(\lambda, b, \infty) \in C(\lambda, b)$ for any λ lying out of this strip, and circles $C(\lambda, b)$ tighten to $C(\lambda, \infty)$ as $b \to \infty$. Therefore functions $\ell_1(\lambda, b, \infty)$ (b is a parameter) form a family of equipotentially bounded holomorphic functions in every finite part of λ-plane entirely lying out of strip (3.11). Due to well-known theorem on the theory of functions of a complex variable it follows that we may choose a sequence $b_n \to \infty$ such that $\ell_1(\lambda, b_n, \cdot)$ converge to some function $m_2(\lambda)$ holomorphic out of strip (3.11) as $n \to \infty$. It is evident that $m_2(\lambda) \in C(\lambda, b)$.

Similarly we may construct the solution

$$\Psi_1(x, \lambda) = \omega_1(x, \lambda) + m_1(\lambda)\, \omega_2(x, \lambda)$$

belonging to $L^2(-\infty, \infty)$. It occurs that

$$\int_{-\infty}^{\infty} \left\{ [\nu - p_2(x)] |\varphi_{12}|^2 + [\nu + p_2(x)] |\varphi_{11}|^2 + q_2(x) [\varphi_{21}\overline{\varphi}_{11} + \varphi_{11}\overline{\varphi}_{21}] \right\} dx = \operatorname{Im} m(\lambda)$$

So we have the following

<u>THEOREM 3.1</u> <u>If condition</u> (3.1) <u>is carried out then equation</u> (3.3) <u>has solutions</u>

$$\begin{aligned} \Psi_1(x, \lambda) &= \omega_1(x, \lambda) + m_1(\lambda)\, \omega_2(x, \lambda) \in L^2(-\infty, 0) \\ \Psi_2(x, \lambda) &= \omega_1(x, \lambda) + m_2(\lambda)\, \omega_2(x, \lambda) \in L^2(0, \infty) \end{aligned} \tag{3.12}$$

<u>where Weil's functions</u> $m_1(\lambda)$, $m_2(\lambda)$ <u>are holomorphic out of strip</u> (3.11) <u>for all</u> λ <u>lying out of this strip.</u> <u>Equalities</u>

$$\int_{-\infty}^{0}\{[\nu - p_2 x]|\psi_{21}|^2+[\nu + p_2(x)]|\psi_{11}|^2+q_2(x)[\psi_{12}\overline{\psi}_{11}+\psi_{11}\overline{\psi}_{12}]\}\,dx = -\operatorname{Im} m_1(\lambda)$$

$$\int_{0}^{\infty}\{[\nu - p_2(x)]|\psi_{22}|^2+[\nu + p_2(x)]|\psi_{21}|^2+q_2(x)[\psi_{22}\overline{\psi}_{21}+\psi_{21}\overline{\psi}_{21}]\}\,dx = \operatorname{Im} m_2(\lambda) \qquad (3.13)$$

<u>are valid under this condition.</u>

Vronskyi's determinant of the solutions of equation (3.3) does not depend on x and may be calculated by condition (2.2):

$$W(\Psi_1, \Psi_2) = \psi_{11}\psi_{22} - \psi_{12}\psi_{21} = m_2(\lambda) - m_1(\lambda). \qquad (3.14)$$

Since the solutions $\omega_1(x,\lambda)$, $\omega_2(x,\lambda)$ form a fundamental system from equalities (3.12) and (3.13) we obtain

<u>COROLLARY 3.1</u> <u>Vronskian</u> $W(\Psi_1, \Psi_2)$ <u>of the solutions</u> $\Psi_1(x,\lambda)$, $\Psi_2(x,\lambda)$ <u>of the equation</u> (3.3) <u>differs from zero, more exactly</u>

$$\operatorname{Im}[m_2(\lambda) - m_1(\lambda)] > 0 \qquad \text{for} \quad \operatorname{Im}\lambda > \tau_1$$

$$\operatorname{Im}[m_2(\lambda) - m_1(\lambda)] < 0 \qquad \text{for} \quad \operatorname{Im}\lambda < \tau_0$$

<u>for all</u> λ <u>lying out of strip</u> (3.11).

We define operator R_λ on all vector functions $f(x) \in L^2(-\infty,\infty)$ by the equality

$$R_\lambda f(x) = \frac{1}{m_1(\lambda) - m_2(\lambda)}\left\{\Psi_2(x,\lambda)\int_{-\infty}^{x}\tilde{\Psi}_1(t,\lambda)\,f(t)\,dt + \Psi_1(x,\lambda)\int_{x}^{\infty}\tilde{\Psi}_2(t,\lambda)\,f(t)\,dt\right\}. \qquad (3.15)$$

From the definition of $R_\lambda f$ it follows that the vector function $y(x,\lambda) = R_\lambda f(x)$ is absolutely continuous in every finite interval of variation of x .

The equality

$$(L - \lambda)\, R_\lambda f(x) = f(x) \tag{3.16}$$

holds for every vector function $f(x) \in L^2(-\infty, \infty)$ and for every λ not belonging to strip (3.11).

Moreover, there exist the equality

$$R_\lambda (L-\lambda)\, f(x) = f(x) \tag{3.17}$$

for every vector function $f(x) \in \mathcal{D}(L)$ from the domain of **definition of the operator** L

It follows from equalities (3.16) and (3.17) that operator R_λ defined by formula (3.15) is the resolvent of operator L.

LEMMA 3.1 If $f(x) \in \mathcal{D}(L)$ then vector function $y(x,\lambda) = R_\lambda f(x)$ admits the following representation :

$$R_\lambda f(x) = \frac{1}{\lambda}\left[f(x) - R_\lambda L f(x) \right]. \tag{3.18}$$

To prove this it is sufficient to substitute

$$\frac{1}{\lambda}\left[B\ \psi'(t,\lambda) + Q(t)\,\psi(t,\lambda) \right]$$

for vector function $\psi_j(t,\lambda)$ under the integral in formula (3.7), and to integrate by parts using equality (3.14).

Let us introduce some notations. We denote reduction of the operator L to finite vector functions by L_0. Let $m_1(\lambda)$, $m_2(\lambda)$ are Weil's functions the existence of which theorem 3.1 establishes. With the help of Weil's functions $m_1(\lambda)$, $m_2(\lambda)$ we introduce functions $m_{jk}(\lambda)$ in the following way:

$$m_{11}(\lambda) = \frac{1}{m_1(\lambda) - m_2(\lambda)}, \quad m_{22}(\lambda) = \frac{m_1(\lambda)\, m_2(\lambda)}{m_1(\lambda) - m_2(\lambda)}, m_{12}(\lambda) = m_{21}(\lambda) = \frac{m_1(\lambda)}{m_1(\lambda) - m_2(\lambda)} \quad (3.19)$$

The following theorem is valid.

THEOREM 2.4 *Every vector function* $f(x) \in \mathcal{D}(L_0)$, *the domain of definition of the operator* L_0, *admits the expansion*

$$f(x) = -\frac{1}{2\pi i} \lim_{\delta \to 0} \sum_{j,k=1}^{2} \left\{ \int_{-\infty + i(\tau_1+\varepsilon)}^{+\infty + i(\tau_1+\varepsilon)} + \int_{+\infty + i(\tau_0-\varepsilon)}^{-\infty + i(\tau_0-\varepsilon)} \right\} e^{-\delta\lambda^2} m_{jk}(\lambda)\, \omega_k(x,\lambda)\, E_f^j(\lambda)\, d\lambda . \quad (3.20)$$

Proof. Let $f(x) \in \mathcal{D}(L_0)$. Then $E_f^j(\lambda)$ are entire functions of parameter λ. By theorem 3.1 and corollary 3.1 we conclude that vector function $y(x,\lambda) = R_\lambda f(x)$ is analytical out of strip (3.11) which does not have any singular point (except an infinitely removed point). Hence, for any $N > 0$ we have identities

$$\int_{-N+i(\tau_1+\varepsilon)}^{N+i(\tau_1+\varepsilon)} y(x,\lambda)\, d\lambda + \int_{C_N^+} y(x,\lambda)\, d\lambda = 0 \quad (3.21)$$

$$\int_{N+i(\tau_0-\varepsilon)}^{-N+i(\tau_0-\varepsilon)} y(x,\lambda)\, d\lambda + \int_{C_N^-} y(x,\lambda)\, d\lambda = 0 \quad (3.22)$$

where C_N^+ (C_N^-) denotes semicircumference lying in the upper (lower) half-plane diameter of which is the segment

$$-N \leq \operatorname{Re}\lambda \leq N, \quad \operatorname{Im}\lambda = \tau_1 + \varepsilon, \quad (-N \leq \operatorname{Re}\lambda \leq N, \ \operatorname{Im}\lambda = \tau_0 - \varepsilon).$$

Moreover, semicircumferences C_N^+, C_N^- go round in positive direction in formulas (3.21) and (3.22). At first we show that

$$\lim_{N \to \infty} \int_{C_N^+} y(x,\lambda)\, d\lambda = \pi i\, f(x) \quad (3.23)$$

$$\lim_{N \to \infty} \int_{C_N^-} y(x,\lambda)\, d\lambda = \pi i\, f(x) .$$

Since proofs are similar we give only the first one. By Lemma 3.1

$$y(x,\lambda) = \frac{1}{\lambda} f(x) - \frac{1}{\lambda} R_\lambda g(x)$$

where $g(x) = Lf(x) \in L^2(-\infty, \infty)$. Therefore

$$\int_{C_N^+} y(x,\lambda)\,d\lambda = f(x)\int_{C_N^+} \frac{d\lambda}{\lambda} - \int_{C_N^+} \frac{1}{\lambda} R_\lambda g(\lambda)\,d\lambda .$$

The first of the integrals in the right side is equal to $\pi i f(x)$. We can show that the second integral has infinitely small value at large N. Then taking in account (3.23) from identities (3.21) and (3.22) we obtain

$$f(x) = -\frac{1}{2\pi i} \lim_{N\to\infty} \left\{ \int_{-N+i(\tau_1+\varepsilon)}^{N+i(\tau_1+\varepsilon)} + \int_{N+i(\tau_0-\varepsilon)}^{-N+i(\tau_0-\varepsilon)} \right\} y(x,\lambda)\,d\lambda \qquad (3.24)$$

where integrals are understood in the sense of principal value.

Using the formula

$$\lim_{N\to\infty} \int_{-N+ia}^{N+ia} \varphi(\lambda)\,d\lambda = \lim_{\delta\to 0} \int_{-\infty+ia}^{\infty+ia} e^{-\delta\lambda^2} \varphi(\lambda)\,d\lambda$$

which is valid if there exists the limit in the left side, the equality (3.24) may be written in the form

$$f(x) = -\frac{1}{2\pi i} \lim_{\delta\to+0} \left\{ \int_{-\infty+i(\tau_1+\varepsilon)}^{\infty+i(\tau_1+\varepsilon)} + \int_{+\infty+i(\tau_0-\varepsilon)}^{-\infty+i(\tau_0-\varepsilon)} \right\} e^{-\delta\lambda^2} y(x,\lambda)\,d\lambda . \qquad (3.25)$$

Now we transform $y(x,\lambda)$. Substituting for vector functions $\psi_j(x,\lambda)$ into formula (3.15) their expression by $\omega_1(x,\lambda)$ $\omega_2(x,\lambda)$ due to formulas (3.12) we obtain

$$y(x,\lambda) = \frac{1}{m_1(\lambda) - m_2(\lambda)} \Big\{ \left[\omega_1(x,\lambda) + m_2(\lambda)\,\omega_2(x,\lambda)\right] \times$$

$$\times \int_{-\infty}^{x} \left[\tilde{\omega}_1(t,\lambda) + m_1(\lambda)\,\tilde{\omega}_2(t,\lambda)\right] f(t)\,dt + \left[\omega_1(x,\lambda) + m_1(\lambda)\,\omega_2(x,\lambda)\right] \times$$

$$\times \int_{x}^{\infty} \left[\tilde{\omega}_1(t,\lambda) + m_2(\lambda)\,\tilde{\omega}_2(t,\lambda)\right] f(t)\,dt \Big\} =$$

$$= \frac{1}{m_1(\lambda)-m_2(\lambda)}\,\omega_1(x,\lambda)\,E_f^1(\lambda) + \frac{m_1(\lambda)}{m_1(\lambda)-m_2(\lambda)}\Big[\omega_2(x,\lambda)\,E_f^1(\lambda) +$$

$$+\ \omega_1(x,\lambda)\,E_f^2(\lambda)\Big] + \frac{m_1(\lambda)\,m_2(\lambda)}{m_1(\lambda)-m_2(\lambda)}\,\omega_2(x,\lambda)\,E_f^2(\lambda) - \mu(x,\lambda)$$

where

$$\mu(x,\lambda) = \omega_2(x,\lambda)\int_{-\infty}^{x}\tilde{\omega}_1(t,\lambda)\,f(t)\,dt + \omega_1(x,\lambda)\int_{x}^{\infty}\omega_2(t,\lambda)\,f(t)\,dt\,. \qquad (3.26)$$

Using notations (3.19) equality (3.25) may be written as

$$f(x) = -\frac{1}{2\pi i}\lim_{\delta\to+0}\sum_{j,k=1}^{2}\Big\{\int_{-\infty+i(\tau_1+\varepsilon)}^{\infty+i(\tau_1+\varepsilon)} + \int_{\infty+i(\tau_0-\varepsilon)}^{-\infty+i(\tau_0-\varepsilon)}\Big\}\, e^{-\delta\lambda^2}\, m_{jk}(\lambda)\,\omega_k(x,\lambda)\,E_f^i(\lambda)\,d\lambda -$$

$$-\lim_{\delta\to+0}\frac{-1}{2\pi i}\Big\{\int_{-\infty+i(\tau_1+\varepsilon)}^{\infty+i(\tau_1+\varepsilon)} + \int_{\infty+i(\tau_0-\varepsilon)}^{-\infty+i(\tau_0-\varepsilon)}\Big\}\, e^{-\delta\lambda^2}\,\mu(x,\lambda)\,dx\,.$$

To end the proof of the theorem it is sufficient to prove that the last integral is zero. By formulas (2.3) we have

$$O(\omega_1(x,\lambda)) = O\big(e^{|\mathrm{Im}\,\lambda x|}\big), \qquad O(\omega_2(x,\lambda)) = O\big(e^{|\mathrm{Im}\,\lambda x|}\big).$$

Since $f(x) \in \mathcal{D}(L_0)$ we can point out number $N > 0$ such that $f(x) = 0$ for $|x| > N$ therefore

$$O\Big\{\int_{-\infty}^{x}\tilde{\omega}_1(t,\lambda)\,f(t)\,dt\Big\} = O\big(e^{|\mathrm{Im}\,\lambda N|}\big)$$

$$O(\omega_1(x,\lambda)) = O\big(e^{|\mathrm{Im}\,\lambda N|}\big)\,.$$

Hence vector function $e^{-\delta\lambda^2}$ which is entire in regard to λ in every strip $|\mathrm{Im}\,\lambda| < C$ tends uniformly to zero as $|\lambda| \to \infty$ for every x. Therefore

$$\lim_{\delta\to+0}\Big\{\int_{-\infty+i(\tau_1+\varepsilon)}^{+\infty+i(\tau_1+\varepsilon)} + \int_{+\infty+i(\tau_0-\varepsilon)}^{-\infty+i(\tau_0-\varepsilon)}\Big\}\, e^{-\delta\lambda^2}\,\mu(x,\lambda)\,d\lambda = 0\,.$$

That proves the theorem.

A common corollary of theorems 2.3 and 3.2 is the following

THEOREM 3.3. Spectral matrix function $\| R_{jn}(\lambda) \|$ of problem (3.3)-(2.2) is given by formulas

$$(R_{jn}, E^j) = \lim_{\delta \to +0} \frac{-1}{2\pi i} \left\{ \int_{-\infty + i(\tau_1+\varepsilon)}^{\infty + i(\tau_1+\varepsilon)} + \int_{\infty + i(\tau_0-\varepsilon)}^{-\infty + i(\tau_0-\varepsilon)} \right\} e^{-\delta\lambda^2} m_{jn}(\lambda) E_f^j(\lambda) d\lambda$$

on all functions $E_f^j(\lambda)$ $(j=1,2)$ being Fourier ω-transformation of vector function $f(x) \in \mathfrak{D}(L_0)$.

The authors are grateful to F.S.Rofe-Beketov for his acquaintance with present work and helpful remarks.

Bibliography

1. Marchenko, V.A., Matematicheskyi sbornik, v.52, N2, 1960
2 Funtakov, V.N., Differentsialnye uravneniya, v.VI, N11, 1970
3 Gasymova, E.M., Izv. Acad. Nauk Azerb. SSR, ser.phys.-math., 1969, N2
4 Maksudov, F.G. and Veliev, S.G., Trudy IMM AN Azerb.SSR, "Investigation on differential equations". Ed. "Elm"AN Azerb SSR, 1971.
5 Gelfand I.M. and Shilov, G.E., "Generalized functions and performances with them". M.Fizmatgiz, 1959
6 Lidskyi, V.B., Trudy Moskovskogo matematicheskogo obschestva, 8, 1959

TOPOLOGICAL ALGEBRAS IN SEVERAL COMPLEX VARIABLES

Anastasios Mallios

0. Introduction

We give in this paper complete proofs of previous results of this author, regarding continuous (algebra) homomorphisms of topological tensor product algebras [18] into a given topological algebra, in what essentially concerns the (canonical) decomposition of such homomorphisms into tensor products of analogous maps of the factors of the tensor product algebras considered (cf., for instance, Theorem 2.2 below). Spaces of such continuous homomorphisms of a given topological algebra (: generalized characters) have been called generalized spectra of the algebras considered [22], a notion motivated by a discussion concerning representations of (topological, actually Banach) algebras of operator-valued holomorphic functions by operators on a given Hilbert space, included in Ref. [28]. As an application, the corresponding results of the last reference are extended to the entirely abstract setting adopted herein. On the other hand, when considering topological algebras with identity elements, the results obtained below have also a special bearing on previous results of this author concerning analogous decomposition questions for the usual spectrum (: Gel'fand space) of a given topological tensor product algebra in terms of the corresponding spectra of the factor algebras (cf., for instance, [15], [18], [19]). Definitive results from this part of the present paper have been previously announced (without proofs) in Ref. [20], [22]. Besides, the results presented in this part of the present paper have been arranged with the concrete applications in mind considered in the second part of this paper.

Thus, we are discussing next, by applying results of the first part of the paper, the particular form which representations (: continuous algebra homomorphisms) of algebras of holomorphic functions into abstract topological algebras take, in case the functions involved are topological algebra-valued as well. The latter case is naturally connected with the notion of topological tensor product algebras, hence the kind of applications considered in connection with the previous part of the paper (cf., for instance, Theorem 2.1 in Part II below). The results obtained also extend analogous considerations in Ref. [28] , which have also been the initial motivation to the present setting.

Finally, another feature, which might be of interest in itself, is the systematic use of the notion of a topological algebra sheaf in getting results in this part of the present paper, and which also seems to be so far one of the best justifications in considering this notion, which on the other hand was motivated by an analogous setting, to that considered in the sequel, given in Ref. [21] . In this concern, a more detailed presentation regarding this point of view has been given in Ref. [24] , [25] .

Partial results of the present paper (cf., for instance, § 3 of Part I), which on the other hand have essentially been incorporated, although in another context, in Ref. [17] , have a special bearing on, by providing besides a better insight into, considerations included in Ref. [27] .

The author is greatly indebted to Professors K.R.Unni and A.Ramakrishnan for the invitation to the Conference, which provided an opportunity to give the lecture on which the present paper is based.

PART I

1. Generalized spectra of topological algebras

In this part of the present paper we essentially give complete proofs of previous results referring to the title of this Section, and which have already been reported in [22].

The topological algebras, we are dealing with in the sequel, will be locally convex ones with a (jointly) continuous multiplication (cf., for instance, [15]). The field of coefficients is the complexes and all topological spaces are assumed to be Hausdorff, unless otherwise indicated.

Thus, by simply referring in the sequel to topological algebras, we shall always mean those indicated above, unless otherwise specified.

Now, given the topological algebras E, F, we denote by $\mathcal{L}_s(E,F)$ the space of continuous linear maps between the respective (locally convex) topological vector spaces, equipped with the topology of simple convergence in E. On the other hand, we denote by $\mathcal{H}om_s(E,F)$ its subspace of continuous algebra homomorphisms, and by $\mathcal{M}(E,F)$ the respective set of the non-zero elements endowed with the relative topology. It is such a set, which we shall call the generalized spectrum of the topological algebra E, with respect to a given (topological) algebra F.

In this respect, given the topological algebras E and F, the generalized Gel'fand transform of an element $x \in E$ is the map $\hat{x} : \mathcal{H}om_s(E,F) \longmapsto F$, defined by $\hat{x}(h) = h(x)$, for every $h \in \mathcal{H}om_s(E,F)$, being continuous for the respective topological spaces.

Now, we first have the following lemma, which is of fundamental importance for the sequel. It extends the analogous situation described in [19; p. 102, Lemma 2.1] (cf. also the discussion in [29; p.112], which was also the initial motivation to the present setting). Thus, we have

LEMMA 1.1. *Let $\hat{E}$ be a locally convex algebra, completion of a locally convex algebra E and let S be a set of generators of E. Moreover, let F be a complete locally convex algebra (with a continuous multiplication), and let the following condition be satisfied:*

(1.1.1) For every $h \in \mathcal{H}om(E, F)$, there exists a neighborhood U of h in $\mathcal{H}om_s(E, F) \subseteq \mathcal{L}_s(E, F)$ such that for every $z \in \hat{E}$ and for every net (x_i) in E converging to z, the net $(\hat{x}_i)$ converges to $\hat{z}$, uniformly on U.

Then, the topology of $\mathcal{H}om(\hat{E}, F) \hookrightarrow \mathcal{L}_s(\hat{E}, F)$ is the "weak topology" defined on it by the maps $\hat{x}$, $x \in S$.

Proof. Let $z \in \hat{E}$, and let $(x_i)_{i \in I}$ be a net in E converging to z in $\hat{E}$. Now, $x_i = P_i(\{a_k\})$, $i \in I$, where P_i denotes a polynomial in a finite number of elements $a_k \in S$. Hence, $x_i(h) = P_i(\{\hat{a}_k(h)\})$ for every $h \in \mathcal{H}om(\hat{E}, F)$. Now, by the cond.(1.1.1) above, $\hat{x}_i \longrightarrow \hat{z}$ locally uniformly in $\mathcal{H}om_s(E, F)$, so that by the continuity of each $\hat{a}_k$ and hence of each $\hat{x}_i$, one obtains that of $\hat{z}$ in $\mathcal{H}om_s(E, F)$, and this proves the assertion.

The following proposition is now an immediate consequence of Lemma 1.1 above, so that we may omit the details of the proof. It is also in the form of Proposition 1.1 below, that we are actually going to apply the preceding lemma in the sequel.

We first fix the terminology applied in this context. Thus, by referring to the topological space $\mathcal{H}om_s(E, F)$, as defined above, we shall say that a subset A of it is *locally equicontinuous*, if, for every $h \in A$, there exists a neighborhood U of h in $\mathcal{H}om_s(E, F)$, which is an equicontinuous subset of $\mathcal{H}om(E, F) \hookrightarrow \mathcal{L}_s(E, F)$ (cf. also [19; p.102, 2]).

We now have the following.

PROPOSITION 1.1. Let E, F be locally convex (topological) algebras (with continuous multiplication), in such a way that $\mathcal{H}om(E, F) \subsetneq \mathcal{L}_s(E, F)$ is a locally equicontinuous subset of the space indicated. Then, the space $\mathcal{H}om_s(E, F)$ satisfies the condition (1.1.1) of Lemma 1.1 above, and hence one has

$$\mathcal{H}om_s(E,F) = \mathcal{H}om_s(\hat{E}, F), \tag{1.1}$$

within a homeomorphism of the topological spaces involved.

The significance of the preceding proposition lies exactly in the establishment of a sufficient condition in order the relation (1.1) above to be valid (in the case under consideration, this is the local equicontinuity of the (generalized) spectra involved), which in general holds true only within a continuous bijection. In this respect, cf. for instance [15; p.173] or [4; p.210]. On the other hand, cf. also [23; p.479, Scholium]. Now, we shall have the occasion to apply presently the same proposition in the next section.

2. Generalized spectra of topological tensor (product) algebras

By a tensor algebra we shall mean in the following a tensor product of (two) algebras with corresponding (ring) multiplication the (linear) extension of that between the decomposable tensors (cf., for example, [2, A III. 33, § 4]). We shall exclusively consider below the case the factor algebras are topological ones and the respective tensor algebra is equipped with a suitable vector space topology making it a topological algebra. We shall call such an object a topological tensor algebra (cf. also [20]). In this case, we shall also speak of an admissible (or compatible) topology on the respective tensor (product) algebra (cf. also [15; p.174, Definition 3.1] and [18; p.4, Definition 3.1]). Thus, given the topological algebras E, F, we denote by $E \underset{\tau}{\otimes} F$ the respective topological tensor algebra, corresponding to a given admissible topology τ on $E \otimes F$, as indicated above.

On the other hand, we usually require for admissible topologies τ as above to satisfy one or both of the following two conditions:

The canonical bilinear map of $E \times F$ into $E \underset{\tau}{\otimes} F$ is separately continuous. (2.1)

For every topological algebra G and for any elements $f \in \mathcal{H}om(E,G)$ and $g \in \mathcal{H}om(F,G)$, one has $f \otimes g \in \mathcal{L}(E \underset{\tau}{\otimes} F, G)$. (2.2)

The projective tensorial topology π on $E \otimes F$ (A.Grothendieck [11]) is an admissible one in the preceding sense, which also satisfies the above two conditions (cf. [15; p.176, Proposition 3.2] and [11; Chap.I, p.30, Proposition 2]). The same remark holds also true in case the topological algebras involved in the preceding two conditions have a separately continuous multiplication (ibid.).

A stronger version of the second condition above is the following one, which we shall also use in the sequel. That is, in the context of the preceding terminology, we have

For any equicontinuous subsets $A \subseteq \mathcal{H}om(E,G)$ and $B \subseteq \mathcal{H}om(F,G)$ one has that $A \otimes B$ is an equicontinuous subset of $\mathcal{L}(E \underset{\tau}{\otimes} F, G)$. (2.3)

The preceding condition is also satisfied in case $\tau = \pi$: Indeed, for any $f \in \mathcal{H}om(E,G)$ and $g \in \mathcal{H}om(F,G)$, $f \times g : E \times F \to G$ is a continuous bilinear map defined by $(f \times g)(x, y) = f(x)\, g(y)$, for any $x \in E$, $y \in F$. (Separate continuity of the multiplication in G would imply the last map to be also separately continuous). Now, $A \times B$ is an equicontinuous subset of $\mathcal{B}(E,F; G)$, the space of continuous bilinear maps of E x F into G, so that the assertion is now obtained by a basic property of topology π (cf. [11; Chap. I, p.30, Proposition 2]). On the other hand, in case the topological algebra G is commutative, then condition (2.3) implies, in particular, that $A \otimes B$ is actually a subset of $\mathcal{H}om(E \underset{\tau}{\otimes} F, G)$, and a similar remark is also in order, concerning the condition (2.2) above.

The subsequent theorem is fundamental for the sequel. In case of algebras with identity elements, it extends the basic formula relating the spectrum of a topological tensor algebra to those of the factor algebras of the tensor product considered (cf., for instance, [18] as well as [15], [19]). It has also been obtained in [28] in the particular case of vector-valued analytic function algebras, which case was also the motivation to the present abstract setting. Thus, we have

THEOREM 2.1. *Let* E *and* F *be topological algebras with identity elements, and let* G *be a topological algebra with continuous multiplication. Moreover, let* τ *be an admissible topology on the respective tensor product algebra* $E \otimes F$ *satisfying the condition (2.1) and (2.2) above. Then, there exists a homeomorphism*

$$\varphi : \mathcal{H}om_s(E \underset{\tau}{\otimes} F, G) \longmapsto \mathcal{H}om_s(E, G) \times \mathcal{H}om_s(F, G) \quad (2.4)$$

whose range are those (f, g) *for which one has* $f(x)\ g(y) = g(y)\ f(x)$ *in* G, *for every* $(x, y) \in E \times F$. *On the other hand, by restricting (2.4) to the non-zero homomorphims, we get a homeomorphism (into)*

$$\psi : \mathfrak{M}(E \underset{\tau}{\otimes} F, G) \longmapsto \mathfrak{M}(E, G) \times \mathfrak{M}(F, G) . \quad (2.5)$$

In particular, when G is commutative, the preceding maps are onto.

Proof. If $h \in \mathcal{H}om\,(E \underset{\tau}{\otimes} F, G)$, the relations

$$f(x) = h(x \otimes 1),\ x \in E \text{ and } g(y) = h(1 \otimes y),\ y \in F, \quad (2.6)$$

where 1 stands for the identity element in E, F respectively, define f and g as (algebra) homomorphisms of E and F respectively into G, which are also continuous by the hypothesis for h and cond. (2.1) above. On the other hand, one has by (2.6),

$$h(x \otimes y) = h(\,(x \otimes 1)\,(1 \otimes y)\,) = h\,(\,(1 \otimes y)\,(x \otimes 1)\,)$$
$$= h(x \otimes 1)\ h(1 \otimes y) = f(x)g(y) = h(1 \otimes y)\ h(x \otimes 1)$$
$$= g(y)\ f(x),$$

so that $h = f \otimes g$, that is, for every $z = \sum_i x_i \otimes y_i$, one has $h(z) = \sum_i f(x_i)\ g(y_i)$. This also proves that the map φ given by (2.4) is injective. Now, let $f \in \mathcal{H}om\,(E, G)$ and $g \in \mathcal{H}om(F,G)$, so that by the cond.(2.2), one has $f \otimes g \in \mathcal{L}(E \underset{\tau}{\otimes} F, G)$, and hence if, moreover, $f(x)\ g(y) = g(y)\ f(x)$, for any elements $x \in E$ and $y \in F$, one obtains that $f \otimes g \in \mathcal{H}om\,(E \underset{\tau}{\otimes} F, G)$, so that $h = f \otimes g = f_1 \otimes g_1$, where f_1, g_1 are given by the relations (2.6) above. Therefore, $f_1(x) = h(x \otimes 1) = (f \otimes g)\ (x \otimes 1) = f(x)$, $x \in E$ and similarly $g_1(y) = (f \otimes g)\ (1 \otimes y) = g(y)$, $y \in F$, so that φ is, in particular, a bijection whose range is the set indicated in the statement of the theorem. Now, the bicontinuity of the map φ can be derived by an obvious modification, for the case under consideration, of the respective argumentation applied in Ref. [15; p.178, Theorem 4.1]. The rest of the proof is now an immediate consequence of the preceding and the theorem is proved.

The topological algebras considered in the preceding theorem might be not necessarily locally convex ones with multiplication separately continuous (except of the algebra G as indicated). In this respect, cf. for instance the considerations in Ref. [18]. On the other hand, to facilitate the arguments, we suppose below that the respective topological algebras are locally convex ones with continuous multiplication, as this was indicated at the beginning of this Part of the present paper.

Thus, let E, F and G be topological algebras as indicated before with G complete, and let τ be an admissible topology on $E \otimes F$. Then, there exists a continuous bijection of $\mathcal{H}om_s(E \hat{\otimes}_\tau F, G)$ into $\mathcal{H}om_s(E \otimes_\tau F, G)$, where $E \hat{\otimes}_\tau F$ denotes the completion of the respective topological algebra $E \otimes_\tau F$, so that by the preceding theorem one obtains a continuous bijection of the first space onto that subset of $\mathcal{H}om_s(E, G) \times \mathcal{H}om_s(F, G)$ indicated by Theorem 2.1 above (cf. the relation (2.4)). Now, motivated by the situation one has in the case of the usual spectrum (: complex continuous (algebra) homomorphisms) of a topological algebra [10], [18], we presently examine the case, the preceding continuous bijection has also a continuous inverse.

The following lemma is justified by Proposition 1.1 in the preceding and Theorem 2.2 below. On the other hand, it can be obtained by applying the argumentation in the first half of the proof of Proposition 2.1 in Ref. [29; p.103], and making the obvious modifications for the present setting in connection with Theorem 2.1 above and cond.(2.3), so that we may omit the details of the proof. Besides, we note that cond. (2.3), with τ being any admissible topology on $E \otimes F$, obviously implies the cond. (2.2) cf. also [18; p.9, cond. (4.3)]). Thus, we have :

LEMMA 2.1. Let E, F, G be topological algebras satisfying the conditions of Theorem 2.1 above, and let τ be an admissible topology on $E \otimes F$, satisfying the conditions (2.1) and (2.3) above. Moreover, suppose that $\mathcal{H}om(E, G) \subseteq \mathcal{L}_s(E, G)$ and $\mathcal{H}om(F, G) \subseteq \mathcal{L}_s(F, G)$ are locally equicontinuous subsets of the spaces indicated. Then, $\mathcal{H}om(E \hat{\otimes}_\tau F, G) \subseteq \mathcal{L}_s(E \hat{\otimes}_\tau F, G)$ is also a locally equicontinuous subset of the respective space.

We are now in the position to state the following fundamental theorem; an early announcement of it has been given in Ref. [20]. On the other hand, a more general version of it can be given in the context of Theorem 4.1 in Ref. [26], by considering topological algebras with approximate identities (cf. also ibid., Scholium 4.1). That is, we have:

THEOREM 2.2. Let E, F, G be locally convex (topological) algebras with continuous multiplication, all having identity elements and let G be complete. Moreover, suppose that $\mathcal{H}om(E, G) \subseteq \mathcal{L}_s(E, G)$ and $\mathcal{H}om(F, G) \subseteq \mathcal{L}_s(F, G)$ are locally equicontinuous subsets of the respective spaces, and let τ be an admissible topology on the tensor product algebra $E \otimes F$ satisfying cond. (2.1) and (2.3) above. Then, the map

$$\varphi : \mathcal{H}om_s(E \hat{\otimes}_\tau F, G) \longmapsto \mathcal{H}om_s(E, G) \times \mathcal{H}om_s(F, G) \quad (2.7)$$

(cf. also Theorem 2.1, the relation (2.4) above) defines a bicontinuous injection between the corresponding spaces, which is a bijection (homeomorphisms), when G is commutative.

Proof. By the proof of Lemma 2.1 above, $\mathcal{H}om(E \otimes_\tau F, G)$ is a locally equicontinuous subset of $\mathcal{L}_s(E \otimes_\tau F, G)$ (cf. also [19; p.103, Proposition 2.1]), so that by Proposition 1.1 in the preceding, one has the relation $\mathcal{H}om_s(E \otimes_\tau F, G) = \mathcal{H}om_s(E \hat{\otimes} F, G)$ within a homeomorphism of the topological spaces involved, and hence the assertion is now reduced to that of the preceding Theorem 2.1, and the proof is complete. (An analogous assertion in the present context concerning the map ψ in the relation (2.5) above is here obviously redundant).

3. Inductive limits of topological tensor algebras.

The purpose of this section is to examine the relation, connecting the spectrum of a topological inductive limit algebra, whose ingredients are topological tensor product algebras, with the spectra of these later algebras involved. The motivations are the applications given in Part II of this paper concerning analytic function algebras (cf., for instance, Lemma 1.1 or Theorem 2.1), and they have been caused by the discussion contained in Ref. [28], regarding similar considerations in case of Banach algebra-valued analytic function algebras.

Thus, suppose that $(E_\alpha, f_{\beta\alpha})$ is an inductive system of topological algebras and let $E = \varinjlim E_\alpha$ be the respective locally convex inductive limit algebra. In this concern, the topological algebra E has a separately or jointly continuous multiplication in so far as this is the case for the individual algebras E_α, $\alpha \in I$: Cf. [16; p. 214, Proposition 2.1], as well as the considerations in Ref. [28; p.490, Section 7.7.11]. On the other hand, one has a different situation in case of locally m-convex (topological) algebras (cf. [16; p.215, Remark 2.1]).

Now, given a topological algebra F, one has the following relation, regarding the generalized spectra of the topological algebras involved (cf. also [16; p.215, Proposition 3.1]):

$$\mathcal{H}om_s(E, F) = \mathcal{H}om_s(\varinjlim E_\alpha, F) = \varprojlim \mathcal{H}om(E_\alpha, F), \tag{3.1}$$

within a homeomorphism. In particular, suppose that the following two conditions are satisfied:

For any $\alpha, \beta \in I$ and for every $g \in \mathcal{M}(E_\alpha, F)$, one has

$$\operatorname{Im}(f_{\beta\alpha}) \cap \complement \operatorname{Ker}(g) \neq \emptyset . \tag{3.1.1}$$

For every $\alpha \in I$, and for every $h \in \mathfrak{M}(E_\alpha, F)$, one has

$$\operatorname{Im}(f_\alpha) \cap \complement \operatorname{Ker}(h) \neq \emptyset . \tag{3.1.2}$$

Then, by the preceding relation (3.1), one obtains:

$$\mathfrak{M}(E, F) = \mathfrak{M}(\varinjlim E_\alpha, F) = \varprojlim \mathfrak{M}(E_\alpha, F) \tag{3.2}$$

within a homeomorphism (cf. also [16; p.128, Proposition 2.2]). On the other hand, we also note that the preceding two conditions are besides necessary in order the relation (3.2) above to be true (loc. cit.). In particular, the said conditions are fulfilled, if the (topological) algebras have identity elements and one considers identity preserving (algebra) homomorphism. In this respect, we note that this is, in particular, the case one usually encounters in the applications concerning the preceding relation (3.2) (cf., for instance, [27]).

Now, suppose that $E = \varinjlim E_\alpha$ and $F = \varinjlim F_\lambda$ are locally convex (topological) inductive limit algebras, and let, for each pair (α, λ) of indices, $E_\alpha \hat{\otimes}_\pi F_\lambda$ be the respective completed topological tensor (product) algebra in the projective tensorial topology π.

Now, one has an algebraic isomorphism onto

$$H = \varinjlim (E_\alpha \otimes_\pi F_\lambda) \longmapsto E \otimes_\pi F \tag{3.3}$$

which is also a continuous map, under the indicated topologies, so that one also obtains, for every locally convex (topological) algebra G a (natural) map:

$$\mathcal{H}om_s(E \otimes_\pi F, G) \longmapsto \mathcal{H}om_s(\varinjlim (E_\alpha \otimes_\pi F_\lambda), G) \tag{3.4}$$

On the other hand, the following lemma asserts that the spaces indicated in the preceding relation (3.4) are actually the same, within a homeomorphism, into a more general context regarding the tensorial topologies involved. That is, we have :

LEMMA 3.1. Let $E = \varinjlim E_\alpha$ and $F = \varinjlim F_\lambda$ be locally convex (topological) inductive limit algebras and let, for every pair (α, λ) of indices, $\tau(\alpha, \lambda)$ be an admissible topology on the tensor (product) algebra $E_\alpha \otimes F_\lambda$, satisfying the conditions (2.1) and (2.2) in the preceding. Moreover, let τ be an admissible topology on the respective tensor product algebra $E \otimes F$ satisfying the same conditions as above, and finally let G be a (locally convex) topological algebra. Then, one has a homeomorphism "from-onto", as follows

$$\varphi : \mathcal{H}om_s(\varinjlim (E_\alpha \underset{\tau(\alpha,\lambda)}{\otimes} F_\lambda), G) \longmapsto \mathcal{H}om_s(E \underset{\tau}{\otimes} F, G), \tag{3.5}$$

whose domain of definition is characterized by the analogous condition to that indicated in Theorem 2.1 in the preceding (cf. also the proof that follows). On the other hand, when G is commutative, one has the relation

$$\mathcal{H}om_s(\varinjlim (E_\alpha \underset{\tau(\alpha,\lambda)}{\otimes} F_\lambda), G) = \mathcal{H}om_s(E \underset{\tau}{\otimes} F, G), \tag{3.6}$$

within a homeomorphism of the topological spaces involved.

Proof. The argumentation is based on a repeated application of the relation (3.1) above as well as of Theorem 2.1 in the preceding, the equalities involved in the sequel of the proof being valid within a homeomorphism of the topological spaces indicated. Thus, we have :

$$\begin{aligned}
\mathcal{H}om_s(\varinjlim (E_\alpha \underset{\tau(\alpha,\lambda)}{\otimes} F_\lambda), G) &= \varprojlim \mathcal{H}om_s(E_\alpha \underset{\tau(\alpha,\lambda)}{\otimes} F_\lambda, G) \\
&\underset{\rightarrow}{\subseteq} \varprojlim (\mathcal{H}om_s(E_\alpha, G) \times \mathcal{H}om_s(F_\lambda, G)) \\
&= (\varprojlim \mathcal{H}om_s(E_\alpha, G)) \times (\varprojlim \mathcal{H}om_s(F_\lambda, G)) \\
&= \mathcal{H}om_s(\varinjlim E_\alpha, G) \times \mathcal{H}om_s(\varinjlim F_\lambda, G) \\
&= \mathcal{H}om_s(E, G) \times \mathcal{H}om_s(F, G),
\end{aligned}$$

so that the first part of the assertion of the lemma is now a consequence of Theorem 2.1 in the preceding and the hypothesis for the topology τ on $E \otimes F$. On the other hand, if moreover the algebra G is commutative, then by the same theorem as above, the last cartesian product space obtained by the preceding relations is homeomorphic to the space $\mathcal{H}om_s(E \underset{\tau}{\otimes} F, G)$ (an equality (: homeomorphism) being also valid in place of the inclusion relation above), which is the second part of the assertion, and this finishes the proof.

On the other hand, motivated by the situation one has in Theorem 2.2 above and the considerations in Proposition 3.2 of Ref. [16; p.216], one obtains the corresponding result of the preceding lemma, when completed (topological) tensor product algebras are also under consideration. Thus, we have the following

THEOREM 3.1. Let $E = \varinjlim E_\alpha$ and $F = \varinjlim F_\lambda$ be locally convex (topological) inductive limit algebras such that the individual (topological) algebras involved all have identity elements. Moreover, let G be a complete topological algebra (with an identity element) in such a way that, for any indices $\alpha \in I$ and $\lambda \in K$, the sets $\mathcal{H}om(E_\alpha, G) \underset{\to}{\subseteq} \mathcal{L}_s(E_\alpha, G)$ and $\mathcal{H}om(F_\lambda, G) \underset{\to}{\subseteq} \mathcal{L}_s(F_\lambda, G)$ are equicontinuous subsets of the spaces indicated. Finally, let $\tau(\alpha, \lambda)$ be an admissible topology on $E_\alpha \otimes F_\lambda$ satisfying the conditions (2.1) and (2.3) in the preceding, for every pair (α, λ) of indices, and let τ be a similar topology on the tensor product algebra $E \otimes F$. Then, in case the algebra G is commutative, one has the relation,

$$\mathcal{H}om_s(E \underset{\tau}{\hat{\otimes}} F, G) = \mathcal{H}om_s(\widehat{\varinjlim}(E_\alpha \underset{\tau(\alpha,\lambda)}{\hat{\otimes}} F_\lambda), G) \qquad (3.7)$$

within a homeomorphism of the topological spaces indicated, the general case, for G non-commutative, being analogously formulated as in the preceding Lemma 3.1.

Proof. We first note that by hypothesis and Proposition 3.2 in Ref. [16; p.216] (by an easy adaptation to the present context), each of the algebras $E = \varinjlim E_\alpha$ and $F = \varinjlim F_\lambda$ is such that the sets $\mathcal{H}om_s(E, G) \subseteq \mathcal{L}_s(E, G)$, $\mathcal{H}om_s(F,G) \subseteq \mathcal{L}_s(F,G)$ are equicontinuous subsets of the respective spaces as indicated. Hence, by hypothesis for the topology τ and Lemma 2.1, in connection with Proposition 1.1 in the foregoing, one has the relation

$$\mathcal{H}om_s(E \underset{\tau}{\otimes} F, G) = \mathcal{H}om_s(E \underset{\tau}{\hat{\otimes}} F, G) \tag{3.8}$$

within a homeomorphism of the spaces indicated. Similarly, by hypothesis for the topology $\tau(\alpha,\lambda)$ and Lemma 2.1 above, each of the algebras $E_\alpha \underset{\tau(\alpha,\lambda)}{\hat{\otimes}} F_\lambda$ is such that the set $\mathcal{H}om(E_\alpha \underset{\tau(\alpha,\lambda)}{\hat{\otimes}} F_\lambda, G) \subseteq \mathcal{L}_s(E_\alpha \underset{\tau(\alpha,\lambda)}{\hat{\otimes}} F_\lambda, G)$ is an equicontinuous subset of the space indicated, so that one has, for every pair (α, λ) of indices, the relation

$$\begin{aligned} \mathcal{H}om_s(E_\alpha \underset{\tau(\alpha,\lambda)}{\hat{\otimes}} F_\lambda, G) &= \mathcal{H}om_s(E_\alpha \underset{\tau(\alpha,\lambda)}{\otimes} F_\lambda, G) \\ &= \mathcal{H}om_s(E_\alpha, G) \times \mathcal{H}om_s(F_\lambda, G), \end{aligned} \tag{3.9}$$

within a homeomorphism of the topological spaces involved (cf. Proposition 1.1 and Theorem 2.1 above). Moreover, by [16; p.216, Proposition 3.2], the (topological) algebra $\varinjlim(E_\alpha \underset{\tau(\alpha,\lambda)}{\hat{\otimes}} F_\lambda)$ has the respective set of continuous G-valued (algebra) homomorphisms of it as an equicontinuous subset of the respective set of continuous linear maps, as above, so that one concludes, by Proposition 1.1 in the preceding, the relation:

$$\begin{aligned} &\mathcal{H}om_s(\widehat{\varinjlim} (E_\alpha \underset{\tau(\alpha,\lambda)}{\otimes} F_\lambda), G) \\ &\quad = \mathcal{H}om_s(\varinjlim(E_\alpha \underset{\tau(\alpha,\lambda)}{\hat{\otimes}} F_\lambda), G) \end{aligned} \tag{3.10}$$

within a homeomorphism of the topological spaces involved, and hence one obtains, within a homeomorphism for the respective equalities involved in the sequel, the relation :

$$\begin{aligned}
&\mathcal{H}om_s\,(\varinjlim\,(E_\alpha \underset{\tau(\alpha,\lambda)}{\hat{\otimes}} F_\lambda),\, G) \\
&= \varprojlim\, \mathcal{H}om_s\,(E_\alpha \underset{\tau(\alpha,\lambda)}{\hat{\otimes}} F_\lambda,\, G) \\
&= (\text{by the rel.}(3.9))\,\varprojlim\,(\,\mathcal{H}om_s(E_\alpha, G) \times \mathcal{H}om_s(F_\lambda, G)) \\
&= (\varprojlim\, \mathcal{H}om_s(E_\alpha, G)) \times (\varprojlim\, \mathcal{H}om_s(F_\lambda, G)) \qquad (3.11)\\
&= \mathcal{H}om_s(\varinjlim E_\alpha, G) \times \mathcal{H}om_s(\varinjlim F_\lambda, G) \\
&= \mathcal{H}om_s(E, G) \times \mathcal{H}om_s(F, G) \\
&= \mathcal{H}om_s(E \underset{\tau}{\otimes} F,\, G) = (\text{by the rel.}(3.8))\, \mathcal{H}om_s(E \underset{\tau}{\hat{\otimes}} F, G),
\end{aligned}$$

and this proves the desired relation (3.7) above. On the other hand, in case the algebra G is not necessarily commutative, following the reasoning in Theorem 2.1 in the preceding, as well as the argumentation in the proof of the preceding Lemma 3.1, one easily gets the respective relation to (3.7) above, the corresponding proof to this case being obvious, and this completes the proof of the theorem.

Remark.3.1. Regarding the hypothesis of the equicontinuity of the (generalized) spectra of the topological algebras involved in the preceding theorem, we remark that one could also consider locally equicontinuous spectra instead, taking into account the situation described in Proposition 3.1 in Ref. [19; p.105] (cf. also [21; p.304, Theorem 2.3] for a similar account). For simplicity's sake however we have restricted ourselves to consider only equicontinuity in the preceding Theorem 3.1.

On the other hand, we could also require local equicontinuity for the (generalized) spectra involved, in case we only were interested in a relation like (3.11) above. This will be, in particular, the case for the topological algebras considered in Part II below (cf., for instance, the situation related to Theorem 5.2 in the sequel). Thus, by applying the terminology of Lemma 3.1 and Theorem 3.1 in the preceding, and based on the corresponding proofs _one obtains respectively a map_

$$\varphi : \mathcal{H}om_s(\underset{\longrightarrow}{\lim}\,(E_\alpha \underset{\tau(\alpha,\lambda)}{\otimes} F_\lambda), G) \mapsto \mathcal{H}om_s(E,G) \times \mathcal{H}om_s(F,G), \tag{3.12}$$

and for locally equicontinuous (generalized) spectra a map

$$\Phi : \mathcal{H}om_s(\underset{\longrightarrow}{\lim}\,(E_\alpha \underset{\tau(\alpha,\lambda)}{\hat{\otimes}} F_\lambda), G) \longmapsto \mathcal{H}om_s(E,G) \times \mathcal{H}om(F,G), \tag{3.13}$$

which are homeomorphisms into, for the topological spaces indicated and are reduced to homeomorphisms (onto) in case the algebra G is commutative.

Now, by applying the preceding considerations, for reasons of later use, we give the following:

THEOREM 3.2. _Let the conditions of Theorem 3.1 be satisfied concerning the topological algebras considered, in such a way that the respective generalized spectra are only locally equicontinuous, so that the relation (3.13) holds true._ _On the other hand, if_

$$h \in \mathcal{H}om_s(\underset{\longrightarrow}{\lim}\,(E_\alpha \underset{\tau(\alpha,\lambda)}{\hat{\otimes}} F_\lambda), G), \qquad \text{and } h = f \otimes g$$

denotes the respective decomposition of h given by (3.13), let

$$\mathcal{M} = \mathrm{Im}(h), \quad \mathcal{M}_f = \mathrm{Im}(f) \quad \text{and} \quad \mathcal{M}_g = \mathrm{Im}(g)$$

be the respective by h, f and g image subalgebras of G. _Then, regarding the commutants of these subalgebras of G, one obtains_

$$\mathcal{M}'' \cap \mathcal{M}' \subseteq \mathcal{M}'' \cap \mathcal{M}'_g \tag{3.14}$$

so that if, moreover, the algebra E is commutative, one gets

$$\mathcal{M}'' \cap \mathcal{M}' = \mathcal{M}'' \cap \mathcal{M}'_g \tag{3.15}$$

<u>In particular, if $\mathcal{M}$ has a trivial center, the $\mathcal{M}_g$ has also a trivial center, and $\mathcal{M}_f \subseteq \mathbb{C} . 1$, where 1 stands for the identity of G, so that the map</u>

$$\chi : x \longmapsto \lambda_x , \tag{3.16}$$

<u>with $\lambda_x \in \mathbb{C}$, defines a character of the algebra E, i.e. one has that $\chi \in \mathcal{M}(E)$.</u>

<u>Proof.</u> If $A \subseteq G$, we denote by A' the <u>commutant</u> of A, that is those elements $z \in G$ for which one has $za = az$, for every $a \in A$. Thus, since both of $\mathcal{M}_f$, $\mathcal{M}_g$ are contained in $\mathcal{M}$, by the relation $h = f \otimes g$, one has that $\mathcal{M}' \subseteq \mathcal{M}_f \cap \mathcal{M}_g$, and on the other hand the converse relation is also true, by the corresponding definitions, so that one actually obtains $\mathcal{M}' = \mathcal{M}_f' \cap \mathcal{M}_g'$ and hence one has :

$$\mathcal{M}'' \cap \mathcal{M}' = \mathcal{M}'' \cap \mathcal{M}_f' \cap \mathcal{M}_g' \tag{3.17}$$

which obviously implies the relation (3.14). Now, suppose that <u>the algebra</u> E <u>is commutative</u>: Then, one obviously has $\mathcal{M}_f \subseteq \mathcal{M}'$, so that $\mathcal{M}'' \subseteq \mathcal{M}_f$, and hence by (3.17) one gets the relation (3.15). On the other hand, we get

$$\mathcal{M}_g \cap \mathcal{M}_g' \subseteq \mathcal{M} \cap \mathcal{M}_g' = \mathcal{M} \cap \mathcal{M}'' \cap \mathcal{M}_g'$$

$$= (\text{by } (3.15))\ \mathcal{M} \cap \mathcal{M}'' \cap \mathcal{M}' = \mathcal{M} \cap \mathcal{M}' ,$$

that is, the center of the algebra $\mathcal{M}_g$ is contained in that of the algebra $\mathcal{M}$, so that if the latter is trivial, the same is true for the first. Moreover, since by hypothesis for the algebra E, $\mathcal{M}_f \subseteq \mathcal{M} \cap \mathcal{M}'$, if $\mathcal{M}$ has a trivial center, then by the last relation, one has

$$\mathcal{M}_f \subseteq \mathcal{M} \cap \mathcal{M}' = \mathbb{C} . 1 \tag{3.18}$$

where 1 denotes the identity of the algebra $\mathcal{U} \subseteq G$, so that one obtains the relation

$$f(x) = \lambda_x . 1 , \tag{3.19}$$

with $\lambda_x \in \mathbb{C}$, for every $x \in E$, and the map $\chi : x \mapsto \lambda_x$ defined by (3.19) is of course a non-zero continuous complex homomorphism of the algebra E, that is one has $\chi \in \mathcal{M}(E)$, and this completes the proof of the theorem.

The preceding theorem extends several aspects of an analogous result in Ref. [28] (ibid.; p.488, Theorem 9.2). Besides, it will be applied, in the context given above, in the second part of this paper concerning (topological) analytic function algebras (cf. Theorem 5.1 in the sequel).

PART II

1. Topological algebra sheaves

We apply in this part of the paper the terminology and notation of Ref. [23] without further explanations. For convenience, the term topological algebra will mean in the sequel a locally convex topological algebra with a jointly continuous multiplication and an identity element. More general situations will be clear from the context in connection with the results in Ref. [25].

Let $(X, \mathcal{A})$ and $(X, \mathcal{B})$ be sheaves of locally convex (topological) algebras on the Hausdorff topological space X([24], [25]). Now, if U varies over the open subsets of X, the relation

$$\Gamma(U, \mathcal{A}) \hat{\otimes}_{\pi} \Gamma(U, \mathcal{B}) \tag{1.1}$$

defines a presheaf of (complete locally convex) topological algebras on X (cf. also [15; p.174, §3] and [11; Chap. I, p.37, §2]).

We denote by $\mathcal{A} \pi \mathcal{B}$ the sheaf on X generated by the presheaf

(1.1) as above (cf. also [25; Definition 2.1]).

In particular, if one of the sheaves $\mathcal{A}$, $\mathcal{B}$ is of a nuclear type, and either one of them determines a topological dual weakly flabby precosheaf [25], then one has the relation:

$$\Gamma(U, \mathcal{A}) \hat{\otimes}_{\pi} \Gamma(U, \mathcal{B}) = \Gamma(U, \mathcal{A} \pi \mathcal{B}) \tag{1.2}$$

within a bijection, for every open set $U \subseteq X$: This can be proved by a similar argument as that applied in the proof of Theorem 2.1 of Ref. [25].

Now, if S denotes an arbitrary subset of X, then by regarding the sheaf $\mathcal{A} \pi \mathcal{B}$, one defines:

$$\Gamma(S, \mathcal{A} \pi \mathcal{B}) = \varinjlim_{U \supseteq S} \Gamma(U, \mathcal{A} \pi \mathcal{B}), \tag{1.3}$$

where U ranges over the open neighbourhoods of S in X. (In this respect, cf. also [8; p.151, Corollaire 1]). Hence, by (1.2) one obtains,

$$\Gamma(S, \mathcal{A} \pi \mathcal{B}) = \varinjlim_{U \supseteq S} (\Gamma(U, \mathcal{A}) \hat{\otimes}_{\pi} \Gamma(U, \mathcal{B})) . \tag{1.4}$$

We realize, by the preceding, that it would be of interest to have the distributivity of the "inductive limit" involved in the relation (1.4) above with respect to the "topological tensor product" indicated, as this is usually the case in the applications. On the other hand, motivated by the analogous situation already considered by A.Grothendieck [11; p.47, Corollaire] , we can also discern the following:

Suppose that the topological algebra sheaves involved are such that the respective topological algebras of local sections define the underlying topological vector spaces as complete barrelled DF-locally convex spaces. Besides, suppose that the topological space X is second countable, so that there exists a denumerable basis of open neighborhoods of S in X. Finally, assume that the corresponding topological vector spaces of local sections of the topological algebra sheaves involved, satisfy the conditions of the Köthe's completeness theorem (cf., for instance, [14; p.164, Corollary]), regarding the inductive limit appearing in the relation (1.3) above. Therefore, the algebra $\Gamma(S, \mathcal{A}\,\pi\,\mathcal{B})$ is complete, and hence one obtains:

$$\begin{aligned}\Gamma(S, \mathcal{A}\ \pi\ \mathcal{B}) &= \varinjlim (\Gamma(U,\mathcal{A}) \underset{\pi}{\hat{\otimes}} \Gamma(U,\mathcal{B})) \\ &= \hat{\varinjlim} (\Gamma(U,\mathcal{A}) \underset{\pi}{\hat{\otimes}} \Gamma(U,\mathcal{B})) \\ &= (\varinjlim \Gamma(U,\mathcal{A})) \underset{\pi}{\hat{\otimes}} (\varinjlim \Gamma(U,\mathcal{B})) \\ &= (\hat{\varinjlim} \Gamma(U,\mathcal{A})) \underset{\pi}{\hat{\otimes}} (\hat{\varinjlim} \Gamma(U,\mathcal{B})) \\ &= \Gamma(S,\mathcal{A}) \underset{\pi}{\hat{\otimes}} \Gamma(S,\mathcal{B}),\end{aligned}$$

within a topological algebraic isomorphism. (In this respect, cf. also [11; Chap.I, p.74 and p.76, Proposition 14] and [10; p.78, Theorem 9]).

As a matter of fact, we shall not use however the last relations obtained by the preceding argument, in their strong form derived within the previous context, since what we are actually going to apply in the sequel is the form of the spectrum of a topological algebra sheaf [25; (1.7)] of the type given by the relation (1.3) above, in such a way that one may then apply, for instance, Theorem 3.1 of Part I of this paper, and hence obtain a more general setting than that suggested by the preceding last relations.

Thus, we are led to formulate the following basic theorem, an initial version of which has been motivated by an analogous result in Ref. [28; p.490, Theorem 9.6] (cf. also the results in the subsequent section below). That is, we have:

THEOREM 1.1. <u>Let $\mathcal{A}$ and $\mathcal{B}$ be topological algebra sheaves over a second countable Hausdorff topological space X, such that one of them is of a nuclear type and either one determines a topological</u>

dual weakly flabby precosheaf [25]. Moreover, suppose that the respective local section algebras of $\mathcal{A}$, $\mathcal{B}$ are Fréchet (locally convex topological) algebras with identity elements and locally equicontinuous generalized spectra with respect to a complete topological algebra E, having an identity element, and besides let the sheaf $\mathcal{A}$ be such that the respective local section algebras are commutative. Finally, let S be an arbitrary subset of the topological space X. Then, for every $h \in \mathcal{H}om(\Gamma(S, \mathcal{A}\,\varepsilon\,\mathcal{B}), E)$, such that Im(h) has a trivial center, one obtains

$$h = \varphi \otimes \rho , \tag{1.5}$$

where $\varphi \in \mathcal{M}(\Gamma(S,\mathcal{A}))$ and $\rho \in \mathcal{H}om(\Gamma(S,\mathcal{B}), E)$, in such a way that the decomposition (1.5) is uniquely defined by a given h, whenever the pair (φ, ρ) as above determines an element h by (1.5) (cf. also Theorem 2.1 and/or Lemma 3.1 in Part I above).

Proof. By hypothesis and Theorem 2.1 in Ref. [25], one has that $(X, \mathcal{A}\,\pi\,\mathcal{B})$ is an abstract topological algebra space, such that we also have $\mathcal{A}\,\pi\,\mathcal{B} = \mathcal{A}\,\varepsilon\,\mathcal{B}$ (ibid., Scholium 2.1). In this respect, we also note that the topological algebra sheaves considered in [25] were not necessarily on the same base space X, the corresponding proofs however being easily adapted to the present context. On the other hand, if U is an open subset of X, then the local section algebra $\Gamma(U, \mathcal{A}\,\varepsilon\,\mathcal{B})$ is a Fréchet topological algebra with an identity element such that the space (: generalized spectrum) $\mathcal{H}om(\Gamma(U, \mathcal{A}\,\varepsilon\,\mathcal{B}), E)$ is locally equicontinuous : This follows by the hypothesis, the relation (1.2), as well as Lemma 2.1 in Part I above. Now, by (1.4), one obtains:

$$\mathcal{H}om(\Gamma(S, \mathcal{A}\,\varepsilon\,\mathcal{B}), E) = \mathcal{H}om(\varinjlim(\Gamma(U,\mathcal{A}) \mathbin{\hat{\otimes}_{\varepsilon}} \Gamma(U,\mathcal{B})), E)$$
$$= \varprojlim \mathcal{H}om(\Gamma(U,\mathcal{A}) \mathbin{\hat{\otimes}_{\varepsilon}} \Gamma(U,\mathcal{B}), E),$$

so that if h belongs to the left hand member of the last relation, one actually has, by applying (3.13) above, that there exist uniquely defined elements (cf. also Theorem 2.1) $\varphi \in \mathcal{H}om(\Gamma(S,\mathcal{A}), E)$ and $\rho \in \mathcal{H}om(\Gamma(S,\mathcal{B}), E)$ with $h = \varphi \otimes \rho$. Now, if $Im(h) \subseteq E$ has a trivial center, then by the "commutativity" of the algebra sheaf $\mathcal{A}$ and Theorem 3.2, one obtains that $Im(\varphi) \subseteq Im(h) \cap (Im(h))' = \mathbb{C}.1$, so that the map φ is given by a relation like (3.16), i.e. one has that $\varphi \in \mathfrak{M}(\Gamma(S,\mathcal{A}))$, and this finishes the proof.

We are now going to apply the preceding considerations to the case of certain topological function algebra sheaves, as this is indicated in the subsequent section.

2. Analytic function algebra sheaves

We specialize in the sequel to the case the topological algebra sheaves involved have local sections constituting algebras of holomorphic functions on complex analytic spaces. The results obtained have a special bearing on previous results in Ref. [28], which have also been the motive to the present part of this paper.

We refer to [12] for the terminology applied herein, regarding complex (analytic) spaces. Thus, let $(X, \mathcal{O})$ be a complex space with X a second countable (Hausdorff topological) space and $\mathcal{O}$ the respective structure sheaf of X. Now, it is well known that $\mathcal{O}$ is a topological algebra sheaf of a Fréchet nuclear type, in the sense that the corresponding local sections of $\mathcal{O}$ determine Fréchet nuclear (locally m-convex) topological algebras (cf., for instance, [12; p.158, Theorem 5], [9; p.21] as well as [25]). On the other hand, it is a consequence of the Oka-Weil approximation theorem that $\mathcal{O}$ determines a topologically dual weakly flabby precosheaf on X (cf. [25; § 1, Example 1.1] for a more general situation).

In particular, we shall also be concerned below with Stein algebra sheaves, i.e. topological algebra sheaves in the preceding sense, corresponding to the structure sheaf $\mathcal{O}$ of a Stein space $(X, \mathcal{O})$ (cf. also [25] and [21; p.307, § 4]).

Thus, if $(X, \mathcal{O})$ is a Stein space, then by definition X is second countable, in such a way that the structure sheaf $\mathcal{O}$ determines by its local section algebras, commutative Fréchet nuclear locally m-convex (topological) algebras with identity elements, whose spectra are locally compact spaces, and hence locally equicontinuous: Indeed, if $\Gamma(U, \mathcal{O})$ is a local section algebra of $\mathcal{O}$, with $U \subseteq X$ open, then its spectrum is $\mathfrak{M}(\Gamma(U, \mathcal{O})) = \tilde{U}$, the respective " envelope of holomorphy" of U, the last relation being valid within a homeomorphism (cf., for instance, [13; p.510, Theorem 2.1 (Rossi)]), which is locally compact, so that the assertion follows by [21; p.303, Theorem 2.1].

As an application of Theorem 1.1 of the preceding section, we are now in the position to prove the following:

THEOREM 2.1. Let $(X, \mathcal{O})$ be a Stein space, and let S be a subset of X such that one has $\mathfrak{M}(\Gamma(S, \mathcal{O})) = S$, within a homeomorphism (cf. also the Scholium that follows). On the other hand, let $\mathbb{E}$ be a Fréchet (locally convex topological) algebra with an identity element and let $\mathbb{F}$ be a complete topological algebra with an identity element. Finally, suppose that the generalized spectra, with respect to the algebra $\mathbb{F}$, of the local section algebras of $\mathcal{O}$, as well as of the algebra $\mathbb{E}$, are locally equicontinuous. Then, for every element $h \in \mathcal{H}om(\Gamma(S, \mathcal{O} \varepsilon \mathbb{E}), \mathbb{F})$, such that $Im(h) \subseteq \mathbb{F}$ has a trivial center, there exist uniquely defined elements $x \in S$ and $\rho \in \mathcal{H}om(\mathbb{E}, \mathbb{F})$, in such a way that, for every $\vec{f} \in \Gamma(S, \mathcal{O} \varepsilon \mathbb{E})$, one has the relation:

$$h(\vec{f}) = \rho(\vec{f}(x)) \tag{2.1}$$

<u>Moreover, the image of ρ in $\mathbb{F}$ has a trivial center.</u>

<u>Proof</u>. One has, by definition, the relation

$$\Gamma(S, \mathcal{O}\, \varepsilon\, \mathbb{E}) = \varinjlim \Gamma(U, \mathcal{O}\, \varepsilon\, \mathbb{E}) = \varinjlim (\Gamma(U, \mathcal{O}) \hat{\otimes}_{\varepsilon} \mathbb{E}), \qquad (2.2)$$

where U runs over a fundamental system of open neighborhoods of S in X. (In this respect, cf. also the relations (1.3) and (1.4) in the preceding, as well as the considerations in Ref. [21;p.307, §4]) Therefore, one has by (2.2) the relation:

$$\mathcal{H}om(\Gamma(S, \mathcal{O}\, \varepsilon\, \mathbb{E}), \mathbb{F}) = \mathcal{H}om(\varinjlim (\Gamma(U, \mathcal{O}) \hat{\otimes}_{\varepsilon} \mathbb{E}), \mathbb{F}), \qquad (2.3)$$

so that, by hypothesis, one has in particular the situation described by the relation (3.13) in Part I above. Hence, by Theorem 1.1 above, one obtains the unique existence of elements $\varphi \in \mathcal{M}(\Gamma(S, \mathcal{O}))$ and $\rho \in \mathcal{H}om(\mathbb{E}, \mathbb{F})$ with $h = \varphi \otimes \rho$, so that by hypothesis the map φ is uniquely determined by the evaluation at a point $x \in S$. Now, if $\vec{f} \in \Gamma(U, \mathcal{O}) \hat{\otimes} \mathbb{E} = \Gamma(U, \mathcal{O}\, \varepsilon\, \mathbb{E})$, with $U \supseteq S$ open, and $\vec{f} = g \otimes \vec{a}$, where $g \in \Gamma(U, \mathcal{O})$ and $\vec{a} \in \mathbb{E}$, then, one obtains:

$$h(\vec{f}) = (\varphi \otimes \rho)(g \otimes \vec{a}) = \varphi(g)\, \rho(\vec{a}) = g(x)\, \rho(\vec{a})$$
$$= \rho(g(x)\vec{a}) = \rho(\vec{f}(x)),$$

which proves the assertion for the particular $\vec{f}$ considered. On the other hand, one gets the general case by continuity and linearity. Besides, the last assertion is immediate by hypothesis for the map h as above and Theorem 3.2 in Part I, and this finishes the proof.

<u>Scholium 2.1</u>. In connection with the situation described in Section 1 above and the preceding theorem, we remark that in case of a complex space $(X, \mathcal{O})$ with X second countable, if S is a subset of X, one can consider a denumerable fundamental system of open neighborhoods of S in X, in such a way that concerning the corresponding to that system local sections of the structure sheaf $\mathcal{O}$, one has the situation occurring in Köthe's completeness theorem

[14; p.164, Corollary] . Therefore, if $\mathbb{E}$ denotes a complete locally m-convex (topological) algebra, for which the respective topological vector space is a DF - (locally convex) space, one has the relation:

$$\Gamma(S, G \varepsilon \mathbb{E}) = \varinjlim \Gamma(U_n, G \varepsilon \mathbb{E})$$
$$= \varinjlim (\Gamma(U_n, G) \hat{\otimes}_\pi \mathbb{E}) = (\varinjlim \Gamma(U_n, G)) \hat{\otimes}_\pi \mathbb{E}$$
$$= \Gamma(S, G) \hat{\otimes}_\pi \mathbb{E}, \qquad (2.4)$$

within a topological algebraic isomorphism (cf. also [11; Chap.I, p.47, Corollaire]). Now, the preceding suggest the following more general remarks :

Suppose that E and F are topological algebras with identity elements, such that the algebra F has also a locally equicontinuous spectrum. Moreover, let $(E_\alpha)_{\alpha \in I}$ be an inductive system of topological algebras having identity elements and locally equicontinuous spectra in such a way that E = lim E_α , as a topological (inductive limit) algebra. On the other hand, suppose that

$$E \hat{\otimes}_\tau F = (\varinjlim E_\alpha) \hat{\otimes}_\tau F = \varinjlim (E_\alpha \hat{\otimes}_{\tau_\alpha} F), \qquad (2.5)$$

within a topological algebraic isomorphism of the topological algebras involved, where τ and τ_α , $\alpha \in I$, denote " admissible (tensorial) topologies" [15, 18] on the tensor (product) algebras indicated. (In this respect, cf. also the situation described by the relation (2.4) above). Now, by considering the spectra of the respective algebras, one obtains:

$$\mathfrak{M}(E \hat{\otimes}_\tau F) = \mathfrak{M}(\varinjlim (E_\alpha \hat{\otimes}_{\tau_\alpha} F))$$
$$= \varprojlim \mathfrak{M}(E_\alpha \hat{\otimes}_{\tau_\alpha} F) = \varprojlim (\mathfrak{M}(E_\alpha) \times \mathfrak{M}(F))$$
$$= \mathfrak{M}(E) \times \mathfrak{M}(F), \qquad (2.6)$$

within a homeomorphism. (In this respect, the case of generalized spectra of the topological algebras in (2.6), with respect to a given topological algebra G, commutative or not, is clear).

Now, an interesting consequence of the preceding relation is that it provides another example for <u>the validity of the basic formula (2.7)</u> in Part I above, with $G = \mathbb{C}$, <u>without the topological algebras involved to have all locally equicontinuous spectra</u> : In this respect, cf. [19; p.103, Proposition 2.1 and p.105, Proposition 3.1], as well as [21; p.304, Theorem 2.3] . On the other hand, cf. also the Scholium in p.479 of Ref. [23] .

By concluding this section, we finally remark, in connection with Theorem 2.1 above, that one has the relation $\mathfrak{M}(\Gamma(S, \mathcal{O})) = S$, within a homeomorphism, if S is for example an open holomorphically convex subset of X, in which case the homeomorphism considered is given by the corresponding evaluation map on S (cf., for instance, [12; p.214, Corollary 9]). In this connection, we also remark that the preceding have a special bearing on the considerations included in Ref. [28] (cf., in particular, p.490, Theorem 9.6 and p.491, Remark).

3. <u>Runge-type results</u>

We apply in the sequel the technique developed hitherto in order to get results of the type indicated in the heading of this section, within the context of the preceding sections of this paper and which also extend similar results in Ref. [1], [7] and [28] .

Thus, suppose that $\mathcal{S}$ is a <u>coherent analytic algebra sheaf</u> [12], [25, 24] on a complex space (X, $\mathcal{O}$). Then, $\mathcal{S}$ is a Fréchet algebra sheaf, which is also of a nuclear type (cf. [3; p.326, and p.327, Proposition 8.1]). Hence, if the space X is, in particular, second countable, then for every Fréchet locally convex algebra, <u>the sheaf</u> $\mathcal{S} \varepsilon E = \mathcal{S} \pi E$ <u>is a topological algebra</u>

sheaf on X (cf. [25. Definition 2.1, Theorem 2.2, as well as Example 1.1]), so that if U is an open subset of X, one obtains the relation:

$$\Gamma(U, \mathcal{S} \varepsilon E) = \Gamma(U, \mathcal{S}) \hat{\otimes}_{\pi} E, \tag{3.1}$$

within a topological algebraic isomorphism of the topological algebras involved (cf. also [21; p.307, Theorem 4.1, rel. (4.2)] , [3; p.328, Proposition 9.2] , as well as [25; Theorem 2.1] or [24; p.218, Theorem 2.1]).

Now, given the complex space (X, $\mathcal{O}$) and an open subset A of X, we shall say that (X, A) is a Runge pair, with respect to the (structure) sheaf $\mathcal{O}$, if the topological algebra $\Gamma(X, \mathcal{O}) \subsetneq \Gamma(A, \mathcal{O})$ is dense in $\Gamma(A, \mathcal{O})$. On the other hand, the respective terminology is quite clear concerning the preceding notion referring, more generally to a coherent analytic (topological) algebra sheaf $\mathcal{S}$ on X or to a (topological algebra) sheaf $\mathcal{S} \varepsilon E$ as above.

In this respect, we are now in the position to state the following theorem, the proof of which can easily be supplied by the preceding considerations and standard argumentation concerning topological tensor products [11] , so that we may omit the details. Thus, we have:

THEOREM 3.1. Let (X, $\mathcal{O}$) be a complex space with X second countable, $\mathcal{S}$ a coherent analytic (topological) algebra sheaf on X, and A an open subset of X, in such a way that (X, A) is a Runge pair with respect to the sheaf $\mathcal{S}$. Moreover, let E be a Fréchet locally convex (topological) algebra. Then, (X, A) is also a Runge pair, with respect to the (Fréchet algebra) sheaf $\mathcal{S} \varepsilon E$.

In particular, one concludes the following generalized form of the classical Runge's Theorem. (cf. also [28; p.491, Proposition 9.8] and [7; p.57, Corollary to Theorem 4]). That is, we have:

COROLLARY 3.1. Let $(X, \mathcal{O})$ be a complex space with X second countable, and let A be an open subset of X, in such a way that (X, A) is a Runge pair with respect to the (structure) sheaf $\mathcal{O}$. Besides, let E be a Fréchet locally convex (topological) algebra. Then (X, A) is a Runge pair with respect to the (Fréchet algebra) sheaf $\mathcal{O} \varepsilon E$, as well.

On the other hand, based on the preceding considerations we also have the following:

THEOREM 3.2. Let $(X, \mathcal{O})$ be a complex space with X second countable, and let $\mathcal{S}$ be a coherent analytic (topological) algebra sheaf on X. Moreover, let K be a subset of X, and let (U_n) be a (denumerable) fundamental system of open neighborhoods of K in X, in such a way that, for each U_n, $n \in \mathbb{N}$, (X, U_n) is a Runge pair with respect to the sheaf $\mathcal{S}$. Furthermore, let E be a Fréchet locally convex (topological) algebra. Then, (X, K) is a Runge pair, with respect to the (Fréchet algebra) sheaf $\mathcal{S} \varepsilon E$, in the sense that the topological algebra $\Gamma(X, \mathcal{S} \varepsilon E)$ is dense in the (topological) algebra $\Gamma(K, \mathcal{S} \varepsilon E)$.

Proof. By definition and the hypothesis (cf. also the relation (1.3) above), one has

$$\Gamma(K, \mathcal{S} \varepsilon E) = \varinjlim_{U_n \supseteq K} \Gamma(U_n, \mathcal{S} \varepsilon E). \tag{3.2}$$

On the other hand, by hypothesis and Theorem 3.1 above, one concludes that the algebra $\Gamma(X, \mathcal{S} \varepsilon E)$ is dense in each one of the algebras $\Gamma(U_n, \mathcal{S} \varepsilon E)$, for $n \in \mathbb{N}$, and hence in the respective topological inductive limit algebra of the second member of (3.2), as well, which proves the assertion, by the same relation, and the proof is finished.

In particular, suppose that $K \subseteq \mathbb{C}^n$ is a compact polynomially convex set [12; p.39]. Now, one concludes to an analogous result as in the preceding theorem, by considering the algebra $E[x]$ of polynomials on K with coefficients in the topological algebra E, as above, in such a manner that one obtains that $E[x]$ is dense in the algebra $\mathcal{H}(K, E)$ of E-valued holomorphic maps on K. (In this respect, cf. also [28; p.491, Proposition 9.8]).

In connection with the preceding considerations, we now conclude with the following comments:

Suppose that K is a compact subset of a Stein space $(X, \mathcal{O})$, and let (A_n) be a decreasing sequence (fundamental system) of open relatively compact neighborhoods of K in X in such a way that one has $K = \bigcap_n A_n$, and moreover suppose that $\mathfrak{M}(\Gamma(A_n, \mathcal{O})) = A_n$, $n \in \mathbb{N}$, within a homeomorphism (cf. also [6; p.311] and [30]). Now, by considering the (topological algebra) sheaf $\mathcal{O} \varepsilon E$, with E a Fréchet locally convex (topological) algebra with an identity element and a locally equicontinuous spectrum, one has, regarding the spectra of the topological algebras involved the following:

$$\mathfrak{M}(\Gamma(K, \mathcal{O} \varepsilon E)) = \mathfrak{M}(\varinjlim \Gamma(A_n, \mathcal{O} \varepsilon E))$$

$$= \varprojlim \mathfrak{M}(\Gamma(A_n, \mathcal{O} \varepsilon E)) = \varprojlim \mathfrak{M}(\Gamma(A_n, \mathcal{O}) \hat{\otimes}_\pi E)$$

$$= \varprojlim (\mathfrak{M}(\Gamma(A_n, \mathcal{O})) \times \mathfrak{M}(E)) = (\varprojlim A_n) \times \mathfrak{M}(E) = (\bigcap A_n) \times \mathfrak{M}(E),$$

that is by the hypothesis for K, one finally has the relation:

$$\mathfrak{M}(\Gamma(K, \mathcal{O} \varepsilon E)) = K \times \mathfrak{M}(E), \tag{3.3}$$

within a homeomorphism of the topological spaces involved.

On the other hand, one can apply, within the context of the previous remarks, the situation described for instance in the preceding Theorem 3.2. In this respect, cf. also the relevent comments in Ref. [28; p.491, Remark] , as well as the respective results in Ref. [30] for the case considered therein.

Addendum.- In connection with Theorem 2.1 in Part II above, we remark that one could get an analogous result to Theorem 1.1, namely within the context of abstract topological algebra sheaves which are not necessarily associated with a Stein space. Besides, it was actually in this form which the said theorem has been presented at the Conference.

References

1 Bishop, E., Analytic functions algebras with values in a Fréchet space. Pacific J. Math. 12 (1962), 1177-1192.

2 Bourbaki, N., Algèbre I, Chap. 1-3. Hermann, Paris, 1970.

3 Bungart, L., Holomorphic functions with values in locally convex spaces and applications to integral formulas. Trans. Amer. Math. Soc. 11 (1964), 317-344.

4 Dietrich, W.E. Jr., The maximal ideal space of the topological algebra C(X, E). Math. Ann. 183 (1969), 201-212.

5 Edwards, R.E., Functional Analysis (Theory and Applications). Holt, Rinehart and Winston, New York, 1965.

6 Forster, O., Primärzerlegung in Steinschen Algebren. Math. Ann. 154 (1964), 307-329.

7 Fujimoto, H., Vector-valued holomorphic functions on a complex space. J. Math. Soc. Japan 17 (1965), 52-66.

8 Godement, R., Topologie algébrique et théorie des faisceaux. Hermann, Paris, 1964.

9 Grauert, H., Ein Theorem der analytischen Garbentheorie und die Modulräume Komplexer Strukturen. Publ. math., No.5. Inst. Hautes Ét. Sci., Paris, 1960.

10 Grothendieck, A., Sur les espaces (F) et (DF). Summa Bras. Math. 3 (1954), 57-123.

11 Grothendieck, A., Produits tensoriels topologiques et espaces nucléaires. Mem. Amer. Math. Soc. Nr. 16 (1955).

12 Gunning, R. and Rossi, H., Analytic Functions of Several Complex Variables. Prentice-Hall, Englewood Cliffs, New Jersey, 1965.

13 Harvey, R. and Wells, R.O. Jr., Compact holomorphically convex subsets of a Stein manifold. Trans. Amer. Math. Soc. 136 (1969), 509-516.

14 Horváth, J., Topological Vector Spaces and Distributions, I. Addison-Wesley, Reading, Mass., 1966.

15 Mallios, A., On the spectrum of a topological tensor product of locally convex algebras. Math. Ann. 154 (1964), 171-180.

16 Mallios, A., Inductive limits and tensor products of topological algebras. Math. Ann. 170 (1967), 214-220.

17 Mallios, A., Note on the spectrum of topological inductive limit algebras. Bull. Soc. Math. Grece 8 (1967), 127-131.

18 Mallios, A., Semi-simplicity of tensor products of topological algebras. Bull. Soc. math. Grece 8 (1967), 1-16.

19 Mallios, A., Spectrum and boundary of topological tensor product algebras. Bull. Soc. math. Grece 8 (1967), 101-115.

20 Mallios, A., On topological tensor algebras. Proc. VII Austrian Math. Congress, Linz, Sept. 1968 (: Nachr. Öster. math. Gesel Nr. 91 (1970), 69-70).

21 Mallios, A., On the spectra of topological algebras. J. Funct. Anal. 3 (1969), 301-309.

22 Mallios, A., On generalized spectra of topological tensor algebras. Prakt. Akad. Athēnōn 45 (1970), 76-81.

23 Mallios, A., On functional representations of topological algebras, J. Funct. Anal. 6 (1970), 468-480.

24 Mallios, A., On topological algebra sheaves of a nuclear type. Studia Math. 38 (1970), 215-220.

25 Mallios, A., On topological algebra spaces (to appear).

26 Mallios, A., Spectral decomposition of representations of topological algebras (to appear).

27 Porta, H. and Recht, L., Spectra of algebras of holomorphic germs. Ill. J. Math. 13 (1969), 515-520.

28 Porta, H. and Schwartz, J.T., Representations of the algebra of all operators in Hilbert space, and related analytic function algebras. Comm. Pure Appl. Math. 20 (1967), 457-492.

29 Rickart, C.E., Banach Algebras. Van Nostrand, Princeton, N.J. 1960.

30 Rossi, H., Holomorphically convex sets in several complex variables. Ann. Math. 74 (1961), 470-493.

Appendix

(Note received.................). Referring to the remarks of the preceding addendum, it is actually proved that the whole of Section 2 above can be considered within the context of the general theory of abstract topological algebra spaces (i.e. "topologically algebraized spaces"), or in particular topological function algebra spaces.

In this regard, by an abstract topological algebra space we mean a pair $(X, \mathcal{A})$ consisting of a Hausdorff topological space X and a topological algebra sheaf $\mathcal{A}$ over X (cf. [24]). On the other hand, a topological function algebra space is a pair $(X, \mathcal{A})$ as before, where now the (local) sections of $\mathcal{A}$ are, in particular, continuous complex-valued functions, i.e. we suppose that $\mathcal{A}$ is, in this case, a subsheaf of the sheaf $\mathcal{C}$ of germs of continuous complex-valued functions on X.

Thus, by arguing within this context the following extension and improvement as well of Theorem 2.1 in Part II above can be formulated. That is, we have

THEOREM. Suppose that $(X, \mathcal{A})$ is a topological function algebra space such that the local section algebras determined by $\mathcal{A}$ are locally convex algebras containing the constants and having locally equicontinuous generalized spectra with respect to a complete topological algebra $\mathbb{F}$ which has an identity element. On the other hand, let $\mathbb{E}$ be a locally convex algebra with an identity element and a locally equicontinuous generalized spectrum with respect to $\mathbb{F}$, and moreover let S be an $\mathcal{A}$-convex subset of X.

Then, the following map

$$h \longmapsto (x, \rho) : \mathcal{H}om(\Gamma(S, \mathcal{A} \tau \mathbb{E}), \mathbb{F}) \longmapsto S \times \mathcal{H}om(\mathbb{E}, \mathbb{F})$$

is one-to-one, in case the image of h in $\mathbb{F}$ *has a trivial center, and in such a way that one has*

$$h = \rho \circ \delta_x \tag{i}$$

(cf. the notation below). Moreover $\operatorname{Im}(\rho) \subseteq \mathbb{F}$ *has also a trivial center.*

Referring to the terminology applied in the preceding theorem, we shall say that a subset S of X is $\mathcal{A}$ -convex whenever one has the relation

$$\mathcal{M}(\Gamma(S, \mathcal{A})) = S,$$

within a homeomorphism. For example, this will be the case if S is a compact holomorphically convex (i.e. $\mathcal{O}$-convex) subset of a Stein manifold (X, $\mathcal{O}$).

On the other hand, one denotes by δ_x in the relation (i) above the image of $x \in S$ by the (continuous, not in general one-to-one) *generalized Dirac transform*

$$\delta : S \longmapsto \mathcal{H}om_S(\Gamma(S, \mathcal{A} \tau \mathbb{E}), \mathbb{E}),$$

so that (i) is equivalently written

$$h(\vec{f}) = \rho(\vec{f}(x)), \tag{ii}$$

for every $\vec{f} \in \Gamma(S, \mathcal{A} \tau \mathbb{E})$.

Besides, the "vectorization" $\mathcal{A} \tau \mathbb{E}$ of the (structure) sheaf $\mathcal{A}$ on X is defined by the presheaf

$$\Gamma(U, \mathcal{A}) \tau \mathbb{E},$$

where U ranges over an open basis of X, and where τ denotes some " compatible topology" [19] on the respective tensor product algebra $\Gamma(U, \mathcal{A}) \otimes \mathbb{E}$ induced on it by the canonical exact sequence

$$0 \longrightarrow \Gamma(U, \mathcal{A}) \otimes \mathbb{E} \longrightarrow \mathcal{L}_{\mathfrak{T}}(\mathbb{E}'_{\mathfrak{S}}, \Gamma(U, \mathcal{A})), \qquad \text{(iii)}$$

for suitable vector space topologies $\mathfrak{T}$ and $\mathfrak{S}$. For example, take $\mathcal{A}$ nuclear [24], $\mathfrak{T} = e$ (: biequicontinuous convergence topology), $\mathfrak{S}$ the Mackey topology on $\mathbb{E}'$ (: topological dual of $\mathbb{E}$), and τ the relative topology on the respective tensor product algebra.

Finally, within this same framework it also seems to fit naturally results of the "Runge type" as these, for example, have been exhibited in Section 3 of Part II above. Details on this, as well as the proof of the preceding theorem will be given in subsequent reports (cf., for instance, [25]).

SPECTRA OF COMPOSITION OPERATORS ON $C[0,1]$.*

Bruce Montador

A composition operator $[f]$ on $C[0,1]$ is generated by a continuous function $f : [0,1] \to [0,1]$; for any complex valued continuous function g on $[0,1]$, $[f]g = g \circ f$. The norm of $[f]$ is always 1, so that the spectrum of $[f]$ is always contained in $D = \{ \lambda \in \mathbb{C} \mid |\lambda| \leq 1 \}$.

Before continuing to discuss the spectra of these operators, we shall need to define two types of functions $f : [0,1] \to [0,1]$.

DEFINITION 1. An eventually surjective function $f : [0,1] \longrightarrow [0,1]$ is a function for which there exist an integer n and a set $E \subset [0,1]$ such that

$$f^n([0,1]) = f^{n+k}([0,1]) = E \quad \text{for all} \quad k \geq 0.$$

DEFINITION 2. An eventually injective function $f : [0,1] \longrightarrow [0,1]$ is a function for which there exists an integer n such that if $f^{n+k}(x) = f^{n+k}(y)$ for two points in $[0,1]$ and a positive integer k, then $f^n(x) = f^n(y)$.

Remark. An eventually surjective function which is also eventually injective is injective on the set E.

Proposition 1. *If f is not eventually surjective, the spectrum of $[f]$ is D.*

Outline of Proof. For each $\lambda \in \mathbb{C}$, $0 < |\lambda| < 1$, we find a sequence $\{g_n\}$ of functions: $[0,1] \longrightarrow \mathbb{C}$ such that $\| \lambda g_n - [f] g_n \| \longrightarrow 0$. Such a λ is thus an approximate proper value and so a spectral value.

*Communication presented to the International Conference on Functional Analysis and its Applications, Madras 1973. Research Supported in part by the Canada Council and by the Ministere de l'Education du Quebec

Proposition 2. If f is a bijection, the spectrum of $[f]$ is

(a) $\{1\}$ if $f(x) = x \quad \forall\, x \in [0,1]$.

(b) $\{1, -1\}$ if $f^2(x) = x \quad \forall\, x \in [0,1]$ and if $\exists\, y \in [0,1]$ such that $f(y) \neq y$.

(c) $S^1 = \{\lambda \in \mathbb{C} \mid |\lambda| = 1\}$ if $\exists\, y \in [0,1]$ such that $f^2(y) \neq y$.

Outline of Proof. If f is a bijection $[f^{-1}]$ is the inverse of $[f]$. Since both operators are of norm 1 the spectrum of $[f]$ is contained in S^1. It is possible to find elements of $C[0,1]$ not in the image of $\lambda I - [f]$ if $|\lambda| = 1$ and if there is a y such that $f^2(y) \neq y$.

Proposition 3. If f is eventually surjective and eventually injective, but is not a bijection, then the spectrum of $[f]$ is

a) $\{1,0\}$ if $f(x) = x \quad \forall\, x \in E$.

b) $\{1,-1,0\}$ if $f^2(x) = x \quad \forall\, x \in E$ and if $f(y) \neq y$ for at least one $y \in E$.

c) S^1 o $\{0\}$ if $f^2(y) = y$ for at least one $y \in E$.

Proposition 4. If f is eventually surjective, but is not eventually injective, then the spectrum of $[f]$ is D.

Outline of Proof. For each λ, $0 < |\lambda| < 1$, it is possible to construct a linear functional μ on $C[0, 1]$ such that $\mu(g \circ f) = \lambda\mu(g)$. μ is thus a proper measure of f and λ is a spectral value.

LINEAR FUNCTIONALS ON VECTOR VALUED KÖTHE SPACES

C.W. Mullins

We first give some notation and terminology associated with vector lattices. Let F be a vector lattice. F is <u>σ - order complete</u> if every non-empty majorized countable subset has a supremum. A subset S <u>exhausts</u> F if every element of F is dominated in absolute value by a positive scalar multiple of some element of S. A linear functional on F is called <u>order continuous</u> (resp. <u>order sequentially continuous</u>) if it maps a net (resp. sequence) in F which order converges to θ onto a net (resp. sequence) of real numbers which converges to 0. The set of all order continuous (resp. order sequentially continuous) linear functionals on F is denoted by F^{o} (resp. F^{so}). The set of all order bounded linear functionals on F is denoted by F^{b}.

Let X be a σ-compact locally compact Hausdorff space equipped with a positive Radon measure μ. Let Λ be a (perfect) Köthe function space on X and Λ^{*} its Köthe dual. Let E be a Banach lattice. If E' is endowed with the dual order structure and norm, then E' is also a Banach lattice. We shall denote by $\Lambda[E]$ the vector space (of equivalence classes modulo μ-null sets) of μ-measurable E-valued functions $\vec{f}$ on X such that $\|\vec{f}\| \in \Lambda$ and similarly for $\Lambda^{*}[E']$. If $\Lambda[E]$ is endowed with the natural (almost everywhere) pointwise order structure, then it is a vector lattice. Moreover, $\Lambda[E]$ is σ-order complete whenever E is. In fact, the pointwise supremum serves as the lattice supremum.

If $\vec{g} \in \Lambda^{*}[E']$ then it follows from $|\langle \vec{f}, \vec{g} \rangle| \leq \|\vec{f}\|\|\vec{g}\|$ that $\langle \vec{f}, \vec{g} \rangle$ is integrable for all $\vec{f} \in \Lambda[E]$. Hence, we may use $\vec{g}$ to define a linear functional ψ on $\Lambda[E]$ by

$$\psi(\vec{f}) = \int \langle \vec{f}, \vec{g} \rangle \, d\mu.$$

THEOREM 1. <u>If E is order complete and $E' \subseteq E^{\circ}$, then ψ is order continuous. Hence, under these conditions, we may write</u>

$$\Lambda^{*}[E'] \subseteq \Lambda[E]^{\circ} \subseteq \Lambda[E]^{so} \subseteq \Lambda[E]^{b}.$$

Các [1] has established the following results:

(a) The mapping $(\vec{f},\vec{g}) \longrightarrow \int \langle \vec{f},\vec{g} \rangle \cdot d\mu$ places the spaces $\Lambda[E]$ and $\Lambda^{*}[E']$ in duality.

(b) If for $g \in \Lambda^{*}$ we define the semi-norm p_g on $\Lambda[E]$ by

$$p_g(\vec{f}) = \int \|\vec{f}\| \; |g| \; d\mu ,$$

then the topology $\mathcal{T}_n$ generated by the set $\{ p_g : g \in \Lambda^{*} \}$ is complete. Moreover, if E' is separable, then this topology is admissible for the dual system $\langle \Lambda[E], \Lambda^{*}[E'] \rangle$.

It is clear that the topology $\mathcal{T}_n$ is a lattice topology. Using the fact that the dual of a complete metrizable lattice is the set of order bounded linear functionals, we obtain our second result.

THEOREM 2. <u>In addition to the conditions of Theorem 1, if E' is separable and Λ^{*} contains a countable exhausting set, then</u>

$$\Lambda^{*}[E'] = \Lambda[E]^{\circ} = \Lambda[E]^{so} = \Lambda[E]^{b}.$$

<u>References</u>.

1 Các, N.P., Generalized Köthe function spaces. Proc. Cambridge Philos. Soc. 65, 601-611 (1969).

2 Mullins, C.W., Order continuous mappings on Köthe spaces, I and II. Proc. Acad. Amsterdam, A70, 404-421 (1967).

A GENERAL VIEW ON UNITARY DILATIONS

Béla Sz.-Nagy (Szeged)

1. It is 20 years ago that I proved the following theorem:

For every contraction T on Hilbert space $\underline{H}$ (i.e. linear operator with norm $\|T\| \leq 1$) there exists a unitary operator U on some Hilbert space $\underline{K}$ containing $\underline{H}$ as a subspace, such that

(1) $$T^n h = P_{\underline{H}} U^n h \qquad (h \in \underline{H};\ n = 0,1,2,\ldots).$$ *)

U is called a unitary dilation of T. One can require that U be minimal, i.e. that the relation

(2) $$\bigvee_{n=-\infty}^{\infty} U^n \underline{H} = \underline{K}$$

hold; U is then determined uniquely (up to an isometric isomorphism).

One of the first applications was an easy proof of the inequality

(3) $$\|p(T)\| \leq \max_{|\lambda| \leq 1} |p(\lambda)| \qquad \text{(J.von Neumann, 1951)}$$

for polynomials. Indeed, (1) implies that $p(T)h = P_{\underline{H}} p(U)h$ $(h \in \underline{H})$, whence $\|p(T)\| \leq \|p(U)\|$, and for unitary operators (3) easily follows from spectral theory. Let us remark that von Neumann's original proof was rather complicated.

My first proof of the existence of a U satisfying (1) started by observing that

(4) $$\mathrm{Re}\left[(h,h) + 2\sum_{n=1}^{\infty} \lambda^n (T^n h, h)\right] \geq 0 \quad \text{for } h \in \underline{H},\ |\lambda| < 1.$$

---

*) P will always denote orthogonal projection onto the subspace indicated in the subscript.

Using the Stieltjes integral representation (due to F.Riesz) of holomorphic functions with non-negative real part in the unit circle, and some well-known facts on quadratic forms on Hilbert space, I showed that

$$T^n = \int_0^{2\pi} e^{int} dB_t \quad (n = 0,1,\ldots)$$

where B is a semi-spectral measure on $[0,2\pi]$. Now, by M.A.Neumark's theorem (1943) the semi-spectral measure B can be 'dilated' to a spectral measure E, i.e. one has

$$B(\sigma)h = P_{\underline{H}} E(\sigma)h \quad (h \in \underline{H};\ \sigma \text{ Borelian}),$$

and the proof concludes by taking $U = \int_0^{2\pi} e^{it} dE_t$.

My second proof also started from (4), but then observed that this implies that the sequence T(n) of operators defined by T^n for $n \geq 0$ and by $T^{*|n|}$ for $n < 0$ is positive definite, i.e. such that

$$\sum_n \sum_m (T(n-m)h_n, h_m) \geq 0$$

for every $\underline{H}$-vector valued finite sequence of vectors $h_n \in \underline{H}$. The proof ended by applying the theorem of Gelfand-Raikov-Godement-Neumark stating that every positive definite function T(x) on a group G such that T(e) = I for the neutral element $e \in G$, can be 'dilated' to a representation U(x) of the group G by unitary operators.

Shortly after these proofs were published, J.J.Schäffer gave (1955) another proof by developing a simple method of P.R.Halmos (1950) for the construction of a unitary U_o such that $Th = P_{\underline{H}}U_o h$ (no higher powers being considered). Schäffer's proof, in its slightly modified form I gave it later (1960) to yield just the

$\underline{\text{minimal}}$ U , runs as follows.

First, one introduces the 'defect operators' D, D_* and 'defect spaces' $\underline{D}$, $\underline{D}_*$ by

$$D = (I - T^*T)^{\frac{1}{2}}, \quad D_* = (I - TT^*)^{\frac{1}{2}}; \quad \underline{D} = \overline{D\underline{H}}, \quad \underline{D}_* = \overline{D_*\underline{H}},$$

and observes that

$$(5) \qquad TD = D_*T$$

(consequence of $T(I-T^*T) = (I-TT^*)T$). Then one constructs the Hilbert space $\underline{K}$ of 'vectors'

$$(6) \qquad k = \langle \ldots, d_{-2}, d_{-1}, \boxed{h}, d_1, d_2, \ldots \rangle$$

with components $h \in \underline{H}$, $d_n \in \underline{D}$ for $n \geq 1$, $d_n \in \underline{D}_*$ for $n \leq -1$, with finite sum of norm-squares. One embeds $\underline{H}$ in $\underline{K}$ by identifying

$$h \sim \langle \ldots, 0, 0, \boxed{h}, 0, 0, \ldots \rangle \qquad (h \in \underline{H}),$$

and defines, for k given in (6),

$$(7) \quad Uk = \langle \ldots, d_{-2}, D_*d_{-1} + Th, \boxed{-T^*d_{-1} + Dh}, d_1, d_2, \ldots \rangle .$$

Using (5) one shows that U is unitary, its inverse being given by

$$(7)' \quad U^{-1}k = \langle \ldots, d_{-2}, d_{-1}, D_*h - Td_1, \boxed{T^*h + Dd_1}, d_2, \ldots \rangle$$

If $k = h \in \underline{H}$, then

$$U^n h = \langle \ldots, 0, \boxed{T^n h}, DT^{n-1}h, DT^{n-2}h, \ldots, Dh, 0, \ldots \rangle$$

whence (1). Denote by $\underline{L}$ and $\underline{L}^*$ the subspaces of $\underline{K}$ formed respectively by the vectors

$$(8) \quad \langle \ldots, 0, \boxed{0}, d_1, 0, \ldots \rangle \text{ and } \langle \ldots, d_{-1}, \boxed{0}, 0, 0, \ldots \rangle .$$

($d_1 \in \underline{D}$, $d_{-1} \in \underline{D}_*$). From (6) - (7)', one deduces that

$$(9) \quad \underline{K} = \ldots \oplus U^{-1}\underline{L}^* \oplus \underline{L}^* \oplus \boxed{\underline{H}} \oplus \underline{L} \oplus U\underline{L} \oplus \ldots .$$

Furthermore, one observes that

$$(10) \qquad \underline{L} = \overline{(U-T)\underline{H}}, \quad \underline{L}^* = \overline{(U^{-1}-T^*)\underline{H}};$$

these relations exhibit the meaning, independent of the actual representation of the dilation, of the subspaces $\underline{L}$ and $\underline{L}^*$, and also show, together with (9), that U satisfies the minimality requirement (2) also.

Note that uniqueness of the minimal unitary dilation U follows from the fact that

$$(U^n h, U^m h') = (T^{n-m}h, h') \text{ for } n \geqslant m, \text{ and for } h, h' \in \underline{H},$$

which shows that the inner products on the left-hand side do not depend on the special choice of U.

Since orthogonality $U^n\underline{L} \perp U^m\underline{L}$ for n, m $\geqslant 0$ ($n \neq m$) implies orthogonality for any integers $n \neq m$, one can form the subspaces

$$M(\underline{L}) = \bigoplus_{-\infty}^{\infty} U^n\underline{L} \text{ and } M(\underline{L}^*) = \bigoplus_{-\infty}^{\infty} U^n\underline{L}^*,$$

on both of which U is a bilateral shift, of multiplicity d and d_*, where

$d = \dim \underline{L} = \dim \underline{D}$ and $d_* = \dim \underline{L}^* = \dim \underline{D}_*$.

One can show that

$M(\underline{L}) = \underline{K}$ if and only if $T \in C_{0\cdot}$, i.e. if $T^n \to 0$ $(n \to \infty)$;

$M(\underline{L}^*) = \underline{K}$ if and only if $T \in C_{\cdot 0}$, i.e. if $T^{*n} \to 0$ $(n \to \infty)$.

Thus in these cases U is a bilateral shift on the whole of $\underline{K}$.

2. By a theorem found independently by H.Langer and C.Foiaş every contraction T on a Hilbert space $\underline{H}$ induces a unique decomposition of $\underline{H}$ into the orthogonal sum of two subspaces, one of which reduces T to a <u>unitary</u> operator, and the other to a <u>completely non-unitary</u> (c.n.u.) contraction (i.e. which has no unitary part in

any non-zero reducing subspace). Now, for unitaries we have at our disposal the classical spectral theorem and functional calculus. For the opposite type of operator, the c.n.u. contractions, it is just the theory of unitary dilations which yields the best means for investigation.

First of all it turns out (by using in particular relation (9)) that

$$M(\underline{L}) \vee M(\underline{L}^*) = \underline{K} \tag{11}$$

holds if and only if T is c.n.u. Thus for c.n.u. T , the minimal unitary dilation U is the (skew) sum of two bilateral shifts. As a consequence, the spectral measure of U is absolutely continuous with respect to Lebesgue measure.

This makes it possible to define a functional calculus for a c.n.u. contraction T which allows all functions of class H^{∞}, i.e. bounded and holomorphic in the open unit disc; notably if $u(\lambda) = a_0 + a_1 \lambda + a_2 \lambda^2 + \ldots$ then the definition

$$u(T) = \lim_{r \to 1-0} \left[a_0 I + r a_1 T + r^2 a_2 T^2 + \ldots \right]$$

is meaningful, with operator norm convergence for the series, and operator strong convergence for $r \to 1 - 0$. This functional calculus is an algebra homomorphism from H^{∞} into the algebra of operators on $\underline{H}$; moreover we have

$$\| u(T) \| \leq \| u \|_{\infty}$$

(extended form of (3)). Some continuity properties also hold, e.g. if $u_n \in H^{\infty}$ and $u_n(e^{it}) \to 0$ boundedly and a.e. on the circle, then $u_n(T) \to 0$ strongly, while if $u_n(\lambda) \to 0$ boundedly in the open unit disc, then $u_n(T) \to 0$ weakly.

Of particular interest are those c.n.u. contractions T for which there exists some non-zero $u \in H^{\infty}$ such that $u(T) = 0$; they are called of <u>class</u> C_0. We have $C_0 \subset C_{00}, (C_{00} = C_{0\cdot} \cap C_{\cdot 0})$. These operators resemble in many respects to operators on finite dimensional spaces. A simple type of operator $T \in C_0$ is the operator $S(m)$ on the space $\underline{H}(m)$ associated with an arbitrary <u>inner</u> function m (i.e., $m \in H^{\infty}$ and $|m(e^{it})| = 1$ a.e.) in the following way:

$$\underline{H}(m) = H^2 \ominus mH^2 \quad , \quad S(m)v = P_{\underline{H}(m)}(\lambda v) \quad \text{for} \quad v \in \underline{H}(m);$$

H^2 is the Hardy subspace of the space L^2 of functions on the unit circle. In fact, m is then the 'minimal' function for $S(m)$, that is, $u(S(m)) = 0$ if and only if u is a multiple of m in H^{∞}.

Operators of class C_0 were studied from different aspects: one of their most striking properties is that they allow a kind of Jordan reduction:

<u>If</u> $T \in C_0$ <u>on</u> $\underline{H}$ <u>is such that</u>

$$\underline{H} = \bigvee_{n=0}^{\infty} T^n \hat{h},$$

<u>where</u> $\hat{h} = \{h_1, \ldots, h_r\}$ <u>is some finite set of vectors in</u> $\underline{H}$, <u>then</u> T <u>is quasi-similar to a unique operator of the form</u>

$$S(m_1) \oplus S(m_2) \oplus \ldots \oplus S(m_s) ,$$

<u>where</u> $m_1, m_2, \ldots, m_s$ <u>are non-constant inner functions, each of which is a divisior</u> (in H^{∞}) <u>of its predecessor and</u> s <u>is equal to the minimal value of</u> r .

(Two operators, say A and B are 'quasi-similar' if there exist operators X and Y, with zero kernel and dense range, such that

$$AX = XB \quad \text{and} \quad BY = YA.)$$

Note that similarity in the usual sense would not do the same.

I do not go into more details about nice properties of operators of class C_0. One recent outgrowth of this subject is a theorem of E. Nordgren on 'quasi-equivalence' of a finite matrix with entries in H^∞, to the diagonal matrix formed by the 'invariant factors'-a by no means immediate extension of a classical result on matrices with polynomial entries.

3. Let us return to the decomposition (9) of the space $\underline{K}$ of minimal unitary dilation of the contraction T. The subspace

$$(12) \qquad \underline{K}_+ = \underline{H} \oplus M_+(\underline{L}) \qquad (M_+(\underline{L}) = \underline{L} \oplus U\underline{L} \oplus U^2\underline{L} \oplus \ldots)$$

is invariant for U and one can easily show that $U_+ = U \mid \underline{K}_+$ is the minimal isometric dilation of T, i.e.

$$(13) \qquad T^n h = P_{\underline{H}} U_+^n h \quad (h \in \underline{H},\ n = 0, 1, \ldots), \text{ and}$$

$$(14) \qquad \bigvee_{n=0}^{\infty} U_+^n \underline{H} = \underline{K}_+ .$$

The isometry U_+ induces a 'Wold decomposition' of $\underline{K}_+$, notably

$$(15) \qquad \underline{K}_+ = M_+(\underline{L}_*) \oplus \underline{R}, \quad (M_+(\underline{L}_*) = \underline{L}_* \oplus U\underline{L}_* \oplus U^2\underline{L}_* \oplus \ldots),$$

where

$$\underline{L}_* = \underline{K}_+ \ominus U_+ \underline{K}_+ \quad \text{and} \quad \underline{R} = \bigcap_{n=0}^{\infty} U_+^n \underline{K}_+ ;$$

one can show that the Wold wandering subspace $\underline{L}_*$ is in the following relation with the wandering subspace $\underline{L}^*$ introduced above:

$$(16) \qquad \underline{L}_* = U\underline{L}^* = U\overline{(U^{-1} - T^*)\underline{H}} = \overline{(I - UT^*)\underline{H}} .$$

Combining (12) and (15) we obtain:

$$\underline{H} = [M_+(\underline{L}_*) \oplus \underline{R}] \ominus M_+(\underline{L}),$$

or denoting by Q orthogonal projection from $M(\underline{L})$ into $M(\underline{L}_*)$,

$$(17). \qquad \underline{H} = [M_+(\underline{L}_*) \oplus \underline{R}] \ominus \{Q\ell \oplus (\ell - Q\ell) : \ell \in M_+(\underline{L})\} .$$

Moreover, if T is c.n.u. we can use relation (10) to get the following expression for $\underline{R}$:

$$\underline{R} = \overline{Q\, M(\underline{L})} \, . \tag{18}$$

The essential in this representation of $\underline{H}$ (and $\underline{R}$) is that it is entirely in terms of the spaces $M_+(\underline{L}_*)$, $M_+(\underline{L})$, and $M(\underline{L})$, in which U is a unilateral or a bilateral shift, and of the operator Q which measures the 'angle' between $M(\underline{L}_*)$ and $M(\underline{L})$.

Assume now that $\underline{H}$ - and therefore $\underline{K}$ also - are separable. Then using the canonical unitary maps of $\underline{L}$ and $\underline{L}_*(=U\underline{L}^*)$ onto the defect spaces $\underline{D}$ and $\underline{D}_*$, resulting from (8) or (9), and performing obvious 'Fourier' transformations

$$M_+(\underline{L}) \to H^2(\underline{D}), \quad M(\underline{L}) \to L^2(\underline{D}), \quad M_+(\underline{L}_*) \to H^2(\underline{D}_*),$$

where, e.g., $L^2(\underline{D})$ denotes the space of $\underline{D}$-vector valued functions on the unit circle (with normalized Lebesgue measure), the operator Q will be transformed into an operator

$$\Theta_T : L^2(\underline{D}) \longrightarrow L^2(\underline{D}_*),$$

which is, in fact, multiplication on $L^2(\underline{D})$ by an operator valued function

$$\Theta_T(e^{it}) : \underline{D} \longrightarrow \underline{D}_* \, .$$

This function is the strong, non-tangential limit, a.e. on the unit circle, of a contractive analytic function $\Theta_T(\lambda) : \underline{D} \to \underline{D}_*$, called the <u>characteristic function</u> of T ; its precise form is:

$$\Theta_T(\lambda) = \left[-T + \lambda \sum_0^\infty \lambda^n D_* T^{*n} D \right] \Big| \underline{D} \, .$$

Denoting by Δ_T multiplication on $L^2(\underline{D})$ by the operator valued function

$$\Delta_T(e^{it}) = \left[I_{\underline{D}} - \Theta_T(e^{it})^* \, \Theta_T(e^{it}) \right]^{1/2},$$

it results from (17) and (18) that the c.n.u. contraction T is unitarily equivalent to the operator $\underline{\underline{T}}$ defined on the space

$$\underline{\underline{H}} = [H^2(\underline{D}_*) \oplus \overline{\Delta_T L^2(\underline{D})}] \ominus \{\Theta_T w \oplus \Delta_T w : w \in H^2(\underline{D})\}$$

by

$$\underline{\underline{T}}(u \oplus v) = P_{\underline{\underline{H}}}(\chi u \oplus \chi v) \qquad (u \oplus v \in \underline{\underline{H}}),$$

where $\chi(e^{it}) \equiv e^{it}$.

The function $\Theta_T(\lambda)$ is purely contractive, i.e. $\|\Theta_T(0)h\| < \|h\|$ for every $h \in \underline{D}$, $h \neq 0$. Conversely, if we give the role of $\Theta_T(\lambda)$ in the above construction to an arbitrarily given, purely contractive analytic function Θ with values operators $\Theta(\lambda): \underline{E} \to \underline{E}_*$ ($\underline{E}$, $\underline{E}_*$ separable Hilbert spaces) then the resulting operator

$$\underline{\underline{T}} = S(\Theta)$$

is a c.n.u. contraction, whose characteristic function equals (up to constant unitary factors) the given function $\Theta(\lambda)$.

Thus the theory of unitary dilations leads in an entirely natural way to a general functional model for c.n.u. contractions.

This model is the counterpart of the classical model of unitary operators, viz. multiplication by χ on an orthogonal sum of spaces L^2_μ with (not-necessarily homogeneous) Borel measures μ on the unit circle.

From the further developments and applications of this theory let us mention the following ones: Mutual relation between invariant subspaces and 'regular' factorizations of the characteristic function; Spectral theory for 'weak' contractions T, i.e. for which $I - T^*T$ is of trace class and the spectrum of T does not cover the unit disc; Criteria for a contraction to be similar to some unitary operator, etc.

4. The continuous analogue of the unitary dilation theorem deals with one parameter weakly continuous semi-groups $T(s)(0 \leq s < \infty)$ of contractions on a Hilbert space $\underline{H}$ and asserts that there exists a one parameter group $U(s)$ $(-\infty < s < \infty)$ of unitary operators on some Hilbert space $\underline{K}$ such that we have

$$(19) \qquad T(s)h = P_{\underline{H}}U(s)h \qquad (h \in \underline{H},\ s \geq 0)$$

and

$$(20) \qquad \bigvee_{s=-\infty}^{\infty} U(s)\underline{H} = \underline{K} .$$

This group $U(s)$ is determined uniquely (i.e. up to an isometric isomorphism) and is strongly continuous.

My original proof (1953) of the existence of a continuous group $U(s)$ satisfying (19) started by proving that

$$\operatorname{Re} \int_0^\infty e^{-sz}(T(s)h,h)ds \geq 0 \quad \text{for} \quad h \in \underline{H}, \text{ and } \operatorname{Re} z > 0$$

and then applied the Stieltjes integral representation

$$F(z) = \int_{-\infty}^{\infty} \frac{1}{z - i\lambda}\, d\alpha(\lambda) \text{ (with bounded, non-decreasing } \alpha(\lambda))$$

of a function holomorphic and with $\operatorname{Re} F(z) \geq 0$ for $\operatorname{Re} z > 0$, and satisfying

$$|x \cdot F(x+iy)| \leq C \quad \text{for} \quad x > 0 .$$

By virtue of inversion theorems for Laplace transforms and by elementary facts on quadratic forms on Hilbert space it resulted that

$$T(s) = \int_{-\infty}^{+\infty} e^{i\lambda s}\, dB_\lambda \qquad (s \geq 0)$$

for a semi-spectral measure B on $(-\infty, \infty)$. The proof concluded by dilation of B to a spectral measure (Neumark's theorem) and by setting

$$U(s) = \int_{-\infty}^{+\infty} e^{i\lambda s} dE_\lambda \qquad (-\infty < s < \infty) .$$

In a second proof I observed that the function T(s) - when extended to negative reals by setting $T(-s) = T(s)^*$ - is positive definite on the additive group of reals.

There are at least two further proofs; one of them uses the infinitesimal cogenerator

$$T = (A + I)(A-I)^{-1}, \quad \text{where} \quad A = \lim_{s\to 0} \frac{1}{s}(T(s)-I);$$

T is a contraction and if

$$U = \int_0^{2\pi} e^{it} dE_t$$

is the minimal unitary dilation of T then

$$U(s) = \int_0^{2\pi} \left(\exp s \frac{e^{it}+1}{e^{it}-1}\right) dE_t = \int_{-\infty}^{\infty} e^{is\lambda} dE_{2 \operatorname{arc} \operatorname{cotg} \lambda}$$

will be the minimal unitary dilation of T(s), i.e. satisfying (19) and (20).

5. Let $\{T_k\}$ be a commutative system of contractions on the same space $\underline{H}$. The question is whether there exists a corresponding commutative system $\{U_k\}$ of unitaries on some space $\underline{K}(\supset \underline{H})$ such that

(21) $$\prod_k T_k^{n_k} h = P_{\underline{H}} \prod_k U_k^{n_k} h \qquad (h \in \underline{H})$$

holds for any finite product of powers with exponents $n_k \geq 0$.

The existence of such a system $\{U_k\}$ was first established under some further restrictions (Sz.-Nagy, Brehmer). Ando proved it in case of a commuting pair $\{T_1, T_2\}$ without any further restriction. It was a surprise when Parrott constructed an example of a commuting triple $\{T_1, T_2, T_3\}$ for which no commuting triple

$\{U_1, U_2, U_3\}$ exists for which

$$T_1 h = P_{\underline{H}} U_1 h, \quad T_2 h = P_{\underline{H}} U_2 h, \quad T_3 h = P_{\underline{H}} U_3 h \qquad (h \in \underline{H})$$

hold simultaneously.

This curious situation has among others the following consequence. Ando's theorem and the joint spectral representation of two commuting unitaries imply that the inequality of von Neumann extends to polynomials of any two commuting contractions, in the form:

$$\| p(T_1, T_2) \| \leq \max_{|\lambda_1|, |\lambda_2| \leq 1} | p(\lambda_1, \lambda_2) | .$$

However, for more than two commuting contractions the unitary dilation method breaks down, and the problem of validity of the von Neumann inequality is in general still unsettled.

6. Another kind of 'norm' of an operator T on $\underline{H}$ is its 'numerical radius' defined by

$$w(T) = \sup_{\|h\| \leq 1} |(Th, h)|; \tag{22}$$

clearly $w(T) \leq \| T \|$. It is a surprising result of C.Berger (1965) <u>that if</u> $w(T) \leq 1$ <u>then there exists a unitary operator</u> U <u>on some space</u> $\underline{K}(\supset \underline{H})$ <u>such that</u>

$$T^n h = 2 \cdot P_{\underline{H}} U^n_2 h \qquad (h \in \underline{H}, \; n = 1, 2, \ldots). \tag{23}$$

(Such a U is called a unitary '2-dilation' of T .)

There are several proofs; one of them starts by observing that the spectrum $\sigma(T)$ cannot have points outside the unit circle, for then it would also have a boundary point z_0 there, and since such a z_0 belongs to the approximate point spectrum this easily leads to a contradiction to the assumption $w(T) \leq 1$. It follows that $I - zT$ is (boundedly) invertible for $|z| < 1$ and satisfies

$$(I-zT)^{-1} = \sum_{0}^{\infty} z^n T^n \qquad \text{(convergence in operator norm)};$$

then using $w(T) \leq 1$ we obtain that

$$\text{Re} \sum_{0}^{\infty} z^n (T^n h, h) \geq 0 \qquad \text{for} \quad |z| < 1,$$

and the proof continues (as in the original proofs for the case $\|T\| \leq 1$) either by applying the Riesz representation or by a positive definiteness argument.

Among the various interesting consequences of (23) we have the von Neumann type inequality

$$\| p(T) \| \leq 2 \max_{|z| \leq 1} |p(z)| \qquad \text{for polynomials such that } p(0) = 0.$$

Also, it leads to the proof of the inequality, conjectured by P.R.Halmos,

$$w(T^n) \leq (w(T))^n \qquad (n = 1, 2, \dots);$$

etc.

A natural problem is then to consider those operators T which possess a unitary ρ-dilation ($\rho > 0$) (2 replaced by ρ in (23)). The class of these operators T is denoted by $\mathcal{C}_\rho$.

We have $\|T^n\| \leq \rho$ for $T \in \mathcal{C}_\rho$, but it turns out that not all 'power bounded' operators belong to some class $\mathcal{C}_\rho$.

Perhaps the most interesting result in this connection is that <u>all operators belonging to some class</u> $\mathcal{C}_\rho$ (thus in particular the operators T with $w(T) \leq 1$) <u>are similar to contractions</u>.

Most of the subject reviewed above are results of joint work with C.Foiaş and are treated in detail in our book. There are still other aspects of the theory which I could not include for it would take us too far to consider them all.

This is an intriguing and growing area of research, whose usefulness is now beyond doubt, but which has not yet been fully exploited.

A few references are given below.

References

1. Neumark, M.A., Positive definite operator functions on a commutative group, Bulletin (Izvestiya) Acad. Sci. URSS (série math.), 7 (1943), 237-244. (Russian, with English summary).
2. Halmos, P.R., Normal dilations and extensions of operators, Summa Brasil. Math., 2 (1950), 125-134.
3. von Neumann, J., Eine Spektraltheorie für allgemeine Operatoren eines unitären Raumes, Math. Nachr., 4 (1951), 258-281.
4. Sz.-Nagy, B., Sur les contractions de l' espace de Hilbert, Acta Sci. Math., 15 (1953), 87-92.
5. Sz.-Nagy, B., Transformations de l'espace de Hilbert, fonctions de type positif sur un groupe, Acta Sci. Math. 15 (1954) 104-114.
6. Sz.-Nagy, B., Prolongements des transformations de l'espace de Hilbert qui sortent de cet espace, (1955). English translation New York, 1960.
7. Sz.-Nagy, B. and Foiaş, C., Harmonic Analysis of operators on Hilbert space, North Holland (1970).
8. Sz.-Nagy, B., Modèle de Jordan pour une classe d'operateurs de l'espace de Hilbert, Acta Sci. Math., 31 (1970), 93-117.
9. Schäffer, J.J., On unitary dilations of contractions, Proc. Amer. Math. Soc., 6 (1955), 322.
10. Halperin, I., The unitary dilation of a contraction operator, Duke J. Math., 28 (1961), 563-571.
11. Ando, T., On a pair of commutative contractions, Acta Sci. Math. 24 (1961), 80-90.
12. Mlak, W., Unitary dilations of contraction operators, Rozprawy mat., 46 (1965), 1-88

QUANTIZATION IN HAMILTONIAN PARTICLE MECHANICS

Ebbe Thue Poulsen

1. Introduction

The formal similarity between the equations of motion of a system of point particles in classical Hamiltonian mechanics and in the Heisenberg formulation of quantum mechanics is well-known. The similarity is striking when the classical equations are written in the following form

$$\frac{dq_{nj}}{dt} = \{ q_{nj},h \} , \qquad \frac{dp_{nj}}{dt} = \{p_{nj},h\} ,$$

where (q_{n1},q_{n2},q_{n3}) resp. (p_{n1},p_{n2},p_{n3}) denote the position resp. momentum of the n'th particle, h the Hamilton-function of the system, and $\{ .,. \}$ the Poisson-bracket (cf. Section 2, Notation).

The equations of motion of a quantum mechanical system of mutually distinguishable point particles without spin have the form

$$\frac{d\,\mathbb{Q}_{nj}}{dt} = -i\left[\mathbb{Q}_{nj}, \mathbb{H}\right], \qquad \frac{d\,\mathbb{P}_{nj}}{dt} = -i\left[\mathbb{P}_{nj}, \mathbb{H}\right].$$

Here the $\mathbb{Q}_{nj}$ and $\mathbb{P}_{nj}$ are selfadjoint operators in a Hilbert space, representing the observables corresponding to the coordinates of position and momentum of the n'th particle, while the Hamilton-operator $\mathbb{H}$ represents the total energy of the system. The bracket $[.,.]$ denotes the commutator of the two operators appearing as its arguments.

A mapping θ from functions h to operators $\mathbb{H} = \theta(h)$ will be called a <u>quantization</u> if, formally, it maps the classical equations of motion into those of quantum mechanics, i.e., if

$$\text{(i)} \qquad \begin{aligned} \theta(\{ q_{nj},h \}) &= -i \left[\theta(q_{nj}), \theta(h)\right] \\ \theta(\{ p_{nj},h \}) &= -i \left[\theta(p_{nj}), \theta(h)\right] . \end{aligned}$$

In addition, we shall require that $\theta(1) = I$, the identity operator, a requirement which, at least partly, plays the role of a normalization condition. It implies that the operators $\mathbb{Q}_{nj} = \theta(q_{nj})$ and $\mathbb{P}_{nj} = \theta(p_{nj})$ satisfy the so-called canonical commutation relations

$$[\mathbb{Q}_{nj}, \mathbb{Q}_{mk}] = 0 = [\mathbb{P}_{nj}, \mathbb{P}_{mk}] ,$$

$$[\mathbb{Q}_{nj}, \mathbb{P}_{mk}] = i I \delta_{nm} \delta_{jk} .$$

A number of concrete realizations of these commutation relations are known, the best known one being the Schrödinger representation, where the underlying Hilbert space is $L^2(\mathbb{R}^{3N})$, and $\mathbb{Q}_{nj}$ resp. $\mathbb{P}_{nj}$ is the operator of multiplication by x_{nj} resp. $-i\partial/\partial x_{nj}$ It is furthermore known that under suitable supplementary conditions, the most stringent one being an irreducibility condition, a set of selfadjoint operators satisfying the canonical commutation relations is unitarily equivalent with those of the Schrödinger representation.

We shall not worry about the above mentioned supplementary conditions or about the time-dependence of the unitary operators which take the time-dependent operators $\mathbb{Q}_{nj}$ and $\mathbb{P}_{nj}$ into their Schrödinger representation. For this, see for instance Leaf [13]. Forgetting about the time-dependence which was essential in the motivation of the problem, we shall in the sequel let $\mathbb{Q}_{nj}$ and $\mathbb{P}_{nj}$ denote the operators of the Schrödinger representation, and study properties of quantizations.

The fundamental difficulty is caused by the fact that the $\mathbb{Q}$'s and $\mathbb{P}$'s do not commute, so that the usual functional calculus is not available.

The first example of a quantization was constructed in 1926 by H. Weyl [33]. His construction, now usually called the Weyl correspondence, lies at the heart of most of the subsequent development, and is also basic for the present paper. It is performed with the aid of two unitary representations of the additive group $\mathbb{R}^{3N}$, and generalizations or extensions primarily based upon this point of view have been considered by, among others, Segal [25], Kastler [10], Shale and Stinespring [26], Aslaksen and Klauder [1].

It has long been realized that Weyl's correspondence does not have the property

(ii) $$\theta(\{f,h\}) = -i[\theta(f),\theta(h)] \quad .$$

Thus, Groenewold [6] among other things gives a formula for

$$\theta^{-1}(-i[\theta(f),\theta(h)])$$

in terms of f and h, a formula now commonly called the Moyal sine-bracket, cf. Moyal [17]. It is normally expressed as a series expansion whose first term is the Poisson bracket. Elsewhere in the same paper Groenewold argues that for an arbitrary quantization the property (ii) implies commutativity of the operators of the form $\theta(f)$. However, his argument is incorrect, and in fact, using a reducible representation of the commutation relations Souriau has constructed a quantization with this property (see Streater [30]).

In attempts to save the property (ii) several authors have investigated the possibility of restricting the Weyl correspondence to suitable classes of functions in such a way that this relation holds for these, cf. van Hove [32], Hermann [8], Joseph [9].

Our approach is similar in that we restrict one of the functions, say f, to a rather small domain, namely first order polynomials in the q's and p's and those polynomials of second order which are components of the angular momenta $q_n \times p_n$.

Now, let us restrict our attention to the conditions (i). A technical difficulty is caused by the fact that in the Hilbert space L_2 the operators $\mathbb{Q}_{nj}$ and $\mathbb{P}_{nj}$ are unbounded and not everywhere defined. In order to overcome this difficulty we require all quantizations $\theta(h)$ to be defined on the common domain $\mathcal{S}$ of all polynomials in the $\mathbb{Q}$'s and $\mathbb{P}$'s. Since (i) involves products of the form $\mathbb{Q}\theta(h)$ and $\mathbb{P}\theta(h)$ too, it is convenient to allow the dual space $\mathcal{S}^*$ of temperate distributions as range of the operators $\theta(h)$. With these conventions we first prove (Theorem 1) that Weyl's correspondence is indeed a quantization, and that it is an isomorphism between the topological vector spaces $\mathcal{S}^{6N*}$ and $L(\mathcal{S}^{3N}, \mathcal{S}^{3N*})$. Thus, where most previous results involving unbounded operators have been based on rather formal arguments, the use of the spaces of distribution theory allows us to introduce topology and use the full force of the methods of analysis. The most satisfactory result hitherto known, although rather elementary, is that the functions in $L^2(\mathbb{R}^{6N})$ correspond precisely to the Hilbert-Schmidt operators in $L^2(\mathbb{R}^{3N})$, c.f. Loupias and Miracle-Solé [14] and Pool [18]. Furthermore, Loupias and Miracle-Solé [15] have proved that the quantization of a function in $\mathcal{S}^{6N}$ is an operator of trace-class, and by a duality argument that every bounded operator in L^2 can be considered the quantization of a temperate distribution.

We next prove (Theorem 2) the existence of a large class of quantizations, parametrized by smooth functions on $\mathbb{R}^{6N}$. It is not to be expected that such general quantizations are of interest, but rather that one can single out various quantizations by imposing suitable additional conditions.

Thus, we prove (Theorem 3) that Weyl's correspondence is uniquely characterized by the additional conditions

$$\theta(q_{nj}h) = \frac{1}{2}(\mathbb{Q}_{nj}\theta(h) + \theta(h)\mathbb{Q}_{nj}),$$

$$\theta(p_{nj}h) = \frac{1}{2}(\mathbb{P}_{nj}\theta(h) + \theta(h)\mathbb{P}_{nj}).$$

That Weyl's correspondence has this property is an immediate consequence of a formula (Moyal's cosine bracket) due to Groenwold [6] for

$$\theta^{-1}\left(\frac{1}{2}(\theta(g)\theta(h) + \theta(h)\theta(g))\right).$$

This formula also shows that in general this inverse image is different from $g \cdot h$. Let us note in this connection that the quantization suggested by Rivier [21] is characterized by

$$\theta(g(q)h(p)) = \frac{1}{2}(\theta(g(q))\theta(h(p)) + \theta(h(p))\theta(g(q))).$$

We also show (Theorem 4) that the usual Wick-ordering or normal ordering is indeed a quantization. For a comparison between these three special quantizations and some of their properties, see Mehta [16] .

Finally, Theorem 5 characterizes those quantizations which satisfy (ii) with the components of the angular momenta in the place of f. As is to be expected, this condition is satisfied provided the parameter function of the quantization has a suitable invariance property relative to rotations.

In the concluding section of the paper it is shown that the quantization by Wick-ordering has a particularly simple appearance in the Fock-Bargman-Segal representation of the commutation relations. Also, attention is called to the important and apparently deep problem of characterizing those functions (distributions) whose quantizations are essentially self-adjoint.

In conclusion it should be mentioned that other approaches to the quantization problem exist, some of them apparaently with little relationship to the one presented here. In addition to the literature cited above we refer to Shewell [27] , Segal [23] ,

Prosser [19], Klauder [11], Souriau [28, 29]. It was noticed by Kohn and Nirenberg [12] that the relationship established between operators and functions by means of a quantization is very similar to the relationship between a pseudodifferential operator and its symbol. This analogy was exploited by Grossman, Loupias and Stein [7], but may yet contain interesting further possibilities.

The author wishes to express his indebtnedness to Konrad Jörgens and Povl Kristensen for inspiration and stimulating discussions.

2. Notation

Euclidean spaces of dimension d will be denoted by $\mathbb{R}^d$, and although often different copies of this space will, formally, play different roles, we shall not distinguish between them in notation. Accordingly, if x and y belong to $\mathbb{R}^d$, their scalar product will be denoted

$$x \cdot y = \sum_{k=1}^{d} x_k y_k$$

whether they belong to the same or to different copies of $\mathbb{R}^d$.

Distribution spaces. Schwartz' space of rapidly decreasing infinitely often differentiable functions on $\mathbb{R}^d$ will be denoted by $\mathcal{S}^d$. Its dual space, i.e. the space of temperate distributions will be denoted by $\mathcal{S}^{d*}$.

For elements of $\mathcal{S}^{d*}$ we shall use the conventional functional notation, writing them as $f(x)$ etc. even though they need not be locally integrable functions. Accordingly, the value of a distribution $f \in \mathcal{S}^{d*}$ on a test function $\varphi \in \mathcal{S}^d$ will be written symbolically as an integral:

$$\langle f, \varphi \rangle = \int \overline{f(x)}\, \varphi(x) dx = \langle \overline{\varphi, f} \rangle .$$

The complex conjugation is introduced in order to facilitate translation into the usual Hilbert space framework.

The space $L(\mathcal{S}^d, \mathcal{S}^{d*})$ of continuous linear operators from $\mathcal{S}^d$ into $\mathcal{S}^{d*}$ will always be equipped with the topology of uniform convergence on bounded sets. By Schwartz' nuclear theorem (Schwartz [22], see also for instance Treves [31]) $L(\mathcal{S}^d, \mathcal{S}^{d*})$ is isomorphic algebraically and topologically with $\mathcal{S}^{2d*}$, the correspondance between an operator $T \in L(\mathcal{S}^d, \mathcal{S}^{d*})$ and its <u>kernel</u> $k(x,y) \in \mathcal{S}^{2d*}$ being given by

$$\langle \psi, T\varphi \rangle = \int \overline{\psi(x)}\, k(x,y) \quad \varphi(y)\, dx dy$$

or, conventionally,

$$T\,\varphi(x) = \int k(x,y)\, \varphi(y)\, dy.$$

For an operator $T \in L(\mathcal{S}^d, \mathcal{S}^{d*})$ we define its <u>adjoint</u> as the operator $T^* \in L(\mathcal{S}^d, \mathcal{S}^{d*})$ ($\mathcal{S}^d$ is reflexive) characterized by

$$\langle T^* \psi, \varphi \rangle = \langle \psi, T\varphi \rangle \qquad \varphi, \psi \in \mathcal{S}^d .$$

The kernel of T^* is

$$k^*(x,y) = \overline{k(y,x)} ,$$

and the mapping $T \longmapsto T^*$ is a conjugate linear homeomorphism of $L(\mathcal{S}^d, \mathcal{S}^{d*})$ onto itself.

<u>Weyl Operators</u>. The operators $\mathbb{Q}_k$ and $\mathbb{P}_k$, $k = 1,2,\ldots,d$ are defined by

$$(\mathbb{Q}_k\, f)\,(x) = x_k\, f(x) ,$$

$$\mathbb{P}_k f = -i \frac{\partial}{\partial x_k} f.$$

Clearly $\mathbb{Q}_k$ and $\mathbb{P}_k$ belong to $L(\mathcal{S}^{d*}, \mathcal{S}^{d*})$ as well as to $L(\mathcal{S}^d, \mathcal{S}^d)$.

For $(\lambda,\mu) \in \mathbb{R}^{2d}$ we define

$$\lambda \cdot \mathbb{Q} + \mu \cdot \mathbb{P} = \sum_{k=1}^{d} (\lambda_k \mathbb{Q}_k + \mu_k \mathbb{P}_k).$$

This operator is the infinitesimal generator of a strongly continuous equicontinuous one-parameter group of operators in $L(\mathcal{S}^{d*}, \mathcal{S}^{d*})$. The operators in this group will be denoted by

$$\exp(it(\lambda \cdot \mathbb{Q} + \mu \cdot \mathbb{P})) = \exp(i(t\lambda \cdot \mathbb{Q} + t\mu \cdot \mathbb{P})).$$

They are given by

$$\exp(i\lambda \cdot \mathbb{Q})\, f(x) = e^{i\lambda\cdot x} f(x),$$

$$\exp(i\mu \cdot \mathbb{P})\, f(x) = f(x+\mu),$$

$$\begin{aligned}\exp(i(\lambda \cdot \mathbb{Q} + \mu \cdot \mathbb{P})) &= e^{\frac{1}{2} i\lambda\cdot\mu} \exp(i\lambda \cdot \mathbb{Q}) \exp(i\mu \cdot \mathbb{P}) \\ &= e^{-\frac{1}{2} i\lambda\cdot\mu} \exp(i\lambda \cdot \mathbb{P}) \exp(i\mu \cdot \mathbb{Q}) \\ &= \exp(\tfrac{1}{2} i\lambda \cdot \mathbb{P}) \exp(i\mu \cdot \mathbb{Q}) \exp(\tfrac{1}{2} i\lambda \cdot \mathbb{P}).\end{aligned}$$

The Hilbert space $L^2(\mathbb{R}^d)$ is clearly invariant under the operators $\exp(i(\lambda \cdot \mathbb{Q} + \mu \cdot \mathbb{P}))$, in fact, on $L^2(\mathbb{R}^d)$ these operators are unitary operators. These unitary operators were introduced by Hermann Weyl [33] , and are now usually called the Weyl operators.

Commutators. For two linear operators A, B we define the commutator by

$$[A,B] = AB - BA$$

on its natural domain.

If $A \in L(\mathcal{S}^d, \mathcal{S}^d) \cap L(\mathcal{S}^{d*}, \mathcal{S}^{d*})$, as is the case for $\mathbb{Q}_k$ and $\mathbb{P}_k$, and if $B \in L(\mathcal{S}^d, \mathcal{S}^{d*})$, then

$$[A,B] \in L(\mathcal{S}^d, \mathcal{S}^{d*}).$$

Poisson brackets. Two particular copies of $\mathbb{R}^d$ will be denoted position space and momentum space, their points being denoted by q and p respectively. If f and g denote distributions on phase space $\mathbb{R}^d \times \mathbb{R}^d$, then the Poisson bracket of f and g is defined as

$$\{f, g\} = \sum_{k=1}^{d} \left\{ \frac{\partial f}{\partial q_k} \frac{\partial g}{\partial p_k} - \frac{\partial f}{\partial p_k} \frac{\partial g}{\partial q_k} \right\}.$$

(It is assumed that the products make sense, for instance that either f or g is a $\mathcal{C}^{\infty}$-function.)

3. Quantizations. Definition and first analysis.

Definition 1. By a quantization we shall understand a continuous linear transformation θ from a $\mathfrak{A}$ -space (see Definition 2) $W \in \mathcal{S}^{2d*}$ into $L(\mathcal{S}^{d}, \mathcal{S}^{d*})$ satisfying the following conditions:

(1) $$\theta(\{q_k, f\}) = -i [\theta(q_k), \theta(f)],$$

(2) $$\theta(\{p_k, f\}) = -i [\theta(p_k), \theta(f)],$$

(3) $$\theta(\bar{f}) = \theta(f)^*,$$

(4) $$\theta(q_k) = \mathbb{Q}_k, \quad \theta(p_k) = \mathbb{P}_k.$$

Note that condition (4) is compatible with (1), (2) and (3), and implies $\theta(1) = I$.

Definition 2. By a $\mathfrak{A}$ -space we shall understand a locally convex topological vector space W satisfying the following conditions:

a) As a vector space W is a subspace of $\mathcal{S}^{2d*}$, and the inclusion mapping is continuous.

b) W is invariant under differential operators with constant coefficients, and under complex conjugation, and these operations are continuous from W into W.

c) The functions $(q,p) \mapsto q_k$ and $(q,p) \longmapsto p_k$ belong to W. So do the functions $e_{\lambda\mu}$ for all $(\lambda,\mu) \in \mathbb{R}^{2d}$, where

$$e_{\lambda\mu}(q,p) = \exp(i(\lambda \cdot q + \mu \cdot p)),$$

and they generate a dense linear subspace of W.

Note that the basic requirements formulated in Definition 2 are that the conditions of Definition 1 be meaningful.

LEMMA 1. *Let* θ *be a quantization defined on a* $\mathfrak{A}$*-space* W. *Then there exists a complex valued function* c *on* $\mathbb{R}^{2d}$ *such that*

$$\theta(e_{\lambda\mu}) = c(\lambda,\mu)\exp(i(\lambda \cdot \mathbb{Q} + \mu \cdot \mathbb{P})).$$

The function c *has the properties*

$$c(-\lambda, -\mu) = \overline{c(\lambda,\mu)},$$

$$c(0,0) = 1.$$

The function c *determines* θ *uniquely.*

Proof. Let $(\lambda,\mu) \in \mathbb{R}^{2d}$ be fixed. Since

$$\{q_k, e_{\lambda\mu}\} = i\mu_k e_{\lambda\mu},$$

$$\{p_k, e_{\lambda\mu}\} = -i\lambda_k e_{\lambda\mu},$$

it follows from the conditions (1), (2) and (4) in Definition 1 that

$$[\mathbb{Q}_k, \theta(e_{\lambda\mu})] = -\mu_k \theta(e_{\lambda\mu}),$$

$$[\mathbb{P}_k, \theta(e_{\lambda\mu})] = \lambda_k \theta(e_{\lambda\mu}).$$

Direct computation shows that the Weyl-operator $T = \exp(i(\lambda \cdot \mathbb{Q} + \mu \cdot \mathbb{P}))$ satisfies these conditions. Hence, if we define $C = T^{-1}\theta(e_{\lambda\mu})$, remembering that $T^{-1} \in L(\varphi^{d*}, \varphi^{d*})$, it follows that

$$[\mathbb{Q}_k, C] = [\mathbb{P}_k, C] = 0 \qquad \text{for all } k.$$

It is well-known, and easily proved, that this implies that C is a multiple of I, say $C = c(\lambda,\mu)\, I$.

The fact that $c(-\lambda,-\mu) = \overline{c(\lambda,\mu)}$ now follows from condition (3) in Definition 1 since

$$\exp(i(\lambda \cdot \mathbb{Q} + \mu \cdot \mathbb{P}))^* = \exp(i(-\lambda . \mathbb{Q} - \mu \cdot \mathbb{P})),$$

and the property $c(0,0) = 1$ follows from the fact, that $\theta(1) = I$, as remarked after Definition 1.

Finally, it follows from the continuity of θ and condition c) of Definition 2 that the function c determines θ uniquely.

4. Weyl's quantization

We shall now consider the case $c(\lambda,\mu) \equiv 1$, which is essentially the case already considered by Weyl [33] and by many others since then.

The main theorem is suggested by the observation that the Weyl operators can be written formally as

$$(5)\quad \exp(i(\lambda.\mathbb{Q} + \mu.\mathbb{P})) = \exp(\tfrac{1}{2}i\mu\,.\,\mathbb{P})\exp(i\lambda.x)\exp(\tfrac{1}{2}i\,\mu\,.\,\mathbb{P})$$

$$= \int \exp(\tfrac{1}{2}i\; z.\,\mathbb{P})\,\delta(z-\mu)\exp(i\lambda.x)\exp(\tfrac{1}{2}i\; z.\,\mathbb{P})dz$$

$$= \int \exp(\tfrac{1}{2}i\; z.\,\mathbb{P})(\mathcal{F}_2 e_{\lambda\mu}(x,z))\exp(\tfrac{1}{2}i\; z.\,\mathbb{P})dz,$$

where $\mathcal{F}_2$ denotes the partial Fourier transform defined by

$$\mathcal{F}_2 f(q,z) = (2\pi)^{-d}\int \exp(-i\; z\,.\,p)f(q,p)dp.$$

Applying (5) to a function $\varphi \in \mathcal{S}^d$, we get (with $x,y,z \in \mathbb{R}^d$)

$$\exp(i(\lambda\,.\,\mathbb{Q} + \mu\,.\,\mathbb{P}))\,\varphi(x) = \int \mathcal{F}_2\, e_{\lambda\mu}(x + \tfrac{1}{2}z, z)\;\varphi(x + z)dz$$

$$= \int \mathcal{F}_2\, e_{\lambda\mu}(\tfrac{1}{2}x + \tfrac{1}{2}y, y-x)\,\varphi(y)dy.$$

This observation suggests the following definition of a quantization.

THEOREM 1. *The formula*

$$(6)\qquad \theta_1(f)\,\varphi(x) = \int \mathcal{F}_2\, f(\tfrac{1}{2}\,x + \tfrac{1}{2}\,y, y-x)\;\varphi(y)dy$$

defines a transformation θ_1 *of* $\mathcal{S}^{2d*}$ *into* $L(\mathcal{S}^d, \mathcal{S}^{d*})$. *The transformation* θ_1 *is a quantization, in the sequel denoted Weyl's quantization. It is an isomorphism of the topological vector space* $\mathcal{S}^{2d*}$ *onto* $L(\mathcal{S}^d, \mathcal{S}^{d*})$.

Proof. First, note that the partial Fourier transform $\mathcal{F}_2$ is an isomorphism of $\mathcal{S}^{2d*}$ onto itself. Secondly, if we define

$$h(x,y) = g(\tfrac{1}{2}x + \tfrac{1}{2}y, y - x),$$

then clearly the transformation $g \longmapsto h$ is an isomorphism of $\mathcal{S}^{2d*}$ onto itself. Finally, Schwartz' nuclear theorem (cf. Section 2), tells us how to interpret the formula defining θ_1, and gives the final statement of the theorem.

It remains to prove that θ_1 is a quantization, since obviosuly $\mathcal{S}^{2d*}$ is a $\mathcal{O}$-space. This is verified by straightforward computation. Let us for instance prove the identity (2). Writing ∂_k^1 for the derivative of a distribution on $\mathbb{R}^d \times \mathbb{R}^d$ w.r.t. the argument with index k in the first copy of $\mathbb{R}^d$, we get

$$-i[\mathbb{P}_k, \theta_1(f)]\,\varphi(x) = -\frac{\partial}{\partial x_k}\int \mathcal{F}_2 f(\tfrac{1}{2}x + \tfrac{1}{2}y, y-x)\,\varphi(y)\,dy +$$

$$+\int \mathcal{F}_2 f(\tfrac{1}{2}x + \tfrac{1}{2}y, y-x)\,\frac{\partial\varphi}{\partial y_k}(y)\,dy$$

$$= -\int\Big[\Big(\frac{\partial}{\partial x_k} + \frac{\partial}{\partial y_k}\Big)\mathcal{F}_2 f(\tfrac{1}{2}x + \tfrac{1}{2}y, y-x)\Big]\varphi(y)\,dy$$

$$= -\int \partial_k^1\,\mathcal{F}_2 f(\tfrac{1}{2}x + \tfrac{1}{2}y, y-x)\,\varphi(y)\,dy$$

$$= -\int \mathcal{F}_2\,\partial_k^1 f(\tfrac{1}{2}x + \tfrac{1}{2}y, y-x)\,\varphi(y)\,dy$$

$$= -\theta_1\Big(\frac{\partial f}{\partial q_k}\Big)\varphi(x) = \theta_1(\{p_k, f\})\,\varphi(x).$$

5. A class of general quantizations.

For the construction of more general quantizations, it is convenient to introduce the inverse Fourier transform in $\mathcal{S}^{2d*}$ defined by

$$(\mathcal{F}^{-1} g)(q,p) = \int \exp(i(\xi \cdot q + \eta \cdot p))\,g(\xi,\eta)\,d\xi d\eta.$$

Now, let θ denote an arbitrary quantization, and put

$$\pi_1 = \theta \circ \mathfrak{F}^{-1} \quad \text{and} \quad \pi = \theta \circ \mathfrak{F}^{-1}.$$

Then we have, denoting the delta-distribution at (λ,μ) by $\delta_{\lambda\mu}$,

$$\begin{aligned} \pi(\delta_{\lambda\mu}) = \theta(e_{\lambda\mu}) &= c(\lambda,\mu) \quad \theta_1(e_{\lambda\mu}) \\ &= c(\lambda,\mu) \quad \pi_1 \ (\delta_{\lambda\mu}) = \pi_1(c(\lambda,\mu)\delta_{\lambda\mu}). \end{aligned}$$

If c is a $\mathcal{C}^{\infty}$-function, so that multiplication of distributions by c is defined, this can be written

$$\pi(\delta_{\lambda\mu}) = \pi_1 \ (c \ \delta_{\lambda\mu}).$$

We now have

<u>THEOREM 2</u>. <u>Let</u> c <u>be a complex valued</u> $\mathcal{C}^{\infty}$<u>-function on</u> $\mathbb{R}^{2d}$ <u>with the properties</u>

$$c(-\lambda,-\mu) = \overline{c(\lambda,\mu)},$$
$$c(0,0) = 1,$$
$$\frac{\partial c}{\partial \lambda_k}(0,0) = \frac{\partial c}{\partial \mu_k}(0,0) = 0 \quad \text{for all} \quad k.$$

<u>Then there exist a quantization</u> θ_c <u>such that</u>

$$\theta_c(e_{\lambda\mu}) = c(\lambda,\mu) \exp(i(\lambda \cdot \mathbb{Q} + \mu \cdot \mathbb{P})).$$

θ_c <u>is given by</u>

$$(7) \qquad \theta_c(f) = \theta_1 \circ \mathfrak{F}^{-1} (c \cdot \mathfrak{F} f)$$

<u>on the</u> $\mathcal{O}\!\!\!/$<u>-space</u>

$$W_c = \left\{ f \in \mathcal{S}^{2d*} \mid c \cdot \mathfrak{F} f \in \mathcal{S}^{2d*} \right\}.$$

Before proving the theorem, we first complete the formulation of it by specifying a topology on the space W_c with which it is a $\mathcal{O}\!\!\!/$-space, and θ_c is continuous.

<u>LEMMA 2</u>. <u>Let</u> c <u>and</u> W_c <u>be as in Theorem 2. Denote by</u> $\hat{W}_c$ <u>the space</u>

$$\hat{W}_c = \{ g \in \mathcal{S}^{2d*} \mid c \cdot g \in \mathcal{S}^{2d*} \},$$

<u>and provide it with the coarsest topology such that both the identical injection and the mapping</u> $g \longmapsto c \cdot g$ <u>are continuous from</u> $\hat{W}_c$ <u>into</u> $\mathcal{S}^{2d*}$.

<u>Define the topology of</u> W_c <u>such that the Fourier transform</u> $\mathcal{F}$ <u>is a homeomorphism of</u> W_c <u>onto</u> $\hat{W}_c$. <u>Then</u> W_c <u>is a</u> $\mathcal{O}\!\!\!+$<u>-space</u>.

<u>Proof</u>. In view of the properties of $\mathcal{F}$, the conditions a), b), c) of Definition 2 are equivalent to

a') $\hat{W}_c$ is a subspace of $\mathcal{S}^{2d*}$, and the inclusion mapping is continuous.

b') $\hat{W}_c$ is invariant under multiplication by polynomials, and under the mapping $f \longmapsto \tilde{f}$, where $\tilde{f}(\xi,\eta) = \bar{f}(-\xi,-\eta)$. These operations are continuous in $\hat{W}_c$.

c') The distributions $\partial\delta/\partial\xi_k$ and $\partial\delta/\partial\eta_k$ belong to $\hat{W}_c$. So do the functions $\delta_{\lambda\mu}$, and they generate a dense linear subspace of $\hat{W}_c$.

The only fact needing proof is the final part of c'). To prove it, note that $\hat{W}_c$ is locally convex and that every continuous linear functional F on it can be represented (not uniquely) in the form

$$\langle F, g\rangle = \int \varphi(\xi,\eta) \; g(\xi,\eta) \, d\xi \, d\eta + \int \psi(\xi,\eta) \; c(\xi,\eta) \; g(\xi,\eta) \, d\xi \, d\eta$$

for some $\varphi, \psi \in (\mathcal{S}^{2d*})^* = \mathcal{S}^{2d}$. Now, if

$$\langle F, \delta_{\lambda\mu}\rangle = 0 \quad \text{for all } \lambda,\mu,$$

then

$$\varphi(\lambda,\mu) + \psi(\lambda,\mu) \, c(\lambda,\mu) \equiv 0,$$

and hence

$$\langle F, g\rangle = 0 \quad \text{for all } g \in \hat{W}_c.$$

<u>Proof of Theorem 2</u>. It is clear that θ_c is a continuous linear transformation from W_c into $L(\mathcal{S}^d, \mathcal{S}^{d*})$, and the

properties of c immediately give that the conditions (3) and (4) of Definition 1 are satisfied. Finally, since

$$\{q_k, f\} = \frac{\partial f}{\partial p_k}$$

and

$$\mathcal{F}^{-1}\left(c \cdot \mathcal{F} \frac{\partial f}{\partial p_k}\right) = \frac{\partial}{\partial p_k} \mathcal{F}^{-1}(c \cdot \mathcal{F} f),$$

the condition (1) (and, equivalently, (2)) follows from the fact that θ_1 is a quantization.

In view of the form of the general quantization θ_c it is immediate to deduce from knowledge of the function c whether θ_c is defined on all of $\mathcal{S}^{2d*}$, is onto all of $L(\mathcal{S}^d, \mathcal{S}^{d*})$, is one-to-one, etc.

6. Quantizations satisfying additional requirements.

In this section we shall only consider quantizations of the type θ_c described in Theorem 2. First let us note that since $\mathcal{S}^{2d*}$ is invariant under differential operators with polynomial coefficients, the same holds for $\hat{W}_c$, and hence, by Fourier transform, for W_c.

THEOREM 3. Weyl's quantization θ_1 is uniquely characterized (among all quantizations θ_c) by the conditions

$$\theta(q_k f) = \tfrac{1}{2}(\mathbb{Q}_k \theta(f) + \theta(f) \mathbb{Q}_k), \tag{8}$$

$$\theta(p_k f) = \tfrac{1}{2}(\mathbb{P}_k \theta(f) + \theta(f) \mathbb{P}_k). \tag{9}$$

Proof. From the definition of the spaces W_c in Lemma 2 immediately follows that the mapping $(\lambda, \mu) \longmapsto e_{\lambda\mu}$ is differentiable from $\mathbb{R}^{2d}$ into an arbitrary space W_c with the derivatives

$$\frac{\partial}{\partial \lambda_k} e_{\lambda\mu} = i q_k e_{\lambda\mu}, \qquad \frac{\partial}{\partial \mu_k} e_{\lambda\mu} = i p_k e_{\lambda\mu}.$$

Also, the mapping

$$(\lambda, \mu) \longmapsto E_{\lambda\mu} = \exp(i(\lambda \cdot \mathbb{Q} + \mu \cdot \mathbb{P}))$$

is differentiable from $\mathbb{R}^{2d}$ into $L(\mathcal{S}^d, \mathcal{S}^{d*})$ with the derivatives

$$\frac{\partial}{\partial \lambda_k} E_{\lambda\mu} = \frac{i}{2} (\mathbb{Q}_k E_{\lambda\mu} + E_{\lambda\mu} \mathbb{Q}_k),$$

$$\frac{\partial}{\partial \mu_k} E_{\lambda\mu} = \frac{i}{2} (\mathbb{P}_k E_{\lambda\mu} + E_{\lambda\mu} \mathbb{P}_k).$$

It follows that if a quantization θ_c satisfies the conditions of the theorem, then

$$\frac{\partial c}{\partial \lambda_k} = \frac{\partial c}{\partial \mu_k} = 0$$

for all k, and hence $c(\lambda,\mu) \equiv 1$. Conversely, if $c \equiv 1$, then the conditions are satisfied for f of the form $e_{\lambda\mu}$, and hence by linearity and continuity for all $f \in \mathcal{S}^{2d*}$.

In view of this theorem the Weyl quantization might be called quantization by symmetrization. Another special quantization which may be of interest for quantum field theory might be called quantization by Wick-ordering. It is described in the following theorem.

<u>THEOREM 4.</u> <u>Among all quantizations</u> θ_c <u>there exists precisely one which satisfies</u>

(10) $$\theta((p_k - iq_k)f) = \theta(f) (\mathbb{P}_k - i\mathbb{Q}_k),$$

(11) $$\theta((p_k + iq_k)f) = (\mathbb{P}_k + i \mathbb{Q}_k) \theta(f),$$

<u>namely the one corresponding to the function</u>

$$c(\lambda , \mu) = \exp(\frac{1}{4} |\lambda|^2 + |\mu|^2) .$$

<u>Proof</u>. The conditions are equivalent with the following:

(12) $$\theta(q_k f) = \frac{1}{2} (\mathbb{Q}_k \theta(f) + \theta(f) \mathbb{Q}_k) - \frac{i}{2} [\mathbb{P}_k,\theta(f)] ,$$

(13) $$\theta(p_k f) = \frac{1}{2} (\mathbb{P}_k \theta(f) + \theta(f) \mathbb{P}_k) + \frac{i}{2} [\mathbb{Q}_k,\theta(f)] .$$

Now, by (7) we have with $\theta = \theta_c$

(14) $$\begin{aligned}\theta(q_k f) &= \theta_1 \mathcal{F}^{-1}(c \mathcal{F}(q_k f)) \\ &= \theta_1 \mathcal{F}^{-1}(i c \frac{\partial}{\partial \xi_k} \mathcal{F} f),\end{aligned}$$

by (7) and (8)

$$(15)\quad \frac{1}{2}\,(\,\mathbb{Q}_k\,\theta(f) + \theta(f)\,\mathbb{Q}_k\,) \;=\; \theta_1(q_k\,\mathcal{F}^{-1}\,(c\,\mathcal{F}f))$$
$$= \theta_1\,\mathcal{F}^{-1}\,(i\,\frac{\partial}{\partial\xi_k}\,(c\,\mathcal{F}f)),$$

and finally by (2) and (7)

$$(16)\quad -\frac{i}{2}\,[\mathbb{P}_k,\,\theta(f)] \;=\; -\frac{1}{2}\,\theta_1\,\mathcal{F}^{-1}\,(c\,\mathcal{F}(\frac{\partial f}{\partial q_k}))$$
$$= \theta_1\,\mathcal{F}^{-1}\,(-\frac{i}{2}\,c\,\xi_k\,\mathcal{F}f).$$

Inserting these expressions in (12) we get

$$\frac{\partial c}{\partial \xi_k} \;=\; \frac{1}{2}\,\xi_k\,c.$$

Similarly, (13) gives

$$\frac{\partial c}{\partial \eta_k} \;=\; \frac{1}{2}\,\eta_k\,c,$$

and the theorem follows.

The quantization by Wick-ordering has a particularly simple form in the Bargman-Segal-representation. We shall briefly comment upon this in Section 7.

In the remaining part of this section we shall consider the case corresponding to N particles moving in $\mathbb{R}^3$, i.e. $d = 3N$. We shall then interpret $\mathbb{R}^{3N}$ as $(\mathbb{R}^3)^N$, and write points $x \in \mathbb{R}^{3N}$ as

$$x \;=\; (x_1, x_2, \ldots, x_N),$$

where

$$x_n \;=\; (x_{n1}, x_{n2}, x_{n3}) \in \mathbb{R}^3 \qquad \text{for} \quad n = 1, 2, \ldots, N.$$

<u>Definition 3.</u> The vector function l_n on phase space $\mathbb{R}^{3N} \times \mathbb{R}^{3N}$ defined by

$$l_n \;=\; q_n \times p_n$$

will be denoted the (classical) <u>angular momentum</u> of the n'th particle. The function

$$l \;=\; \sum_{n=1}^{N} l_n$$

will be denoted the (classical) <u>total angular momentum</u>.

The vector valued operator

$$\mathbb{L}_n \in L(\mathcal{S}^{3N}, \mathcal{S}^{3N})^3 \times L(\mathcal{S}^{3N*}, \mathcal{S}^{3N*})^3$$

defined in a natural way by the expression

$$\mathbb{L}_n = \mathbb{Q}_n \times \mathbb{P}_n$$

will be denoted the (quantum) <u>angular momentum operator</u> of the n'th particle. The operator

$$\mathbb{L} = \sum_{n=1}^{N} \mathbb{L}_n$$

will be denoted the (quantum) total <u>angular momentum operator</u>.

The coordinates of l_n will be called its <u>components</u> and denoted (l_n^1, l_n^2, l_n^3), and correspondingly for l, $\mathbb{L}_n$ and $\mathbb{L}$.

<u>LEMMA 3.</u> <u>Let</u> θ <u>denote an arbitrary quantization. The quantization of an arbitrary angular momentum component</u> l_n^j <u>has the form</u>

$$\theta(l_n^j) = \mathbb{L}_n^j + \alpha_n^j I,$$

<u>where</u> α_n^j <u>is a scalar constant.</u>

<u>Proof.</u> Since each Poisson bracket $\{q_{mi}, l_n^j\}$ and $\{p_{mi}, l_n^j\}$ is a linear combination of q_{nk}'s and p_{nk}'s, one can check directly that

$$\theta(\{q_{mi}, l_n^j\}) = -i[\mathbb{Q}_{mi}, \mathbb{L}_n^j]$$

and

$$\theta(\{p_{mi}, l_n^j\}) = -i[\mathbb{P}_{mi}, \mathbb{L}_n^j].$$

It follows that $\theta(l_n^j) - \mathbb{L}_n^j$ commutes with all $\mathbb{Q}_{mi}$ and all $\mathbb{P}_{mi}$, and hence is a multiple of the identity.

<u>DEFINITION 4.</u> A quantization θ is said to be <u>compatible with rotations</u> if and only if

$$\theta(\{l^j, f\}) = -i[\theta(l^j), \theta(f)]$$

for j = 1,2,3.

It is said to be <u>fully compatible with rotations</u> if and only if

$$\theta(\{l_n^j, f\}) = -i[\theta(l_n^j), \theta(f)]$$

for $n = 1,2,\ldots,N$, $j = 1,2,3$.

<u>Definition 5.</u> By a <u>rotation</u> we shall understand an orthogonal transformation in $\mathbb{R}^3$ with determinant $+1$. The group of all rotations is denoted $SO(3)$.

For every $\gamma \in SO(3)$ we denote by γ^N the orthogonal transformation in $\mathbb{R}^{3N}$ defined by

$$\gamma^N(x_1, x_2, \ldots, x_N) = (\gamma x_1, \gamma x_2, \ldots, \gamma x_n).$$

THEOREM 5. *A quantization* θ_c *is compatible with rotations if and only if* c *is invariant under simultaneous rotation of all* (λ_n, μ_n), *i.e. if and only if*

$$c(\lambda, \mu) = c(\gamma^N\lambda, \gamma^N\mu) \quad \text{for all} \quad \gamma \in SO(3) .$$

θ_c *is fully compatible with rotations if and only if* c *is invariant under separate rotation of each* (λ_n, μ_n), *i.e. if and only if*

$$c(\lambda_1, \ldots, \lambda_N, \mu_1, \ldots, \mu_N) = c(\lambda_1, \ldots, \gamma\lambda_n, \ldots, \gamma\mu_n, \ldots, \mu_N)$$

for all $\gamma \in SO(3)$, $n = 1,2,\ldots,N$.

<u>Proof.</u> In view of Lemma 3, θ is fully compatible with rotations if and only if

$$\theta(\{l_n^j, f\}) = -i[\mathbf{L}_n^j, \theta(f)] \tag{17}$$

for all $n = 1,2,\ldots,N$, $j = 1,2,3$. An analogous reformulation of compatibility with rotations is immediate.

From (14) and (15) we get

$$\theta(q_{nk}\, g) = \tfrac{1}{2}(\mathbf{Q}_{nk}\,\theta(g) + \theta(g)\,\mathbf{Q}_{nk}) - i\, e_1\, \mathcal{F}^{-1}\left(\frac{\partial c}{\partial \xi_{nk}}\, \mathcal{F} g\right).$$

This relation and the analogous expression for $\theta(p_{nk}\, g)$ together with the fact that θ is a quantization give

$$\theta(\{l_n^1, f\}) = \theta(p_{n3}\frac{\partial f}{\partial p_{n2}} + q_{n3}\frac{\partial f}{\partial q_{n2}} - p_{n2}\frac{\partial f}{\partial p_{n3}} - q_{n2}\frac{\partial f}{\partial q_{n3}})$$

$$= \tfrac{1}{2}(-i\,\mathbb{P}_{n3}[\mathbb{Q}_{n2},\theta(f)] - i[\mathbb{Q}_{n2},\theta(f)]\,\mathbb{P}_{n3} +$$

$$+ i\,\mathbb{Q}_{n3}[\mathbb{P}_{n2},\theta(f)] + i[\mathbb{P}_{n2},\theta(f)]\,\mathbb{Q}_{n3} +$$

$$+ i\,\mathbb{P}_{n2}[\mathbb{Q}_{n3},\theta(f)] + i[\mathbb{Q}_{n3},\theta(f)]\,\mathbb{P}_{n2} -$$

$$- i\,\mathbb{Q}_{n2}[\mathbb{P}_{n3},\theta(f)] - i[\mathbb{P}_{n3},\theta(f)]\,\mathbb{Q}_{n2} -$$

$$- i\,\theta_1\mathcal{F}^{-1}(\frac{\partial c}{\partial \eta_{n3}}\mathcal{F}\frac{\partial f}{\partial p_{n2}} + \frac{\partial c}{\partial \xi_{n3}}\mathcal{F}\frac{\partial f}{\partial q_{n2}} -$$

$$- \frac{\partial c}{\partial \eta_{n2}}\mathcal{F}\frac{\partial f}{\partial p_{n3}} - \frac{\partial c}{\partial \xi_{n2}}\mathcal{F}\frac{\partial f}{\partial q_{n3}})$$

$$= - i[\mathbb{L}_n^1,\theta(f)] +$$

$$+ \theta_1\mathcal{F}^{-1}((\xi_{n2}\frac{\partial c}{\partial \xi_{n3}} - \xi_{n3}\frac{\partial c}{\partial \xi_{n2}} + \eta_{n2}\frac{\partial c}{\partial \eta_{n3}} -$$

$$- \eta_{n3}\frac{\partial c}{\partial \eta_{n2}})\,\mathcal{F}f).$$

If (17) is to be satisfied for $j = 1$ and for all $f \in W_c$, we must clearly have

$$\xi_{n2}\frac{\partial c}{\partial \xi_{n3}} - \xi_{n3}\frac{\partial c}{\partial \xi_{n2}} + \eta_{n2}\frac{\partial c}{\partial \eta_{n3}} - \eta_{n3}\frac{\partial c}{\partial \eta_{n2}} = 0,$$

which means that for all rotations γ that leave the 1-axis invariant we have

$$c(\xi_1,\ldots,\gamma\xi_n,\ldots,\xi_N,\eta_1,\ldots,\gamma\eta_n,\ldots,\eta_N) = c(\xi,\eta).$$

Similarly, if

$$\theta(\{l^1,f\}) = -i[\mathbb{L}^1,\theta(f)] ,$$

we must have

$$c(\gamma^N\xi,\gamma^N\eta) = c(\xi,\eta)$$

for all such γ, and conversely.

The condition (17) with $j = 2$ and 3 corresponds to invariance of c under rotations that leave the 2-axis and 3-axis invariant, and the theorem follows.

We note that, in particular, Weyl's quantization and the quantization of Theorem 4 are fully compatible with rotations.

7. Concluding remarks

The preceding results all refer to a particular representation of the commutation relations, namely the one built over the spectrum of the position operators such that these are realized as multiplication operators. As remarked by Grossmann, Loupias and Stein [7] it may well be of interest to investigate the form of quantizations in other familiar representations of the commutation relations, both reducible and irreducible representations. (Note that some of the correspondences considered by these authors do not satisfy the condition (3) of Definition 1, and hence are not quantizations in the sense of the present paper).

A particularly simple and important example concerns the form of the quantization by Wick-ordering in the Fock-Bargman-Segal representation.

The underlying Hilbert space $\mathcal{F}^d$ consists of those entire functions h on $\mathbb{C}^d$ for which

$$\int |h(z)|^2 \, d\mu < \infty,$$

where the measure μ on $\mathbb{C}^d$ is the 2d-dimensional Lebesgue measure multiplied by the density function

$$\rho(z) = \pi^{-d} e^{-|z|^2} .$$

This space was introduced by Fock [5] , but with the scalar product given by a series expansion in the Taylor coefficients. The integral representation was introduced by Segal [24] and Bargman [2] , who studied in detail the relationship of this representation with the Schrödinger representation. Also the spaces $\tilde{\varphi}^d$ and $\tilde{\varphi}^{d*}$ corresponding to φ^d and φ^{d*} have been studied by Bargman [3]. They too are spaces of entire functions on $\mathbb{C}^d$ and may be characterized as follows:

$$\tilde{\varphi}^d = \bigcap_{n=1}^{\infty} \left\{ h \;\middle|\; |z|^n \, |h(z)| \, e^{-\frac{1}{2}|z|^2} \longrightarrow 0 \text{ for } |z| \longrightarrow \infty \right\},$$

$$\tilde{\varphi}^{d*} = \bigcup_{n=1}^{\infty} \left\{ h \;\middle|\; |z|^{-n} \, |h(z)| \, e^{-\frac{1}{2}|z|^2} \longrightarrow 0 \text{ for } |z| \longrightarrow \infty \right\}.$$

In this framework the commutation relations are realized by the operators $\tilde{\mathbb{Q}}_k$, $\tilde{\mathbb{P}}_k$, $k = 1,\dots,d$, where

$$\tilde{\mathbb{Q}}_k h(z) = \frac{1}{2}\sqrt{2}\left(z_k + \frac{\partial}{\partial z_k}\right) h,$$

$$\tilde{\mathbb{P}}_k h(z) = \frac{i}{2}\sqrt{2}\left(z_k - \frac{\partial}{\partial z_k}\right) h.$$

A corresponding form of the nuclear theorem is the following: The topological vector spaces $\tilde{\varphi}^{2d*}$ and $L(\tilde{\varphi}^d, \tilde{\varphi}^{d*})$ are isomorphic; for every kernel $k(z,w) \in \tilde{\varphi}^{2d*}$ the corresponding linear operator $K \in L(\tilde{\varphi}^d, \tilde{\varphi}^{d*})$ is defined by

$$Kh(z) = \int k(z,\bar{w}) \, h(w) \, d\mu(w).$$

For the present discussion it is more convenient to replace each kernel $k(z,w)$ by a function $\mathcal{K}$, which is essentially the function called the Wick-symbol of the operator K by Berezin [4]. Thus, we put

$$\mathcal{K}(z,w) = e^{-z \cdot w} k(z,w),$$

so that

$$Kh(z) = \int \mathcal{K}(z,\bar{w}) e^{z \cdot \bar{w}} h(w) \, d\mu(w).$$

It is now immediate to derive from identities proved by Bargman and Berezin that the correspondence

$$\mathcal{K} \longleftrightarrow K$$

has the following properties

$$1 \longleftrightarrow I,$$

$$z_j \mathcal{K} \longleftrightarrow 2^{-\frac{1}{2}}(\tilde{\mathbb{Q}}_j - i\, \tilde{\mathbb{P}}_j)\, K,$$

$$w_j \mathcal{K} \longleftrightarrow K\, 2^{-\frac{1}{2}}(\tilde{\mathbb{Q}}_j + i\, \tilde{\mathbb{P}}_j),$$

$$\frac{\partial \mathcal{K}}{\partial z_j} \longleftrightarrow [2^{-\frac{1}{2}}(\mathbb{Q}_j + i\, \tilde{\mathbb{P}}_j), K],$$

$$\frac{\partial \mathcal{K}}{\partial w_j} \longleftrightarrow -[2^{-\frac{1}{2}}(\tilde{\mathbb{Q}}_j - i\, \tilde{\mathbb{P}}_j), K].$$

Now let f denote a function on the phase space $\mathbb{R}^{2d}$, and consider the function $\mathcal{K}$ defined on a real 2d-dimensional subspace D of $\mathbb{C}^{2d}$ by the definition

$$\mathcal{K}\,(2^{-\frac{1}{2}}(q - ip),\, 2^{-\frac{1}{2}}(q + ip)) = f(q,p).$$

If f is real-analytic, then $\mathcal{K}$ has a unique analytic extension to a neighborhood of D. It is now immediate to check that on the set of functions f, for which the corresponding $\mathcal{K}$ is the Wick-symbol of an operator K, the mapping $f \longmapsto K$ is analogous to the quantization by Wick-ordering described in Theorem 4.

A most interesting open problem is that of characterizing those classical Hamilton-functions whose quantization is essentially self-adjoint in the Hilbert space L^2. One might conjecture that this essential selfadjointness, which corresponds to unitarity of the time evolution of the quantum system, is linked with unitarity of the flow associated with the time evolution of the classical system. Recent results by Rauch and Reed [20] show that this attractive conjecture is false in its broadest generality. However, for the Weyl quantization sufficiently regular Hamilton functions of the important type considered by these authors, the statement of the conjecture is correct.

References

1. Aslaksen, Erik W; Klauder, John R, Continuous representation theory using the affine group, J. Mathematical Phys. 10(1969), 2267-2275.
2. Bargman, V., On a Hilbert space of analytic functions and an associated integral transform, Comm. Pure Appl. Math. 14 (1961), 187-214.
3. Bargman, V., On a Hilbert space of analytic functions and an associated integral transform. Part II. A family of related function spaces. Application to distribution theory, Comm. Pure Appl. Math. 20 (1967), 1-101.
4. Berezin, F.A., Wick and Anti-Wick operator symbols, Mat. Sb. (N.S.) 86 (128) (1971), 578-610.
5. Fock, V., Verallgemeinerung und Losung der Diracschen Statistichen Gleichung, Z. Physik 49 (1928), 339-357.
6. Groenewold, H.J., On the principles of elementary quantum mechanics, Physica 12 (1946), 405-460.
7. Grossmann, A., Loupias, G. and Stein, E.M., An algebra of pseudo-differential operators and quantum mechanics in phase space, Ann. Inst. Fourier (Grenoble) 18 (1968), fasc. 2, 343-368, (1969).
8. Hermann, Robert., Lie groups for physicists, W.A.Benjamin, Inc., New York - Amsterdam, 1966.
9. Joseph, A., Derivations of Lie brackets and canonical quantization, Comm. Math. Phys. 17 (1970), 210-232.
10. Kastler, D., The C*-algebras of a free boson field. I. Discussion of the basic facts, Comm. Math. Phys. 1 (1965), 14-48.
11. Klauder, John R., Continuous representation theory. III. On functional quantization of classical systems, J. Mathematical Phys. 5 (1964), 177-187.

12. Kohn, J.J., and Nirenberg, L., An algebra of pseudo-differential operators, Comm. Pure Appl. Math. 18 (1965), 269-305.

13. Leaf, Boris, Weyl transform in nonrelativistic quantum dynamics, J. Mathematical Phys. 9 (1968), 769-781.

14. Loupias, G., Miracle-Solé, S., C*-algèbres des systèmes canoniques. I, Comm. Math. Phys. 2 (1966), 31-48.

15. Loupias, G. and Miracle-Solé, S., C*-algèbres des systèmes canoniques. II, Ann. Inst. H. Poincaré Sect. A (N.S.) 6 (1967), 39-58.

16. Mehta, C.L., Phase-space formulation of the dynamics of canonical variables, J. Mathematical Phys. 5 (1964), 677-686.

17. Moyal, J.E., Quantum mechanics as a statistical theory, Proc. Cambridge Philos. Soc. 45 (1949), 99-124.

18. Pool, James C.T., Mathematical aspects of the Weyl correspondence, J. Mathematical Phys. 7 (1966), 66-76.

19. Prosser, R.T., Segal's quantization procedure, J. Mathematical Phys. 5 (1964), 701-707.

20. Rauch, Jeffrey, and Reed, Michael, Two examples illustrating the differences between classical and quantum mechanics, mimeographed, to appear.

21. Rivier, D.C., On a one-to-one correspondence between infinitesimal canonical transformations and infinitesimal unitary transformations, Phys. Rev. (2) 83 (1951), 862-863.

22. Schwartz. L., Séminaire Schwartz de la faculté des sciences de Paris, 1953/54. Produits tensoriels topologiques d'espaces vectoriels topologiques. Espaces vectoriels topologiques nucléaires. Applications, Secrétariat mathématique, Paris, 1954.

23. Segal, I.E., Quantization of nonlinear systems, J. Mathematical Phys. 1 (1960), 468-488.

24. Segal, I.E., Mathematical characterization of the physical vacuum for a linear Bose-Einstein field, Illinois J. Math. 6 (1962), 500-523.

25. Segal, I.E., Transforms for operators and symplectic automorphisms over a locally compact abelian group, Math. Scand. 13 (1963), 31-43.

26. Shale, David, Stinespring, W.Forrest, The quantum harmonic oscillator with hyperbolic phase space, J. Functional analysis 1 (1967), 492-502.

27. Shewell, John Robert., On the formation of quantum mechanical operators, Amer. J. Phys. 27, (1959), 16-21.

28. Souriau, J.-M., Quantification géométrique, Comm. Math. Phys. 1 (1966), 374-398.

29. Souriau, J.-M., Quantification géométrique. Applications, Ann. Inst. H.Poincaré Sect. A (N.S.) 6 (1967), 311-341.

30. Streater, R.F., Canonical quantization, Comm. Math. Phys. 2 (1966), 354-374.

31. Treves, François, Topological vector spaces, distributions and kernels, Academic Press, New York-London, 1967.

32. Van Hove, Léon, Sur certaines représentations unitaires d'un groupe infini de transformations, Acad. Roy. Belg. Cl. Sci. Mém. Coll. in - 8° . 26 (1951), fasc. 6, 102 pp.

33. Weyl, Hermann, Quantenmechanik und Gruppentheorie, Z. Phys. 46 (1927), 1-46.

THE LEBESGUE CONSTANTS FOR POLYNOMIAL INTERPOLATION

T.J.Rivlin

Introduction

Let X denote an infinite triangular matrix whose k^{th} row is

$$x(k) = (x_{1,k},\dots,x_{k,k})$$

where

$$-1 \leq x_{k,k} < \dots < x_{1,k} \leq 1.$$

Let $\mathcal{P}_n$ be the set of real polynomials of degree at most n. For each real-valued function f, defined on X and for each $k = 1,2,\dots$ let $L_k(f,X)$ denote the unique element of $\mathcal{P}_{k-1}$ such that

$$L_k(f,X;\, x_{i,k}) = f(x_{i,k}), \quad i = 1,\dots,k.$$

That is, L_k is the polynomial of degree at most k-1 which interpolates f at $x(k)$. If we put

$$\omega_k(x) = \omega_k(X;x) = (x-x_{1,k})\dots(x-x_{k,k})$$

and

$$\ell_{i,k}(x) = \ell_{i,k}(X;x) = \frac{\omega_k(x)}{x-x_{i,k}} \, \frac{1}{\omega_k'(x_{i,k})} \tag{1}$$

$i = 1,\dots,k$, then

$$L_k(f,X;x) = \sum_{i=1}^{k} f(x_{i,k})\, \ell_{i,k}(x). \tag{2}$$

Suppose

$$|f|_p = \left(\sum_{i=1}^{k} |f(x_{i,k})|^p \right)^{\frac{1}{p}}, \quad p \geq 1,$$

with the understanding that

$$|f|_{\infty} = \max_{i=1,\dots,k} |f(x_{i,k})| .$$

The mapping $f \xrightarrow{L_k} L_k(f,X)$ is a linear operator, and if $\mathcal{P}_{k-1}$ is normed by $\|\cdot\|$, the norm of the operator is

$$\| L_k \| = \max_{|f|_p \leq 1} \| L_k(f,X) \| = \Lambda_k(X) \qquad (3)$$

$\Lambda_k(X)$ is the Lebesgue constant of order k with respect to p and $\|\cdot\|$. Our purpose is to present a selective survey of what is known about the size of the Lebesgue constants as functions of the parameters k, X, p and $\|\cdot\|$.

Since, as a general rule, in applications small Lebesgue constants are desirable it is of interest to observe that there exists X^* such that

$$\Lambda_k(X^*) \leq \Lambda_k(X)$$

for all X. For, if

$$\inf_X \Lambda_k(X) = b \geq 0$$

there exists a convergent sequence of vectors $x^{(m)}(k) \to x^*(k)$. If the components of $x^*(k)$ are distinct, then the required X^* exists. But, by choosing m sufficiently large and $f_j(x^{(m)}_{i,k}) = \delta_{ij}, j=1,\dots,k$ we conclude that

$$\| \ell_j(x) \| \leq b + 1, \quad j = 1,\dots,k \; ^*$$

and hence by the comparability of norms on a finite dimensional space

$$|\ell_j(x)| \leq c, \quad j = 1,\dots,k.$$

But

$$c \geq \max_{-1 \leq x \leq 1} |\ell_j(x)| = \max_{-1 \leq x \leq 1} \prod_{\substack{i=1 \\ i \neq j}}^{k} \left| \frac{x - x^{(m)}_{i,k}}{x^{(m)}_{j,k} - x^{(m)}_{i,k}} \right| \geq \frac{2^{2-k}}{\left(x^{(m)}_{j,k} - x^{(m)}_{j+1,k}\right) 2^{k-2}}$$

*When quantities obviously depend on k or X we often omit the explicit notation of dependency.

so that

$$x^{(m)}_{j,k} - x^{(m)}_{j+1,k} \geqslant \frac{1}{4^{k-2}c},$$

for $j = 1,2,\ldots,k-1$ and all m. A minimal Lebesgue constant exists.

We shall restrict our attention to the integral q-norms on $\mathcal{P}_{k-1}$, $q \geqslant 1$, that is, for $g \in \mathcal{P}_{k-1}$

$$\|g\|_q = \left[\int_{-1}^{1} |g(x)|^q \, dx\right]^{1/q},$$

with

$$\|g\|_\infty = \|g\| = \max_{-1 \leqslant x \leqslant 1} |g(x)| .$$

The case $q = \infty$ is of particular interest since it covers most of the problems which have been treated in the literature. Indeed, when $q = \infty$ and $|f|_p \leqslant 1$.

$$\begin{aligned}
\left\| \sum_{i=1}^{k} f(x_i)\, \ell_i(x) \right\| &\leq \max_{-1 \leq x \leq 1} \sum_{i=1}^{k} |f(x_i)|\,|\ell_i(x)| \\
&\leq \left[\sum_{i=1}^{k} |f(x_i)|^p\right]^{1/p} \max_{-1 \leq x \leq 1} \left[\sum_{i=1}^{k} |\ell_i(x)|^{p'}\right]^{1/p'} \\
&\leq \max_{-1 \leq x \leq 1} \left[\sum_{i=1}^{k} |\ell_i(x)|^{p'}\right]^{1/p'} \qquad (4)
\end{aligned}$$

where

$$\frac{1}{p'} + \frac{1}{p} = 1,$$

in view of Hölder's inequality. By appropriate choice of f the inequalities in (4) can be replaced by equalities and we conclude that

$$\Lambda_k(X) = \max_{-1 \leqslant x \leqslant 1} \left[\sum_{i=1}^{k} |\ell_i(x)|^{p'}\right]^{1/p'} \qquad (5)$$

If $q < \infty$, a similar application of Hölder's inequality yields

$$\Lambda_k(X) \leq \left[\int_{-1}^{1} \left(\sum_{i=1}^{k} |\ell_i(x)|^{p'} \right)^{q/p'} dx \right]^{1/q} \tag{6}$$

If $p = 1$ we have, by Minkowski's inequality

$$\| \sum_{i=1}^{k} f(x_i)\,\ell_i(x) \|_q \leq \sum_{i=1}^{k} |f(x_i)|\, \| \ell_i \|_q \leq \sum_{i=1}^{k} |f(x_i)| \max_i \| \ell_i \|_q ,$$

with equality if $f(x_i) = \delta_{ij}$ where j is any index such that

$$\max_i \| \ell_i \|_q = \| \ell_j \|_q .$$

Thus for $p = 1$

$$\Lambda_k(X) = \max_i \| \ell_i \|_q .$$

It is worth mentioning that, for example, (5) has an application in the theory of errors. Namely, if the value $f(x_i)$ has a possible error whose magnitude does not exceed $\varepsilon_i, i=1,\ldots,k$. Then the magnitude of the difference between $L_k(x)$ and the value at x of the interpolating polynomial to the true values does not exceed

$$\left[\sum_{i=1}^{k} \varepsilon_i^{\,p} \right]^{1/p} \Lambda_k(X)$$

for $-1 \leq x \leq 1$.

The first section of this paper is devoted to the most widely studied case, $p = q = \infty$, while the concluding section presents a scattering of results for other p and q. Instead of encumbering the notation for the Lebesgue constants with p and q we specify p and q at the beginning of each discussion and omit further mention of these parameters when no confusion can arise.

1. $p = q = \infty$

This is the most widely studied case. The mapping $f \longrightarrow L_k(f,X)$ is now a projection of $C[-1,1]$ (with the uniform norm) onto

$\mathcal{P}_{k-1}$, and its norm is

$$\Lambda_k(X) = \max_{-1 \leq x \leq 1} \sum_{i=1}^{k} |\ell_i(x)| . \tag{7}$$

Since

$$1 = \sum_{i=1}^{k} \ell_i(x) \leq \sum_{i=1}^{k} |\ell_i(x)| ,$$

obviously $\Lambda_k \geq 1$ and for $k > 1$, $\Lambda_k > 1$.

If $f \in C[-1,1]$ and $E_n(f) = \| f-p_n \|$ where p_n is the polynomial of degree n of best approximation to f on $[-1,1]$, then if I denotes the identity operator

$$\| f-L_k(f) \| = \| (f-p_{k-1}) - L_k(f-p_{k-1}) \| = \| (I-L_k)(f-p_{k-1}) \|$$

$$\leq E_{k-1}(f) \| I-L_k \| .$$

Furthermore,

$$\| I-L_k \| = 1 + \| L_k \| = 1 + \Lambda_k(X). \tag{8}$$

((8) is a special case of a theorem of Daugavet. See Cheney and Price [3] . But the proof of (8) alone is so simple that we give it here.) To see this, fix X and suppose $k > 1$. (The case $k = 1$ is obvious.)

Let

$$\Lambda_k = \sum_{i=1}^{k} |\ell_i(\bar{x})| , \qquad -1 \leq \bar{x} \leq 1.$$

Observe that $\bar{x} \neq x_{i,k}$, $i = 1,\dots,k$. Choose $g \in C[-1,1]$ such that 1) $g(x_i) = \operatorname{sgn} \ell_i(\bar{x})$; 2) $g(\bar{x}) = -1$ and 3) $\| g \| = 1$. Then

$$1 + \| L_k \| \geq \| I-L_k \| \geq \| g-L_k(g) \| \geq | g(\bar{x}) - \sum_{i=1}^{k} |\ell_i(\bar{x})| =$$

$$= 1 + \| L_k \| .$$

Thus,

$$\| f - L_k(f) \| \leq (1 + \Lambda_k(X))\, E_{k-1}(f) \qquad (9)$$

and we see that the size of the Lebesgue constant provides a bound on how good a uniform approximation is provided by interpolation in x(k). (9) reveals that with estimates of $E_{k-1}(f)$, as provided say by Jackson's theorem, the size of the Lebesgue constants has implications for the convergence of $L_k(f)$ to f on $[-1,1]$. We shall not discuss any further the important topic of the convergence properties of $L_k(f)$ but refer the reader to Askey [1].

We turn next to estimates of the size of the Lebesgue constants (7). Erdös [5] has shown that for some positive constant c, and every X and k,

$$\Lambda_k(X) > \frac{2}{\pi} \log k - c. \qquad (10)$$

$T_k(x) = \cos k\theta$, $x = \cos\theta$ is the Chebyshev polynomial of degree k. Let

$$x_{i,k} = \cos\frac{(2i-1)\pi}{2k}, \quad i = 1,\dots,k; \quad k=1,2,\dots$$

the zeros of $T_k(x)$. Call the resulting infinite matrix of nodes, T. If

$$x_{i,k} = \cos\frac{(i-1)\pi}{k-1}, \quad i = 1,\dots,k; \quad k = 2,3,\dots,$$

the extrema of $T_{k-1}(x)$, the resulting matrix of nodes ($x_{1,1} = 1$, say) we call U. We wish to show that

$$\Lambda_k(T) \leq \frac{2}{\pi} \log k + 1, \quad k = 1,2,\dots .$$

Ehlich and Zeller [4] showed that for $-1 \leq x \leq 1$

$$\sum_{i=1}^{k} |\ell_{i,k}(T;x)| \leq \sum_{i=1}^{k} |\ell_{i,k}(1)| = \Lambda_k(T) \qquad (11)$$

with equality holding only for $x = \pm 1$. An easy computation reveals then that

$$\Lambda_k(T) = \frac{1}{k} \sum_{i=1}^{k} \cot \frac{(2i-1)\pi}{4k}. \tag{12}$$

We now prove

<u>Theorem 1.</u> <u>The sequence</u>

$$t_k = \Lambda_k(T) - \frac{2}{\pi} \log k$$

<u>is strictly monotone decreasing with</u> $t_1 = 1$ <u>and</u>

$$\lim_{k \to \infty} t_k = \frac{2}{\pi}(\log \frac{8}{\pi} + \gamma) = .9625\ldots,$$

where γ is Euler's constant.

Proof.

$$\frac{\pi}{2} t_k = \frac{\pi}{2k} \sum_{i=1}^{k} \left[\cot \frac{(2i-1)\pi}{4k} - \frac{4k}{(2i-1)\pi}\right] + \left[2 \sum_{i=1}^{k} \frac{1}{2i-1} - \log k\right]$$

$$= a_k + u_k \tag{13}$$

The a_k are a sequence of Riemann sums of the integral

$$\int_0^{\pi/2} (\cot x - \frac{1}{x})dx = \log \frac{2}{\pi},$$

hence

$$\lim_{k \to \infty} a_k = \log \frac{2}{\pi}$$

while

$$\lim_{k \to \infty} u_k = \gamma + \log 4.$$

It remains, therefore, only to show that $(\pi/2)t_k$ is strictly monotone decreasing. The following Lemma due to D.J.Newman and the speaker is the key to the proof.

<u>Lemma 1.</u> <u>If</u> f'' <u>and</u> f''' <u>are both non-negative in</u> $[0,1]$ <u>then</u>

the Riemann sums

$$s_k = \frac{1}{k} \sum_{i=1}^{k} f\left(\frac{2i-1}{2k}\right)$$

are monotone increasing as k *increases*. (*The same conclusion holds if* $f'' \geqslant 0$ *and* $f''' \leqslant 0$.)

Proof. Integrating by parts three times yields

$$s_k = \int_0^1 f(t)dt - \frac{f''(0)}{24k^2} - \int_0^1 \left\{ \frac{4(kt-[kt+\frac{1}{2}])^3 + [kt+\frac{1}{2}]}{24k^3} \times f'''(1-t)\right\}dt, \tag{14}$$

($[\ ]$ here is the integer part notation.) Since $f''(0) \geqslant 0$, the sequence $-f''(0)/(24k^2)$ is monotone increasing, hence since $t^3 f'''(1-t) \geqslant 0$, it suffices to show that the function

$$\frac{4(kt - [kt+\frac{1}{2}])^3 + [kt+\frac{1}{2}]}{24(kt)^3}$$

decreases as k increases. Thus it is enough to show that

$$\frac{4(x - [x+\frac{1}{2}])^3 + [x+\frac{1}{2}]}{x^3}$$

is a decreasing function for $x > 0$. This function is continuously differentiable for $x > 0$, hence it suffices to verify that its derivative is negative for $n - (1/2) < x < n+(1/2)$. But in this interval the function is

$$\frac{4(x-n)^3 + n}{x^3}$$

whose derivative is

$$\frac{12n}{x^4}\left((x-n)^2 - \frac{1}{4}\right),$$

which is, indeed, negative throughout the interval. The final parenthetical remark can be established by putting $1 - t = u$ in (14). This concludes the proof of the lemma.

Returning now to the proof of the theorem, we apply Lemma 1 with

$$f(x) = \frac{1}{\frac{\pi}{2}x} - \cot\frac{\pi}{2}x. \qquad (15)$$

Since

$$\frac{1}{z} - \cot z = \sum_{j=1}^{\infty} c_{2j-1} z^{2j-1}, \quad |z| < \pi$$

which $c_{2j-1} > 0$, it is evident that f'' and f''' are non-negative in $[0,1]$, and therefore $s_k = -(2/\pi)a_k$ is monotone increasing. Hence a_k is monotone decreasing. But

$$u_k - u_{k+1} = \log\left(1 + \frac{1}{k}\right) - \frac{2}{2k+1},$$

is positive when $k = 1$ and tends to 0 as k tends to infinity.

Since

$$h(x) = \log\left(1 + \frac{1}{x}\right) - \frac{2}{2x+1}$$

has as its derivative

$$h'(x) = \frac{-1}{x(x+1)(2x+1)^2} < 0, \quad x > 0,$$

$$u_{k+1} < u_k \qquad k = 1,2,\dots,$$

and $(\pi/2)t_k$ is strictly monotone decreasing.

Ehlich and Zeller [4] also show that

$$\Lambda_k(U) = \begin{cases} \Lambda_{k-1}(T), & k = 2,4,6,\dots \\ \Lambda_{k-1}(T) - \varepsilon_{k-1}, & k = 3,5,7,\dots \end{cases}$$

where

$$0 < \varepsilon_k < \frac{1}{k^2}.$$

Consequently, if

$$\tau_k = \Lambda_k(U) - \frac{2}{\pi}\log k,$$

$$\frac{\pi}{2}\tau_k = a_{k-1} + 2\sum_{i=1}^{k-1}\frac{1}{2i-1} - \log k - \delta_{k-1}$$

$$= a_{k-1} + v_{k-1} - \delta_{k-1},$$

where a_k is defined in (13) and

$$\delta_{k-1} = \begin{cases} 0, & k \text{ even} \\ \frac{\pi}{2}\varepsilon_{k-1}, & k \text{ odd}, \quad k \geq 3. \end{cases}$$

Suppose $f(x)$ is defined by (15). Then $f''(0) = 0$ and (14) implies that

$$0 < s_k - s_{k-1} = \frac{1}{24}\int_0^1 \{G(k-1,t) - G(k,t)\} f'''(1-t)dt,$$

where

$$G(k,t) = \frac{4(kt - [kt+\frac{1}{2}])^3 + [kt+\frac{1}{2}]}{k^3}.$$

Now,

$$-\frac{1}{2} \leq kt - [kt+\frac{1}{2}] < \frac{1}{2}$$

and $[kt+\frac{1}{2}] \leq k$, so

$$G(k-1,t) \leq \frac{\frac{1}{2} + [(k-1)t+\frac{1}{2}]}{(k-1)^3}, \; G(k,t) \geq \frac{-\frac{1}{2} + [(k-1)t+\frac{1}{2}]}{k^3}$$

and

$$s_k - s_{k-1} < \frac{1}{24k^3(k-1)^3}\int_0^1 \{k^3 + (k^3-(k-1)^3)k\} f'''(1-t)dt$$

$$< \frac{1}{6(k-1)^3} f''(1) = \frac{2}{3\pi}\frac{1}{(k-1)^3}.$$

Thus

$$a_k - a_{k-1} > -\frac{1}{3(k-1)^3},$$

while

$$v_k - v_{k-1} = \frac{2}{2k-1} - \log(1 + \frac{1}{k}) > \frac{2}{2k-1} - \frac{1}{k} = \frac{1}{k(2k-1)} .$$

Hence

$$(a_k + v_k) - (a_{k-1} + v_{k-1}) > \frac{1}{k(2k-1)} - \frac{1}{3(k-1)^3} > 0,$$

for $k \geq 3$. Thus we have proved:

<u>Theorem 2.</u> <u>The sequence</u>

$$\tau_k = \Lambda_k(U) - \frac{2}{\pi} \log k, \quad k = 2,4,6,\ldots,$$

<u>is strictly monotone increasing.</u> <u>Also,</u>

$$\lim_{k \to \infty} \tau_k = \frac{2}{\pi} (\log \frac{8}{\pi} + \gamma)$$

<u>and</u>

$$\tau_k < \frac{2}{\pi} (\log \frac{8}{\pi} + \gamma), \quad k = 2,3,4,5,\ldots .$$

<u>Remark.</u> We conjecture that τ_k, $k \geq 2$ is strictly monotone increasing ($\tau_2 = 1 - (2/\pi) \log 2 = .559\ldots$), and believe this result can be obtained by closer scrutiny of the quantities ϵ_k (Cf. Ehlich and Zeller [4]).

Szegö [12, p.336] shows that if X is made up of the zeros of the Jacobi polynomials with parameters (α,β), $\alpha > -1$, $\beta > -1$, and $\gamma = \max(\alpha,\beta)$ then

$$\Lambda_k = \begin{cases} O(k^{\gamma + \frac{1}{2}}), & \gamma > -\frac{1}{2}, \\ O(\log k), & \gamma \leq -\frac{1}{2}. \end{cases} \tag{16}$$

Fejér [9] introduced the notion of "normal" matrices. It is the following. Let $H_k(x)$ denote the unique polynomial of degree at most $2k-1$ satisfying

$$H_k(x_{i,k}) = f_i \; ; \; H_k'(x_{i,k}) = g_i, \quad i = 1,\ldots,k$$

for given f and g. Then

$$H_k(x) = \sum_{j=1}^{k} (h_j(x) f_j + k_j(x) g_j)$$

where

$$h_j(x) = v_{j,k}(x) \; \ell_{j,k}(x), \quad j = 1,\ldots,k$$

and

$$v_{j,k}(x) = 1 - \frac{\omega_k''(x_j)}{\omega_k'(x_j)} (x - x_j), \quad j = 1,\ldots,k.$$

If

$$v_{j,k}(x) \geqslant \rho \geqslant 0; \quad -1 \leqslant x \leqslant 1, \quad j = 1,\ldots,k,$$

for $k = 1,2,\ldots,$ we say that X is ρ-normal (0-normal being simply "normal"). By choosing $f_i = 1$, $g_i = 0$, $i = 1,\ldots,k$ we obtain

$$\sum_{j=1}^{k} v_j(x) \; \ell_j^2(x) \equiv 1, \tag{17}$$

from which we conclude that if X is ρ-normal, $\rho > 0$

$$\sum_{j=1}^{k} \ell_j^2 (x) \leqslant \frac{1}{\rho}, \tag{18}$$

and hence

$$\Lambda_k(X) \leqslant \frac{k^{1/2}}{\rho^{1/2}} .$$

There is a unique 1-normal matrix (ρ can never exceed 1). We call it F after Fejér and Fekete. Its kth row consists of the zeros of $(1-x^2)\, P_{k-1}'(x)$, which are also the zeros of $P_k - P_{k-2}$ and

$$\int_{-1}^{x} P_{k-1}(t)\,dt,$$

where P_k is the Legendre polynomial of degree k. Thus

$$\Lambda_k(F) \leqslant k^{1/2}. \tag{19}$$

In the case of equally spaced nodes the Lebesgue constants grow geometrically. Let E denote the matrix defined by

$$x_{i,k} = 1 - \frac{2(i-1)}{k-1} , \quad i = 1,\ldots,k$$

then according to Golomb [10] , as k runs through odd integers

$$e^{1/2} \leq \overline{\lim_{k \to \infty}} (\Lambda_k(E))^{1/k} \leq 2.$$

Next let us consider the problem of a minimal Lebesgue constant. Neither X^* nor $\Lambda_k(X^*)$, $k \geq 4$, are known. For $k \geq 3$

$$\lambda_k(x) = \sum_{i=1}^{k} |\ell_i(x)|$$

is a spline function whose values are never less than 1, taking the value 1 only at the x_i, having a single maximum ("hump") between consecutive nodes, monotone decreasing and convex in $(-1,x_k)$ and monotone increasing and convex in $(x_1,1)$.

There is a conjecture, going back at least to S. Bernstein [2] , that for X^* $\lambda_k(x)$ has k+1 <u>equal</u> humps, and conversely. It is known (Cf. Luttmann and Rivlin [11]) that X^* is not unique, and it is always possible to find an X^* with $x_{1,k} = 1$, $x_{k,k} = -1$. Perhaps X^* is unique if end points must be included. The Bernstein conjecture would then call for k-1 equal humps.

Some numerical computations (Cf. Luttman and Rivlin [11]) suggest that (19) can be improved to

$$\Lambda_k(F) < \Lambda_k(T).$$

Note that O(log k) result (16) does not cover F, since $(1-x^2)P'_{k-1}$ is not a Jacobi polynomial by Szegö's definition.

Finally, it is an open question as to what the best constant c is in (10). Does (10) continue to hold if c = 0 ?

2. Other choices of p,q.

2.1. $q = \infty$, $1 \leq p < \infty$.

Suppose X is ρ- normal with $\rho > 0$, and $p' = 2$. Then we have, in view of the identity (17)

$$1 = \left(\sum_{j=1}^{k} v_j \ell_j^2 \right) \geq \rho \sum_{j=1}^{k} \ell_j^2$$

and so, according to (5),

$$\Lambda_k(X) \leq \rho^{-1/2} . \tag{20}$$

Jensen's inequality (see Hardy, Littlewood, Polya, " Inequalities") shows that (20) holds for $p' \geq 2$. Since, clearly

$$\Lambda_k(X) \geq 1,$$

we conclude that $X^* = F$ is the (unique) minimal set of nodes for $q = \infty$, $1 \leq p \leq 2$.

When $p = 1$

$$\Lambda_k(X) = \max_{-1 \leq x \leq 1} \max_i |\ell_i(x)| .$$

When the nodes are the zeros of certain ultraspherical polynomials, i.e., orthogonal polynomials with respect to the weight function $(1-x^2)^\alpha$ $\alpha > -1$ some results are known ($p = 1$). In the case of T ($\alpha = -\frac{1}{2}$), Erdös and Grünwald [8] showed that

$$\sup_k \Lambda_k(T) = \frac{4}{\pi},$$

while for $\alpha = (1/2)$ Webster [13] showed that

$$\sup_k \Lambda_k(X(\tfrac{1}{2})) = 2.$$

Also, if $-(1/2) \leq \alpha \leq (1/2)$ Webster [14] showed

$$\lim_{k \to \infty} \ell_{1,k}(1) = (\tfrac{1}{2} j_1)^{\alpha-1} |\Gamma(1+\alpha) J_{1+\alpha}(j_1)|^{-1},$$

where j_1 is the first positive zero of the Bessel Function $J_\alpha(x)$. For the same α, Erdös [7] showed that

$$\Lambda_k(X(\alpha)) = \ell_{1,k}(1) = \ell_{k,k}(-1),$$

and thought that it could be shown that the Λ_k were monotone increasing.

2.2. $q < \infty$, $1 \leq p \leq \infty$.

Results in these cases are quite skimpy, and the reader should see Erdös [6] for $q = p' = 1,2$. The case of $p = 1$ has some interest. Of course,

$$\Lambda_k(F) \leq 2^{1/q},$$

but it would be interesting to have better estimates.

References:

1 Askey, R., Mean convergence of orthogonal series and Lagrange interpolation, Acta Mathematica (Budapest). To appear.

2 Bernstein, S., Sur la limitation des valeurs d'un polynome $P_n(x)$ de degré n sur tout un segment par ses valeurs en $(n+1)$ points du segment, Izv. Akad. Nauk S.S.R. Classe des Sciences Math. et Naturelles, 1931, 1025-1050.

3 Cheney, E.W., Price, K., Minimal projections, Approximation Theory, Proceedings of a Symposium held at Lancaster, July 1969, ed. A. Talbot, Academic Press, London, 1970, 261-289.

4 Ehlich, H., and Zeller, K., Answertung der Normen von Interpolations-operatoren, Math. Ann., 164, 1966, 105-112.

5 Erdös, P., Problems and results on the theory of interpolation, II, Acta Math. Acad. Sci. Hung., 12, 1961, 235-244.

6 Erdös, P., Problems and results on the convergence and divergence properties of the Lagrange interpolation polynomials and some extremal problems, Mathematica, 10, 1968, 65-73

7 Erdös, P., On the maximum of the fundamental functions of the ultra-spherical polynomials, Annals of Math., 45, 1944, 335-339

8 Erdös, P., Grünwald, G., Note on an elementary problem of interpolation, Bull. A.M.S., 44, 1938, 515-518.

9 Fejér, L., Lagrangesche Interpolation und die zugehorigen konjugierten Punkte, Math. Ann., 106, 1932, 1-55

10 Golomb, M., Lectures on Theory of Approximation, Argonne Nat. Lab., Argonne, Ill., U.S.A., 1962.

11 Luttmann, F., and Rivlin, T., Some numerical experiments in the theory of polynomial interpolation, IBM J. of Research and Development, 9, 1965, 187-191.

12 Szegö, G., Orthogonal Polynomials, Amer. Math. Soc., N.Y., 1959

13 Webster, M., Note on certain Lagrange interpolation polynomials Bull. Am. M. S. 45, 1939, 870-873.

14 Websterm M., Maximum of certain fundamental Lagrange interpolation polynomials, Bull. A.M.S., 47, 1941, 71-73.

APPROXIMATION THEOREMS FOR POLYNOMIAL SPLINE OPERATORS

Karl Scherer

1. Introduction

The purpose of this paper is to present some approximation theorems for polynomial spline operators, the emphasis lying on the connections between them. A general theorem is established in Section 2 for (linear or non-linear) spline operators T_Δ with range in the class $Sp(n,m,\Delta)$ of polynomial splines of degree n and deficiency $n+1-m$ with respect to the partition Δ. It expresses L_p-error bounds for the difference $f^{(\ell)} - T_\Delta^{(\ell)}(f)$ and for $T_\Delta^{k}(f)$ in terms of the difference $f - T_\Delta(f)$ and a modulus of continuity $\omega_r(t;f)_p$ of f of arbitrary order r. As a consequence, a unified approach to many concrete results on simultaneous approximation and error bounds for derivatives of approximating splines is possible.

This is carried out in Section 3 for several types of polynomial spline operators, for example, for interpolating splines of odd degree and for smooth interpolating splines of even degree. Thereby error estimates in case of the existence of first or second integral relations as well as stability results of Swartz-Varga [25] are improved and extended. In performing this, we carry on recent investigations in Scherer [20] and Demko-Varga [11].

In Section 3 assertions inverse to those of Sections 2 and 3 are considered, the final aim being the use of linear spline operators, in particular interpolating splines, for characterizations of the generalized Lipschitz spaces

$$\mathrm{Lip}(\psi, r, q; p) = \Big\{ f \in L_p(a,b) : \int_0^1 [\psi(t)^{-1} \omega_r(t; f)_p]^q dt/t < \infty, \ 1 \le q < \infty$$

$$\operatorname*{ess.sup}_{0<t<1} \psi(t)^{-1} \omega_r(t; f)_p < \infty, \ q = \infty \Big\},$$

the particular choice $\psi(t) = t^{\alpha}$, $0 < \alpha \le r$ giving the familiar Besov spaces. Here the general theory for sequences of linear commutative operators in Banach spaces, as developed in Butzer-Scherer [10] , can only be applied in part. Essential modifications, especially for the case $1 \le q < \infty$, are needed, due to the fact that e.g. for interpolating spline operators their commutativity depends on the underlying meshes. Another critical point is the lack of smoothness of the spline functions to be considered.

The resulting equivalence Theorem 3 yields, among others, characterizations of the (periodic) Lipschitz spaces $\mathrm{Lip}(\psi, r, q; p)$ expressed by the degree of approximation by (periodic) interpolating splines (Theorem 4). Here an extra derivation of a direct theorem for the case $1 \le p < 2$ is necessary (Lemma 4). Finally we remark that similar characterizations by means of best approximation by polynomial splines are established in [19] .

2. Connections between Jackson-type inequalities for polynomial spline approximation.

Let $W_p^k(a,b)$ denote the Sobolev space of functions with absolutely continuous $(k-1)$th derivative and k-th derivative belonging to $L_p(a,b)$, $1 \le p \le \infty$, $k \in \mathbb{N}$ *), and let $W_p^0(a,b) = L_p(a,b)$, the norm being denoted by $\|\cdot\|_p$. We shall

*) $\mathbb{P}$ denotes the set of all non-negative integers and $\mathbb{N}$ the set of all positive integers.

be concerned with the spline classes $(n \geq m \geq 0)$

$$Sp(n,m,\Delta) = \{ s(x) ; \; s(x) \in W_\infty^m(a,b) ; \; s(x) \in P_n , \; x \in (x_i, x_{i+1}) , \; 0 \leq i \leq N-1 \} , \tag{2.1}$$

where P_n denotes the class of algebraic polynomials of degree $\leq n$ and Δ is a partition

$$\Delta : a = x_0 < x_1 < \ldots < x_N = b \tag{2.2}$$

of the interval $[a,b]$ with $\overline{\Delta} = \max_{0 \leq i \leq N-1} (x_{i+1} - x_i)$ and $\underline{\Delta} = \min_{0 \leq i \leq N-1} (x_{i+1} - x_i)$.

As a measure for the jumps of the j-th derivative of a spline s in $Sp(n,m,\Delta)$ let us introduce the quantity

$$[s]_p^j = \begin{cases} \left\{ \sum_{i=1}^{N-1} |s^{(j)}(x_i + 0) - s^{(j)}(x_i - 0)|^p (x_{i+1} - x_{i-1}) \right\}^{1/p} , & 1 \leq p < \infty \\ \sup_{1 \leq i \leq N-1} |s^{(j)}(x_i + 0) - s^{(j)}(x_i - 0)| , & p = \infty \end{cases} \tag{2.3}$$

which vanishes for $j < m$. Furthermore, let us make the convention that $s^{(j)}(x)$ is that function in $L_\infty(a,b)$ which is obtained by differentiating $s \in Sp(n,m,\Delta)$ piecewise on each segment (x_{i-1}, x_i), no matter what the continuity index m may be.

The modulus of continuity is defined by $(t > 0 , r \in \mathbb{N})$

$$\omega_r(t;f)_q = \begin{cases} \sup_{|h| \leq t} \left\{ \int_{x, x+rh \in (a,b)} |\Delta_h^r f(x)|^q dx \right\}^{1/q} , & 1 \leq q < \infty \\ \sup_{|h| \leq t} \; \operatorname{ess\,sup}_{x, x+rh \in (a,b)} |\Delta_h^r f(x)| , & q = \infty , \end{cases} \tag{2.4}$$

where $\Delta_h^r f(x) = \sum_{i=0}^{r} (-1)^i \binom{r}{i} f(x + ih)$ is the r-th (forward) difference of f. In addition, we set $\omega_0(t;f)_q \equiv \| f \|_q$.

We begin with two auxiliary lemmas.

LEMMA 1: *With* $k \in \mathbb{P}$, $1 \leq q \leq \infty$, *let* U *be a map defined on* $W_q^k(a,b)$ *into the spline class* $Sp(n,m,\Delta)$ *such that* $\|U(f)\|_p$ *is a sublinear functional on* $W_q^k(a,b)$. *If the inequalities* $(1 \leq p \leq \infty,\ \ell > k)$

$$\|U(f)\|_p \leq A\, \|f^{(k)}\|_q \qquad (f \in W_q^k(a,b)),$$

$$\|U(g)\|_p \leq B\, \|g^{(\ell)}\|_q \qquad (g \in W_q^{\ell}(a,b))$$

hold for some constants $A,B > 0$ *then*

$$\|U(f)\|_p \leq C^{*)} \max(A, B\,\bar{\Delta}^{k-\ell})\,\omega_{\ell-k}(\bar{\Delta}; f^{(k)})_q . \tag{2.5}$$

Proof: The argument follows along the standard method of the theory of intermediate spaces (cf. [8], [15]).
Since $\|U(f)\|_p$ is sublinear in f one has for arbitrary $g \in W_q^{\ell}(a,b)$

$$\|U(f)\|_p \leq \|U(f-g)\|_p + \|U(g)\|_p$$
$$\leq A\,\|f^{(k)} - g^{(k)}\|_q + B\,\|g^{(k+\ell-k)}\|_q .$$

Taking the infimum over g yields

$$\|U(f)\|_p \leq \max(A, B\,\bar{\Delta}^{k-\ell})\, K(\bar{\Delta}^{\ell-k}, f^{(k)}; L_q(a,b), W_q^{\ell-k}(a,b)) .$$

---

*) C denotes a constant which may change its value from line to line. It does not depend on $\bar{\Delta}$ or the functions considered but in general on n,m,a,b and the various norms and orders of the derivatives considered, as well as upon the various constants connected with mesh quantities such as $\bar{\Delta}/\underline{\Delta}$. Sometimes this dependence is made clear by stating certain of these quantities explicitly.

Since the well-known K-functional satisfies the inequality

$$K(\bar{\Delta}^{\ell-k}, f^{(k)}; L_q(a,b), W_q^{\ell-k}(a,b)) \leq C\, \omega_{\ell-k}(\bar{\Delta}; f^{(k)})_q$$

(see [17]), the desired inequality follows immediately.

<u>LEMMA 2:</u> <u>For</u> $r \in \mathbb{N}$, $1 \leq p \leq \infty$, <u>there exists a linear map</u> $\mathcal{R}_{r,\Delta}: L_p(a,b) \longrightarrow Sp(r,1,\Delta)$ <u>satisfying</u>

$$\| \mathcal{R}_{r,\Delta}(f) - f \|_p \leq C\, \omega_{r+1}(\bar{\Delta}; f)_p . \tag{2.6}$$

<u>Proof:</u> There exists a linear smoothing operator $G_r: L_p(a,b) \to W_p^{r+1}(a,b)$ such that

$$\| f - G_r(f) \|_p \leq C\, \omega_{r+1}(\bar{\Delta}; f)_p;\ \bar{\Delta}^{r+1} \| G_r^{(r+1)}(f) \|_p \leq C\, \omega_{r+1}(\bar{\Delta}; f)_p \tag{2.7}$$

(The construction of G_{r+1} is carried out by means of a linear extension operator due to Whitney and Hestenes by performing the r-th Steklov means of this extension, see Johnen [17]). Then set $\mathcal{R}_{r,\Delta}(f) = L_{r,\Delta}(G_r(f))$ where $L_{r,\Delta}(g;x)$ denotes the spline function in $Sp(r,1,\Delta)$ which coincides on each subsegment (x_i, x_{i+1}) with the Lagrange interpolation polynomial $L_{r,i}(g)$ of g at the nodes $x_{i,k} = x_i + (x_{i+1} - x_i)k/r$, $0 \leq k \leq r$. Then by Corollary 4.2 of Swartz-Varga [25] and by (2.7) it follows that

$$\begin{aligned}\| \mathcal{R}_{r,\Delta}(f) - f \|_p &\leq \| L_{r,\Delta}(G_r(f)) - G_r(f) \|_p + \| G_r(f) - f \|_p \\ &\leq C\, \bar{\Delta}^{r+1} \| G_r^{(r+1)}(f) \|_p + \| G_r(f) - f \|_p \\ &\leq C\, \omega_{r+1}(\bar{\Delta}; f)_p .\end{aligned}$$

Note that an 'ordinary' interpolating spline cannot be employed to prove the lemma since it is defined for continuous functions only. The element of best approximation to f can neither be used since it is in general not linear in f . Another linear map which may be used for the proof of Lemma 2 can be found in a recent paper of Demko-Varga [11, Thm.3.5].

The main result of this section now reads

THEOREM 1: *Let $T_\Delta(f)$ be a (not necessarily linear) operator on $W_q^k(a,b)$ into $Sp(n,m,\Delta)$ with $k \in \mathbb{P}$, $1 \le q \le \infty$. The following inequalities hold:*

a) *If for some* $j \in \mathbb{P}$ *with* $j \geq k$, p *with* $1 \le p \le \infty$ *and* $r \in \mathbb{N}$

$$\int_0^1 u^{-j+\min(0,\ 1/p\ -\ 1/q)-1}\ \omega_r(u;\ f)_q\, du\ <\ \infty \qquad (2.8)$$

then $f \in W_p^j(a,b)$, *and*

$$\| T_\Delta^{(j)}(f) - f^{(j)} \|_p \le C\ \underline{\Delta}^{-j+\min(0,\ 1/p-1/q)} \Big\{ \| T_\Delta(f) - f \|_q + \qquad (2.9)$$
$$+ \int_0^{\bar{\Delta}} (t\,\bar{\Delta}^{-1})^{-j+\min(0,\ 1/p\ -\ 1/q)}\ \omega_r(t;\ f)_q\, dt/t \Big\} .$$

b) *If* $f \in W_q^\ell(a,b)$, $\ell \geq k$, *then for any* $r \in \mathbb{P}$

$$\| T_\Delta^{(\ell)}(f) - f^{(\ell)} \|_q \le C\ \underline{\Delta}^{-\ell} \Big\{ \| T_\Delta(f) - f \|_q + \bar{\Delta}^{\ell}\, \omega_r(\bar{\Delta};\ f^{(\ell)})_q \Big\}, \qquad (2.10)$$

and for $0 \le j < \ell$ (*if* $j = \ell - 1$ *the case* $p = \infty$, $q = 1$ *is excluded*)

$$\| T_\Delta^{(j)}(f) - f^{(j)} \|_p \le C\ \underline{\Delta}^{-j+\min(0,\ 1/p\ -\ 1/q)} \Big\{ \| T_\Delta(f) - f \|_q + \qquad (2.11)$$
$$+ \bar{\Delta}^{\ell}\, \omega_r(\bar{\Delta};\ f^{(\ell)})_q \Big\} ,$$

c) For any $j \in \mathbb{N}$

$$\| T_{\Delta}^{(j)}(f) \|_p \leq C \, \underline{\Delta}^{-j+\min(0,1/p-1/q)} \{ \| T_{\Delta}(f) - f \|_q + \omega_j(\bar{\Delta}; f)_q \}, \quad (2.12)$$

and for arbitrary $r \in \mathbb{N}$

$$[T_{\Delta}(f)]_p^j \leq C(\bar{\Delta}/\underline{\Delta})^{1/p} \underline{\Delta}^{-j+\min(0,1/p-1/q)} \{ \| T_{\Delta}(f) - f \|_q + \omega_r(\bar{\Delta}; f)_q \} \quad (2.13)$$

Proof: To prove (2.9) first note that (2.8) implies $f \in W_p^j(a,b)$ (see e.g. [19]). Furthermore, denoting by S an element of best approximation to f from $Sp(r-1, r-1, \Delta)$, one has by a Bernstein-type inequality for splines (cf. [19])

$$\| T_{\Delta}^{(j)}(f) - f^{(j)} \|_p \leq \| T_{\Delta}^{(j)}(f) - S^{(j)} \|_p + \| S^{(j)} - f^{(j)} \|_p \leq$$

$$\leq c \left\{ \underline{\Delta}^{-j+\min(0,1/p-1/q)} [\| T_{\Delta}(f) - f \|_q + \| S - f \|_q] + \right.$$

$$\left. + \| S^{(j)} - f^{(j)} \|_p \right\} .$$

By results in [19] on best approximation by splines these terms can be estimated in the form (2.9). The proof of (2.12) is similar using [19] and observing that

$$\| T_{\Delta}^{(j)}(f) \|_p \leq \| T_{\Delta}^{(j)}(f) - S^{(j)} \|_p + \| S^{(j)} \|_p .$$

Inequality (2.11) follows from (2.9) by noting that for $f \in W_q^{\ell}(a,b)$ and $0 \leq j < \ell$

$$\int_0^{\bar{\Delta}} u^{-j+\min(0,1/p-1/q)-1} \omega_{r+\ell}(u; f)_q \, du \leq$$

$$\leq \bar{\Delta}^{-j+\ell+\min(0,1/p-1/q)} \omega_r(\bar{\Delta}; f^{(\ell)})_q ,$$

the case $p = \infty$, $q = 1$ being excluded for $j = \ell - 1$.

To show (2.10) we employ the linear map $h_{r+\ell,\Delta}(f;x)$ constructed in Lemma 2 to deduce

$$\|T_\Delta^{(\ell)}(f)-f^{(\ell)}\|_q \le$$

$$\le \|T_\Delta^{(\ell)}(f)-h^{(\ell)}_{r+\ell,\Delta}(f)\|_q+\|h^{(\ell)}_{r+\ell,\Delta}(f)-f^{(\ell)}\|_q \qquad (2.14)$$

$$\le C\,\underline{\Delta}^{-\ell}\{\|T_\Delta(f)-f\|_q+\|h_{r+\ell,\Delta}(f)-f\|_q\}+\|h^{(\ell)}_{r+\ell,\Delta}(f)-f^{(\ell)}\|_q .$$

Next we apply (2.11) and (2.12) to this operator to obtain for $g\in W_q^{r+\ell}(a,b)$ by (2.6)

$$\|h^{(\ell)}_{r+\ell}(g)-g^{(\ell)}\|_q \le C\,\underline{\Delta}^{-\ell}\{\|h_{r+\ell}(g)-g\|_q+\bar{\Delta}^{r+\ell}\|g^{(r+\ell)}\|_q\}$$
$$\le C(\bar{\Delta}/\underline{\Delta})^{\ell}\,\bar{\Delta}^{r}\,\|g^{(r+\ell)}\|_q ,$$

and for $f\in W_q^{\ell}(a,b)$

$$\|h^{(\ell)}_{r+\ell}(f)\|_q \le C\,\underline{\Delta}^{-\ell}\{\|h_{r+\ell}(f)-f\|_q+\omega_\ell(\bar{\Delta};f)\}\le C(\bar{\Delta}/\underline{\Delta})^{\ell}\,\|f^{(\ell)}\|_q .$$

By Lemma 1 the latter inequalities imply in view of the linearity of $h^{(\ell)}_{r+\ell}(f)$

$$\|h^{(\ell)}_{r+\ell}(f)-f^{(\ell)}\|_q \le C(\bar{\Delta}/\underline{\Delta})^{\ell}\,\omega_r(\bar{\Delta};f^{(\ell)})_q .$$

Substituted in (2.14) this yields together with (2.6) the inequality (2.10).

Finally, to prove (2.13) we proceed similarly as in Lemma 5 of [19]. Let S^* be an element of best approximation to f from $Sp(r-1,r-1,\Delta^*)$, where Δ^* is the partition of [a,b] consisting of the nodes $x_i \pm \underline{\Delta}/3$, the x_i's being given by (2.2). Then S^* and its derivatives have no jumps at the nodes of Δ so that $(1 \le p < \infty)$

$$[T_\Delta(f)]_p^j \le \Big\{ \sum_{i=1}^{N-1} |T_\Delta^{(j)}(f; x_i+0) - S^{*(j)}(x_i)|^p \, 2\bar{\Delta} \Big\}^{1/p} +$$

$$+\Big\{ \sum_{i=1}^{N-1} |T_\Delta^{(j)}(f; x_i-0) - S^{*(j)}(x_i)|^p \, 2\bar{\Delta} \Big\}^{1/p}.$$

Proceeding further as in [19] one obtains

$$[T_\Delta(f)]_p^j \le C(\bar{\Delta}/\underline{\Delta})^{1/p} \underline{\Delta}^{-j} \| T_\Delta(f) - S^* \|_p \le$$

$$\le C(\bar{\Delta}/\underline{\Delta})^{1/p} \underline{\Delta}^{-j+\min(0,1/p-1/q)} \| T_\Delta(f) - S^* \|_q$$

from which (2.13) follows by the estimate for the best approximation $\| S^* - f \|_q$ given in [19].

This theorem presents a unified approach to assertions on simultaneous approximation by splines of a given function f and its derivatives as well as to estimates for the derivatives (or jumps) of the approximating spline. The latter assertions are often called of Zamansky-type (cf. [9]). An actual application of Theorem 1 to a particular approximating spline $T_\Delta(f)$ requires only an error bound for $T_\Delta(f) - f$ itself. Such error bounds have been proved in many cases, see e.g. [5], [2], [22], [3], [4], [6], [7], [12] [13], [18], [21], [23], [25], [20], [11] so that they may be completed by those of Theorem 1 if this has not yet been carried out in the concrete case by other means. In the following these statements shall be illustrated by considering several particular types of approximating splines.

3. Applications to particular types of approximating splines.

Before proceeding let us remark that essentially the same method of proof can be used to establish certain 'local' versions of the inequalities (2.9) - (2.12). We state those related with (2.11) and (2.12) in

COROLLARY 1: *Let* $\| g \|_{p,i}$ *denote the* L_p*-norm with respect to the interval* (x_i, x_{i+1}), i.e. $\| g \|_{p,i} = \{ \int_{x_i}^{x_{i+1}} |g(x)|^p dx \}^{1/p}$, *with corresponding modulus of continuity* $\omega_r(t ; f)_{q,i}$ *where* $0 \le i \le N-1$. *The following inequalities are valid for* $T_\Delta(f)$ *defined for* $f \in W_q^\ell(a,b)$, $1 \le q \le \infty$, *if* $\ell \ge 1$ *or* $f \in C[a,b]$ *if* $\ell = 0$:

a) *For any* $p, \bar{p}$ *with* $1 \le p, \bar{p} \le \infty$ *and any* $r \in \mathbb{P}$, *if* $0 \le j < \max(1, \ell)$

$$\| T_\Delta^{(j)}(f) - f^{(j)} \|_{p,i} \le C \, \Delta_i^{-j + 1/p} \{ \Delta_i^{-1/\bar{p}} \| T_\Delta(f) - f \|_{\bar{p},i} + \Delta_i^{\ell - 1/q} \omega_r(\Delta_i ; f^{(\ell)})_{q,i} \};$$

b) For any $j \in \mathbb{N}$ and $p, \bar{p}$ with $1 \le p, \bar{p} \le \infty$

$$\| T_\Delta^{(j)}(f) \|_{p,i} \le C \, \Delta_i^{-j + 1/p} \{ \Delta_i^{-1/\bar{p}} \| T_\Delta(f) - f \|_{\bar{p},i} + \Delta_i^{-1/q} \omega_j(\Delta_i ; f)_{q,i} \}.$$

For the proof of a) we use Markov's inequality and a further inequality for algebraic polynomials (see Timan [26, p.236]. Indeed, together with Hölder's inequality one has

$$\| T_\Delta^{(j)}(f) - f^{(j)} \|_{p,i} \le$$

$$\le C \{ \Delta_i^{-j} \| T_\Delta(f) - L_{r,\Delta}(f) \|_{p,i} + \| L_{r,\Delta}^{(j)}(f) - f^{(j)} \|_{p,i} \}$$

$$\le C \, \Delta_i^{1/p} \{ \Delta_i^{-j} \| T_\Delta(f) - L_{r,\Delta}(f) \|_{\infty,i} + \| L_{r,\Delta}^{(j)}(f) - f^{(j)} \|_{\infty,i} \}$$

$$\le \Delta_i^{1/p} \{ \Delta_i^{-j - 1/\bar{p}} \| T_\Delta(f) - L_{r,\Delta}(f) \|_{\bar{p},i} + \| L_{r,\Delta}^{(j)}(f) - f^{(j)} \|_{\infty,i} \}$$

$$\leq C \, \Delta_i^{1/p - 1/\bar{p} - j} \| T_\Delta(f) - f \|_{\bar{p}, i} +$$

$$+ C \, \Delta_i^{1/p} \{ \Delta_i^{-j} \| f - L_{r,\Delta}(f) \|_{\infty, i} + \| L_{r,\Delta}^{(j)}(f) - f^{(j)} \|_{\infty, i} \} .$$

Using a result of Swartz-Varga [25, Corollary 4.2] it follows that

$$\| f^{(j)} - L_{r,\Delta}^{(j)}(f) \|_{\infty, i} \leq C \, \Delta_i^{\ell - j - 1/q} \| f^{(\ell)} \|_{q, i}$$

and

$$\| g^{(j)} - L_{r,\Delta}^{(j)}(g) \|_{\infty, i} \leq C \, \Delta_i^{r - j - 1/q} \| g^{(r)} \|_{q, i} .$$

Lemma 1 then completes the proof of a).

Next we specialize Theorem 1 to linear operators satisfying Jackson-type inequalities.

COROLLARY 2: With $k \in \mathbb{P}$ and $1 \leq q \leq \infty$ let T_Δ be a linear operator on $W_q^k(a,b)$ into the spline class $Sp(n,m,\Delta)$ satisfying the two Jackson type inequalities ($k' \in \mathbb{N}$, $k' > k$)

i) $\| T_\Delta(f) - f \|_q \leq C \, \bar{\Delta}^k \| f^{(k)} \|_q$

ii) $\| T_\Delta(f) - f \|_q \leq C \, \bar{\Delta}^{k'} \| f^{(k')} \|_q \qquad (f \in W_q^{k'}(a,b))$.

Then the following inequalities are valid for $f \in W_q^\ell(a,b)$, $k' \geq \ell \geq k$, $1 \leq p \leq \infty$*)

a) $\| T_\Delta^{(j)}(f) - f^{(j)} \|_p \leq C \, \bar{\Delta}^\ell \, \underline{\Delta}^{-j + \min(0, 1/p - 1/q)} \omega_{k'-\ell}(\bar{\Delta}; f^{(\ell)})_q, \quad (0 \leq j < \ell)$,

b) $\| T_\Delta^{(\ell)}(f) - f^{(\ell)} \|_q \leq C \, (\bar{\Delta}/\underline{\Delta})^\ell \, \omega_{k'-\ell}(\bar{\Delta}; f^{(\ell)})_q \qquad (j = \ell)$,

---

*) As above the case $q = 1$, $p = \infty$ is excluded if $j = \ell - 1$.

c) $\| T_\Delta^{(j)}(f) \|_p \leq C \, \overline{\Delta}^{\ell} \, \underline{\Delta}^{-j + \min(0, 1/p - 1/q)} \omega_{\min(j,k') - \ell}(\overline{\Delta}; f^{(\ell)})_q, (j \geq \ell)$,

d) $[T_\Delta(f)]_p^j \leq C(\overline{\Delta}/\underline{\Delta})^{1/p} \, \overline{\Delta}^{\ell} \, \underline{\Delta}^{-j + \min(0, 1/p - 1/q)} \omega_{k' - \ell}(\overline{\Delta}; f^{(\ell)})_q \; (j \geq m)$.

The proof is an immediate consequence of Theorem 1, relations (2.10) (2.13), and Lemma 1.

We note that under the stronger hypothesis that there is given a <u>sequence</u> of linear operators T_Δ part a) can also be obtained by a general Theorem in [10], however not the limiting case b).

A particular class of examples to Corollary 2 is given by linear interpolating splines of odd degree satisfying the first and second integral relation. Extending results in [20] we have here

<u>COROLLARY 3</u>: Let $T_\Delta(f)$ be a spline in $Sp(2n-1, 2n-1, \Delta)$ interpolating $f \in W_q^k(a,b)$ with $1 \leq p \leq \infty$ and $2n \geq \ell \geq n \geq 2$, which satisfies the first and second integral relations*) and is linear in f. Then, defining $\sigma \equiv \max_{1 \leq i \leq N-1} ((x_i - x_{i-1})(x_{i+1} - x_i)^{-1} - 1)$, there hold the following inequalities for σ and $\overline{\Delta}$ small enough, and $\max(2,q) \leq p \leq \infty$

a) $\| T_\Delta^{(j)}(f) - f^{(j)} \|_p \leq C \, \overline{\Delta}^{\ell - j + 1/p - 1/q} \omega_{2n-\ell}(\overline{\Delta}; f^{(\ell)})_q, (0 \leq j < \ell)$,

b) $\| T_\Delta^{(\ell)}(f) - f^{(\ell)} \|_q \leq C \, \omega_{2n-\ell}(\overline{\Delta}; f^{(\ell)})_q$,

c) $\| T_\Delta^{(j)}(f) \|_p \leq C \, \overline{\Delta}^{\ell - j + 1/p - 1/q} \omega_{j-\ell}(\overline{\Delta}; f^{(\ell)})_q \quad (j \geq \ell)$,

d) $[T_\Delta(f)]_p^{2n-1} \leq C \, \overline{\Delta}^{\ell - j + 1/p - 1/q} \omega_{2n-\ell}(\overline{\Delta}; f^{(\ell)})_q$,

---

*) By this we mean that the boundary terms obtained by partial integration in these relations must vanish, see e.g. [3, Chap. V] .

where the constants depend on the ratio $\bar{\Delta}/\underline{\Delta}$.

This in an immediate consequence of Corollary 2 since Jackson-type inequalities are satisfied with k' = 2n and k = n (see [20]).

As a final application of Theorem 1 error bounds for perturbed interpolating splines in [25], [20] are improved and extended. In a somewhat different form this is also carried out in Demko-Varga [11].

THEOREM 2: *For* $f \in C[a,b]$ *set* $f_i = f(x_i)$ *if* m *is odd and* $(x_{i+1} - x_i)f_i = \int_{x_i}^{x_{i+1}} f(x)\,dx$ *if* m *is even. Let* T_Δ *be the spline of class* $Sp(m,m,\Delta)$, $m \geq 2$, *satisfying for a given set of numbers* r_i *the conditions*

$$T_\Delta(x_i) = r_i \qquad (1 \leq i \leq N-1) \quad (3.1)$$

if m *is odd, or*

$$\int_{x_i}^{x_{i+1}} T_\Delta(x)\,dx = r_i(x_{i+1} - x_i) \quad (3.2)$$

if m is even. Furthermore, let T_Δ satisfy the boundary conditions

$$T_\Delta^{(j_i)}(a) = \alpha_i, \quad T_\Delta^{(j_i)}(b) = \beta_i \qquad (1 \leq i \leq m' = [(m+1)/2]) \quad (3.3)$$

where α_i, β_i are given reals, and the j_i, k_i are integers satisfying $0 \leq j_1 < \ldots < j_{m'} \leq 2m'-1$, $0 \leq k_1 < \ldots < k_{m'} \leq 2m'-1$ such that

$$\int_a^b [g^{(m')}(x)]^2\,dx = (-1)^{m'} \int_a^b g^{(2m')}(x)\,dx$$

is fulfilled for all $g \in W_2^{2m'}(a,b)$ satisfying the homogeneous conditions (3.1), (3.3) or (3.2), (3.3).

Now assume that $f \in W_q^{\ell}(a,b)$, $1 \le q \le \infty$, for some ℓ with $1 \le \ell \le m+1$, or $f \in C[a,b]$ if $\ell = 0$ and let f have continuous derivatives of orders $j_{i'}$, and $k_{i''}$ at the points a and b, respectively, such that $k \equiv \min(j_{i'}, k_{i''}) \geq \max(1,\ell)$. Then one has with $\bar{\Delta}$, $\underline{\Delta}$, σ and p as in Corollary 3 for $0 \le j < \max(1,\ell)$

$$\begin{aligned}\|T_\Delta^{(j)} - f^{(j)}\|_p \le\ & C\left\{\bar{\Delta}^{1/p-1/q+\ell-j}\omega_{k-\ell}(\bar{\Delta}; f^{(\ell)})_q\right\} + \\ & + C\left\{\sum_i (f_i - r_i)^q\right\}^{1/q} + \\ & + C\,\bar{\Delta}^{1/p}\left\{\sup_{1 \le i \le i'} \bar{\Delta}^{j_i}|f^{(j_i)}(a) - \alpha_i| + \sup_{i' \le i \le m'} \bar{\Delta}^{j_i}|\alpha_i|\right\} + \\ & + C\,\bar{\Delta}^{1/p}\left\{\sup_{1 \le i \le i''} \bar{\Delta}^{k_i}|f^{(k_i)}(b) - \beta_i| + \sup_{i'' \le i \le m'} \bar{\Delta}^{k_i}|\beta_i|\right\}.\end{aligned} \tag{3.4}$$

There also hold estimates similar to the cases b), c) and d) of Corollary 3.

<u>Proof</u>: It suffices to prove the case j = 0 of (3.4) since the other cases follow by Theorem 1. We restrict ourselves to the case m is even.

A direct application of Corollary 2 is not possible since T_Δ does not need to be linear in f. It is therefore necessary to insert a suitable linear spline operator. To this end let $S_\Delta(g;x)$ be the spline in $Sp(2m'+1, 2m'+1, \Delta)$ interpolating $g \in C[a,b]$, i.e. $S(g;x_i) = g(x_i)$, $0 \le i \le N$, and satisfying the boundary conditions

$$S^{(j_i+1)}(g;a) = L_{m,1}^{(j_i+1)}(g;a), \quad S_\Delta^{(k_i+1)}(g;b) = L_{m,N}^{(k_i+1)}(g;b) \qquad (1 \le i \le m') \tag{3.5}$$

where $L_{m,1}(g;x)$, $L_{m,N}(g;x)$ denote the Lagrange interpolation polynomials of g with respect to the nodes $a + i(x_1 - a)/m$ and

$b - i(b - x_{N-1})/m$ with $0 \leq i \leq m$, respectively. By results in [20] one derives in the same manner as in Swartz-Varga [25]

$$\| S_\Delta(g) - g \|_p \leq \begin{cases} C\, \bar{\Delta}^{1/p - 1/q + 1} \| g' \|_q & (g \in W_q^1(a,b)) \\ C\, \bar{\Delta}^{1/p - 1/q + m + 2} \| g^{(m+2)} \|_q & (g \in W_q^{m+2}(a,b)) \end{cases} \tag{3.6}$$

Then one considers, setting $F(x) = \int_a^x f(u)\, du$, the inequality

$$\| T_\Delta - f \|_p \leq \| T_\Delta - S'_\Delta(F) \|_p + \| S'_\Delta(F) - F' \|_p . \tag{3.7}$$

Results in [20] on even-degree splines yield for the first term of (3.7) in view of $\int_{x_i}^{x_{i+1}} S'_\Delta(F; x)\, dx = f_i(x_{i+1} - x_i)$

$$\| T_\Delta - S'(F) \|_p \leq$$

$$\leq C \left\{ \bar{\Delta}^{1/p - 1/q + m} \| S_\Delta^{(m+1)}(F) \|_q \right\} + C \left\{ \sum_i (f_i - r_i)^q \right\}^{1/q} +$$

$$+ C\, \bar{\Delta}^{1/p} \sup_{1 \leq i \leq m'} \left\{ \bar{\Delta}^{j_i} | S_\Delta^{(j_i+1)}(F; a) - \alpha_i | + \bar{\Delta}^{k_i} | S_\Delta^{(k_i+1)}(F; b) - \beta_i | \right\} . \tag{3.8}$$

Now we apply Corollary 2 to the second term of (3.7) and to the first of (3.8) which is possible by the Jackson-type inequalities (3.6). This gives by (3.5)

$$\| T_\Delta - f \|_p \leq C \left\{ \bar{\Delta}^{1/p - 1/q + \ell} \omega_{m-\ell}(\bar{\Delta}; F^{(\ell+1)})_q + \right.$$

$$\left. + \left[\sum_i (f_i - r_i)^q (x_{i+1} - x_i)^{-q} \right]^{1/q} \right\} +$$

$$+ C\, \bar{\Delta}^{1/p} \sup_{1 \leq i \leq m'} \left\{ \bar{\Delta}^{j_i} | L_{m,1}^{(j_i+1)}(F; a) - \alpha_i | + \bar{\Delta}^{k_i} | L_{m,N}^{(k_i+1)}(F; b) - \beta_i | \right\} . \tag{3.9}$$

In order to obtain the desired estimate (3.4) for $j = 0$ it remains therefore only to estimate the latter two terms in (3.9). For $i < i$

one has

$$\bar{\Delta}^{j_i} | L_{m,1}^{(j_i+1)}(F;a) - \alpha_i | \leq \bar{\Delta}^{j_i} | L_{m,1}^{(j_i+1)}(F;a) - F^{(j_i+1)}(a) | + \\ + \bar{\Delta}^{j_i} | f^{(j_i)}(a) - \alpha_i | .$$

Since $F^{(j_{i'}+1)}(x)$ is continuous in x = a by assumption, for $\bar{\Delta}$ small enough Rolle's theorem yields for the first term on the right the bound $C\,\bar{\Delta}^{j_{i'}} \sup_{x\in[a,x_1]} | L_{m,1}^{(j_{i'}+1)}(F;x) - F^{(j_{i'}+1)}(x) |$. It follows that

$$\begin{aligned} \bar{\Delta}^{j_i} | L_{m,1}^{(j_i+1)}(F;a) - \alpha_i | &\leq C \Big\{ \bar{\Delta}^{j_i} | f^{(j_i)}(a) - \alpha_i | + \\ &+ \bar{\Delta}^{j_{i'}} \sup_{x\in[a,x_1]} | L_{m,1}^{(j_{i'}+1)}(F;x) - F^{(j_{i'}+1)}(x) | \Big\} + \\ &+ C \Big\{ \bar{\Delta}^{j_{i'}} | f^{(j_{i'})}(a) - \alpha_{i'} | + \bar{\Delta}^{j_{i'}} |\alpha_{i'}| \Big\} . \end{aligned} \tag{3.10}$$

For $i \geqslant i'$ one uses simply the estimate

$$\begin{aligned} \bar{\Delta}^{j_i} | L_{m,1}^{(j_i+1)}(F;a) - \alpha_i | &\leq \bar{\Delta}^{j_i} | L_{m,1}^{(j_i+1)}(F;a) | + \bar{\Delta}^{j_i} |\alpha_i| \\ &\leq C\,\bar{\Delta}^{j_{i'}} \sup_{x\in[a,x_1]} | L_{m,1}^{(j_{i'}+1)}(F;x) | + \bar{\Delta}^{j_i} |\alpha_i| , \end{aligned} \tag{3.11}$$

which follows by Markov's inequality.

By known error bounds on Lagrange interpolation (cf. [25]) one has

$$\bar{\Delta}^{j_{i'}} \sup_{x\in[a,x_1]} | L_{m,1}^{(j_{i'}+1)}(g;x) | \leq \begin{cases} C\,\bar{\Delta}^{\ell-1/q} \| g^{(\ell+1)} \|_q \\ \\ C\,\bar{\Delta}^{j_{i'}-1/q} \| g^{(j_{i'}+1)} \| \end{cases}$$

which implies by Lemma 1, since $L_{m,1}^{(j_{i'}+1)}$ is a linear operator, that

$$\bar{\Delta}^{j_{i'}} \sup_{x\in[a,x_1]} |L_{m,1}^{(j_{i'}+1)}(F;x)| \le C\,\bar{\Delta}^{\ell-1/q}\,\omega_{j_{i'}-\ell}(\bar{\Delta};f^{(\ell)})_q. \quad (3.12)$$

The estimates (3.10) - (3.12) together with the corresponding ones for the point x = b now give the desired inequality (3.4) for j = 0 after substitution in (2.20).

4. Equivalence theorems for approximation by linear polynomial Spline operators.

The object here is to complete the different types of error bounds presented in Sec. 2 and 3, especially in Corollary 2, in such a way that they can serve as characterizations of Besov spaces. In the case of best approximation such a characterization theorem has been established in [19] using splines with equidistant nodes. In order to do this for linear spline operators one could apply the general theory for sequences of linear commutative operators in Banach spaces as developed in Butzer-Scherer [9,10]. But then some limiting cases due to the lack of smoothness of the approximating splines are not covered. Another crucial point presenting difficulties is that the commutativity of linear spline operators depends upon the underlying meshes.

Thus a somewhat different approach is necessary using the more concrete results on best approximation by splines with equidistant nodes in [19]. Let $E_N^{(n,m)}(f;p)$ denote the best approximation to $f\in L_p(a,b)$, $1\le p<\infty$, or $f\in C[a,b]$ if $p=\infty$, by splines of the class $Sp(n,m,\Delta_N)$, where Δ_N is the equidistant partition of [a,b] into segments of length (b-a)/N. Then the main result we need states that for N odd, $N \geq 3$ (cf. [19])

$$\omega_{n+1}(1/N;f)_p \leq C\left\{E_N^{(n,m)}(f;p)+E_{N+1}^{(n,m)}(f;p)\right\},\ (n\geq m\geq 0) \quad (4.1)$$

we need further an auxiliary lemma which is connected with well-known inequalities of Hardy (cf. [14], [16], [10]).

<u>LEMMA 3: Let</u> $\psi(t)$ <u>be a positive monotone increasing function on</u> (0,1] <u>satisfying</u>

$$\int_0^t \psi(u)\,u^{-1}\,du = O[\psi(t)], \qquad \int_t^1 \psi^{-1}(u)\,u^{-1}\,du = O[\psi^{-1}(t)] \quad (4.2)$$

<u>for</u> $t \to 0+$. <u>Then for any sequence</u> $\{a_n\}_{n=1}^{\infty}$ <u>of non-negative numbers and</u> $1 \leq q \leq \infty$ <u>there holds</u>

$$\left\{\sum_{n=1}^{\infty}\left[\psi(1/n)^{-1} n \sum_{j=0}^{\infty} a_{n2^j}\right]^{q}n^{-1}\right\}^{1/q} \leq C\left\{\sum_{n=1}^{\infty}\left[\psi(1/n)^{-1} n a_n\right]^{q} n^{-1}\right\}^{1/q}, \quad (4.3)$$

<u>provided the right-side is finite.</u>

<u>Proof</u>. We establish the above inequality for q = 1 and $q = \infty$ whence the general case follows by the interpolation theorem of Riesz-Thorin (cf.[10]). For q = 1 one has

$$\sum_{n=1}^{\infty}\psi(1/n)^{-1}\sum_{j=0}^{\infty}a_{n2^j} = \sum_{m=1}^{\infty}a_m\sum_{n\in\mathbb{N}_m}\psi(1/n)^{-1}$$

where $\mathbb{N}_m = \{k \in \mathbb{N} : k = m2^{-j},\ j = 1,2,\ldots\}$. It is easy to see that

$$\sum_{n\in\mathbb{N}_m}\psi(1/n)^{-1} \leq 2\left\{\psi^{-1}(1/m) + \int_{1/m}^{1}\psi^{-1}(u)\,u^{-1}\,du\right\}$$

so that inequality (4.3) follows for $q = 1$ by (4.2). For $q = \infty$ one has

$$\sup_{n\in\mathbb{N}} \psi^{-1}(1/n)\, n \sum_{j=0}^{\infty} a_{n2^j} \leq$$

$$\leq \sup_{n\in\mathbb{N}} a_n \psi^{-1}(1/n)\, n \sup_{m\in\mathbb{N}} \psi^{-1}(1/m)\, m \sum_{j=0}^{\infty} \psi(m^{-1}2^{-j})\, m^{-1} 2^{-j} .$$

Here the estimate

$$\sum_{j=0}^{\infty} \psi(m^{-1}2^{-j})\, m^{-1} 2^{-j} \leq 2\Big\{\psi(1/m)\, m^{-1} + \int_0^{1/m} u\, \psi(u)\, u^{-1} du\Big\} ,$$

gives (4.3) for $q = \infty$, again by (4.2).

In order to formulate the following theorem we recall (cf. [19]) that the generalized Lipschitz space $\mathrm{Lip}(\psi, q, r; p)$ is defined as the set of functions which belong $L_p(a,b)$ for $1 \leq q \leq \infty$ and to $C[a,b]$ for $p = \infty$, and for which the semi-norm

$$|f|_p^{\psi,q,r} = \begin{cases} \Big\{ \int_0^1 [\psi(t)^{-1}\, \omega_r(t;f)_p]^q \, dt/t \Big\}^{1/q} < \infty , & 1 \leq q < \infty \\ \operatorname*{ess.sup}_{0<t<1} \big[\psi(t)^{-1} \omega_r(t;f)_p\big] < \infty & , q = \infty \end{cases}$$

is finite, where $\psi(t)$ is a bounded positive function on $(0,1]$ with $\lim_{t\to 0+} \psi(t) = 0$.

<u>THEOREM 3</u>: <u>With</u> $k \in \mathbb{P}$, $1 \leq p \leq \infty$ <u>let</u> $\{S_N(f)\}_{n=1}^{\infty}$ <u>be a sequence of linear operators on</u> $W_p^k(a,b)$ <u>into the spline class</u> $Sp(n,m,\Delta_N)$ <u>satisfying the two Jackson type inequalities</u> $(n+1 \geq k' > k)$

$$\| S_N(f) - f \|_p \leq C\, N^{-k} \| f^{(k)} \|_p , \tag{4.4}$$

$$\| S_N(f) - f \|_p \leq C\, N^{-k'} \| f^{(k')} \|_p , \tag{4.5}$$

(*the space* $L_\infty(a,b)$ *being replaced by* $C[a,b]$). *Furthermore, let* $\psi(t)$ *be a monotonely increasing function on* $[0,1]$ *satisfying* (4.2) *and*

$$\int_t^1 u^{-k'+k-1}\psi(u)\,du = O[t^{-k'+k}\psi(t)],$$

$$\int_0^t u^{k'-k-1}\psi^{-1}(u)\,du = O[t^{k'-k}\psi^{-1}(t)] \tag{4.6}$$

for $t\to 0+$. *Then the following two assertions are equivalent for* $1\leq q<\infty$:

(i) $\left\{\sum_{N=1}^{\infty}[N^k\psi^{-1}(1/N)\,\|S_N(f)-f\|_p]^q\,1/N\right\}^{1/q}<\infty,$

(ii) $f^{(k)}\in \mathrm{Lip}(\psi,q,k'-k;p),$

with a corresponding modification for $q=\infty$ (cf. [19]).

If instead of (4.2) *the (stronger) conditions*

$$\int_0^t \psi(u)u^{-\ell+k-1}\,du = O[t^{-\ell+k}\psi(t)],$$

$$\int_t^1 \psi^{-1}(u)\,u^{\ell-k-1}\,du = O[t^{\ell-k}\psi^{-1}(t)] \tag{4.7}$$

are satisfied for some $\ell\in\mathbb{N}$ *with* $k<\ell<k'$, *then these assertions are equivalent to*

(iii)
$$f\in W_p^{\ell}(a,b),$$
$$\left\{\sum_{N=1}^{\infty}[N^{k-\ell}\psi^{-1}(1/N)\,\|S_N^{(\ell)}(f)-f^{(\ell)}\|_p]^q N^{-1}\right\}^{1/q}<\infty.$$

If in addition $S_N S_{2N} = S_{2N} S_N$ *for all* $n\in\mathbb{N}$ (*assuming* $m\geq k$)

then these assertions are equivalent to

(iv)
$$\Big\{\sum_{N=1}^{\infty} (N^{k-j}\psi^{-1}(1/N)\,[S_N]_p^j)^q N^{-1}\Big\}^{1/q} < \infty \quad , \qquad m \leq j < k' \;,$$
$$\Big\{\sum_{N=1}^{\infty} (N^{k-k'}\psi^{-1}(1/N)\,\| S_N^{(k')} \|_p)^q N^{-1}\Big\}^{1/q} < \infty \;^{*)} \qquad (j=k').$$

In case $k'=n+1$ the equivalence of these assertions remains true without condition (3.6) on $\psi(t)$.

Proof: One half of the theorem follows directly from Corollary 2, namely the implications (ii) $\Rightarrow$ (i), (i) $\Rightarrow$ (iii) and (i) $\Rightarrow$ (iv). Concerning the converses one first observes that assertion (i) implies $f \in \mathrm{Lip}(\psi_{-k}, q, n+1; p)$ with $\psi_{-k}(t) = \psi(t)t^k$ by $E_N^{(n,m)}(f;p) \leq \|S_N(f)-f\|_p$ and (4.1). In view of (4.2) and (4.5) results in [19] on generalized Lipschitz spaces apply giving first that $f \in \mathrm{Lip}(\psi_{-k}, q, k'; p)$ and also $f^{(k)} \in \mathrm{lip}(\psi, q, k'-k; p)$, thus (ii). Similarly (iii) implies $f^{(\ell)} \in \mathrm{lip}(\psi_{\ell-k}, q, n-\ell+1; p)$ by (4.7), hence $f^{(k)} \in \mathrm{Lip}(\psi, q, n-k+1; p)$ and also $f^{(k)} \in \mathrm{Lip}(\psi, q, k'-k; p)$ by (4.2), (4.6).

The equivalence of all assertions above is therefore complete provided one shows the implication (iv) $\Rightarrow$ (i). To this end we observe that the commutativity assumption allows by a known argument (cf. [10]) to estimate

$$\| S_N(f) - S_{2N}(f) \|_p \leq$$

$$\leq \| S_N(S_{2N}(f)) - S_{2N}(f) \|_p + \| S_{2N}(S_N(f)) - S_N(f) \|_p \qquad (4.8)$$

$$\leq C\, N^{-k} \{\omega_{k'-k}(1/N;\, S_N^{(k)}(f))_p + \omega_{k'-k}(1/N;\, S_{2N}^{(k)}(f))_p\} ,$$

*) In case $k'=n+1$ the second condition in (iv) is dropped. In this connection recall that $S_N^{(k')}$ is defined as that function which is obtained by k'-fold piecewise differentiation of S_N.

Corollary 2 being used.

Next set $F(x) = \int_a^x f(u)du$ and $T_N(f;x) = \int_a^x S_N(f;u)du + t_N(x)$, where with $x_i = a+i(b-a)/N$, $0 \leq i \leq N$,

$$t_N(x) = \int_a^x [f(u) - S_N(f;u)]\,du. \quad (x \in [x_i, x_{i+1}))$$

It therefore follows that $T_N(f;x_i) = F(x_i)$ for each i, $0 \leq i \leq N-1$. Hence by Rolle's theorem

$$\| T_N(f) - T_{2N}(f) \|_p \leq C\, N^{-1} \| S_N(f) - S_{2N}(f) \|_p. \quad (4.9)$$

Since

$$\| T_N(f) - F \|_p \leq (b-a)^{1/p} \| S_N(f) - f \|_1 + \| t_N \|_p$$

$$\leq 2(b-a)^{1/p} \| S_N(f) - f \|_1 \leq 2(b-a) \| S_N(f) - f \|_p$$

it follows by (4.4), (4.8) and (4.9) that

$$\| T_N(f) - F \|_p \leq \sum_{j=0}^{\infty} \| T_{N2^j}(f) - T_{N2^{j+1}}(f) \|_p$$

$$\leq C \sum_{j=0}^{\infty} (N2^j)^{-k-1}\, \omega_{k'-k}(N^{-1}2^{-j};\, S^{(k)}_{N2^j})_p .$$

Then we employ Lemma 3 to obtain

$$\Big\{ \sum_{N=1}^{\infty} [\psi^{-1}(1/N)\, N^{k+1} \| T_N(f) - F \|_p]^{q_N} N^{-1} \Big\}^{1/q}$$

$$\leq C \Big\{ \sum_{N=1}^{\infty} [\psi^{-1}(1/N)\, \omega_{k'-k}(N^{-1};\, S^{(k)}_N)_p]^{q_N} N^{-1} \Big\}^{1/q} .$$

By a result in [19] we have furthermore

$$N^{-k}\, \omega_{k'-k}(N^{-1};\, S_N^{(k)})_p \leq C\, N^{-m}\, \omega_{k'-m}(N^{-1};\, S_N^{(m)})_p$$

$$\leq \sum_{j=m}^{k'-1} N^{-j} [S_N]^j_p + N^{-k'} \| S_N^{(k')} \|_p ,$$

where the first sum is dropped if $k' \leq m$ and the last term if $k'=n+1$. Thus from (iv) it follows by (4.1) and the characterization of generalized Lipschitz spaces by best approximation by splines in [19] that $F \in Lip(\psi_{-k-1}, q, n+2; p)$, and consequently $f^{(k)} \in Lip(\psi, q, k'-k; p)$ by the reduction assertions already mentioned.

This completes the proof of the theorem by observing that in case $k'=n+1$ one needs only one reduction assertion not involving condition (4.6).

Let us remark that also another version of Theorem 3 may be formulated for sequences of operators on $W_p^k(a,b)$ into the spline classes $Sp(n,m,\Delta_k)$ where this time the sequence $\{\Delta_k\}_{k=1}$ consists of nested partitions, i.e. $\Delta_k \subset \Delta_{k+1}$, which satisfy $\overline{\Delta}_k \leq C \underline{\Delta}_{k+1}$. Its proof follows more closely to that of the general theorem for sequences of commutative linear operators given in [10].

There are many special instances of sequences of linear commutative spline operators S_N which satisfy the assumptions of Theorem 3. In particular, interpolating splines of odd (or even) degree defined for equidistant meshes and with suitable boundary conditions can be considered. We mention e.g. natural interpolating splines, splines obtained by piecewise Hermite interpolation and periodic splines. Only the latter ones shall be studied here in more detail. To this end we first extend known error bounds on periodic splines of odd degree.

LEMMA 4. a) Let f be continuous with period b-a, and let $S_{N,2n-1}(f;x)$ be the periodic spline (with period b-a) in $Sp(2n-1,2n-1,\Delta_N)$, $n \geq 2$, interpolating f at the nodes of Δ_N, i.e. $S_{N,2n-1}(f,x_{i,N})=f(X_{i,N})$ for $x_{i,N} = a+i(b-a)/N$, $0 \leq i \leq N$. Then

$$\| S_{N,2n-1}(f) - f \|_p \leq C N^{-k} \| f^{(k)} \|_p \tag{4.10}$$

<u>for periodic</u> $f \in W_p^k(a,b)$, <u>where</u> $1 \leq k \leq 2n$ <u>if</u> $1 \leq p < \infty$ <u>and</u> $0 \leq k \leq 2n$ <u>if</u> $p = \infty$.

b) <u>Let</u> f <u>be a periodic function (period</u> b-a) <u>in</u> $L_p(a,b)$, $1 \leq p \leq \infty$, <u>and let</u> $S_{N,2n}(f;x)$ <u>be the periodic spline in</u> $Sp(2n,2n,\Delta_N)$ <u>satisfying</u> $\int_{x_i}^{x_{i+1}} S_{N,2n}(f;u)\,du = \int_{x_i}^{x_{i+1}} f(u)\,du$. <u>Then</u>

$$\| S_{N,2n}(f) - f \|_p \leq C\, N^{-k} \| f^{(k)} \|_p \tag{4.11}$$

<u>for periodic</u> $f \in W_p^k(a,b)$, $0 \leq k \leq 2n+1$.

<u>Proof</u>. Let us first remark that the interpolating splines are uniquely defined (concerning b) see e.g. [27]) and that the estimates of part a) are well known for $2 \leq p \leq \infty$ (see e.g. [2],[12] , [15]).

In order to establish (4.10) for $1 \leq p \leq 2$, observe that if one considers both f and $S_{N,2n-1}(f)$ as functions on $(-\infty,\infty)$, $S_{N,2n-1}(f)$ is the unique spline with bounded (2n-2)-th derivative interpolating f at the nodes $x_{i,N}$, modulo b-a. Setting $M_i = S_{N,2n-1}^{(2n-2)}(f;x_i)$ one has

$$S_{N,2n-1}^{(2n-2)}(f;x) = M_i(x - x_{i-1,N})h^{-1} + M_{i-1}(x_{i,N} - x)h^{-1}, \tag{4.12}$$

$$(h = (b-a)/N)$$

for $x_{i-1,N} \leq x \leq x_{i,N}$. Now in Ahlberg-Nilson [1] the relation

$$M_i = (2n-1)!\, h^{-2n+2} \sum_{j=-\infty}^{\infty} a_j(n)\, \Delta_h^{2n-2} f(x_{i+j} - (n-1)h) \tag{4.13}$$

is proved where the coefficients $a_j(n)$ satisfy

$$\sum_{j=-\infty}^{\infty} | a_j(n) | < C . \tag{4.14}$$

Using these facts it follows that

$$\|S^{(2n-2)}_{N,2n-1}(f)\|_p = \Big\{\sum_{i=1}^{N}\int_{x_{i-1}}^{x_i} |M_i(x-x_{i-1,N})+M_{i-1}(x_{i,N}-x)|^p h^{-p}dx\Big\}^{1/p}$$

$$\leq 2\Big\{\sum_{i=1}^{N} |M_i|^p h\Big\}^{1/p}$$

Furthermore one has ($x_{i+j,N}-(n-1)h = x_{i+j-n+1,N}$)

$$|\Delta_h^{2n-2} f(x_{i+j-n+1})| \leq h^{2n-3}\int_{x_{i+j-n+1,N}}^{x_{j+i+n-1,N}} |f^{(2n-2)}(u)|du$$

$$\leq h^{2n-2-1/p}\Big\{\int_{x_{i+j-n+1,N}}^{x_{i+j+n-1,N}} |f^{(2n-2)}(u)|^p du\Big\}^{1/p},$$

and by (3.13), (3.14) and Minkowski's inequality

$$\|S^{(2n-2)}_{N,2n-1}(f)\|_p \leq C\Big(\sum_{i=1}^{N}\Big[\sum_{j=-\infty}^{\infty}|a_j(n)|\Big\{\int_{x_{i+j+n-1,N}}^{x_{i+j+n-1,N}} |f^{(2n-2)}(u)|^p du\Big\}^{1/p}\Big]\Big)^{1/p} \tag{4.15}$$

$$\leq C\sum_{j=-\infty}^{\infty}|a_j(n)|\Big\{\sum_{j=1}^{N}\int_{x_{i+j-n+1,N}}^{x_{i+j+n-1,N}} |f^{(2n-2)}(u)|^p du\Big\}^{1/p}$$

$$\leq C\,\|f^{(2n-2)}\|_p .$$

Then by Rolle's theorem and an argument of de Boor (see [24,p.19]) using the polygonal line interpolating $f^{(2n-2)}$ it follows from (4.15) that

$$\|S_{N,2n-1}(f)-f\|_p \leq C\,N^{-2n+2}\|S^{(2n-2)}_{N,2n-1}(f)-f^{(2n-2)}\|_p$$

$$\leq C\,N^{-2n}\|f^{(2n)}\|_p ,$$

establishing (4.10) for $k = 2n$. To obtain (4.10) for the other values of k, let H(f) be the Hermite spline in $Sp(4n+1, 2n+1, \Delta_N)$ defined by

$$H^{(j)}(f; x_{i,N}) = \begin{cases} f^{(j)}(x_{i,N}), & 0 \le j \le \max(1,k)-1 \\ 0, & \max(1,k) \le j \le 2n, \end{cases}$$

where f is given as in (4.10). Then, by known estimates for H(f) (see Swartz-Varga [25]), it follows that

$$\| S_{N,2n-1}(f) - f \|_p = \| S_{N,2n-1}(H) - f \|_p$$

$$\le \| S_{N,2n-1}(H) - H(f) \|_p + \| H(f) - f \|_p$$

$$\le C \{ N^{-2n} \| H^{(2n)} \|_p + N^{-k} \| f^{(k)} \|_p \}$$

$$\le C\ N^{-k} \| f^{(k)} \|_p .$$

To prove part b), assume first $\int_a^b f(u)du = 0$, and set $F(x) = \int_a^x f(u)du$ which is therefore a continuous function on [a,b] with period b-a. Then $S'_{N,2n+1}(F)$ is a periodic spline of class $Sp(2n, 2n, \Delta_N)$ satisfying

$$\int_{x_i}^{x_{i+1}} S'_{N,2n+1}(F;x)\,dx = S_{N,2n+1}(F;x_{i+1}) - S_{N,2n+1}(F;x_i)$$

$$= F(x_{i+1}) - F(x_i)$$

$$= \int_{x_i}^{x_{i+1}} f(u)\,du .$$

By the uniqueness of its definition this means that $S'_{N,2n+1}(F) = S_{N,2n}(f)$.

Using this fact one derives by Theorem 1

$$\| S_{N,2n}(f) - f \|_p = \| S'_{N,2n+1}(F) - F' \|_p$$
$$\le C\ N \{ \| S_{N,2n+1}(F) - F \|_p + N^{-1} \omega_{2n}(N^{-1}; f)_p \},$$

whence (4.11) follows in case $\int_a^b f(u)du = 0$ by applying (4.10). If $\int_a^b f(u)du \neq 0$ define $\bar{f}(x) = f(x)-(b-a)^{-1}\int_a^b f(u)du$, so that $\int_a^b \bar{f}(u)du = 0$. This yields

$$\| S_{N,2n}(f) - f \|_p = \| S_{N,2n}(\bar{f}) - \bar{f} \|_p \leq C N^{-k} \| \bar{f}^{(k)} \|_p .$$

The proof is completed by observing that $\bar{f}^{(k)} = f^{(k)}$ for $k \in \mathbb{N}$ and $\| \bar{f} \|_p \leq C \| f \|_p$.

In view of this lemma one can apply Theorem 3 giving ($k'=2n+1$, $k=0$, $m=2n$)

THEOREM 4: Let f and $S_{N,2n}(f)$ be given as in Lemma 4, part b) and let $\psi(t)$ be a positive monotonely increasing function on $(0,1]$ satisfying (4.2). Then the following assertions are equivalent for $1 \leq q \leq \infty$:

i) $\left\{ \sum_{N=1}^{\infty} [\psi^{-1}(1/N) \| S_{N,2n}(f) - f \|_p]^q N^{-1} \right\}^{1/q} < \infty$,

ii) $f \in \mathrm{Lip}(\psi, q, 2n+1; p)$,

iii) $\left\{ \sum_{N=1}^{\infty} [\psi^{-1}(1/N)\, N^{-2n} [S_{N,2n}(f)]_p^{2n}]^q N^{-1} \right\}^{1/q} < \infty$.

If in addition $\psi(t)$ satisfies ($\ell \leq 2n$)

$$\int_0^t \psi(u)\, u^{-\ell-1} du = O[t^{-\ell}\psi(t)], \quad \int_t^1 \psi^{-1}(u)\, u^{\ell-1} du = O[t^{\ell}\psi^{-1}(t)],$$

then these assertions are further equivalent to

$$f \in W_p^{\ell}(a,b),$$

(iv)

$$\left\{ \sum_{N=1}^{\infty} [N^{-\ell}\psi^{-1}(1/N) \| S_{N,2n}^{(\ell)}(f) - f^{(\ell)} \|_p]^q N^{-1} \right\}^{1/q} < \infty .$$

A similar chain of equivalences holds for the splines $S_{N,2n-1}(f)$ defined in Lemma 4, part a).

This theorem can be considered not only as a collection of various approximation statements on the spline operators $S_{N,2n}(f)$ and $S_{N,2n-1}(f)$ but also as a characterization theorem of the (periodic) generalized Lipschitz spaces by periodic interpolating splines. The corresponding characterizations with an element of best approximation replacing the interpolating spline have been given in [19]. In this connection we reemphasise the fact that all generalized Lipschitz spaces are characterized in one and the same setting in contrast to the case of approximation by trigonometric polynomials where best approximation and linear operators on the one hand, and saturation and non-optimal approximation order on the other hand, must be treated separately.

The author thanks Professor Dr.P.L.Butzer and Dozent Dr.E.Görlich for the critical reading of the manuscript.

References

1 Ahlberg, J.H. and Nilson, E.N., Polynomial splines on the real line. J. Approximation Theory 3 (1970), 398-409.

2 Ahlberg, J.H., Nilson, E.N. and Walsh, J.L., Best approximation and convergence properties of higher-order spline approximation. J. Math. Mech. 14 (1965), 231-244.

3 Ahlberg, J.H., Nilson, E.N. and Walsh, J.L., The Theory of Splines and their applications, Academic Press, New York 1967.

4. Birkhoff, G., Local spline approximation by moments, J. Math. Mech. 16, (1967), 987-990.

5 Birkhoff, G. and deBoor, C., Error bounds for spline interpolation J. Math. Mech. 13 (1964), 827-836.

6 deBoor, C., On local spline approximation by moments, J. Math. Mech. 17 (1968), 729-735.

7 deBoor, C., On the convergence of odd-degree spline interpolation. J. Approximation Theory 1 (1968), 452-463.

8 Butzer, P.L. and Berens, H., Semi-Groups of operators and approximation, Springer, New York 1967.

9. Butzer, P.L. and Scherer, K., On the fundamental approximation theorems of D.Jackson, S.N.Bernstein and theorems of M.Zamansky and S.B.Stečkin, Aequationes Math. 3 (1969), 170-185.

10 Butzer, P.L. and Scherer, K., Jackson and Bernstein-type inequalities for families of commutative operators in Banach spaces. J. Approximation Theory 5, (1972), 308-342.

11 Demko, S. and Varga, R.S., Extended L_p-error bounds for spline and L-spline interpolation (to appear).

12 Golomb, M., Approximation by periodic spline interpolants on uniform meshes, J. Approximation Theory 1 (1968), 26-65.

13 Hall, C.A., On error bounds for spline interpolation, J. Approximation Theory 1 (1968), 209-218.

14 Hardy, G.H., Littlewood, J.E. and Polya, G., Inequalities, Cambridge University Press, Cambridge, 1934.

15 Hedstrom, G.W. and Varga, R.S., Application of Besov spaces to spline approximation, J. Approximation Theory 4 (1971), 295-327.

16 Isumi, M., Izumi, S. and Petersen, G.M., On Hardy's inequality and its generalization, Tôhoku Math. J. 21 (1969), 601-613.

17 Johnen, H., Inequalities connected with the moduli of smoothness. Mat. Vesnik 9 (24) (1972), 289-303.

18 Meir, A. and Sharma, A., Convergence of a class of interpolatory splines, J. Approximation Theory 1 (1968), 243-250.

19 Scherer, K., Characterization of generalized Lipschitz classes by best approximation with Splines, To appear in SIAM J. of Num. Analysis.

20 Scherer, K., A comparison approach to direct theorems for polynomial spline approximation. To appear in the Proceedings of 'Conference on Theory of Approximation', Poznan, 22-26 -8 -1972

21 Schultz, M.H., Error bounds for polynomial spline interpolation. Math. Comp. 24 (1970), 507-515.

22 Sharma, A. and Meir, A., Degree of approximation of spline interpolation, J. Math. Mech. 15 (1966), 759-767.

23 Subbotin, Yu.N., On a linear method of approximation of differentiable functions. Math. Notes 7 (1970), 256-260.

24 Swartz, B.K., $O(h^{k-j}\,\omega(D^k f, h))$ bounds on some spline interpolation errors, Los Alamos Scientific Laboratory Report LA-4477, 1970.

25 Swartz, B.K. and Varga, R.S., Error bounds for spline and L-spline interpolation, J. Approximation Theory 6 (1972), 6-49.

26 Timan, A.F., Theory of Approximation of functions of a Real variable, Pergamon Press, New York 1963.

27 Varga, R.S., Error bounds for spline interpolation, In 'Approximations with special emphasis on Spline functions' (ed. by I.J.Schoenberg). Academic Press, New York 1969, pp.367-388.

CHARACTERIZATION OF THE BARRELLED, d-BARRELLED AND σ-BARRELLED SPACES OF CONTINUOUS FUNCTIONS

J.Schmets

Abstract. Let X be a completely regular and Hausdorff topological space, υX its realcompactification and $C_c(X)$ [resp. $C_{\mathfrak{P}}(X)$] the space of all continuous functions on X equipped with the topology of compact convergence [resp. of uniform convergence on a convenient family of subsets of X]. We give here the generalization to the case of the space $C_{\mathfrak{P}}(X)$, of the characterization of the barrelled, d-barrelled and σ-barrelled $C_c(X)$ spaces.

1. Let X be a completely regular and Hausdorff topological space and $\mathscr{C}(X)$ be the space of all continuous functions on X.

We denote by υX (to be read 'upsilon X') the realcompactification of X, which can be looked as the set of all characters of $\mathscr{C}(X)$, considered as a subspace of the weak algebraic dual of $\mathscr{C}(X)$; it is also a completely regular and Hausdorff topological space.

In the sequel we shall as usual not only identify $x \in X$ with its associated Dirac measure (which makes X appear as a topological subspace of υX) but also $f \in \mathscr{C}(X)$ with its unique continuous extension to υX (which identifies $\mathscr{C}(X)$ and $\mathscr{C}(\upsilon X)$ as algebras).

If E is a locally convex topological vector space we denote by E^* its (topological) dual and by E^*_s its weak dual.

2. A subset B of X defines a semi-norm of uniform convergence

$$\| f \|_B = \sup_{x \in B} | f(x)|, \ \forall f \in \mathscr{C}(X),$$

on $\mathscr{C}(X)$ if and only if every $f \in \mathscr{C}(X)$ is bounded on B; such sets are called <u>bounded</u> in X.

With Noureddine, to define locally convex topological vector spaces out of $\mathcal{C}(X)$ by consideration of the set of the semi-norms of uniform convergence on the elements of a family $\mathcal{P}$ of bounded sets in X, we shall only consider the families $\mathcal{P}$ such that $\bigcup_{B\in\mathcal{P}} B \supset X$. Thus we define the space $C_{\mathcal{P}}(X)$ and we may as well suppose that $\mathcal{P}$

-contains $\overline{B}$ if it contains B,

-contains $B_1 \cup B_2$ if it contains B_1 and B_2,

-contains B' if it contains B with $B' \subset B$,

since this gives an equivalent system of semi-norms on $\mathcal{C}(X)$.

For instance the families

$$\mathcal{A}(X),\ \mathcal{K}(X) \text{ and } \mathcal{B}(X)$$

determined respectively by all finite, compact and bounded subsets in X are convenient and for them we shall go back to the usual notations:

$$C_s(X),\ C_c(X) \text{ and } C_b(X).$$

3. Let us recall some results.

a) As far as the barrelledness is concerned we have this situation.

In 1954, Nachbin [4] and Shirota [6] proved the following result.

THEOREM 1. <u>The space</u> $C_c(X)$ <u>is barrelled if and only if every bounded set in</u> X <u>is relatively compact.</u>

Later on Buchwalter [1] introduced the space μX, which is the smallest subspace of υX which contains X and where every bounded subset is relatively compact, and described the barrelled space associated to $C_c(X)$. If E is a locally convex topological vector space, the barrelled space associated to E is the linear space E equipped with the coarsest locally convex topology which at the same time is stronger than the one of E and makes E barrelled.

THEOREM 2. The barrelled space associated to $C_c(X)$ is $C_c(\mu X)$.

Finally Buchwalter and myself [2] noticed that $C_c(\mu X)$ is even the barrelled space associated to $C_s(X)$. So the general result can be stated as follows.

THEOREM 3. For every locally convex topology on $\mathscr{C}(X)$ finer than the one of $C_s(X)$ and coarser than the one of $C_c(\mu X)$, the barrelled space associated is $C_c(\mu X)$.

b) Now let us go to the case of d-barrelledness and σ-barrelledness.

A locally convex topological vector space E is

-d-barrelled if every countable union of equicontinuous subsets of E^* is equicontinuous if it is weakly bounded.

-σ-barrelled if every denumerable and bounded subset of E^*_s is equicontinuous.

We have given previously [5] the characterization of the d-barrelled and σ-barrelled $C_c(X)$ spaces. Our purpose is to do it in the general case of the $C_{\mathfrak{P}}(X)$ spaces.

4. The key is given in the following considerations.

Let $\mathfrak{P}$ be a family of bounded subsets in X as described in paragraph 2 and let us associate to it

$$\mathfrak{P}'' = \left\{ B'' \subset \upsilon X : \exists\, B \in \mathfrak{P} \text{ such that } B'' \subset \overline{B}^{\upsilon X} \right\}$$

and

$$X''_{\mathfrak{P}} = \bigcup_{B'' \in \mathfrak{P}''} B''.$$

Then it is easy to see that the spaces $C_{\mathfrak{P}}(X)$ and $C_{\mathfrak{P}''}(X''_{\mathfrak{P}})$ are equal [let us recall that we have identified $\mathscr{C}(X)$ with $\mathscr{C}(\upsilon X)$ and therefore with $\mathcal{C}(X''_{\mathcal{P}})$] since we have

$$\| f \|_B = \| f \|_{\overline{B}^{\upsilon X}}$$

for every $f \in \mathcal{C}(X)$ and every $B \subset \upsilon X$.

For the space $C_{\mathcal{P}''}(X''_{\mathcal{P}})$ we are in the situation

$$\mathcal{A}(X''_{\mathcal{P}}) \subset \mathcal{P}'' \subset \mathcal{K}(X''_{\mathcal{P}})$$

since every bounded subset in X is bounded in υX and hence relatively compact in υX: in fact every bounded subset in υX is easily seen to be precompact in the weak algebraic dual of $\mathcal{C}(X)$ and υX is complete in that space.

Therefore for every continuous linear functional $\mathcal{T}$ on $C_{\mathcal{P}}(X)$ there exists a smallest compact subset $[\mathcal{T}]$ of X'' such that

$$f \in \mathcal{C}(X),\ f = 0 \text{ on } [\mathcal{T}] \Rightarrow \mathcal{T}(f) = 0$$

and there exists a constant $C > 0$ such that

$$|\mathcal{T}(f)| \leq C \| f \|_{[\mathcal{T}]},\ \forall f \in C_{\mathcal{P}}(X).$$

This set $[\mathcal{T}]$ is called the <u>support of</u> $\mathcal{T}$. Moreover one has $[c\mathcal{T}] = [\mathcal{T}]$ for every complex number $c \neq 0$.

If $\mathcal{B}$ is a subset of $[C_{\mathcal{P}}(X)]^*$ let us call <u>support of</u> $\mathcal{B}$ the set

$$[\mathcal{B}] = \overline{\bigcup_{\mathcal{T} \in \mathcal{B}} [\mathcal{T}]}^{X''_{\mathcal{P}}}.$$

PROPOSITION 1. <u>If</u> $\mathcal{B} \subset [C_{\mathcal{P}}(X)]^*$ <u>is equicontinuous, then</u> $[\mathcal{B}]$ <u>is compact and belongs to</u> $\mathcal{P}''$.

<u>Proof</u>. In fact there exists $B \in \mathcal{P}$ and $C > 0$ such that

$$\sup_{\mathcal{T} \in \mathcal{B}} |\mathcal{T}(f)| \leq C \| f \|_B = C \| f \|_{\overline{B}^{\upsilon X}},\ \forall f \in C_{\mathcal{P}}(X),$$

and therefore one has $[\mathcal{T}] \subset \overline{B}^{\upsilon X}$ for every $\mathcal{T} \in \mathcal{B}$. So $[\mathcal{B}]$ is contained in $\overline{B}^{\upsilon X}$ which is compact and belongs to $\mathcal{P}''$.

PROPOSITION 2. <u>If $\mathfrak{B}$ is a bounded subset of $[C_{\mathcal{P}}(X)]_s^*$, then every real positive, lower semi-continuous and locally bounded function on $X''_{\mathcal{P}}$ is bounded on $[\mathfrak{B}]$.</u>

<u>In particular, $[\mathfrak{B}]$ is bounded in $X''_{\mathcal{P}}$.</u>

<u>Proof.</u> Let us suppose there exists a real positive, lower semi-continuous function f on $X''_{\mathcal{P}}$ which is unbounded on $[\mathfrak{B}]$.

Let us consider the sets

$$\Omega_n = \{ x \in X''_{\mathcal{P}} : |f(x)| > n \} \ (n \in \mathbb{N}).$$

Since f is lower semi-continuous they are open. Moreover since f is locally bounded, for every compact subset K of $X''_{\mathcal{P}}$, there exists $n \in \mathbb{N}$ such that $K \subset X''_{\mathcal{P}} \setminus \Omega_n$ and we have $\overline{\Omega}_n^{X''_{\mathcal{P}}} \downarrow \emptyset$. Finally, by construction, $\Omega_n \cap [\mathfrak{B}] \neq \emptyset$ for every $n \in \mathbb{N}$.

At that time it is possible to construct a subsequence n_k of n, a sequence $\mathcal{T}_k \in \mathfrak{B}$ and a sequence $f_k \in C_{\mathcal{P}}(X)$ such that

$$\Omega_{n_k} \cap [\mathcal{T}_k] \neq \emptyset, \ \mathcal{T}_k(f_k) = 1 \text{ and } f_k(X''_{\mathcal{P}} \setminus \Omega_{n_k}) = 0, \ \forall k.$$

Let us define $c_1 = 1$ and

$$c_k = k^2 - \sum_{j=1}^{k-1} c_j \mathcal{T}_k(f_j), \quad \forall k > 1.$$

Then, for every $B \in \mathcal{P}$, there exists $k_B \in \mathbb{N}$ such that

$$\overline{B}^{X''_{\mathcal{P}}} \subset X''_{\mathcal{P}} \setminus \Omega_{n_{k_B}}$$

because $\overline{B}^{X''_{\mathcal{P}}}$ is compact. Therefore $\| f_k \|_B = 0$ for every $k \geq n_{k_B}$ and so the sequence f_k converges to 0 in $C_{\mathcal{P}}(X)$ and for every sequence c'_k of complex numbers, $\sum_{k=1}^{\infty} c'_k c_k f_k$ is a Cauchy series in $C_{\mathcal{P}}(X)$.

Therefrom we deduce that for every sequence c_k' of complex numbers such that $\sum_{k=1}^{\infty} |c_k'| \leq 1$, the series $\sum_{k=1}^{\infty} c_k' c_k f_k$ converges in $C_{\mathcal{P}}(X)$. In fact, for $x_o \in X$, there exists k_o such that $x_o \notin \overline{\Omega}_{k_o}^{X''_{\mathcal{P}}}$. So there exists a neighbourhood of x_o in $X''_{\mathcal{P}}$ which is disjoint from $\overline{\Omega}_{k_o}^{X''_{\mathcal{P}}}$. In that neighbourhood, if $k_o < n_k$, the series is equal to $\sum_{j=1}^{k-1} c_j' c_j f_j$ and so converges to a continuous function on X.

So the set

$$\mathcal{K} = \left\{ \sum_{k=1}^{\infty} c_k' c_k f_k : \sum_{k=1}^{\infty} |c_k'| \leq 1 \right\}$$

is an absolutely convex and compact subset of $C_{\mathcal{P}}(X)$.

Now we have

$$\sup_{g \in \mathcal{K}} \sup_{\mathcal{T} \in \mathcal{B}} |\mathcal{T}(g)| \geq \sup_n | \mathcal{T}_n (\sum_{j=1}^{n} \frac{c_j}{n} f_j | ,$$

and

$$\mathcal{T}_n (\sum_{j=1}^{n} \frac{c_j}{n} f_j) = n, \quad \forall n \in \mathbb{N} .$$

Hence we obtain a contradiction since every bounded subset of $[C_{\mathcal{P}}(X)]_s^*$ is also bounded on the absolutely convex and compact subsets of $C_{\mathcal{P}}(X)$.

PROPOSITION 3. <u>If $\mathcal{B}$ is a bounded subset of $[C_{\mathcal{P}}(X)]_s^*$ and if $[\mathcal{B}]$ belongs to $\mathcal{P}''$, then $\mathcal{B}$ is equicontinuous.</u>

<u>Proof.</u> The set $\mathcal{A}$ of the linear functionals $\mathcal{T}_{[\mathcal{B}]}$ defined on $C_c([\mathcal{B}])$ by

$$\mathcal{T}_{[\mathcal{B}]}(f) = \mathcal{T}(\tilde{f}), \quad \forall f \in C_c([\mathcal{B}]) ,$$

where $\mathcal{T} \in \mathcal{B}$ and where $\tilde{f}$ is a continuous extension of f to X, is of course bounded in $[C_c([\mathcal{B}])]^*_s$. Since $C_c([\mathcal{B}])$ is a Banach space, $\mathcal{A}$ is equicontinuous. Hence the conclusion by use of the continuous extension theorem again.

5. From these results it is possible to deduce in the same way as in [5] the following results.

THEOREM 4.

a) <u>The space</u> $C_{\mathcal{P}}(X)$ <u>is barrelled if and only if every bounded subset in</u> $X''_{\mathcal{P}}$ <u>belongs to</u> $\mathcal{P}''$.

b) <u>The space</u> $C_{\mathcal{P}}(X)$ <u>is d-barrelled if and only if every bounded subset in</u> $X''_{\mathcal{P}}$ <u>which is the closure in</u> $X''_{\mathcal{P}}$ <u>of a countable union of elements of</u> $\mathcal{P}''$ <u>belongs to</u> $\mathcal{P}''$.

c) <u>The space</u> $C_{\mathcal{P}}(X)$ <u>is</u> σ<u>-barrelled if and only if every bounded subset in</u> $X''_{\mathcal{P}}$ <u>which is the closure in</u> $X''_{\mathcal{P}}$ <u>of a countable union of supports of</u> $\mathcal{T} \in [C_{\mathcal{P}}(X)]^*$ <u>belongs to</u> $\mathcal{P}''$.

<u>Proof of the necessity.</u>

a) If B is a bounded subset in $X''_{\mathcal{P}}$, it may be considered as a subset of $[C_{\mathcal{P}}(X)]^*$; let us call it $\mathcal{B}$ at that time to avoid confusion. This set $\mathcal{B}$ is then a bounded subset of $[C_{\mathcal{P}}(X)]^*_s$, hence equicontinuous. Therefore $\overline{B}^{X''_{\mathcal{P}}} = [\mathcal{B}]$ is compact and belongs to $\mathcal{P}''$ by Proposition 1.

b) Let B''_n $(n \in \mathbb{N})$ be a sequence of elements of $\mathcal{P}''$ whose union is bounded in X''. Then the subsets $\mathcal{B}_n$ of $[C_{\mathcal{P}}(X)]^*$ they define are equicontinuous and their union $\mathcal{B}$ is a bounded subset of $[C_{\mathcal{P}}(X)]^*_s$, so, by Proposition 1,

$$[\mathcal{B}] = \overline{\bigcup_{n=1}^{\infty} [\mathcal{B}_n]}^{X''_{\mathcal{P}}} = \overline{\bigcup_{n=1}^{\infty} B''_n}^{X''_{\mathcal{P}}}$$

is compact and belongs to $\mathcal{P}''$.

c) Let $\mathcal{T}_n$ $(n \in \mathbb{N})$ be a bounded sequence of elements of $[C_{\mathcal{P}}(X)]^*_s$, such that $\overline{\bigcup_{n=1}^{\infty}[\tau_n]}^{X''_{\mathcal{P}}}$ be bounded in $X''_{\mathcal{P}}$. For every n, there exists a complex number $c_n \neq 0$ such that

$$|c_n \tau_n(f)| \leq \sup_{x \in [\tau_n]} |f(x)|, \qquad \forall f \in C_{\mathcal{P}}(X).$$

Since $[c_n \tau_n] = [\tau_n]$ for all n, the sequence $c_n \tau_n$ is bounded in $[C_{\mathcal{P}}(X)]_s^*$ since

$$\sup_n |c_n \tau_n(f)| \leq \sup_{x \in \bigcup_{n=1}^{\infty}[\tau_n]} |f(x)|, \qquad \forall f \in C_{\mathcal{P}}(X).$$

Since $C_{\mathcal{P}}(X)$ is σ-barrelled, the sequence $c_n \tau_n$ is equicontinuous and, by Proposition 1,

$$\left[\bigcup_{n=1}^{\infty} \tau_n\right] = \overline{\bigcup_{n=1}^{\infty}[\tau_n]}^{X''_{\mathcal{P}}}$$

is compact and belongs to $\mathcal{P}''$.

<u>Proof of the sufficiency</u>.

a) If $\mathcal{B}$ is a bounded subset of $[C_{\mathcal{P}}(X)]_s^*$, by Proposition 2, $[\mathcal{B}]$ is bounded in $X''_{\mathcal{P}}$, hence compact by hypothesis. The conclusion then follows from Proposition 3.

b) If $\mathcal{B}_n$ $(n \in \mathbb{N})$ are equicontinuous subsets of $[C_{\mathcal{P}}(X)]_s^*$ whose union $\mathcal{B}$ is bounded in that space, by Proposition 1, every $[\mathcal{B}_n]$ is compact and belongs to $\mathcal{P}''$ and, by Proposition 2, $\bigcup_{n=1}^{\infty}[\mathcal{B}_n] \subset [\mathcal{B}]$ is bounded in $X''_{\mathcal{P}}$. By hypothesis $[\mathcal{B}] = \overline{\bigcup_{n=1}^{\infty}[\tau_n]}^{X''_{\mathcal{P}}}$ is compact and belongs to $\mathcal{P}''$. Hence the conclusion by Proposition 3.

c) The proof goes on as in b).

<u>Remark</u>. These results give of course the way to obtain the d-barrelled and the σ-barrelled space associated to $C_{\mathcal{P}}(X)$.

References

1 Buchwalter, H., Parties bornées d'un espace topologique complètement régulier Sém. Choquet, 9 (1969-70)nº14, 15 p.

2 Buchwalter, H. and Schmets, J., Sur quelques propriétés de l'espace $C_s(T)$, J. Math. Pures et Appl., to appear in 1973.

3 De Wilde, M., Garnir, H.G. and Schmets, J., Analyse Fonctionnelle I, Birkhäuser Verlag, Basel, 1968.

4 Nachbin, L., Topological vector spaces of continuous functions, Proc. Nat. Acad. Sc. U.S.A., 40 (1954), 471-474.

5 Schmets, J., Espaces C(X) tonnelé, infra-tonnelé et σ-tonnelé, Ann. Coll. Int. Bordeaux, avril 1971, à paraitre.

6 Shirota, T., On locally convex vector spaces of continuous functions, Proc. Japan Acad., 30 (1954), 294-298.

CARDINAL SPLINE INTERPOLATION AND THE EXPONENTIAL EULER SPLINES

I.J.Schoenberg

1. Introduction. I wish to thank Professors Alladi Ramakrishnan and K.R.Unni for asking me to take part in this splendid international conference. Before leaving for Madras I sent to the Publishers the monograph [5] which deals with the problem of cardinal spline interpolation and its ramifications. Here I describe a few of the results of [5] and also the main results of the notes [6] and [2] that were written since [5] went to the Printers.

Let

$$(y_\nu), \quad (\nu = 0, \pm 1, \pm 2, \dots), \tag{1.1}$$

be a prescribed biinfinite sequence of real or complex numbers. A problem of cardinal interpolation is to find a function $f(x)$ in some prescribed function space such that

$$f(\nu) = y_\nu \quad \text{for all integer } \nu. \tag{1.2}$$

The analogue of the Lagrange interpolation formula for this problem is the so-called cardinal series

$$f(x) = \sum_{-\infty}^{\infty} y_\nu \frac{\sin \pi (x-\nu)}{\pi (x-\nu)}. \tag{1.3}$$

The piecewise linear analogue of (1.3) is well known: If $M(x)$ is the so-called roof-function such that $M(x) = x + 1$ in $[-1,0]$, $M(x) = 1 - x$ in $[0,1]$, and $M(x) = 0$ elsewhere, then

$$S_1(x) = \sum_{-\infty}^{\infty} y_\nu M(x-\nu) \tag{1.4}$$

Sponsored by the United States Army under Contract No.DA-31-124-ARO-D-462.

is clearly the piecewise linear interpolant of the sequence (1.1). The purpose of <u>cardinal spline interpolation</u> is to bridge the gap between the linear spline (1.4) and the cardinal series (1.3). It aims at retaining some of the sturdiness and simplicity of (1.4), at the same time capturing some of the smoothness and sophistication of (1.3).

We need some notations. Let n be a natural number and let

$$\mathcal{S}_n = \{S(x)\} \tag{1.5}$$

denote the class of functions satisfying the following two conditions:

$$S(x) \in C^{n-1}(\mathbb{R}), \tag{1.6}$$

$$S(x) \in \pi_n \text{ in each of the intervals } (\nu, \nu+1), \text{ for all integer } \nu. \tag{1.7}$$

The functions $S(x)$ are called <u>cardinal spline functions</u> of degree n. We also need the class

$$\mathcal{S}^*_n = \{S(x);\ S(x+\tfrac{1}{2}) \in \mathcal{S}_n\}. \tag{1.8}$$

Its elements are also cardinal spline functions of degree n except that their 'knots' are now halfway between the integers. Finally, let

$$\tilde{\mathcal{S}}_n = \begin{cases} \mathcal{S}_n \text{ if } n \text{ is odd}, \\ \mathcal{S}^*_n \text{ if } n \text{ is even}. \end{cases} \tag{1.9}$$

Let $n > 1$. It is readily seen that for an arbitrary sequence (1.1) there are infinitely many functions $f(x)$ satisfying (1.2) such that $f(x) \in \mathcal{S}_n$, and also infinitely many satisfying (1.2) and $f(x) \in \mathcal{S}^*_n$. Thus <u>existence</u> never fails. The main problem is to insure <u>uniqueness</u> by means of appropriate 'boundary conditions' at infinity. In this direction a main result of [5, Lecture 4] is as follows.

THEOREM 1. *If* (1.1) *is such that*

$$y_\nu = O(|\nu|^\gamma) \quad \text{as} \quad \nu \longrightarrow \pm\infty, \text{ for some } \gamma \geq 0, \tag{1.10}$$

then there is a unique

$$S(x) \in \tilde{\mathcal{S}}_n \tag{1.11}$$

such that

$$S(x) = O(|x|^\gamma) \text{ as } x \longrightarrow \pm\infty \tag{1.12}$$

and satisfying

$$S(\nu) = y_\nu \quad \text{for all } \nu. \tag{1.13}$$

We refer to sequences and splines satisfying (1.10) and (1.12), respectively, as being of *power growth*. If the exponent $\gamma = 0$ we have the case of *bounded* sequences and splines. An example of such a sequence is

$$y_\nu = (-1)^\nu \quad \text{for all } \nu. \tag{1.14}$$

By Theorem 1 there is a unique interpolating spline that we denote by $\mathcal{E}_n(x)$; this function is uniquely characterized by the requirement of being bounded on $\mathbb{R}$ and satisfying

$$\mathcal{E}_n(x) \in \tilde{\mathcal{S}}_n, \text{and } \mathcal{E}_n(\nu) = (-1)^\nu \text{ for all } \nu. \tag{1.15}$$

This is a well-known function that plays an important role in approximation theory. We call $\mathcal{E}_n(x)$ the *Euler spline of degree* n because its polynomial components (i.e. its restrictions in intervals between consecutive knots) are, up to trivial changes of scale, identical with the Euler polynomial $E_n(x)$.

2. *The exponential Euler splines.* No sooner have we stated Theorem1 than the following interpolation problem arises that exceeds its competence. Let t be a real or complex number $\neq 0$ and $\neq 1$.

We wish to interpolate the geometric progression

$$y_\nu = t^\nu \quad \text{for all } \nu \tag{2.1}$$

by a spline $S(x)$ of the class $\mathcal{S}_n$, where n is even or odd. Here Theorem 1 is applicable only if n is odd, and if $|t| = 1$, when the sequence (2.1) is bounded.

To arrive at a general solution we follow Euler and introduce the polynomials $\Pi_n(t)$ defined by the generating function

$$\frac{t-1}{t-e^z} = \sum_0^\infty \frac{\Pi_n(t)}{(t-1)^n} \cdot \frac{z^n}{n!} . \tag{2.2}$$

We find that $\Pi_0(t) = \Pi_1(t) = 1$, $\Pi_2(t) = t + 1$, $\Pi_3(t) = t^2 + 4t + 1$, $\Pi_4(t) = t^3 + 11t^2 + 11t + 1$, $\Pi_5(t) = t^4 + 26t^3 + 66t^2 + 26t + 1$. The polynomial $\Pi_n(t)$ is a monic reciprocal polynomial of degree $n - 1$, having integer coefficients. This polynomial is also the subject of extensive combinatorial investigations (See [1]). Frobenius has shown in 1910 that its zeros are all simple, negative and, of course, reciprocal in pairs. We conclude that if $n = 2m-1$ is odd, then

$$\Pi_{2m-1}(-1) \neq 0 . \tag{2.3}$$

We call $\Pi_n(t)$ the Euler-Frobenius polynomial of degree $n - 1$.

Again following Euler we define an Appell sequence of polynomials $A_n(x; t)$ by

$$\frac{t-1}{t-e^z} e^{xz} = \sum_0^\infty \frac{A_n(x;t)}{n!} z^n . \tag{2.4}$$

In terms of

$$a_n(t) = \frac{\Pi_n(t)}{(t-1)^n} \tag{2.5}$$

we may write

$$A_n(x;t) = x^n + \binom{n}{1} a_1(t)\, x^{n-1} + \binom{n}{2} a_2(t)\, x^{n-2} + \\ + \ldots + a_n(t). \quad (2.6)$$

<u>We assume the parameter</u> t <u>to be such that</u> $\Pi_n(t) \neq 0$, hence by (2.5), (2.6), that

$$A_n(0\,;\,t) \neq 0 . \quad (2.7)$$

We define the <u>exponential Euler spline</u> $S_n(x\,;\,t)$ of degree n as follows:
<u>We set</u>

$$S_n(x\,;t) = \frac{A_n(x\,;t)}{A_n(0\,;t)} \quad \text{in the interval } 0 \le x < 1 \quad (2.8)$$

<u>and extend its definition to all real</u> x <u>by the functional equation</u>

$$S_n(x+1;t) = t\, S_n(x;t). \quad (2.9)$$

<u>The structure of the polynomials</u> (2.6) <u>is such that the resulting function</u> $S_n(x;t)$ <u>has</u> $n-1$ <u>continuous derivatives for all real</u> x, <u>hence</u>

$$S_n(x;t) \in \mathcal{S}_n . \quad (2.10)$$

For details we refer to [4, §§1 to 4].

From (2.8) $S_n(0;t) = 1$ and now (2.9) shows that

$$S_n(\nu;\,t) = t^\nu \quad \text{for all integer } \nu. \quad (2.11)$$

<u>The spline</u> $S_n(x;t)$ <u>is therefore an interpolant of the sequence</u> (2.1).

By (2.3) we may choose the value $t = -1$ if $n = 2m - 1$ is odd; now the sequence (2.1) reduces to (1.14) and therefore

$$S_{2m-1}(x;-1) = \mathcal{E}_{2m-1}(x) . \tag{2.12}$$

Strange as it may seem, the exponential Euler spline $S_n(x;t)$ seems to have appeared for the first time in [4], except for the case $t = -1$, when the functions

$$S_{2m-1}(x;-1) \text{ and } S_{2m}(x + \tfrac{1}{2};-1)/S_{2m}(\tfrac{1}{2};-1)$$

are identical with the Euler splines $\mathcal{E}_{2m-1}(x)$ and $\mathcal{E}_{2m}(x)$, respectively, and go back to Hermite and Sonin.

For the remainder of this note we shall assume that

$$n = 2m - 1 \tag{2.13}$$

is odd. We do this because (2.3) allows us to choose t anywhere on the unit circle giving us all <u>bounded</u> exponential Euler splines. Accordingly, let

$$t = e^{iu}, \quad -\pi \leq u \leq \pi . \tag{2.14}$$

We shall describe below several applications of the following

<u>THEOREM 2</u>. <u>Assuming</u> (2.13) <u>and</u> (2.14) <u>to hold, the following estimate is valid</u>:

$$|e^{iux} - S_{2m-1}(x;e^{iu})| \leq 3\left(\frac{|u|}{\pi}\right)^{2m} \text{ for all real } x . \tag{2.15}$$

For a proof we refer to [5, Lecture 3, § 6]. For the case when $S_{2m-1}(x;e^{iu})$ is a periodic function of x, which happens if and only if e^{iu} is a root of unity, M. Golomb was the first to derive an estimate of the type (2.15). S.D. Silliman [7] recognized its role in the approximation of Fourier transforms.

3. The approximate evaluation of Fourier Transforms. We consider the Fourier transform

$$F(u) = \int_{-\infty}^{\infty} e^{iux} f(x)dx \tag{3.1}$$

which is to be evaluated numerically. We assume that $f(x)$ has the properties:

$$f^{(\nu)}(x) \ (\nu = 0,1,\ldots, 2m-1) \text{ are absolutely continuous, and} \longrightarrow 0 \text{ as } x \longrightarrow \pm\infty, \tag{3.2}$$

$$f^{(2m)}(x) \in L_1(\mathbb{R}). \tag{3.3}$$

We proceed as follows. We choose a step $h > 0$ and consider the unique bounded cardinal spline function $S_h(x)$ of the degree $2m-1$, with knots at the points νh $(-\infty < \nu < \infty)$, that interpolates $f(x)$ at these knots. As the approximation of (3.1) we take the transform of the interpolant, hence

$$F_h(u) = \int_{-\infty}^{\infty} e^{iux} S_h(x)dx. \tag{3.4}$$

A closer analysis [5, Lecture 10, §§ 1 and 2] reveals that

$$F_h(u) = \frac{\psi_{2m}(hu)}{\phi_{2m}(hu)} \cdot h \sum_{\nu=-\infty}^{\infty} f(\nu h)e^{i\nu uh}, \tag{3.5}$$

where

$$\psi_{2m}(u) = \left(\frac{2 \sin \frac{u}{2}}{u}\right)^{2m} \tag{3.6}$$

and

$$\phi_{2m}(u) = \sum_{j=-\infty}^{\infty} \psi_{2m}(u + 2\pi j). \tag{3.7}$$

This last function is a periodic function of period 2π, in fact a cosine polynomial of order $m - 1$ with rational coefficients. Some explicit expressions are

$$\phi_4(u) = \frac{1}{3}(2 + \cos u), \quad \phi_6(u) = \frac{1}{60}(33+26 \cos u + \cos 2u). \tag{3.8}$$

The right side of (3.5) is easy to evaluate, e.g. for cubic or for quintic splines, except for the series (the Riemann sum for (3.1)) which requires some such device as the so-called fast Fourier transform.

We now set $K(x) = e^{iux} - S_{2m-1}(x;e^{iu})$ and apply repeated integrations by parts to the integral

$$Rf = \int_{-\infty}^{\infty} K(x) \, f^{(2m)}(x)\,dx .$$

If we combine the final result with Theorem 2, after a preliminary change of step, we obtain

THEOREM 3. (S.D.Silliman). *The value of* u *having been chosen, if we select* $h > 0$ *such that*

$$-\frac{\pi}{h} \leq u \leq \frac{\pi}{h}, \tag{3.9}$$

then

$$|F(u) - F_h(u)| \leq 3\left(\frac{h}{\pi}\right)^{2m} \|f^{(2m)}\|_{L(\mathbb{R})} \tag{3.10}$$

Silliman originally derived this result in a different way. See [7] also for numerical examples. For the present approach see [5, Lecture 10, §§ 1,2] .

4. *The convergence of cardinal spline interpolation as the degree of the spline tends to infinity; Sufficient conditions.* As a second application of the exponential Euler splines we consider the following problem. Let $f(x)$ be a bounded function from $\mathbb{R}$ to $\mathbb{C}$, and let $S_{2m-1}(x)$ be the unique bounded cardinal spline of degree $2m-1$ that interpolates $f(x)$ at the integers. *Under what conditions do we have*

$$\lim_{m\to\infty} S_{2m-1}(x) = f(x) \text{ uniformly for all real } x \ ? \quad (4.1)$$

The simplest function $f(x)$ such that (4.1) holds is $f(x) = \cos \pi x$. Indeed, at the integers this function produces the sequence (1.14), while on the other hand we have the well known Fourier series

$$\mathcal{E}_n(x) = \sum_{r=1}^{\infty} (2r-1)^{-n-1} \cos(2r-1)x / \sum_{r=1}^{\infty} (2r-1)^{-n-1},$$

valid for all integers n. Since $S_{2m-1}(x) = \mathcal{E}_{2m-1}(x)$ in our case, we see that

$$\mathcal{E}_{2m-1}(x) = \cos \pi x + O(3^{-2m}) \text{ as } m \to \infty . \quad (4.2)$$

In contrast, for the choice $f(x) = \sin \pi x$ we find that $S_{2m-1}(x)$ vanishes identically and (4.1) does not hold. For further results and references see [5, Lecture 9, § § 3,4,5] and [6] . Here we discuss the main result of [6] .

<u>THEOREM 4</u>. <u>Let</u> $f(x)$ <u>be an entire function defined by the Fourier - Stieltjes integral</u>

$$f(x) = \int_{-\pi}^{\pi} e^{iux} \, d\,\alpha(u) , \quad (4.3)$$

<u>where</u> $\alpha(u)$ <u>is a function of bounded variation in</u> $[-\pi, \pi]$ <u>such that</u>

$$\alpha(-\pi + 0) - \alpha(-\pi) = \alpha(\pi) - \alpha(\pi - 0) . \quad (4.4)$$

<u>If</u> $S_{2m-1}(x)$ <u>is the cardinal spline interpolant of</u> $f(x)$, <u>then</u>

$$\lim_{m\to\infty} S_{2m-1}(x) = f(x) \text{ uniformly for all real } x \quad (4.5)$$

We present no proofs in this expository note, but the application of Theorem 2 is so simple and direct that we must at least sketch it. Let

$$\alpha_0(u) = \begin{cases} \alpha(-\pi + 0) & \text{if } u = -\pi, \\ \alpha(u) & \text{if } -\pi < u < \pi, \\ \alpha(\pi - 0) & \text{if } u = \pi, \end{cases} \tag{4.6}$$

and let A be the common value of the equal jumps of $\alpha(u)$ described by (4.4). We claim that <u>the interpolating spline may be expressed by</u>

$$S_{2m-1}(x) = \int_{-\pi}^{\pi} S_{2m-1}(x;e^{iu})\,d\,\alpha_0(u) + 2A\,\mathcal{E}_{2m-1}(x). \tag{4.7}$$

This is indeed an element of $\mathcal{S}_{2m-1}$, while for integer $x = \nu$ we find that

$$S_{2m-1}(\nu) = \int_{-\pi}^{\pi} S_{2m-1}(\nu;e^{iu})\,d\,\alpha_0(u) + 2A(-1)^\nu = \int_{-\pi}^{\pi} e^{i\nu u}d\alpha_0(u) + 2A(-1)^\nu$$

$$= \int_{-\pi}^{\pi} e^{i\nu u}\,d\,\alpha(u) = f(\nu),$$

and our claim is established. Subtracting the relation (4.7) from

$$f(x) = \int_{-\pi}^{\pi} e^{iux}\,d\,\alpha_0(u) + 2A\cos\pi x$$

and applying the triangle inequality we find that

$$|f(x) - S_{2m-1}(x;e^{iu})| \leq \int_{-\pi}^{\pi} |e^{iux} - S_{2m-1}(x;e^{iu})| \cdot |d\,\alpha_0(u)| +$$

$$+ 2|A|\,|\cos\pi x - \mathcal{E}_{2m-1}(x)|,$$

and using Theorem 2 we find that

$$\| f(x) - S_{2m-1}(x;e^{iu}) \|_\infty \leq \int_{-\pi}^{\pi} (|u|/\pi)^{2m}\,|d\,\alpha_0(u)| +$$

$$+ 2|A|\,\|\cos\pi x - \mathcal{E}_{2m-1}(x)\|_\infty .$$

The right side $\to 0$ as $m \to \infty$ because of the continuity of $\alpha_0(u)$ at $\pm\pi$ and the relation (4.5) is established.

5. <u>Necessary conditions.</u> The conditions to be described were given in [2] . The main tool is a cardinal spline analogue of the follow-

ing classical theorem of the brothers Markov (See [3]).

THEOREM 5 (A.A. and W.Markov). We use the supremum-norm in the interval $[-1,1]$. If

$$P(x) \in \pi_n \quad \text{and} \quad \|P\| \leq 1, \tag{5.1}$$

then

$$\|P^{(\nu)}\| \leq \|T_n^{(\nu)}\| = T_n^{(\nu)}(1), \text{ for } \nu = 1,\ldots,n, \tag{5.2}$$

where $T_n(x)$ denotes the Chebyshev polynomial. Here (5.2) gives the best constants, because $T_n(x)$ satisfies (5.1).

The cardinal spline analogue is

THEOREM 6. Now we use the supremum-norm on $\mathbb{R}$. If

$$S(x) \in \tilde{\mathcal{S}}_n \quad \text{and} \quad \|S\| \leq 1 \tag{5.3}$$

then

$$\|S^{(\nu)}\| \leq \|\mathcal{E}_n^{(\nu)}\| = \begin{cases} \mathcal{E}_n^{(\nu)}(0) & \text{if } \nu \text{ is even} \\ \mathcal{E}^{(\nu)}(\tfrac{1}{2}) & \text{if } \nu \text{ is odd} \end{cases} \quad (\nu = 1,\ldots,n) \tag{5.4}$$

Here (5.4) gives the best constants because $\mathcal{E}_n(x)$ satisfies (5.3).

From this we derive in [2] our last result.

THEOREM 7. If $f(x)$ is bounded on $\mathbb{R}$ and

$$\lim_{m\to\infty} S_{2m-1}(x) = f(x) \text{ uniformly on } \mathbb{R} \tag{5.5}$$

then $f(x)$ is the restriction to $\mathbb{R}$ of an entire function $f(z)$ satisfying the inequality

$$|f(z)| \leq A\, e^{\pi |\mathrm{Im}\, z|}, \quad (z \in \mathbb{C}), \text{ (A constant)}. \tag{5.6}$$

Let $f(x)$ be bounded on $\mathbb{R}$. We have derived for (5.5) to hold, the sufficient conditions (4.3), (4.4), and also the necessary condition (5.6). Observe that (4.3) easily implies (5.6), since

$$|f(x+iy)| = \left| \int_{-\pi}^{\pi} e^{iux} e^{-uy} d\alpha(u) \right| \leq e^{\pi|y|} \int_{-\pi}^{\pi} |d\alpha(u)|.$$

However, there is yet a wide gap between our sufficient conditions and our necessary condition. I hope that some day this gap will be closed, and two young mathematicians, A.Cavaretta and A.M.Fink are trying to do that.

References.

1 Foata, D. and Schützenberger,M.-P.,Téorie géometrique des polynômes Eulérians,Lecture Notes in Math.,No.138,Springer, Berlin,1970

2 Richards,F.B. and Schoenberg,I.J.,Notes on spline functions IV. A cardinal spline analogue of the theorem of the brothers Markov. MRC T.S.Report # 1330,April 1973,Madison,Wisconsin. To appear in Israel J. of Math.

3 Schaeffer,A.C. and Duffin,R.J., On some inequalities of S. Bernstein and W.Markoff for derivatives of polynomials. Bull. Amer.Math.Soc., 44(1938), 289-297.

4 Schoenberg,I.J., Cardinal interpolation and spline functions IV. The exponential Euler splines. in 'Linear Operators and Approximation', edited by P.L.Butzer, J.-P.Kahane and B.Sz.-Nagy, Proc.of the Oberwolfach Conf. Aug. 14-22,1971,ISNM,Vol. 20 (1972), 382-404 .

5 _______________, Cardinal spline interpolation. CBMS Regional Conference Monograph No.12, 125 pages, SIAM, Philadelphia 1973, To appear.

6 Schoenberg,I.J., Notes on spline functions III. On the convergence of the interpolating cardinal splines as their degree tends to infinity. MRC T.S.Report # 1326, April 1973,Madison, Wisconsin, To appear in Israel J. of Math.

7 Silliman, S.D., The numerical evaluation by splines of the Fourier transform and the Laplace transform. Ph.D. Thesis, University of Wisconsin-Madison, June 1971, MRC T.S.Report # 1183, January 1972, Madison, Wisconsin

APPROXIMATION OF ANALYTIC FUNCTIONS IN HAUSDORFF METRIC

Bl. Sendov

Our aim in this paper is to estimate the best approximation with algebraic polynomials relative to Hausdorff distance of analytic functions which have singularities at the end-points of the interval of approximation. One typical example is the function $\varphi_\alpha(x) = (1 + x)^\alpha$, $0 < \alpha < 1$, examined on the interval $[-1,1]$. While the best approximation of φ_α relative to the uniform distance depends essentially upon α, the best approximation of φ_α, relative to Hausdorff distance can be estimated in order with $\left(\frac{\ln n}{n}\right)^2$ for every $\alpha > 0$. We shall prove that the last assertion can be carried over for wider class of analytic functions.

Similar questions have been examined concerning approximation with rational functions relative to the uniform distance [1], [2].

1. Hausdorff Distance

The definition and some properties of Hausdorff distance can be found in our review paper [3], but all necessary definitions and properties will be given here in order to make this paper self-contained.

Let $\rho(A,B)$ be an arbitrary Minkowski distance between two points A and B of the Euclidean plane R_2, and let $F \subset R_2$, $G \subset R_2$ be two closed point sets. The Hausdorff distance (or H-distance) $\tau(F,G)$ between F and G (generated by $\rho(\cdot,\cdot)$) is defined through

$$\tau(F,G) = \max \left\{ \max_{A\in F} \min_{B\in G} \rho(A,B), \max_{A\in G} \min_{B\in F} \rho(A,B) \right\} .$$

We shall specify $\rho(A,B)$ as

$$\rho(A,B) = \rho(A(x_1,y_1), B(x_2,y_2)) = \max (|x_1 - x_2|, |y_1 - y_2|) .$$

Immediate consequences of the definition of H-distance are the following two lemmas.

LEMMA 1. *If for every point* $A \in F$ *there exists such a point* $B \in G$ *that* $\rho(A,B) \leq \delta$ *and conversely for every* $A \in G$ *there is a point* $B \in F$ *with* $\rho(A,B) \leq \delta$, *then* $\tau(F,G) \leq \delta$.

LEMMA 2. *If there is an* $A \in F$, *such that for every* $B \in G$ *we have* $\rho(A,B) > \delta$, *then* $\tau(F,G) > \delta$.

Let further $\Delta = [a,b]$ be a closed interval in the real line R_1. We denote by F_Δ the class of all bounded and closed point subsets of R_2, which are convex with respect to the y-axis and whose projections on the x-axis coincide with Δ.

Let f be a bounded real function defined in Δ. The subset of R_2 which is the intersection of all elements $F \in F_\Delta$ containing the graph of f (considered also as a point set in R_2) is called the complete graph of f and is denoted by $\bar{f}$. The complete graph of a continuous function coincides with its graph. Hausdorff distance (or H-distance) $\tau(f,g)$ between two bounded real functions defined on Δ is by definition the H-distance between their complete graphs, i.e. $\tau(f,g) = \tau(\bar{f},\bar{g})$.

For H-distances between functions, Lemma 1 can be improved in the following way:

LEMMA 3. *If* f *and* g *are bounded real functions defined on the interval* Δ *and for every* $x_0 \in \Delta$ *there is a point* $(x_1,y_1) \in \bar{g}$ *so that*

$$\max\left(|x_0 - x_1|, |f(x_0) - y_1|\right) \leq \delta,$$

and a point $(x_2,y_2) \in \bar{f}$ *so that*

$$\max\left(|x_0 - x_2|, |g(x_0) - y_2|\right) \leq \delta,$$

then $\tau(f,g) \leq \delta$.

The following theorem connects the H-distance with the uniform distance between continuous functions [3].

THEOREM 1. *Let* f *and* g *be continuous functions defined on* Δ *and denote their modulii of continuity by* $\omega(f;\delta)$ *and* $\omega(g;\delta)$. *If* $R(f,g) = \max_{x\in\Delta} |f(x) - g(x)|$ *is the uniform distance between* f *and* g *and* $\omega(\delta) = \min\ (\omega(f;\delta), \omega(g;\delta))$, *then we have the inequality*

$$\tau(f,g) \leq R(f,g) \leq \tau(f,g) + \omega(\tau(f,g)).$$

The proof of the theorem follows from the definition of H-distance.

Let H_n be the set of algebraic polynomials of degree $\leq n$. If $p \in H_n$, then the same notation will be used also for the graph of p in Δ. Let further F be a closed subset of R_2. Consider the best approximation

$$E_{n,\tau}(F) = \inf_{p \in H_n} \tau(F,p).$$

The following assertion [3] holds.

THEOREM 2. *For every* $F \in F_\Delta$

$$E_{n,\tau}(F) = O\left(\frac{\ln n}{n}\right).$$

The exact constant before $\frac{\ln n}{n}$ in the former theorem has been found in [4] .

2. Auxiliary Assertions

LEMMA 4. *Let* C_1 *and* C_2 *be two non-negative numbers and* $C = \max\ (C_1, C_2)$. *For every positive integer* n *there exists an algebraic polynomial* $\psi_n(x)$ *of degree* n, *such that*

$$|\psi_{(n)}(x)| \leq C\,n^{-2} \quad \text{for } -1 \leq x \leq 1 - \nu_n,$$

$$\min_{1-\nu_n \leq x \leq 1} \psi_n(x) = -C_1 - C\,n^{-2},$$

$$\max_{1-\nu_n \leq x \leq 1} \psi_n(x) = C_2 + C\,m^{-2},$$

where

$$\nu_n = 32\left(\frac{\ln n}{n}\right)^2 . \tag{1}$$

Proof. Without any restrictions we can suppose that $C_1 \le C_2 = C$. Furthermore, the assertion is not trivial for $\nu_n < 2$, hence, we can suppose that

$$\frac{\ln n}{n} < \frac{1}{4} \tag{2}$$

Let us examine the polynomial

$$g_n(x) = C\, n^{-2}\,(1-x)\, T_{n-1}\left(\frac{x+\alpha}{1-\alpha}\right); \quad \alpha > 0,$$

where $T_m(x) = \cos(m \arc\cos x)$ is Chebyshev's polynomial. It is evident we have

$$|g_n(x)| \le C\, n^{-2} \text{ for } -1 \le x \le 1-2\alpha .$$

We shall prove that α can be chosen in the interval $[0, \frac{1}{4}\nu_n]$ in such a way that,

$$\xi(\alpha) = \max_{1-2\alpha \le x \le 1} g(x) = C_2 + C\, n^{-2}.$$

Really,

$$\xi(\alpha) \geq g_n(1-\alpha) = C\, n^{-2}\, \alpha\, T_{n-1}\left(\frac{1}{1-\alpha}\right)$$

$$\geq C\, n^{-2}\, \alpha\, T_{n-1}\left(\frac{1+\alpha/2}{1-\alpha/2}\right) \geq \frac{\alpha C}{2n^2}\left(\frac{1+\sqrt{\alpha/2}}{1-\sqrt{\alpha/2}}\right)^{n-1}$$

$$\geq \frac{\alpha C}{2n^2}\left(\frac{1-\sqrt{\alpha/2}}{1+\sqrt{\alpha/2}}\right) e^{n\sqrt{2\alpha}} .$$

When $\alpha = \frac{1}{4}\nu_n$ from (1), assuming (2), we obtain

$$\xi\left(\frac{1}{4}\nu_n\right) \geq C(\ln n)^2 \geq 2C.$$

Since $\xi(\alpha)$ is a continuous function of α, $\xi(\frac{1}{4}\nu_n) \geq 2C$, $\xi(0) = 0$, then there exists $\alpha_0 \in [0, \frac{1}{4}\nu_n]$, such that

$$\xi(\alpha_0) = \max_{1-2\alpha_0 \le x \le 1} g_n(x) = C_2 + C\, n^{-2}.$$

We fix $\alpha = \alpha_0$ in the definition of $g_n(x)$.

Now we shall prove that there exists $\beta \in [0, \frac{1}{4}\nu_n]$ such that

$$\eta(\beta) = \min_{1 \leq x \leq 1 + 2\beta} g(x) = - C_1 - C\, n^{-2} .$$

Really, $\eta(\beta)$ is a continuous function of β and $\eta(0) = 0$. On the other hand

$$\eta(\alpha_o) = - C\, n^{-2} \alpha_o \, T_{n-1} \left(\frac{1 + 2\alpha_o}{1 - \alpha_o} \right) ,$$

and since

$$T_{n-1} \left(\frac{1 + 2\alpha_o}{1 - \alpha_o} \right) \geqslant T_{n-1} \left(\frac{1}{1 - \alpha_o} \right) ,$$

then

$$\eta(\alpha_o) \leq - C_2 - C\, n^{-2} \leqslant - C_1 - C\, n^{-2} .$$

But then for some $\beta_o \in [\,0,\alpha_o\,] \subset [\,0,\frac{1}{4}\nu_n\,]$ we will have

$$\eta(\beta_o) = - C_1 - C\, n^{-2} .$$

It is seen immediately that if we take

$$\psi_n(x) = g_n((\,1 + \beta_o)x\,),$$

then ψ_n will satisfy all the conditions of the lemma, that completes its proof.

We need the following S.N.Bernstein's assertion, carried over without proof [5] .

<u>LEMMA 5.</u> <u>Let the function</u> f <u>possess Taylor's expansion</u>

$$f(x) = \sum_{k=0}^{\infty} a_k x^k ,$$

<u>then the expansion of the function in Chebyshev's polynomials is</u>

$$f(x) = \sum_{m=0}^{\infty} A_m T_m(x),$$

<u>where</u>

$$A_m = 2^{1-m} \sum_{k=0}^{\infty} 2^{-2k} \binom{m + 2k}{k} a_{m+2k}. \tag{3}$$

The lemma above was used by S.N.Bernstein for estimation of the best uniform approximation

$$E_n(f) = \inf_{p \in H_n} R(f,p)$$

of analytic functions with algebraic polynomials.

LEMMA 6. Let f be an analytic function in the circle $|z| < \rho$, $|f(z)| \leq M$ for $|z| \leq \rho$ and have real values on the interval $[-\rho, \rho]$, $\rho > 1$. Then for the best approximation of f with algebraic polynomials in the interval $[-1,1]$ the inequality (4) holds

$$E_n(f) \leq \frac{M}{\rho - 1} \left(\rho + \sqrt{\rho^2 - 1} \right)^{-n}. \tag{4}$$

Proof. From the conditions of the lemma there follows that

$$f(x) = \sum_{k=0}^{\infty} a_k x^k$$

where

$$|a_k| \leq M \rho^{-k}.$$

Using Lemma 5 for the coefficients (3) of the expansion of $f(x)$ in Chebyshev's polynomials, we obtain

$$|A_m| \leq \frac{M\rho^{-m}}{2^{m-1}} \sum_{k=0}^{\infty} \binom{m+2k}{k} (2\rho)^{-2k}. \tag{5}$$

Using the well-known identity

$$\sum_{k=0}^{\infty} \binom{m+2k}{k} x^k = \frac{1}{\sqrt{1-4x}} \left(\frac{2}{1+\sqrt{1-4x}} \right)^m,$$

that holds when $|x| < \frac{1}{4}$, from (5) we have

$$|A_m| \leq \frac{2M}{\sqrt{\rho^2 - 1}} \left(\rho + \sqrt{\rho^2 - 1} \right)^{-m} \tag{6}$$

But, on the other hand

$$E_n(f) \leq \sum_{m=n+1}^{\infty} |A_m|$$

and using (6), we obtain (4).

Let us denote by $AM_{[-1,1]}$ the set of functions f, analytic in the circle $|z| < 1$, $|f(z)| \leq M$ for $|z| \leq 1$ and having real values in the interval $[-1, 1]$.

LEMMA 7. Let $f \in AM_{[-1,1]}$ and denote

$$f_n(x) = f(\lambda_n x),$$

where

$$\lambda_n = \left[1 + 8\left(\frac{\ln n}{n}\right)^2 \right]^{-1} .$$

Then for the best uniform approximation of f_n with algebraic polynomials on the interval $[-1,1]$ the following inequality holds

$$E_n(f_n) \leqslant Mn^{-2} .$$

Proof. According to Lemma 6,

$$E_n(f_n) \leqslant \frac{M}{\lambda_n^{-1} - 1} \left(\lambda_n^{-1} + \sqrt{\lambda_n^{-2} - 1} \right)^{-n}$$

$$\leqslant \frac{Mn^2}{8(\ln n)^2} \left(1 + 4 \frac{\ln n}{n} \right)^{-n}$$

$$\leqslant \frac{Mn^2}{8(\ln n)^2} \left(1 + 4 \frac{\ln n}{n} \right) e^{-4 \ln n} ,$$

and for $n \geqslant 2$ we obtain

$$E_n(f_n) \leqslant Mn^{-2} .$$

Since for $n = 1$ the assertion is trivial, with this the lemma is completely proved.

3. Main Theorem

Now we begin the proof of the basic result of this paper.

THEOREM 3. If $f \in A\, M_{[-1,1]}$, then for the best approximation of f with algebraic polynomials relative to Hausdorff distance the following inequality holds

$$E_{n,\tau}(f) \leqslant \left(\frac{8 \ln n}{n}\right)^2 + \frac{5M}{n^2} .$$

Proof. Let us denote by $p_n(x)$ the polynomial from H_n of best uniform approximation of the function $f_n(x) = f(\lambda_n x)$, defined in Lemma 7. Hence, we have

$$|f_n(x) - p_n(x)| \leqslant Mn^{-2} \quad \text{for } |x| \leqslant 1. \tag{7}$$

According to Lemma 4., two polynomials $q_n(x)$ and $q_n^*(x)$ from

H_n can be found, such that for the polynomial

$$S_n(x) = q_n(x) + q_n^*(x) + p_n(x)$$

the inequalities (8) and (9) will hold:

$$\left| \sup_{x \in \Delta_i} f(x) - \max_{x \in \Delta_i'} S_n(x) \right| \leq 4\, Mn^{-2} ; \quad i = 1;2, \tag{8}$$

$$\left| \inf_{x \in \Delta_i} f(x) - \min_{x \in \Delta_i'} S_n(x) \right| \leq 4\, Mn^{-2} ; \quad i = 1,2; \tag{9}$$

where

$$\Delta_1 = [-1,-1+2\nu_n] , \quad \Delta_2 = [1-2\nu_n, 1] ,$$

$$\Delta_1' = [-1,-1+\nu_n] , \quad \Delta_2' = [1-\nu_n, 1] ,$$

and moreover

$$| q_n(x) + q_n^*(x) | \leq 4\, Mn^{-2} \quad \text{for } |x| \leq 1-\nu_n . \tag{10}$$

This choice of the polynomials q_n and q_n^* is possible, because

$$| f(1-\nu_n) - p_n(1) | \leq Mn^{-2},$$

and

$$| f(-1+\nu_n) - p_n(-1) | \leq Mn^{-2}.$$

Let us now estimate the H-distance between f and S_n, using Lemma 3.

Let x_0 be an arbitrary number belonging to the interval $[-1,1]$. We have to prove, that there exist $x_1 \in [-1,1]$ such that

$$\max(| x_0 - x_1 | , | f(x_0) - S_n(x_1) |) \leq 2\nu_n + 5\, Mn^{-2} ,$$

and a point $(x_2, y_2) \in \bar{f}$ such that

$$\max(| x_0 - x_2 | , | y_2 - S_n(x_0) |) \leq 2\nu_n + 5\, Mn^{-2}.$$

If $x_0 \in [-1,-1+2\nu_n]$ or $x_0 \in [1-2\nu_n, 1]$, the mentioned x_1 and a point (x_2, y_2) exist evidently according to the inequalities (8) and (9).

Let $x_0 \in (-1+2\nu_n , 1-2\nu_n)$. Then for x_1 we can take

$x_1 = x_0/\lambda_n$. Really,

$$x_1 = x_0/\lambda_n \in [-1 + \nu_n , 1 - \nu_n] ,$$

and according to (7) and (10)

$$|f(x_0) - S_n(x_1)| \leq |f(x_0) - p_n(x_1)| + 4 Mn^{-2}$$
$$\leq |f(x_0) - f_n(x_0/\lambda_n)| + 5 Mn^{-2} = 5 Mn^{-2}.$$

On the other hand

$$|x_1 - x_0| = |x_0| \left(\frac{1}{\lambda_n} - 1 \right) \leq \frac{\nu_n}{4} < \nu_n$$

For x_2 and $y_2 = f(x_2)$ we can choose

$$x_2 = \lambda_n x_0 \in [-1 + 2\nu_n , 1 - 2\nu_n] .$$

Then we have

$$|S_n(x_0) - f(\lambda_n x_0)| \leq |p_n(x_0) - f(\lambda_n x_0)| + 4 Mn^{-2}$$
$$\leq |f_n(x_0) - f(\lambda_n x_0)| + 5 Mn^{-2} = 5 Mn^{-2}$$

and moreover

$$|x_2 - x_0| = |x_0| (1 - \lambda_n) \leq \frac{\nu_n}{4 + \nu_n} < \nu_n .$$

With this the theorem's proof is completed.

From Theorem 3 and Theorem 1 the following corollary can be obtained:

THEOREM 4. If $f \in AM_{[-1,1]}$ and has the modulus of continuity $\omega(f;\delta)$, then for the best uniform approximation of f with algebraic polynomials on the interval $[-1,1]$ the equation

$$E_n(f) = O\left(\omega\left(f ; \left(\frac{\ln n}{n}\right)^2\right)\right)$$

holds.

We shall show that the estimate in Theorem 3 is exact in order.

THEOREM 5. For the function

$$\tau(x) = \begin{cases} 1 & \text{for } x = -1, 1 , \\ 0 & \text{for } -1 < x < 1 , \end{cases}$$

the following inequality holds

$$E_{n,\tau}(\tau) \geq \left(\frac{\ln n}{2n} \right)^2 .$$

Proof. Let us suppose that for some positive integer

$$E_{n,\tau}(\tau) < \left(\frac{\ln n}{2n} \right)^2 = \sigma_n^2 .$$

This means that an algebraic polynomial $p_n \in H_n$ exists, such that

$$| p_n(x) | < \sigma_n^2 \quad \text{for} \quad |x| \leq 1 - \sigma_n^2$$

and

$$p_n(1) > 1 - \sigma_n^2 . \tag{11}$$

But according to the extremal property of Chebyshev's polynomial

$$p(1) \leq \left(\frac{\ln n}{2n} \right)^2 T_n \left(\frac{1}{1 - \sigma_n^2} \right) . \tag{12}$$

On the other hand, since

$$\frac{1}{1 - \sigma_n^2} < \frac{1 + \sigma_n^2}{1 - \sigma_n^2}$$

then

$$T_n \left(\frac{1}{1 - \sigma_n^2} \right) < \left(\frac{1 + \sigma_n}{1 - \sigma_n} \right)^n < e^{2n\sigma_n} = n. \tag{13}$$

Hence, from (11), (12) and (13) we obtain the following inequality

$$1 - \left(\frac{\ln n}{2n} \right)^2 < \frac{(\ln n)^2}{4n} ,$$

that can not be satisfied by any positive integer n. The obtained contradiction proves the theorem.

We finish with the formulation of a hypothesis. It is interesting to find out if the estimate in Theorem 2 can be improved, if f is a graph of a convex function. Our hypothesis is that if f is convex, then

$$E_{n,\tau}(f) = O(n^{-1}).$$

References

1. A.A. Gončar, On the degree of rational approximation of continuous functions with characteristic singularity, (Russian), Math. Sbornik, 73(115),(1967), 630-638.

2. J. Szabados, Rational approximation of analytic functions with finite number of singularities in the real axis, Acta Math. Acad. Sci. Hungarical, 20(1-2),(1969),159-167.

3. Bl. Sendov, Some problems in the theory of approximation of functions and point sets in Housdorff metric, (Russian), Uspehi mat. nauk. 24, no.5, (1969), 141-178.

4. Bl. Sendov and V.Popov, On a generalization of Jackson's theorem for best approximation, J. Approx. Theory (to appear)

5. S.N. Bernstein, Sur la valeur asymptotique de la meilleure approximation des fonctions analytiques, Comptes rendus, Paris, 155 (1912), 1062-1065.

INVARIANT MEANS AND ALMOST CONVERGENCE IN NON-ARCHIMEDEAN ANALYSIS

A.H.Siddiqi

1.

Invariant means and almost convergence in classical analysis have been studied extensively. But only few results are known concerning means in non-Archimedean analysis. In the present talk we discuss the results concerning invariant means and almost convergence in non-Archimedean analysis. In the first part we mention main results concerning almost convergence in classical analysis. In the second part we discuss the results in non-Archimedean analysis.

Part I

2.1

Definition 2.1 [2]. A Banach limit is a linear functional L defined on the vector space of all bounded sequences of real numbers m, such that the following conditions are satisfied:

(i) $L(x) \geqslant o$ if $x_n \geqslant o$, for all n,

(ii) $L(x) = L(\Omega x)$, where $\Omega x = \Omega(x_1, x_2 \dots\dots x_n \dots) = (x_2, x_3 \dots\dots)$

(iii) $L(x) = 1$ if $x = (1,1,1\dots\dots)$.

It is known [2] that Banach limit always exists. G.G.Lorentz [8] introduced and studied the notion of almost convergence. Some of his results are as follows:

Definition 2.2 $x \in m$ is called almost convergent and the number s is called the F-limit of x if $L(x) = s$ for all Banach limits L.

THEOREM A [8]. x *is almost convergent and the F-limit of* x *is* s *iff*

$$\lim_{p\to\infty} \sum_{k=n}^{n+p-1} x_k = s$$

holds uniformly in n .

THEOREM B [8]. Let N be the semigroup of additive positive integers. Then an infinite matrix A on N is strongly regular iff A is regular and

$$\sum_{m} |A(n,m) - A(n,m+1)| = o .$$

M.M.Day [3] extended the notion of almost convergence to amenable semigroups and gave its characterization. S.A.Douglass [5] extended Th.A, to countable amenable semigroup with identity in which both cancellation laws hold.

Definition 2.3 [3]. Let S be an amenable semigroup and m(S) denote the vector space of bounded functions on S. An element $f \in m(S)$ is called almost convergent if all invariant means on m(S) coincide at f .

THEOREM 5. Let G be a countable amenable semigroup with identity e in which both cancellation laws hold. A necessary and sufficient condition that $f \in m(G)$ almost converges to s, is that for any summing sequence $\{S_n\}$ for G,

$$\lim_{n} |S_n|^{-1} \sum_{g \in S_n} f(gh) = s \qquad \text{and}$$

$$\lim_{n} |S_n|^{-1} \sum_{g \in S_n} f(hg) = s \quad \text{uniformly in } h.$$

2.2

Recently Mah [9] has proved several interesting results, which generalize earlier results in this direction by Lorentz [Th.B], King [6] and Schaefer [16]. Some of his results are as follows:

Definition 2.4. Let S be a left amenable semigroup, then a function $f \in m(S)$ is said to be left almost convergent to k if $\varphi(f) = \psi(f) = k$ for any left invariant means φ and ψ .

Definition 2.5. Let S be a left amenable semigroup, A be an infinite matrix on S and Af is defined on S by the relation $Af = \sum_t A(s,t)f(t)$, whenever the sum on the right hand side converges for each $s \in S$.

(1) A is called almost regular if Af is left almost convergent to k whenever $\lim_s f(s) = k$.

(2) A is called strongly regular if $\lim A f(s) = k$, whenever f is left almost convergent to k.

THEOREM D. Let S be left amenable semigroup. An infinite matrix A on S is almost regular iff the following conditions are satisfied:

(i) $\text{Sup} \sum_t |A(s,t)| < M$ for some $M > o$

(ii) $A(s,t)$, as a function of s, is left almost convergent to o for each $t \in S$

(iii) $\sum_t A(s,t)$, as a function of s, is left almost convergent to 1.

THEOREM E. Let S be a left cancellative left amenable semigroup generated by $B \subset S$. The following conditions are necessary and sufficient for an infinite matrix A to be strongly regular:

(i) $\text{Sup}_s \sum_t |A(s,t)| < M$ for some $M > o$.

(ii) $\lim_s \sum_t A(s,t) = 1$.

(iii) $\lim_s \sum_t |A(s,t) - A(s,at)| = o$ for each $a \in B$.

2.3

The following research papers have been also published connected with the notion of almost convergence:

Deeds [4], Kurtz [7], Julefa Ahmad [1], Mazhar and Siddiqi, A.H. [10,11], Petersen [14], Siddiqi, J.A. [21], Siddiqi, A.H. [18,19,20]

Part II

3.1

A.F.Monna, who is creator of the theory of non-Archimedean analysis has surveyed [13] the results known so far in this field. Mainly W.H.Schikhof [17] and A.C.M.Van Rooij [15] have studied problems in the domain of non-Archimedean harmonic analysis. Rooij has shown that there does not exist any mean on the space B(N) of bounded K-valued sequences, where K is a field complete with non-Archimedean valuation and N is the semigroup of natural numbers. He has investigated the following problems:

(i) For what abelian semigroups S there does exist a mean on the space B(S) of all bounded K-valued functions on S?

(ii) Given S and K do there exist means on suitable non-trivial subspaces of B(S)?

He has also investigated that most important property of K relevant to the existence of a mean with values in H, for a given S, is the characteristic of the residue class field of K.

After going through the results of Part I, naturally one would like to know what would be the notion of almost convergence and connected notions in non-Archimedean analysis. In view of the fact that there does not exist any mean on the space B(N), the parallel study of the notion of almost convergence in non-Archimedean analysis is not possible. However a notion of this type is possible as there exist means on subspaces of B(N).

3.2 Definitions and notations:

In the sequel K is a field with a non-Archimedean valuation $|\cdot|$. The characteristic of its residue class field is p. We assume that K is complete. We assume that $N_E \subseteq N_K$, where $N_K = \{|a| ;\ a \in K\}$, $N_E = \{\|x\|;\ x \in E$, where E is a n.a. Banach space$\}$. S is an abelian semigroup with operation + . B(S) is the set of all bounded maps $S \longrightarrow K$. For $f \in B(S)$ we put $\|f\| = \sup \{f(s) \mid s \in S\}$. With $\|\cdot\|$ as norm B(S) is a n.a. Banach space. N denotes the semigroup of positive integers. The symbol I_K indicates both unit element of K and constant function on S whose value is I_K.. I is identity map on B(S). An element $a \in S$ induces a shift $\Omega a : B(S) \rightarrow B(S)$, by $\Omega a f(y) = f(a+y)$, $f \in B(S)$, $y \in S$. $e(n) = n\, I_K$, $n \in N$.

Definition 3.1 Let $L \subseteq B(S)$. L is called orthogonal to $g \in B(N)$

if $\|\alpha f+\beta g\| = \max(\|\alpha f\|, \|\beta g\|)$ for all $\alpha, \beta \in K$ and all $f \in L$.

In such case we write $L \perp g$.

Definition 3.2 [15] . Let L be an invariant linear subspace of B(S) which also contains the constant functions, and if S is any subset of S, then a linear map $\mu : L \longrightarrow K$ on L into K is called a S-mean on L if it satisfies the following conditions:

(1) $\|\mu\| \leq 1$

(2) $\mu(I_K) = I_K$

(3) $\mu \circ \Omega a = \mu$ for every $a \in S$.

A mean on L is an S-mean on L .

Definition 3.4 [15]. Let $f \in B(S)$, $\varepsilon > 0$, $a \in S$. A number $m \in N$ is called an a-ε-period of f if there exists a $k \in N$ such that $\|\Omega^k a f - \Omega^{k+m} a f\| < \varepsilon$.

Definition 3.5 [15]. $f \in B(S)$ is called weakly almost periodic if for all $a \in S$ and $\varepsilon > 0$ there exists an a-ε-period, which is not divisible by p.

The set of all weakly almost periodic functions is denoted by WAP(S).

Definition 3.6 [15] Let $a \in S$. The period of a is o if $x+ma = x$ for all $x \in S$ and $m \in N$. Otherwise the period is the smallest $m \in N$ for which there exists an $x \in S$ with $x+ma = x$. S is called p-free if no element of S has a period which is a multiple of p.

Definition 3.7 [15] A semigroup S is called reasonably amenable if there is at least one invariant subspace of B(S) which has a mean.

Definition 3.8 Let S be a reasonably amenable semigroup and I(L) denote the set of all means on invariant subspace L of B(S)..

An element $f \in L \subseteq B(S)$ is called <u>almost convergent</u> to s if $\mu(f) = \mu'(f) = s$, for each $\mu, \mu' \in I(L)$.

<u>Remarks</u>. (i) In view of Th.6.1 [15] and the fact that every convergent sequence is orthogonal to e, it is clear that N is reasonably amenable.

By Th.4.1 [15] any p-free semigroup is reasonably amenable.

(ii) By Th.6.1 [15] and the note [p.226,15] the following hold:

(a) The necessary condition for an element of B(N) to be almost convergent is that it is orthogonal to e.

(b) If L is a maximal subspace of B(N) orthogonal to e then every element of L is almost convergent.

(c) Every element of WAP(N) is almost convergent.

Here we restrict ourselves to the case S = N, although some results may be extended to reasonably amenable semigroups.

3.3

The following result gives the relation between convergence and almost convergence.

THEOREM 3.1 <u>Every convergent sequence is almost convergent but the converse is not true</u>.

<u>Proof of Th.3.1</u> Let $a = (a_1, a_2, \ldots . a_n \ldots .)$ be a convergent sequence. Choose $\varepsilon > 0$ and $s \in N$, in order to show that a is almost convergent, it is sufficient to show that $a \in$ WAP(N). For this we have to find an a-ε-period of a. We claim that 1 is such period. In fact, take $k \in N$ so that $|a_n - \lim_{\ell \to \infty} a_\ell| \leq \frac{1}{2}\varepsilon$ for all $n \geq k$. Then $|a_n - a_{n'}| \leq \frac{\varepsilon}{2}$ for all $n, n' \geq k$. In particular,

$$|a_{\ell+ks} - a_{\ell+(k+m)s}| \leq \frac{1}{2}\varepsilon \quad \text{for all} \quad \ell \in N, \text{ so}$$

$$\varepsilon > \sup |a_{\ell+ks} - a_{\ell+(k+m)s}| = \| \Omega_s^k a - \Omega_s^{k+m} a \| .$$

Hence $a \in WAP(N)$.

In order to show that converse is not true, we construct an invariant subspace L of B(N) which has unique mean but it does not contain convergent sequence. If $p \neq 2$, let L be the linear hull of a and b, where $a = (1,o,1,o,1\ldots\ldots)$. $b = (o,1,o,1,\ldots\ldots)$. L is invariant, $I_K \in L$ and $\mu : \alpha a + \beta b \to \frac{1}{2}\alpha + \frac{1}{2}\beta$ is a mean on L.

This mean is unique on L, because for any mean V, we have

$$1 = V(I_K) = V(a) + V(\Omega a) \quad \text{or} \quad V(a) = \frac{1}{2}.$$

Similarly $V(b) = \frac{1}{2}$. By Theorem 6.1 [15] $L \perp e$. Trivially L contains non-convergent elements. If $p = 2$ then $p \neq 3$, and we show the above fact by considering the linear hull of (1,o,o,1,o,o1...) (o,1,o,o1,o,o,1....),(o,o,1,oo1,o...).

The characterization of almost convergence in the classical case given by Theorem A, is not valid in the present case.

3.4

We can prove the following theorems concerning the notion of almost regular and strongly regular matrices in non-Archimedean analysis.

THEOREM 3.2 <u>An infinite matrix A is almost regular if and only if the following conditions are satisfied:</u>

(i) $\sup_{m,n} |A(m,n)| < M$ <u>for some</u> $M > o$

(ii) $A(m,n)$, <u>as a function of</u> m, <u>is almost convergent to</u> o <u>for each</u> n.

(iii) $\sum_n A(m,n)$, <u>as a function of</u> m, <u>is almost convergent to</u> 1.

THEOREM 3.3 <u>The infinite matrix A is strongly regular if and only if</u>

(i) $\sup_{m,n} |A(m,n)| < M$, <u>for some</u> $M > 0$

(ii) $\lim_{m} \sum_{n} A(m,n) = 1$

(iii) $\lim_{m} \sum_{n} |A(m,n)-A(m,n+1)| = 0$ <u>hold</u>.

3.5

Mah [9] has given the following definition of F_A - summability in classical analysis:

THEOREM 3.9 <u>A function</u> f <u>is said to be</u> F_A<u>-summable to</u> ℓ <u>if</u>

$$\lim_{s} \sum_{t} A(s,t)f(tb) = \ell, \text{ uniformly in } b, \text{ where } b \in S.$$

If we take this as definition of F_A-summability in non-Archimedean analysis then $f \in B(N)$ is always F_A-summable to o, if matrix A is regular. For unabounded sequence, F_A-limit may not exist even if A is a convergence preserving matrix.

I am thankful to the authorities of Tabriz university and specially Dr.B.A.Saleemi and Dr.M.Noor Khalichi for providing me with facilities to attend this conference.

Bibliography

1 Ahmad, J., Delta, Journal of Wisconsin University, 1970, MR 1971.

2 Banach, S., Théorie des operations lineaires, Warszawa, 1932, 33-34.

3 Day, M.M., Amenable semigroups, Illinois J. Math. 1 (1957) 509-544

4 Deeds, J.B., Summability of vector sequences, Studia Mathematica, XXX(1968), 361-372.

5 Douglass, S.A., On a concept of summability in amenable semigroups, Math. Scand. 23 (1968) 96-102.

6 King, J.P., Almost summable sequences, Proc. Amer. Math. Soc. 17 (1966), 1219-1225.

7 Kurtz, J.C., Almost convergent vector sequences, Tohoku Math. Journal 22 (1970), 493-498.

8 Lorentz, G.G., A contribution to the theory of divergent sequences, Acta Math. 80 (1948), 167-190.

9 Mah, P.M., Summability in amenable semigroups, Trans. Amer. Math. Soc. 156 (1971) 391-403.

10 Mazhar, S.M. and Siddiqi, A.H., On the almost summability of a Trigonometric sequence, Acta Math. Acad. Sc. 20 (1969) 21-24.

11 Mazhar, S.M. and Siddiqi, A.H., On the F_A and F_B summability of a Trigonometric sequence, Ind. Journ. Math. 9 (1967) 461-466.

12 Monna, A.F., Sur le theoreme de Banach-Steinhaus, Proc. Kond. Ned. Akad. V.Wetensch A66 (1963) 121-131.

13 Monna, A.F., Analyse Non-Archimedean, Springer-Verlag, 1970.

14 Petersen, G.M., Summability of sequences, Proc. Amer. Math. Soc. 11 (1960), 469-477.

15 Van Rooij, A.C.M., Invariant means with values in a non-Archimedean valued field, Proc. Kon. Ned. Akad. V.Wetensch 70 (1967) 220-228.

16. Schaefer, P., Almost covergent and almost summable sequences, Proc. Amer. Math. Soc. 20 (1969) 51-54.

17 Schikhof, W.H., Non-Archimedean Harmonic Analysis, Ph.D. Thesis, Nijmegen (1967).

18 Siddiqi, A.H., On F_A regular matrices (unpublished).

19 Siddiqi, A.H., On the lacunary partial sums of Walsh Fourier series, Rendiconti del Circolo di Math. Xviii (1969) 313-318.

20 Siddiqi, A.H., On the summability of a sequence of Walsh functions, J. Austral. Math. Soc. 10 (1969) 385-394.

21 Siddiqi, J.A., Coefficients properties of certain Fourier series Math. Ann. 181 (1969) 385-394.

PARAMEASURES AND MULTIPLIERS OF SEGAL ALGEBRAS

K.R. Unni

In this paper we introduce the notion of a parameasure as an element of the dual of a certain internal inductive limit of Banach spaces and show that a bounded translation invariant operator on a Segal algebra $S(G)$, where G is a locally compact abelian group, can be represented as a convolution with a parameasure. The Fourier transforms are also defined for such parameasures.

1. Definition and properties of Segal algebras.

Throughout this paper G will denote a locally compact abelian group with Haar measure dx and $\hat{G}$ will denote its character group. $\mathcal{K}(G)$ will stand for the class of all continuous functions on G with compact support. For each compact subset C of G, let us denote by $\mathcal{K}_C(G)$ the subclass of those functions in $\mathcal{K}(G)$, whose supports are contained in C. If $f \in L^1(G)$, its Fourier transform is denoted by $\hat{f}$. We set

$$B(G) = \left\{ f \in L^1(G) : \hat{f} \in \mathcal{K}(\hat{G}) \right\}.$$

If $\hat{L}$ is a compact subset of $\hat{G}$, we write

$$B_{\hat{L}}(G) = \left\{ f \in B(G) : \hat{f} \in \mathcal{K}_{\hat{L}}(\hat{G}) \right\}.$$

Definition 1. A linear subspace $S(G)$ on $L^1(G)$ is called a Segal algebra if the following four conditions are fulfilled.

(a) $S(G)$ is dense in $L^1(G)$.

(b) $S(G)$ is a Banach space under some norm $\|\cdot\|_S$ and

$$\|f\|_S \geqslant \|f\|_1 \qquad f \in S(G)$$

(c) Let $y \in G$ and τ_y denote the translation operator. Then for each $f \in S(G)$, $\tau_y f$ belongs to $S(G)$ and the mapping $y \longrightarrow \tau_y f$ is continuous from G into $S(G)$.

(d) $\| \tau_y f \|_S = \| f \|_S$ for all $f \in S(G)$ and all $y \in G$.

We shall now collect below in the form of lemmas various properties of a Segal algebra that are needed in the sequel.

LEMMA 1. *For every* $f \in S(G)$ *and arbitrary* $h \in L^1(G)$ *the vector valued integral* $\int h(y)\, \tau_y f dy$ *exists as an element of* $S(G)$ *and*

$$h * f = \int h(y)\, \tau_y f\, dy .$$

Moreover

$$\| h * f \|_S \leq \| h \|_1 \| f \|_S .$$

LEMMA 2. *Let* μ *be a bounded complex valued measure on* G. *Then for every* $f \in S(G)$, *the vector valued integral* $\int \tau_y f d\mu(y)$ *exists as an element of* $S(G)$ *and*

$$\mu * f = \int \tau_y f d_\mu(y) .$$

Further,

$$\| \mu * f \|_S \leq \| . \mu \| \, \| f \|_S .$$

Notice that the left hand side expressions in Lemmas 1 and 2 are the usual convolution product.

LEMMA 3. $B(G) \subset S(G)$.

LEMMA 4. *To every compact subset* $\hat{K}$ *of* $\hat{G}$ *there is a constant* $C_{\hat{K}} > 0$ *such that for every* $f \in S(G)$ *whose Fourier transform vanishes outside* $\hat{K}$ *satisfies*

$$\| f \|_S \leq C_{\hat{K}} \| f \|_1 .$$

LEMMA 5. *Given any* $f \in S(G)$ *there is for every* $\varepsilon > 0$ *a* $\nu \in S(G)$ *such that the Fourier transform has compact support and*

$$\| \nu * f - f \|_S < \varepsilon .$$

LEMMA 6. Every Segal algebra has approximate units of L^1-norm 1.

The proofs of these lemmas can be found in Reiter ([4] pp.128-129 and [5] pp.18-20 and p.37).

2. Parameasures and their properties.

Let K be a compact subset of G and $\hat{L}$ a compact subset of $\hat{G}$. We define $E_{K,\hat{L}}(G)$ as the space of all functions u which can be represented as

$$u = \sum_{k=1}^{\infty} f_k * g_k \qquad f_k \in \mathcal{K}_K G, \; \hat{g}_k \in \mathcal{K}_{\hat{L}}(\hat{G})$$

with $\sum_{k=1}^{\infty} \|f_k\|_\infty \|g_k\|_1 < \infty$. We supply it with a norm

$$\|u\|_{K,\hat{L}} = \inf \sum_{k=1}^{\infty} \|f_k\|_\infty \|g_k\|_1$$

where the infimum is taken over all representations of u as an element of $E_{K,\hat{L}}(G)$. If ρ_K denotes the measure of the set K, then it is easy to verify the following relations

$$\|u\|_1 \leq \rho_K \|u\|_{K,\hat{L}}$$

$$\|u\|_S \leq \rho_K C_{\hat{L}} \|u\|_{K,\hat{L}}$$

and

$$\|u\|_\infty \leq \|u\|_{K,\hat{L}}$$

where $C_{\hat{L}}$ is given by Lemma 4. $E_{K,\hat{L}}(G)$ then becomes a Banach space under $\|\cdot\|_{K,\hat{L}}$ and the proof is similar to that of Gaudry ([1] , Theorem 2.4). E(G) is now defined as the internal inductive limit of the Banach space $\{E_{K,\hat{L}}(G)\}$, that is E(G) is the vector space $\bigcup_{K,\hat{L}} E_{K,\hat{L}}(G)$ with the topology which has for a neighbourhood base at the origin open sets of the form

$$U_r = \bigcup_{K,\hat{L}} \{f \in E_{K,\hat{L}}(G) : \|f\|_{K,\hat{L}} < r\}.$$

PROPOSITION 1. (i) E(G) is dense in B(G) both in the L^1-norm and in Segal norm.

(ii) E(G) <u>is a dense subspace of</u> S(G)

<u>Proof</u>. Let $h \in B(G)$. Then there is a compact set $\hat{L} \subset \hat{G}$ such that $\hat{h} \in \mathcal{K}_{\hat{L}}(\hat{G})$. Let $\{e_\alpha\}$ be an approximate identity in $L^1(G)$ such that $\|e_\alpha\|_1 = 1$ and $e_\alpha \in \mathcal{K}(G)$. We can even assume that there is a fixed compact set K_0 such that $e_\alpha \in \mathcal{K}_{K_0}(G)$ for all α. Let $\{\eta_\beta\}$ be an approximate identity in $L^1(G)$ such that $\hat{\eta}_\beta \in \mathcal{K}(\hat{G})$ and $\|\eta_\beta\|_1 = 1$. Then $h * e_\alpha * \eta_\beta = e_\alpha * h * \eta_\beta \in E(G)$ and

$$\|h * e_\alpha * \eta_\beta - h\|_1 \leq \|(h * e_\alpha * \eta_\beta) - (h * e_\alpha)\|_1 + \|h * e_\alpha - h\|_1 \to 0.$$

This shows that E(G) is dense in B(G) in L^1-norm.

Now since $\hat{h}$ and $\hat{h}\,\hat{e}_\alpha\,\hat{\eta}_\beta$ both belong to $\mathcal{K}_{\hat{L}}(\hat{G})$ we see that $h * e_\alpha * \eta_\beta - h \in B_{\hat{L}}(G)$ and by Lemma 4, there exists a constant $C_{\hat{L}} > 0$ such that

$$\|h * e_\alpha * \eta_\beta - h\|_S \leq C_{\hat{L}} \|h * e_\alpha * \eta_\beta - h\|_1$$

which shows that E(G) is dense in B(G) in the Segal norm also. Since B(G) is dense in S(G) it follows that E(G) is a dense subspace of S(G).

<u>Definition 2</u>. The space of continuous linear functionals on E(G) is denoted by W(G). We denote the pairing between E(G) and W(G) by

$$\langle u,\sigma \rangle \qquad u \in E(G) \quad \sigma \in W(G).$$

Then $\sigma \in W(G)$ if and only if σ is linear and σ restricted to $E_{K,\hat{L}}(G)$ is continuous in the topology of $E_{K,\hat{L}}(G)$ for each pair of compact sets $(K,\hat{L})$. The elements of W(G) are termed <u>para - measures</u>.

PROPOSITION 2. <u>If</u> S(G)' <u>denotes the dual of</u> S(G), <u>then</u> S(G)' ⊂ W(G).

Proof. Let $u = \sum_{k=1}^{\infty} f_k * g_k \in E_{K,\hat{L}}(G)$.

If $\varphi \in S(G)'$, then

$$|\langle u,\varphi\rangle_S| \leq \|u\|_S \|\varphi\| \leq \rho_K C_{\hat{L}} \|u\|_{K,\hat{L}} \|\varphi\| .$$

Thus φ defines a continuous linear functional on $E_{K,\hat{L}}(G)$ and hence on all of $E(G)$. Hence $\varphi \in W(G)$.

PROPOSITION 3. If $M(G)$ denotes the algebra of all bounded measures on G, then $M(G) \subset W(G)$.

Proof. Let $\mu \in M(G)$ and $u = \sum_{k=1}^{\infty} f_k * g_k \in E_{K,\hat{L}}(G)$.

Then

$$\left| \int_G u(x)\, d\mu(x) \right| \leq \int \|u\|_\infty |d\mu(x)| \leq \|\mu\| \|u\|_{K,\hat{L}} .$$

PROPOSITION 4. Let $h \in B(G)$. If $u \in E(G)$, then $\tilde{h} * u \in E(G)$ and the mapping $u \to \tilde{h} * u$ is continuous from $E(G)$ to $E(G)$ where $\tilde{h}(x) = h(-x)$.

Proof. If $u \in E(G)$ and $h \in B(G)$, it is clear that $\tilde{h} * u \in E(G)$. We shall now show that the restriction of the mapping $u \to \tilde{h} * u$ to each $E_{K,\hat{L}}(G)$ is continuous from $E_{K,\hat{L}}(G)$ to $E(G)$. But this is immediate since $\|u_n - u\|_{K,\hat{L}} \to 0$ implies

$$\|\tilde{h} * u_n - \tilde{h} * u\|_{K,\hat{L}} \leq \|\tilde{h}\|_1 \|u_n - u\|_{K,\hat{L}} \to 0 .$$

Definition 3. Let $f \in B(G)$ and $\sigma \in W(G)$. We define $\sigma * f$ to be that element of $W(G)$ such that

$$\langle u, \sigma * f\rangle = \langle \tilde{f} * u, \sigma\rangle \qquad u \in E(G).$$

Remark. Proposition 4 asserts that this definition is meaningful.

3. Multipliers on the Segal algebra S(G).

Definition 4. A multiplier on $S(G)$ is a bounded linear operator on $S(G)$ which commutes with translations. The space of all multipliers on $S(G)$ is denoted by $\mathfrak{M}(S(G))$.

THEOREM 1. *If* $T \in \mathfrak{M}(S(G))$, *then* $T(f*g) = Tf * g$ *for* $f, g \in S(G)$.

Proof. Let $T \in \mathfrak{M}(S(G))$. Then $T\tau_y = \tau_y T$ for all $y \in G$. We shall show that T commutes with convolutions. Let $f \in S(G)$ and $\varphi \in S(G)'$. Then if $\|T\|$ denotes the operator norm of T,

$$|\langle Tf, \varphi\rangle| \leq \|Tf\|_S \|\varphi\| \leq \|T\| \|\varphi\| \|f\|_S$$

shows that the mapping $f \longrightarrow \langle Tf, \varphi\rangle$ is a bounded linear functional on $S(G)$ and hence there exists a unique $\psi \in S(G)'$ such that

$$\langle f, \psi\rangle = \langle Tf, \varphi\rangle \quad \text{for all } f \in S(G).$$

It is known that if $f \in L^1(G)$, $g \in S(G)$ and $\varphi \in S(G)'$, then

$$\langle f * g, \varphi\rangle = \int f(y)\langle\tau_y g, \varphi\rangle dy$$

holds. (See Reiter [5] p. 53). Now, let $f, g \in S(G)$. Then

$$\langle Tf * g, \varphi\rangle = \int g(y)\langle\tau_y Tf, \varphi\rangle dy = \int g(y)\langle T\tau_y f, \varphi\rangle dy$$

$$= \int g(y)\langle\tau_y f, \psi\rangle dy = \langle f * g, \psi\rangle = \langle T(f * g), \varphi\rangle.$$

Thus

$$\langle T(f * g), \varphi\rangle = \langle Tf * g, \varphi\rangle$$

for each $\varphi \in S(G)'$. Hence by Hahn-Banach theorem, $Tf * g = T(f*g)$ for all $f, g \in S(G)$.

THEOREM 2. *Let* $T \in \mathfrak{M}(S(G))$. *Then there exists a unique parameasure* $\sigma \in W(G)$ *such that*

$$Tf = \sigma * f \qquad f \in B(G)$$

<u>Proof</u>. Suppose $T \in \mathfrak{M}(S(G))$. If $u = \sum_{k=1}^{\infty} f_k * g_k \in E_{K,\hat{L}}(G)$ then

$$\left| \sum_{k=1}^{\infty} f_k * Tg_k(0) \right| \leq \sum_{k=1}^{\infty} | f_k * Tg_k(0) | \leq \sum_{k=1}^{\infty} \| f_k \|_\infty \| Tg_k \|_1$$

$$\leq \sum_{k=1}^{\infty} \| f_k \|_\infty \| Tg_k \|_S \leq \sum_{k=1}^{\infty} \| f_k \|_\infty \| T \| \, \| g_k \|_S$$

$$\leq \| T \| \, C_{\hat{L}} \sum_{k=1}^{\infty} \| f_k \|_\infty \| g_k \|_1 .$$

Thus $\sum_{k=1}^{\infty} f_k * Tg_k(0)$ converges uniformly since $\sum_{k=1}^{\infty} \| f_k \|_\infty \| g_k \|_1 < \infty$. Define $\omega(u) = \sum_{k=1}^{\infty} f_k * Tg_k(0)$.

First we show that ω is well defined. It is enough to show that if $\sum_{k=1}^{\infty} f_k * g_k$ is a representation of 0 as an element of $E_{K,\hat{L}}(G)$, then $\sum_{k=1}^{\infty} f_k * Tg_k(0) = 0$.

Let $\{ e_\beta \}$ be an approximate identity in $L^1(G)$ such that $\| e_\beta \|_1 = 1$ and $\hat{e}_\beta \in \mathcal{K}(\hat{G})$. Then for each k, it is easy to verify that

$$| f_k * T(e_\beta * g_k)(0) - f_k * Tg_k(0) | \leq C_{\hat{L}} \| T \| \, \| f_k \|_\infty \| e_\beta * g_k - g_k \|_1$$

so that

$$f_k * T(e_\beta * g_k)(0) \longrightarrow f_k * Tg_k(0) .$$

Further,

$$\Big|\sum_{k=1}^{\infty} f_k * T(e_\beta * g_k)(0)\Big| \le \sum_{k=1}^{\infty} \|f_k\|_\infty \|T(e_\beta * g_k)\|_1$$

$$\le \sum_{k=1}^{\infty} \|f_k\|_\infty \|T(e_\beta * g_k)\|_S$$

$$\le C_{\hat{L}} \|T\| \sum_{k=1}^{\infty} \|f_k\|_\infty \|e_\beta * g_k\|_1$$

$$\le C_{\hat{L}} \|T\| \sum_{k=1}^{\infty} \|f_k\|_\infty \|g_k\|_1 .$$

Hence $\sum_{k=1}^{\infty} f_k * T(e_\beta * g_k)(0)$ converges uniformly with respect to β. Hence,

$$\lim_\beta \sum_{k=1}^{\infty} f_k * T(e_\beta * g_k)(0) = \sum_{k=1}^{\infty} f_k * Tg_k(0) .$$

Now let $Te_\beta = \eta_\beta$. Then $T(e_\beta * g_k) = Te_\beta * g_k = \eta_\beta * g_k$ so that

$$f_k * T(e_\beta * g_k)(0) = f_k * \eta_\beta * g_k(0) = f_k * g_k * \eta_\beta(0) .$$

Since $\sum_{k=1}^{\infty} f_k * g_k$ converges uniformly in $C_0(G)$, the space of continuous functions on G vanishing at infinity, we have

$$\sum_{k=1}^{\infty} f_k * T(e_\beta * g_k)(0) = \sum_{k=1}^{\infty} \int (f_k * g_k)(-x)\, \eta_\beta(x)\, dx$$

$$= \int \sum_{k=1}^{\infty} (f_k * g_k)(-x)\, \eta_\beta(x)\, dx = 0 .$$

Hence $\sum_{k=1}^{\infty} f_k * Tg_k(0) = 0$ and ω is well defined. Moreover if $u \in E_{K,\hat{L}}(G)$, then

$$|\omega(u)| = \Big|\sum_{k=1}^{\infty} f_k * Tg_k(0)\Big| \le C_{\hat{L}} \|T\| \sum_{k=1}^{\infty} \|f_k\|_\infty \|g_k\|_1 .$$

Thus $|\omega(u)| \leq C_{\hat{L}} \|T\| \|u\|_{K,\hat{L}}$ and ω restricted to $E_{K,\hat{L}}(G)$ is a continuous linear functional on $E_{K,\hat{L}}(G)$. Hence $\omega \in W(G)$.

We now assert that $Tf = \tilde{\omega} * f$ for each $f \in B(G)$ where $\tilde{\omega} \in W(G)$ is defined by $\langle u,\tilde{\omega} \rangle = \langle \tilde{u},\omega \rangle$ for all $u \in E(G)$. If $f \in B(G)$ and $u \in E(G)$ we have

$$\langle u,\tilde{\omega} * f \rangle = \langle \tilde{f} * u,\tilde{\omega} \rangle = \langle f * \tilde{u},\omega \rangle = \tilde{u} * Tf(o) = \langle u,Tf \rangle.$$

Since this holds for each $u \in E(G)$, we conclude that $\tilde{\omega} * f = Tf$ We set $\sigma = \tilde{\omega}$.

We now claim that σ is unique. Suppose there exist $\sigma, \sigma' \in W(G)$ such that $Tf = \sigma * f = \sigma' * f$ for all $f \in B(G)$. Then if $f \in B(G)$ and $u \in E(G)$ we have

$$\langle f * u, \sigma \rangle = \langle u,\sigma * \tilde{f} \rangle = \langle u,\sigma' * \tilde{f} \rangle = \langle f * u,\sigma' \rangle .$$

Since $\{ f * u \mid f \in B(G),\ u \in E(G) \}$ is dense in $E(G)$ we conclude that $\sigma = \sigma'$ and the uniqueness is proved.

4. Pseudomeasures and parameasures.

We shall now show that the class of parameasures includes all pseudomeasures $P(G)$ as well. Let $A(G)$ denote the space of Fourier transforms of functions integrable over $\hat{G}$. The topology on $A(G)$ is given by

$$\|\hat{f}\|_{A(G)} = \|f\|_{L^1(\hat{G})} .$$

The dual space of $A(G)$ with the above norm is the space of pseudomeasures denoted by $P(G)$. We now claim $P(G) \subset W(G)$.

First we show that $E(G) \subset A(G)$. Suppose that $u \in E(G)$. Then there exists a pair $(K,\hat{L})$ of compact sets such that $u \in E_{K,\hat{L}}(G)$. Let

$$u = \sum_{k=1}^{\infty} f_k * g_k , \qquad f_k \in \mathcal{K}_K(G) , \; \hat{g}_k \in \mathcal{K}_{\hat{L}}(\hat{G})$$

with

$$\sum_{k=1}^{\infty} \| f_k \|_{\infty} \| g_k \|_1 < \infty .$$

If ρ_K and $\lambda_{\hat{L}}$ denote the measures of the sets K and $\hat{L}$ respectively, we follow the proof of Gaudry ([1] Theorem 2.5) to show that $u \in A(G)$ and $\| u \|_{A(G)} \leq (\rho_K \lambda_{\hat{L}})^{1/2} \| u \|_{K,\hat{L}}$. Hence $E(G) \subset A(G)$.

If $\sigma \in P(G)$ and $u \in E_{K,\hat{L}}(G)$, then

$$|\langle u, \sigma \rangle| \leq \| u \|_{A(G)} \| \sigma \|_{P(G)} \leq \| \sigma \|_{P(G)} (\rho_K \lambda_{\hat{L}})^{1/2} \| u \|_{K,\hat{L}}$$

which implies that σ when restricted to $E_{K,\hat{L}}(G)$ defines a continuous linear functional on $E_{K,\hat{L}}(G)$. Thus we conclude that $\sigma \in W(G)$ and that every pseudomeasure is also a parameasure.

5. Fourier transforms of parameasures.

We have shown that the space of multipliers on a Segal algebra $S(G)$ is isomorphic to a subspace of the space of parameasures. We shall now define the Fourier transform for this subspace quite analogous to the definition given by Gaudry [2] for quasi-measures. We need a few preliminaries. Let $M_c(G)$ denote the space of all measures on G with compact support.

LEMMA 7. *If $\mu \in M_c(G)$ and $h = \sum_{k=1}^{\infty} f_k * g_k \in E_{K,\hat{L}}(G)$, then the vector valued integrals*

$$H = \int_{\hat{G}} \gamma^{-1} h \, d\mu(\gamma) = \sum_{k=1}^{\infty} \int_{\hat{G}} (\gamma^{-1} f_k) * (\gamma^{-1} g_k) \, d\mu(\gamma)$$

exists as an element of $E(G)$ *and for each* $x \in G$

$$H(x) = \sum_{k=1}^{\infty} \hat{\mu}(x) f_k * g_k(x) = \hat{\mu}(x)\, h(x) .$$

<u>Moreover, the mapping</u> $h \longrightarrow \hat{\mu} h$ <u>from</u> $E_{K,\hat{L}}(G)$ <u>to</u> $E(G)$ <u>is continuous.</u>

<u>Proof</u>. We first notice that if $\hat{g}_k \in \mathcal{K}_{\hat{L}}(\hat{G})$ then $\widehat{\gamma^{-1} g_k} \in \mathcal{K}_{\hat{L}+\gamma}(\hat{G})$ for each $\gamma \in \hat{G}$. Hence if the support of the measure μ is denoted by $\hat{N}$ then it is easy to see that $h \in E_{K,\hat{L}}(G)$ implies $\gamma^{-1} h \in E_{K,\hat{L}+\hat{N}}(G)$. With this observation, the proof of Gaudry [2] goes through to show that $\int_{\hat{G}} \gamma^{-1} h \, d\mu(\gamma)$ belongs to $E_{K,\hat{L}+\hat{N}}(G)$ and hence in $E(G)$.

Since

$$\|\hat{\mu} h\|_{K,\hat{L}+\hat{N}} = \left\| \int_{\hat{G}} \gamma^{-1} h \, d\mu(\gamma) \right\|_{K,\hat{L}+\hat{N}}$$

$$\leq \int_{\hat{G}} \|\gamma^{-1} h\|_{K,\hat{L}+\hat{N}} \, |d\mu(\gamma)|$$

$$= \int_{\hat{G}} \|h\|_{K,\hat{L}} \, |d\mu(\gamma)| = \|h\|_{K,\hat{L}} \, \|\mu\|$$

the mapping $h \longrightarrow \hat{\mu} h$ is continuous.

<u>Definition 4</u>. If $\mu \in M_c(G)$ and we define $\hat{\mu}\sigma$ to be that element in $W(G)$ such that

$$\langle h, \hat{\mu}\sigma \rangle = \langle \tilde{\hat{\mu}} h, \sigma \rangle = \langle \check{\mu} h, \sigma \rangle \qquad h \in E(G)$$

where

$$\check{\mu}(x) = \int_{\hat{G}} (x, \gamma) \, d\mu(\gamma) .$$

We shall now show how to define the parameasure $f\sigma$ where $f \in \mathcal{K}(G)$ and $\sigma \in W(G)$. Let $f \in \mathcal{K}(G)$ and let $\{u_\beta\}$ be an approximate identity in $L^1(G)$ such that $\hat{u}_\beta \in \mathcal{K}(\hat{G})$ and $\|u_\beta\|_1 = 1$.

Let $f_\beta = f * u_\beta$. Then $f_\beta \in E(G)$ and $\hat{f}_\beta \in M_c(\hat{G})$. Hence $f_\beta\,\sigma$ is given by definition 4. Now, we claim that $\{f_\beta\,\sigma\}$ is a cauchy-net in W(G). We first recall that a net $\{\sigma_\alpha\} \subset W(G)$ converges to $\sigma \in W(G)$ if and only if $\langle h,\sigma_\alpha\rangle \longrightarrow \langle h,\sigma\rangle$ for all $h \in E(G)$. Now the relation

$$|\langle h, f_\beta\sigma\rangle - \langle h, f_\alpha\sigma\rangle| = |\langle h, (f_\beta - f_\alpha)\sigma\rangle|$$

$$= |\langle(\tilde{f}_\beta - \tilde{f}_\alpha) h, \sigma\rangle|$$

$$\leq \lambda \|h\|_{K,\hat{L}} \|f * u_\beta - f * u_\alpha\|_1$$

where λ is independent of α and β, holds for all $h \in E_{K,\hat{L}}(G)$ which shows that $\{f_\beta\,\sigma\}$ is a Cauchy net in W(G).

<u>Definition 5.</u> Let $f \in \mathcal{K}(G)$ and $\sigma \in W(G)$. We define $f\,\sigma$ to be that element of W(G) given by

$$\langle h, f\,\sigma\rangle = \lim_\beta \langle h, f_\beta\,\sigma\rangle \qquad h \in E(G)$$

<u>Definition 6.</u> Let $\sigma \in W(G)$ and let U be an open subset of G. Then the restriction of σ to U, denoted by σ_U, is defined as the continuous linear functional on the internal inductive limit $\bigcup_{K\subset U,\hat{L}\subset\hat{G}} E_{K,\hat{L}}(G)$ determined by the equation

$$\langle h, \sigma_U\rangle = \langle h, \sigma\rangle \qquad (h \in E_{K,\hat{L}}(G),\ K \subset U).$$

LEMMA 8. <u>If</u> $\{U_\alpha\}$ <u>is a locally finite cover of</u> G <u>by open sets with compact closures then there exists a locally finite family</u> $\{f_\alpha\} \subset A(G) \cap \mathcal{K}(G)$ <u>such that</u>

(1) <u>the support of</u> f_α <u>is contained in</u> U_α.

(2) $0 \leq f_\alpha(x) \leq 1$ <u>for each</u> $x \in G$ <u>and all</u> α.

(3) $\sum_\alpha f_\alpha(x) = 1$ <u>for each</u> $x \in G$.

(See Larsen [3] p.206).

It is the property of a locally finite family of functions $\{f_\alpha\}$ that if U is any open set with compact closure then all but a finite number of the elements f_α vanish identically on $\overline{U}$.

LEMMA 9. <u>Let</u> $U \subset \hat{G}$ <u>be an open subset with compact closure.</u> <u>Then there exist</u> $k \in B(G)$ <u>and</u> $g \in A(G) \cap \mathcal{K}(G)$ <u>such that</u> $\check{k}\check{g} = 1$ <u>on</u> $\overline{U}$, <u>where</u> $\check{f} = \hat{\hat{f}} = \tilde{\hat{f}}$

(See Larsen [3] p.208)

We are now in a position to define Fourier transforms for elements of $\mathfrak{M}(S(G))$. Let $\sigma \in \mathfrak{M}(S(G))$. If $h = p * q$ with $p \in \mathcal{K}_{\hat{L}}(\hat{G})$ and $\hat{q} \in \mathcal{K}_K(G)$ where $K \subset U$, define

$$\langle h, \hat{\omega}_U \rangle = \langle h, \hat{g}(\sigma * k)^\wedge \rangle.$$

Since $k \in B(G)$ it follows that $\sigma * k \in S(G) \subset L^1(G)$. Hence the Fourier transform $(\sigma * k)^\wedge$ exists and belongs to $W(\hat{G})$ and hence $\hat{g}(\sigma * k)^\wedge \in W(\hat{G})$ by Lemma 7. Now we show that $\hat{\omega}_U$ is well defined ie, independent of the choices of k and g. If $\{u_\alpha\} \subset B(G)$ is an approximate identity for $L^1(G)$ such that $\|u_\alpha\|_1 = 1$ then the proof given in Larsen ([3] p.209) goes through to show that

$$\langle h, \hat{\omega}_U \rangle = \lim_\alpha \langle h, (\sigma * U_\alpha)^\wedge \rangle = \lim_\alpha \langle \check{h}, \sigma * U_\alpha \rangle .$$

Since this limit is independent of the choices of k and g, $\hat{\omega}_U$ is well defined. Now $\hat{\omega}_U$ is extended linearly to elements of the form $\sum p_i * q_i$ ie. to $E_{\hat{L},K}(\hat{G})$ and hence defines a continuous linear functional on $\bigcup_{\hat{L} \subset U, K \subset G} E_{\hat{L},K}(\hat{G})$. Moreover if U_1 and U_2 are two open sets with compact closures such that $U_1 \cap U_2 \neq \emptyset$ and $h \in E_{\hat{L},K}(\hat{G})$ where $\hat{L} \subset U_1 \cap U_2$, then we also have

$$\langle h, \hat{\omega}_{U_1} \rangle = \langle h, \hat{\omega}_{U_1 \cap U_2} \rangle = \langle h, \hat{\omega}_{U_2} \rangle .$$

Thus $\hat{\omega}_{U_1} = \hat{\omega}_{U_2}$ on $U_1 \cap U_2$.

Let $\{U_\alpha\}$ be a locally finite cover of G by open sets with compact closures and let $\{f_\alpha\} \subset A(G) \cap \mathcal{K}(G)$ be a locally finite family of functions satisfying the conclusions of Lemma 8. If $h = p * q$ as before, we define

$$\langle h, \hat{\sigma} \rangle = \langle p * q, \hat{\sigma} \rangle = \sum_\alpha \langle f_\alpha p * q, \hat{\omega}_{U_\alpha} \rangle .$$

Since all but a finite number of the f_α's vanish identically on the support of p, the sum is finite. Now, the support of $f_\alpha p$ is contained in U_α and so $\langle f_\alpha p * q, \hat{\omega}_{U_\alpha} \rangle$ is well defined. Now extend $\hat{\sigma}$ to $E_{\hat{L},K}(\hat{G})$ linearly to obtain a linear functional on $E(\hat{G})$. This extension is possible since $\sum_{j=1}^{\infty} \| p_j \|_\infty \| q_j \|_1 < \infty$. We can also write if $h = \sum_{j=1}^{\infty} p_j * q_j$

$$\langle h, \hat{\sigma} \rangle = \sum_j \sum_\alpha \langle f_\alpha p_j * q_j , \hat{\omega}_{U_\alpha} \rangle$$

$$= \sum_\alpha \sum_j \langle f_\alpha p_j * q_j , \hat{\omega}_{U_\alpha} \rangle$$

$$= \sum_\alpha \langle (\sum_j f_\alpha p_j * q_j), \hat{\omega}_{U_\alpha} \rangle .$$

Now suppose $\hat{L} \subset \hat{G}$ is compact and $\{h_\beta\} \subset E_{\hat{L},K}(\hat{G})$ is such that $\| h_\beta \|_{\hat{L},K} \to 0$. If $h_\beta = \sum_j p_j^\beta * q_j^\beta$ then, since $0 \leq f_\alpha \leq 1$ for each α we have

$$\| \sum_j f_\alpha p_j^\beta * q_j^\beta \|_{\hat{L},K} \leq \sum_j \| f_\alpha p_j^\beta \|_\infty \| q_j^\beta \|_1 \leq$$

$$\leq \sum_j \| p_j \|_\infty \| q_j \|_1$$

which implies that $\| \sum_j f_\alpha p_j^\beta * q_j^\beta \|_{\hat{L},K} \leq \| h_\beta \|_{\hat{L},K}$.

Hence, $\lim_\beta \langle h_\beta, \hat{\sigma} \rangle = 0$. This shows that σ is a continuous linear functional on $E_{\hat{L},K}(\hat{G})$ and hence is a parameasure.

$\hat{\sigma}$ is called the <u>Fourier transform</u> of σ.

If $U \subset \hat{G}$ is an open set with compact closure then $\hat{\sigma}$ restricted to U is equal to $\hat{\omega}_U$. In fact if $h \in E_{\hat{L},K}(\hat{G})$ where $\hat{L} \subset U$, then for $h = \sum p_j * q_j$

$$\begin{aligned}\langle h, \hat{\sigma}_U \rangle = \langle h, \hat{\sigma} \rangle &= \sum_\alpha \langle (\sum_j f_\alpha p_j * q_j), \hat{\omega}_{U_\alpha} \rangle \\ &= \sum_\alpha \langle (\sum_j f_\alpha p_j * q_j), \hat{\omega}_{U_\alpha \cap U} \rangle \\ &= \sum_\alpha \langle (\sum_j f_\alpha p_j * q_j), \hat{\omega}_U \rangle \\ &= \langle \sum_j \sum_\alpha f_\alpha p_j * q_j, \hat{\omega}_U \rangle \\ &= \langle \sum_j p_j * q_j, \hat{\omega}_U \rangle = \langle h, \hat{\omega}_U \rangle\end{aligned}$$

Hence $\hat{\sigma}_U = \hat{\omega}_U$.

<u>Definition 7</u>. Let $f \in \mathcal{K}(G)$ and $\sigma \in W(G)$. Then we define $\sigma * f$ to be that element of $W(G)$ such that

$$\langle h, \sigma * f \rangle = \langle \tilde{f} * h, \sigma \rangle \qquad h \in E(G).$$

If $\sigma \in \mathfrak{M}(S(G))$ corresponds to the transformation T on $S(G)$, we write $\|\sigma\| = \|T\|$.

LEMMA 10. <u>Let</u> $f \in \mathcal{K}(G)$. <u>If</u> $\sigma \in \mathfrak{M}(S(G))$ <u>then</u> $\sigma * f \in \mathfrak{M}(S(G))$ <u>and</u>

$$\|\sigma * f\| \leq \|\sigma\| \; \|f\|.$$

<u>Proof</u>. Straightforward.

<u>Remark</u>. If $\sigma \in \mathfrak{M}(S(G))$ and $g \in B(G)$ then it is easy to verify that $(\sigma * f) * g = (\sigma * g) * f$.

LEMMA 11. <u>If</u> $f \in \mathcal{K}(G)$ <u>and</u> $\sigma \in \mathfrak{M}(S(G))$ <u>then</u> $(\sigma * f)^\wedge = \hat{\sigma}\,\hat{f}$.

<u>Proof</u>. Let $h \in E_{\hat{L},K}(\hat{G})$ and let U be an open neighbourhood of $\hat{L}$ with compact closure. Choose k and g as before so that $\check{k}\,\check{g} = 1$ on U. Then we have

$$\langle h, \hat{\sigma}\,\hat{f}\rangle = \langle \hat{\tilde{f}}\,h, \hat{\sigma}\rangle = \langle \hat{\tilde{f}}\,h, \hat{g}(\sigma * k)^\wedge\rangle$$
$$= \langle \hat{\tilde{g}}\,\hat{\tilde{f}}\,h, (\sigma * k)^\wedge\rangle = \langle \tilde{g} * \tilde{f} * \check{h}, \sigma * k\rangle$$

and

$$\langle h, (\sigma * f)^\wedge\rangle = \langle h, \hat{g}(\sigma * f * k)^\wedge\rangle = \langle \hat{\tilde{g}}\,h, (\sigma * f * k)^\wedge\rangle$$
$$= \langle \tilde{g} * \check{h}, \sigma * f * k\rangle = \langle \tilde{g} * \check{h}, \sigma * k * f\rangle$$
$$= \langle \tilde{g} * \tilde{f} * \check{h}, \sigma * k\rangle .$$

Thus $\langle h, (\sigma * f)^\wedge\rangle = \langle h, \hat{\sigma}\,\hat{f}\rangle$ for all $h \in E\,(G)$ and the conclusion of the lemma follows.

Now let $T \in \mathfrak{M}(S(G))$. We shall show that there exists a unique parameasure $\hat{\sigma} \in W(\hat{G})$ such that $\widehat{Tf} = \hat{\sigma}\,\hat{f}$ for each $f \in B(G)$.

Let us first notice that $\hat{f} \in \mathcal{K}(\hat{G})$ and the parameasure of $\hat{\sigma}\,\hat{f}$ is defined earlier. Let $\{\mu_\beta\}$ be an approximate identity in $L^1(\hat{G})$ such that $\check{\mu}_\beta$ has compact support and $\|\mu_\beta\|_1 = 1$. Set $F_\beta = \hat{f} * \mu_\beta$. Then by definition 5, we have if $h \in E\,(G)$

$$\langle h, \hat{\sigma}\,\hat{f}\rangle = \lim_\beta \langle h, \hat{\sigma}\, F_\beta\rangle .$$

Now let $h \in E_{\hat{L},K}(\hat{G})$. Then by Lemma 11, $\hat{\sigma}\, F_\beta = (\sigma * f\,\check{\mu}_\beta)^\wedge$.

Choose U, g and k as in the proof of Lemma 11. Then

$$\langle h, (\sigma * f)^\wedge \rangle - \langle h, \hat{\sigma} F_\beta \rangle = \langle h, (\sigma * f)^\wedge \rangle - \langle h, (\sigma * f \check{\mu}_\beta)^\wedge \rangle$$

$$= \langle h, \hat{g}(\sigma * f * k)^\wedge \rangle - \langle h, \hat{g}(\sigma * f \check{\mu}_\beta * k)^\wedge \rangle$$

$$= \langle \hat{\tilde{g}} h, (\sigma * f * k)^\wedge \rangle - \langle \hat{\tilde{g}} h, (\sigma * f \check{\mu}_\beta * k)^\wedge \rangle$$

$$= \langle \tilde{g} * \check{h}, \sigma * f * k \rangle - \langle \tilde{g} * \check{h}, \sigma * f \check{\mu}_\beta * k \rangle$$

$$= \langle \tilde{g} * \check{h}, \sigma * k * f \rangle - \langle \tilde{g} * \check{h}, \sigma * k * f \check{\mu}_\beta \rangle$$

$$= \langle \tilde{g} * \check{h}, \sigma * k * (f - f \check{\mu}_\beta) \rangle$$

$$= \int_G (\tilde{g} * \check{h})(x) \, (\sigma * k * (f - f \check{\mu}_\beta))(x) \, dx$$

$$= \int_G (\tilde{f} - \widetilde{f \check{\mu}_\beta})(x) \, (\sigma * k * g * \tilde{\check{h}})(x) \, dx .$$

Hence

$$|\langle h, (\sigma * f)^\wedge \rangle - \langle h, \hat{\sigma} F_\beta \rangle| \leq \| f - f \check{\mu}_\beta \|_\infty \| \sigma * k * g * \tilde{\check{h}} \|_1$$

$$\leq \| \hat{f} - \hat{f} * \mu_\beta \|_1 \| \sigma * k \|_1 \| g * \tilde{\check{h}} \|_1 \longrightarrow 0$$

taking the limit over β. Hence we obtain for each $f \in B(G)$, $h \in E(G)$ and $\sigma \in \mathfrak{M}(S(G))$

$$\langle h, \hat{\sigma} \hat{f} \rangle = \lim_\beta \langle h, \hat{\sigma} F_\beta \rangle = \langle h, (\sigma * f)^\wedge \rangle .$$

Thus we conclude that

$$(\sigma * f)^\wedge = \hat{\sigma} \hat{f} .$$

Thus the representation $T f = \sigma * f$ for $f \in B(G)$ is equivalent to $\widehat{T} f = \hat{\sigma} \hat{f}$.

Now we show that the mapping $\sigma \longrightarrow \hat{\sigma}$ is one to one. Suppose $\hat{\sigma} = 0$. Then if $h \in E(G)$

$$\langle h, (\sigma * f)^{\wedge} \rangle = \langle \hat{h}, \hat{\sigma} \hat{f} \rangle = \lim_{\beta} \langle h, \hat{\sigma} F_{\beta} \rangle = \lim_{\beta} \langle \tilde{F}_{\beta} h, \hat{\sigma} \rangle = 0.$$

Hence $(\sigma * f)^{\wedge} = 0$. Since the Fourier transform is one to one on $L^1(G)$ we conclude that $\sigma * f = 0$ for each $f \in B(G)$. Hence $\sigma = 0$. Thus we have proved the following:

THEOREM 3. <u>The Fourier transform</u> $\sigma \longrightarrow \hat{\sigma}$ <u>defines a linear one to one transformation from</u> $\mathfrak{M}(S(G))$ <u>onto a linear subspace of</u> $W(G)$. <u>Further if</u> $T \in \mathfrak{M}(S(G))$ <u>then there exists a unique para-measure</u> $\hat{\sigma} \in W(G)$ <u>such that</u> $\widehat{Tf} = \hat{\sigma} \hat{f}$ <u>for each</u> $f \in B(G)$.

<u>References</u> .

1 Gaudry, G.I., Quasimeasures and operators commuting with convolution, Pacific J.Math. 18(3) (1966) 461-476.

2 ------------, Multipliers of type (P,Q), Pacific J.Math. 18(3) (1966), 477-488.

3 Larsen, R., The multiplier problem, Lecture notes in mathematics No.105, Springer-Verlag, 1969.

4 Reiter, H., Classical analysis and locally compact groups, Oxford Mathematical Monographs, 1968.

5 -----------, L^1-algebras and Segal algebras, Lecture notes in mathematics No. 231, Springer-Verlag, 1971.

6 Rudin, W., Fourier analysis on groups, Interscience Pub. 1962

SEGAL ALGEBRAS OF BEURLING TYPE

K.R.Unni

The purpose of this paper is to show that on a nondiscrete locally compact noncompact abelian group G the space of functions in $L^1(G)$ whose Fourier transform satisfies certain conditions of Beurling is a Segal algebra and show that each multiplier on this algebra is a bounded measure, if G is noncompact.

1. Beurling spaces

Let G be a locally compact abelian group with Haar measure dx. Along with Beurling [1] we consider a normed family Ω of strictly positive functions ω on G, which are measurable, summable with respect to dx, together with the norm $N(\omega)$, satisfying the following conditions

(1) For each $\omega \in \Omega$, $N(\omega)$ is finite and

$$0 < \int \omega \, dx \leq N(\omega)$$

(2) If $\omega \in \Omega$, then $\frac{1}{\omega}$ is locally L^∞ on G

(3) For each positive number λ and each $\omega \in \Omega$ we have $\lambda\omega \in \Omega$ and

$$N(\lambda\omega) = \lambda N(\omega)$$

(4) If $\omega_1, \omega_2 \in \Omega$, then the sum $\omega_1 + \omega_2$ is in Ω and

$$N(\omega_1 + \omega_2) \leq N(\omega_1) + N(\omega_2)$$

(5) Ω is complete under the norm N. That is, for any sequence $\{\omega_n\}_{n=1}^{\infty} \subset \Omega$ such that $\sum_{n=1}^{\infty} N(\omega_n) < \infty$ then $\omega = \sum_{n=1}^{\infty} \omega_n$ is in Ω and

$$N(\omega) \leq \sum_{n=1}^{\infty} N(\omega_n) .$$

Let Ω_0 be the subset of Ω consisting of those ω such that

$$N(\omega) = 1 .$$

Let $1 < p < \infty$ and let q be the conjugate index of p given by

$$\frac{1}{p} + \frac{1}{q} = 1 .$$

For a fixed p, we set

$$\omega' = \frac{1}{\omega^{p-1}} .$$

For each $\omega \in \Omega_0$, we obtain the Banach spaces $L^p_{\omega'}(G)$ and $L^q_{\omega}(G)$ of functions measurable on G and having the norms

$$\| f \|_{L^p_{\omega'}} = \left\{ \int |f|^p \omega' dx \right\}^{1/p}$$

and

$$\| \varphi \|_{L^q_{\omega}} = \left\{ \int |\varphi|^q \omega \, dx \right\}^{1/q}$$

respectively. We set

$$\Lambda^p(G) = \bigcup_{\omega \in \Omega_0} L^p_{\omega'}(G)$$

and

$$Ⓗ^q(G) = \bigcap_{\omega \in \Omega_0} L^q_{\omega}(G) .$$

Then we have the following theorem of Beurling [1] .

Let $1 < p < \infty$ and $q = \frac{p}{p-1}$. Then both $\Lambda^p(G)$ and $Ⓗ^q(G)$ are Banach spaces if they are supplied with the norms given by

$$\| f \| = \| f \|_{\Lambda^p} = \inf_{\omega \in \Omega_0} \| f \|_{L^p_{\omega'}} \tag{1}$$

and

$$\| \varphi \| = \| \varphi \|_{\Theta^q} = \sup_{\omega \in \Omega_0} \| \varphi \|_{L^q_\omega} \tag{2}$$

respectively. Moreover $\Theta^q(G)$ is the dual of $\Lambda^p(G)$ in the sense that each continuous linear functional L on $\Lambda^p(G)$ has the representation

$$L(f) = \int \varphi f \, dx \tag{3}$$

where φ is a unique element of $\Theta^q(G)$ and

$$\sup_{f \in \Lambda^p(G)} \frac{|\int \varphi f \, dx|}{\| f \|} = \| \varphi \| . \tag{4}$$

We remark that $\Lambda^p(G)$ is a subset of $L^1(G)$ and $L^\infty(G)$ is a subset of $\Theta^q(G)$ we also have the relations

$$\| f \|_1 \leq \| f \| \qquad f \in \Lambda^p(G)$$

and

$$\| \varphi \| \leq \| \varphi \|_\infty \qquad \varphi \in \Theta^q(G)$$

Beurling's theorem remains valid even in the case $p = 1$.

From the definitions, we can conclude that the spaces $\Lambda^p(G)$ and $\Theta^q(G)$ satisfy the following inclusion relations. If $1 < p_1 < p$, then

$$\Lambda^p(G) \subset \Lambda^{p_1}(G) ; \quad \Theta^q(G) \supset \Theta^{q_1}(G) . \tag{5}$$

If $\mathcal{K}(G)$ denotes the space of all continuous functions on G with compact support, then it follows that

$$\mathcal{K}(G) \subset \Lambda^p(G) \tag{6}$$

since we have assumed that $\frac{1}{\omega}$ is locally L^{∞} on G.

We also observe that if $g \in L^{\infty}(G)$ and $f \in \Lambda^p(G)$ then $g f \in \Lambda^p(G)$ and

$$\| g f \| \leq \| g \|_{\infty} \| f \| . \tag{7}$$

2. <u>The algebras</u> $S^p(G)$.

Let $\hat{G}$ denote the character group of G. The Haar measure $d\gamma$ on $\hat{G}$ is so chosen that the Fourier inversion theorem is valid. Now suppose that $1 \leq p < \infty$. Let $S^p(G)$ denote the translation invariant subspace of $L^1(G)$ consisting of all those f in $L^1(G)$ such that its Fourier transform $\hat{f}$ belongs to $\Lambda^p(\hat{G})$. For each p, we introduce a norm on $S^p(G)$ by setting

$$\| f \|_S = \| f \|_1 + \| \hat{f} \| \qquad f \in S^p(G) \tag{8}$$

It is easy to verify that $\| \cdot \|_S$ is a norm on $S^p(G)$. Further if $1 < p_1 < p$ then

$$S^p(G) \subset S^{p_1}(G)$$

THEOREM 1. <u>Let</u> G <u>be a locally compact nondiscrete abelian group and</u> $1 \leq p < \infty$. <u>Then</u> $S^p(G)$ <u>is a Segal algebra. That is, it satisfies the following conditions</u>

(a) $S^p(G)$ <u>is dense in</u> $L^1(G)$

(b) $S^p(G)$ <u>is a Banach space under the norm</u> $\| \cdot \|_S$ <u>and</u>

$$\| f \|_S \geq \| f \|_1 \qquad f \in S^p(G).$$

(c) <u>Let</u> $y \in G$ <u>and</u> τ_y <u>denote the translation operator</u>. <u>For each</u> $f \in S^p(G)$, $\tau_y f$ <u>belongs to</u> $S^p(G)$ <u>and the mapping</u> $y \longrightarrow \tau_y f$ <u>is continuous from</u> G <u>into</u> $S^p(G)$.

(d) $\| \tau_y f \|_S = \| f \|_S$ <u>for all</u> $f \in S^p(G)$ <u>and all</u> $y \in G$.

Proof. If

$$B(G) = \left\{ f \in L^1(G) : \hat{f} \in \mathcal{K}(\hat{G}) \right\}$$

it is a consequence of (6) that

$$B(G) \subset S^p(G) \subset L^1(G).$$

Since $B(G)$ is norm dense in $L^1(G)$ it follows that $S^p(G)$ is norm dense in $L^1(G)$ also.

We shall now show that $S^p(G)$ is a Banach space under the norm $\|\cdot\|_S$. Let $\{f_n\}_{n=1}^{\infty} \subset S^p(G)$ be a sequence such that $\sum_{n=1}^{\infty} \|f_n\|_S < \infty$. We shall show that there exists an $f \in S^p(G)$ such that

$$\lim_{n\to\infty} \left\| \sum_{k=1}^{n} f_k - f \right\|_S = 0. \tag{9}$$

By the definition of the norm $\|\cdot\|_S$, it follows that $\sum_{n=1}^{\infty} \|f_n\|_1 < \infty$ and $\sum_{n=1}^{\infty} \|\hat{f}_n\| < \infty$. Using the completeness of the spaces $L^1(G)$ and $\Lambda^p(\hat{G})$ we obtain $f \in L^1(G)$ and $g \in \Lambda^p(\hat{G})$ such that

$$\lim_{n\to\infty} \left\| \sum_{k=1}^{n} f_k - f \right\|_1 = 0 \tag{10}$$

and

$$\lim_{n\to\infty} \left\| \sum_{k=1}^{n} \hat{f}_k - g \right\| = 0. \tag{11}$$

Since (11) holds, closely following the proof given by Beurling [1,p.5-6] for the completeness of his space, it is possible to find an $\omega \in \Omega$ such that

$$\lim_{n\to\infty} \int \frac{\left|\sum_{k=1}^{n} \hat{f}_k - g\right|^p}{\omega^{p-1}}\, dx = 0$$

Hence $\sum_{k=1}^{n} \hat{f}_k$ converges to g almost everywhere. But

$$\lim_{n\to\infty} \|\sum_{k=1}^{n} \hat{f}_k - \hat{f}\|_\infty \leq \lim_{n\to\infty} \|\sum_{k=1}^{n} f_k - f\|_1 = 0 .$$

Thus $\hat{f} = g$ and $f \in S^p(G)$. It is clear that (9) is satisfied. We have thus proved (a) and (b). (d) is trivial.

To complete the proof we have only to show that given $f \in S^p(G)$ and $\varepsilon > 0$, there is a neighbourhood U of 0 such that

$$\|\tau_y f - f\|_S < \varepsilon \qquad \text{if} \quad y \in U. \tag{12}$$

There exists $\omega \in \Omega_0$ such that $\hat{f} \in L^p_{\omega'}(\hat{G})$. Now the functions $f \in L^1(G)$ such that $\hat{f} \in L^p_{\omega'}(\hat{G})$ form a Segal algebra with the norm $|||f||| = \|f\|_1 + \|\hat{f}\|_{L^p_{\omega'}}$ (See Reiter [3], §5 Example (vi) pp. 25-26) and hence there exists a neighbourhood U of 0 such that

$$|||\tau_y f - f||| < \varepsilon \qquad \text{for all} \quad y \in U. \tag{13}$$

(12) is now a consequence of (13).

The algebras $S^p(G)$ are termed Segal algebras of Beurling type.

3. Multipliers on $S^p(G)$.

A multiplier on $S^p(G)$ is a bounded linear operator on $S^p(G)$ which commutes with translations. The space of multipliers on $S^p(G)$ is denoted by $\mathfrak{M}(S^p(G))$.

THEOREM 2. *Let G be a nondiscrete locally compact abelian and let $1 \leq p < \infty$. If $T \in \mathfrak{M}(S^p(G))$ then there exists a unique pseudomeasure σ such that $Tf = \sigma * f$ for each $f \in S^p(G)$. If in addition G is non-compact then $\mathfrak{M}(S^p(G))$ is isometrically isomorphic to $M(G)$, the space of bounded Borel measures on G.*

Proof. The first part is proved by the author in [4] for any Segal algebra on a locally compact abelian group G. First we observe that $B(G) \subset L^p(G) \cap C_o(G)$ for $1 \leq p < \infty$ where $C_o(G)$ is the space of all continuous functions on G which vanish at infinity. We now assume that $2 \leq p < \infty$ and let $T \in \mathfrak{M}(S^p(G))$. Let $f \in B(G)$. Choose an $\omega \in \Omega_0$ such that $\hat{f} \in L^p_{\omega'}$. Then if $\| T \|$ denotes the operator norm of T, we have

$$\| Tf \|_1 \leq \| Tf \|_S \leq \| T \| (\| f \|_1 + \| \hat{f} \|)$$

$$\leq \| T \| (\| f \|_1 + \| \hat{f} \|_{L^p_{\omega'}}) . \quad (14)$$

Let K denote the support of f. Then there exists a constant $C(K,\omega)$ which depends on K and ω such that

$$\| \hat{f} \|_{L^p_{\omega'}} \leq C(K, \omega) \| \hat{f} \|_p . \quad (15)$$

If q is given by $\frac{1}{p} + \frac{1}{q} = 1$, then $1 < q \leq 2$. Since $f \in B(G) \subset L^q(G)$, using the Hausdorff-Young inequality, we obtain from (14) and (15)

$$\| Tf \|_1 \leq \| T \| (\| f \|_1 + C(K,\omega) \| f \|_q) . \quad (16)$$

Since G is noncompact for each $f \in L^p(G)$ we have

$$\lim_{y \to \infty} \| f + \tau_y f \|_p = 2^{1/p} \| f \|_p \quad (17)$$

where τ_y denotes the translation operator. For f in $B(G)$ the Fourier transforms of f and $\tau_y f$ have the same support so that we can deduce, from (16) by virtue of (17), that

$$\| Tf \|_1 \leq \| T \| (\| f \|_1 + C(K, \omega) 2^{1/q - 1} \| f \|_q) .$$

Repeating the same argument n times we have

$$\| Tf \|_1 \leq \| T \| \left(\| f \|_1 + C(K,\omega)\, 2^{n(\frac{1}{q}-1)} \| f \|_q \right).$$

Since $\frac{1}{q} - 1 < 0$ we conclude that

$$\| Tf \|_1 \leq \| T \| \, \| f \|_1 \qquad f \in B(G)$$

and thus T is bounded on $B(G)$ which is norm dense in $L^1(G)$ and hence can be extended uniquely as a bounded linear operator from $L^1(G)$ to $L^1(G)$ and it is easy to see that this extended T is also a multiplier on $L^1(G)$. Hence there exists a unique $\mu \in M(G)$ such that

$$T f = \mu * f \, , \qquad f \in B(G) \tag{18}$$

and $\| \mu \| \leq \| T \|$. By standard arguments, (18) holds for all $f \in S^p(G)$.

When $p = 1$, the algebra $S^1(G)$ is the same as $A^1(G)$, consisting of those functions f in $L^1(G)$ such that $\hat{f} \in L^1(\hat{G})$ with the norm $\| f \|_S = \| f \|_1 + \| \hat{f} \|_1$, and the conclusion of our theorem is well known.

Now suppose that $1 < p \leq 2$. Then $S^2(G) \subset S^p(G)$. Since the algebras $S^p(G)$ are semisimple and if $T \in \mathfrak{M}(S^p(G))$ there exists a bounded continuous function φ on $\hat{G}$ such that $\widehat{Tf} = \varphi \hat{f}$ for $f \in S^p(G)$ and hence $\widehat{Tf} = \varphi \hat{f}$ for each $f \in S^2(G)$ and $T \in \mathfrak{M}(S^2(G))$ and the representation (18) is immediate.

On the other hand, if T is given by $Tf = \mu * f$ for $f \in S^p(G)$ where $\mu \in M(G)$ then it is easy to see that $T \in \mathfrak{M}(S^p(G))$ and $\| T \| \leq \| \mu \|$. Hence $\| T \| = \| \mu \|$ and the proof of our theorem is completed.

References

1 Beurling, A., Construction and analysis of some convolution algebras, Ann. Inst. Fourier, Grenoble 14, 2 (1964) 1-32.

2 Larsen, R., The multiplier problem, Lecture notes in Mathematics No.105, Springer-Verlag, 1969.

3 Reiter, H., L^1-algebras and Segal algebras, Lecture notes in Mathematics No.231, Springer-Verlag, 1971

4 Unni, K.R., A note on multipliers on a Segal algebra, Studia Mathematica, to appear.

SPLINES IN HILBERT SPACES *

Vimala Walter

1. Introduction

In 1964, I.J. Schoenberg ([13] , [14]) established two extremal properties of polynomial splines which became the starting point of a series of investigations for splines in Hilbert spaces. Marc Atteia [3] introduced both interpolating and smoothing splines in Hilbert spaces. If X and Y are two real Hilbert spaces, T is a continuous linear transformation of X onto Y and Φ is a given constraint set, then an interpolating spline σ of Φ relative to T is defined, if it exists, by

$$\| T\sigma \|_Y = \min_{\phi \in \Phi} \| T\phi \|_Y .$$

Various choices of X,Y,T and Φ have been considered by different authors. (see for example Atteia ([3] , [4] , [5]), Anselone and Laurent ([2]), Laurent [11] and Jerome and Schumaker [10]).

In this paper we study the properties of splines in a Hilbert space under weaker assumptions than those used by the earlier authors. We prove the existence of two distinct classes of splines and not just one class, as is generally supposed. Various interesting and new results for both interpolating and smoothing splines are also obtained.

Sections 2 and 4 deal with the existence and properties of two classes of interpolating and smoothing splines. In section 3 we characterise the interpolating splines belonging to different types of constraint sets. In section 5, we shall consider some special cases.

* Supported by a grant from the C.S.I.R. under the scheme 'Approximation by Spline functions and Related Topics'

Throughout this paper we use the following notation: If H is a Hilbert space, then $\|\cdot\|_H$, $\langle\ ,\ \rangle_H$ and θ_H will stand for the norm, inner product and the zero element respectively. If M is a closed subspace of H, then $M^\perp$ denotes its orthogonal complement and P_M the projection operator on M. $N(T)$ and $K(T)$ denote the kernel and co-kernel respectively of a transformation T on H. The empty set will be denoted by $\mathcal{N}$.

2. Interpolating spline functions.

Let X be a real Hilbert space. Suppose that A and B are two closed subspaces of X such that $A^\perp + B^\perp$ is closed. (In fact, this is true if either $A^\perp$ or $B^\perp$ is of finite dimension (see [9])).

If $a \in A$ and $b \in B$, we set

$$\Phi_a = a + A^\perp,$$

$$\Psi_b = b + B^\perp.$$

We consider two real Hilbert spaces Y and Z isomorphic to A and B respectively. If I_A and I_B denote the corresponding isomorphisms, it is easy to see that the transformations $T = I_A P_A$ and $\tau = I_B P_B$ are linear and continuous from X onto Y and Z with $N(T) = A^\perp$ and $N(\tau) = B^\perp$. Then we have the following

THEOREM 2.1. *There exist two sets* S_a *and* Σ_b *of elements in* X *satisfying*

$$\|\tau s_a\|_Z = \min_{\phi \in \Phi_a} \|\tau \phi\|_Z \quad \text{for all } s_a \in S_a$$

and

$$\|T\sigma_b\|_Y = \min_{\psi \in \Psi_b} \|T\psi\|_Y \quad \text{for all } \sigma_b \in \Sigma_b .$$

The sets S_a *and* Σ_b *reduce to single elements if and only if* $A^\perp \cap B^\perp = \{\theta_X\}$.

Proof. Since $A^{\perp} + B^{\perp}$ is closed, by a result of Atteia ([5],p.195), it follows that $\tau A^{\perp}$ and $TB^{\perp}$ are closed subspaces. Hence

$$\| \tau a - P_{\tau A^{\perp}}(\tau a) \|_Z = \min_{a^{\perp} \in A^{\perp}} \| \tau a - \tau a^{\perp} \|_Z$$

and

$$\| Tb - P_{TB^{\perp}}(Tb) \|_Y = \min_{b^{\perp} \in B^{\perp}} \| Tb - Tb^{\perp} \|_Y .$$

The required sets S_a and Σ_b are then given by

$$S_a = a - A_a^{\perp}$$

and

$$\Sigma_b = b - B_b^{\perp}$$

where

$$A_a^{\perp} = \{ a^{\perp} \in A^{\perp} \mid \tau a^{\perp} = P_{\tau A^{\perp}}(\tau a) \} ,$$

$$B_b^{\perp} = \{ b^{\perp} \in B^{\perp} \mid T b^{\perp} = P_{TB^{\perp}}(Tb) \} .$$

The proof of the last part of the theorem is obvious.

Definition 2.1. An element of S_a is called an interpolating spline of Φ_a relative to τ and an element of Σ_b an interpolating spline of Ψ_b relative to T.

THEOREM 2.2. *If* $S = \bigcup_{a \in A} S_a$, *then the class S of splines has the following properties*:

1) $\tau S = (\tau A^{\perp})^{\perp}$,

2) $S = (\tau^{*} \tau A^{\perp})^{\perp}$,

3) $\tau^{*} \tau S = A \cap B$

where τ^{*} *is the adjoint operator of* τ .

Proof. We first notice that the adjoint operator τ^{*} of τ exists and is a linear, continuous, one to one map of Z onto B.

Let $s \in S$. Then there exists $a \in A$ such that $s \in S_a$. By the definition of S_a, it follows that $\tau s \in (\tau A^\perp)^\perp$. This proves that $\tau S \subset (\tau A^\perp)^\perp$. To prove the converse, we proceed as follows:

Let $z \in (\tau A^\perp)^\perp$. Since there exists $x \in X$ such that $z = \tau x$, we have $\langle \tau x, \tau a^\perp \rangle_Z = 0$, for all $a^\perp \in A^\perp$. Putting $x = a_x + a_x^\perp$ with $a_x \in A$ and $a_x^\perp \in A^\perp$, it is seen that $\langle \tau a_x + \tau a_x^\perp, \tau a^\perp \rangle_Z = 0$ for all $a^\perp \in A^\perp$. This implies that $(-\tau a_x^\perp)$ is the projection of τa_x on $\tau A^\perp$ so that $z = \tau x \in \tau S_{a_x} \in \tau S$. Hence $(\tau A^\perp)^\perp \subset \tau S$.

We shall now prove (2). Let $s \in S$. Since $\tau S = (\tau A^\perp)^\perp$ (by (1)), we have

$$\langle \tau s, \tau a^\perp \rangle_Z = 0 \quad \text{for all } a^\perp \in A^\perp$$

or equivalently,

$$\langle s, \tau^* \tau a^\perp \rangle_X = 0 \quad \text{for all } a^\perp \in A^\perp.$$

Hence $s \in (\tau^* \tau A^\perp)^\perp$. This proves $S \subset (\tau^* \tau A^\perp)^\perp$. If $x \in (\tau^* \tau A^\perp)^\perp$, then $\langle x, \tau^* \tau a^\perp \rangle_X = 0$ for all $a^\perp \in A^\perp$, so that $\langle \tau x, \tau a^\perp \rangle_Z = 0$ for all $a^\perp \in A^\perp$, from which it follows that $x \in S$. Hence $(\tau^* \tau A^\perp)^\perp \subset S$.

If $s \in S$, then $\tau s \in (\tau A^\perp)^\perp$ and we have

$$\langle \tau s, \tau a^\perp \rangle_Z = 0 \quad \text{for all } a^\perp \in A^\perp$$

which implies that

$$\langle \tau^* \tau s, a^\perp \rangle_X = 0 \quad \text{for all } a^\perp \in A^\perp$$

so that $\tau^* \tau S \subset A$. On the other hand, τ^* maps Z onto B. Thus $\tau^* \tau S \subset A \cap B$. Conversely, let $x \in A \cap B$. τ^* being a continuous, linear, 1 - 1 map of Z onto B we can find a unique $\mathfrak{z} \in Z$ such that $\tau^* \mathfrak{z} = x$. Let $\mathfrak{z} = \tau x_0$. Then $\tau^* \tau x_0 \in A$ and so

$$\langle \tau^* \tau x_0 , a^\perp \rangle_X = 0 \quad \text{for all } a^\perp \in A^\perp$$

which is equivalent to

$$\langle \tau x_0 , \tau a^\perp \rangle_Z = 0 \quad \text{for all } a^\perp \in A^\perp .$$

Thus $x_0 \in S$ and hence $A \cap B \subset \tau^* \tau S$.

Similarly we can prove the following

THEOREM 2.3. *If* $\Sigma = \bigcup_{b \in B} \Sigma_b$, *then*

1) $T \Sigma = (T B^\perp)^\perp$,

2) $\Sigma = (T^* T B^\perp)^\perp$,

3) $T^* T \Sigma = B \cap A$

where T^* *is the adjoint operator of* T.

Remarks.

1) S and Σ are closed subspaces of X and they are connected by the relation

$$\tau^* \tau S = T^* T \Sigma .$$

Furthermore, $B^\perp \subset S$ and $A^\perp \subset \Sigma$.

2) If $s \in S$, then there exists $a \in A$ such that $s \in S_a$ and

$$\| \tau \phi - \tau s \|_Z^2 = \| \tau \phi \|_Z^2 - \| \tau s \|_Z^2 \quad \text{for all } \phi \in \Phi_a .$$

This is an analog of 'the first integral relation' for polynomial splines (see [1]).

3) τS is a closed subspace of Z and hence given

$$x = a_x + a_x^\perp \in X, \; a_x \in A, \; a_x^\perp \in A^\perp,$$

there exists $S_x \subset S$ satisfying

$$\| \tau x - \tau s_x \|_Z = \min_{s \in S} \| \tau x - \tau s \|_Z \text{ for all } s_x \in S_x$$

with $S_x = a_x - A^{\perp}_{a_x}$.

4) Results analogous to (2) and (3) also hold for the class Σ .

3. Existence and characterisation of interpolating splines belonging to certain constraint sets.

We shall assume throughout this section that the condition $A^{\perp} \cap B^{\perp} = \{\theta_X\}$ is satisfied by the subspaces A and B so that S_a and Σ_b reduce to single elements. We shall now consider the problem of finding the minimal element when the constraint set is of the form $\Phi_C = \bigcup_{a \in C} \Phi_a$ where C is a compact, convex subset of A. This includes, in particular, the problems considered by Atteia ([4]) and Jerome and Schumaker (p. 45, [10]). It may be mentioned that similar problems with different assumptions have been considered by Daniel and Schumaker (pg. 17, [6]) and Atteia (pg. 195, [5]).

We prove here the following

THEOREM 3.1. If C is a compact, convex subset of A, then there exists $a_o \in C$ such that if s_{a_o} denotes the unique spline belonging to Φ_{a_o} then

$$\| \tau s \|_Z = \min_{\phi \in \Phi_C} \| \tau \phi \|_Z \qquad (*)$$

for all s in the set $\Delta = [(s_{a_o} + B^{\perp}) \cap \Phi_C]$ and $\tau(\Delta) = z_o$ is the unique element of Z satisfying (*). Moreover, there exists a $\sigma_{b_o} \in \Sigma$ such that

$$\| T \sigma_{b_o} \|_Y \leq \| T s \|_Y \qquad \text{for all } s \in \Delta .$$

Further, Δ reduces to a single element if and only if $T(\Delta)$ reduces to a single element.

Proof. Corresponding to each $\Phi_a \subset \Phi_C$, there exists the interpolating spline $s_a \in \Phi_a$ such that

$$\| \tau s_a \|_Z = \min_{\phi \in \Phi_a} \| \tau \phi \|_Z .$$

Let $\bar{\Delta} = \{ s_a \mid a \in C \}$ and define $f(a) = \| \tau s_a \|_Z$ for $a \in C$. Then f is a continuous function defined on the compact set C and hence it attains its minimum value so that there exists $a_o \in C$ satisfying

$$\| \tau s_{a_o} \|_Z = \min_{a \in C} \| \tau s_a \|_Z = \min_{s_a \in \bar{\Delta}} \| \tau s_a \|_Z .$$

Since

$$\| \tau s_a \|_Z = \min_{\phi \in \Phi_a} \| \tau \phi \|_Z$$

it follows that

$$\| \tau s_{a_o} \|_Z = \min_{\phi \in \Phi_C} \| \tau \phi \|_Z . \tag{1}$$

Every element of the set $\Delta = [(s_{a_o} + B^{\perp}) \cap \Phi_C]$ also satisfies (1) and by virtue of remark 1 of § 2 every element of Δ is an interpolating spline. If we set $z_o = \tau s_{a_o}$, then z_o is the unique element of minimal norm in $\tau \Phi_C$ since Z is a real Hilbert space and C is convex. This completes the proof of the first part.

If $\Psi_{z_o} = \{ \phi \in X \mid \tau \phi = z_o \}$ then $\Delta \subset \Psi_{z_o}$.

Further, $\Psi_{z_o} = k_{z_o} + N(\tau)$, $k_{z_o} \in K(\tau)$ and $\tau k_{z_o} = z_o$

$$= b_o + B^{\perp}, \qquad b_o = k_{z_o} \in B .$$

Since $A^{\perp} + B^{\perp}$ is closed and $A^{\perp} \cap B^{\perp} = \{\theta_X\}$, there exists a spline $\sigma_{b_o} \in \Psi_{z_o}$ satisfying

$$\| T\sigma_{b_o} \|_Y = \min_{\varphi \in \Psi_{z_o}} \| T\varphi \|_Y .$$

In particular,

$$\| T\sigma_{b_o} \|_Y \leq \| T s \|_Y \qquad \text{for all } s \in \Delta .$$

Now Δ consists of a single element if and only if $P_A(\Delta) = \{ a_o \}$.
Since $T = I_A P_A$,

$$P_A(\Delta) = \{ a_o \} \Longleftrightarrow T(\Delta) = \{ y_o \} \text{ where } y_o = Ta_o .$$

Remarks.

Let $\{ e_i \}_{i \in I}$ be an orthonormal basis for the subspace A. The index set I is finite or infinite according as the dimension of A is finite or infinite. Since C is compact, for each $i \in I$, there exist $c_i^{(1)}, c_i^{(2)} \in C$ such that

$$\langle c_i^{(1)}, e_i \rangle_X \leq \langle a, e_i \rangle_X \leq \langle c_i^{(2)}, e_i \rangle_X \text{ for all } a \in C .$$

Hence

$$\Phi_C = \{ \phi \in X \mid \langle e_i, c_i^{(1)} \rangle_X \leq \langle e_i, \phi \rangle_X \leq \langle e_i, c_i^{(2)} \rangle_X, i \in I \} \tag{i}$$

Let $\Gamma = T(C)$. Then Γ is compact and convex. Now Φ_C can also be represented as

$$\Phi_C = \{ \phi \in X \mid T\phi \in \Gamma \} \tag{ii}$$

and

$$\Phi_C = \{ \phi \in X \mid P_A \phi \in C \} . \tag{iii}$$

The representation of Φ_C in the three equivalent forms (i),(ii) and (iii) allows for flexibility in the study of the interpolating splines belonging to the set Δ. The first representation is useful in characterising the splines of Δ, the second in studying the case when Δ contains just one element and the third in studying the minimisation problem of Theorem 3.1 in a dual form.

2) For any subset E of X define a function χ on X as follows:

$$\chi_E(x) = \begin{cases} 0 & \text{if } x \in E \\ +\infty & \text{if } x \notin E \end{cases}$$

and set

$$p(x) = \frac{1}{2} \| \tau x \|_Z^2 + \chi_C(P_A x) .$$

If g is a convex function defined on X, its dual g^* is defined by

$$g^*(v) = \sup_{u \in X} [\langle u, v \rangle_X - g(u)] .$$

Now set $\alpha(x) = \frac{1}{2} \| \tau x \|_Z^2$, $\beta(x) = \chi_C(P_A x)$ and $q(x) = -[\alpha^*(x) + \beta^*(x)]$. Using techniques similar to those in [5] , we can prove that

$$q(x) = \begin{cases} -\infty & \text{if } x \notin A \cap B \\ -\frac{1}{2} \| \tau^{*-1} x \|_Z^2 - \sup\limits_{c \in C} \langle x, c \rangle_X & \text{if } x \in A \cap B \end{cases}$$

and

$$\sup_{v \in X} q(v) = \inf_{\omega \in X} p(\omega) .$$

Thus the extremal problem of Theorem 3.1 can be viewed in any one of the following three equivalent forms:

(1) Minimise the norm of $\tau\phi$ for $\phi \in \Phi_C$

(2) Minimise $p(x) = \frac{1}{2} \| \tau x \|_Z^2 + \chi_C(P_A x)$ for $x \in X$

(3) Maximise $q(x) = -[\alpha^*(x) + \beta^*(x)]$ for $x \in A \cap B$.

The last part of Theorem 3.1 gives a necessary and sufficient condition for z_o to be the image of a unique spline. Since $\tau^* \tau S = T^* T \Sigma$, there exists a unique $y_o \in T \Sigma$ such that $\tau^* z_o = T^* y_o$. Now y_o is the image of a set of interpolating splines relative to T, i.e., there exists an interpolating spline $\sigma_{y_o} \in \Sigma$ such that $T(\sigma_{y_o} + A^{\perp}) = y_o$. We shall now give in terms of y_o a condition which is both necessary and sufficient for z_o to be the image of a unique spline.

THEOREM 3.2. <u>A necessary and sufficient condition for Δ to reduce to a single element is that the hyperplane of support of Γ with the equation</u>

$$\langle y_o, y \rangle_Y = \min_{\gamma \in \Gamma} \langle y_o, \gamma \rangle_Y$$

<u>meet Γ at a unique point.</u>

The proof of Theorem 3.2 runs along the same lines as the proof of a similar theorem in [5] (pg.199).

We shall now study in more detail the minimisation problem of Theorem 3.1 for a particular choice of the constraint set Φ_C. Suppose that the index set $I = \{1, 2, \ldots, 2N\}$, i.e., the dimension of A is 2N and the compact set Φ_C is such that

$$\langle e_i, c_i^{(1)} \rangle_X = \langle e_{i+N}, c_i^{(1)} \rangle_X \text{ and } \langle e_i, c_i^{(2)} \rangle_X = \langle e_{i+N}, c_i^{(2)} \rangle_X .$$

Set $k_i = e_i \ (1 \le i \le N)$ and $\ell_i = e_{i+N} \ (1 \le i \le N)$.

The set $\{k_i - \ell_i\}_1^N$ is a linearly independent set in A and spans a subspace, say $A_{k-\ell}$ of A. Consider the set

$$\Phi_{k\ell} = \{\phi \in X \mid \eta_i \le \langle k_i, \phi \rangle_X = \langle \ell_i, \phi \rangle_X \le \zeta_i, 1 \le i \le N\}$$

where $\langle c_i^{(1)}, e_i \rangle_X = \eta_i = \langle c_i^{(1)}, e_{i+N} \rangle_X, \quad 1 \le i \le N$

and $\langle c_i^{(2)}, e_i \rangle_X = b_i = \langle c_i^{(2)}, e_{i+N} \rangle_X$, $1 \leq i \leq N$.

Now $\Phi_{k\ell} = \Phi_C \cap A^{\perp}_{k-\ell}$. It is however, a particular case of a set of the form Φ_C itself and thus there exists a minimal element in $\tau \Phi_{k\ell}$ which is the image of a set of interpolating splines (see Theorem 3.1). Let $\tilde{z}$ be the unique element of $\tau \Phi_{k\ell}$ such that

$$\| \tilde{z} \|_Z = \min_{\phi \in \Phi_{k\ell}} \| \tau \phi \|_Z \qquad (**)$$

and $\tilde{\Delta} = \{ s \in X \mid \tau s = \tilde{z} \}$.

We shall now characterise the set $\tilde{\Delta}$ of interpolating splines. First we observe that

$$\langle \tau \phi - \tilde{z}, \tilde{z} \rangle_Z \geqslant 0 \qquad \text{for all } \phi \in \Phi_{k\ell}$$

$$\langle \tau^* \tilde{z}, \phi - s \rangle_X \geq 0 \text{ for all } \phi \in \Phi_{k\ell}, \text{ for all } s \in \tilde{\Delta}.$$

Now by theorem 2.2, $\tau^* \tilde{z} \in A \cap B$ and hence

$$\tau^* \tilde{z} = \sum_{i=1}^{N} \lambda_i k_i + \sum_{i=1}^{N} \mu_i \ell_i$$

with

$$\sum_{i=1}^{N} \lambda_i k_{i,B^\perp} + \sum_{i=1}^{N} \mu_i \ell_{i,B^\perp} = \theta_X$$

where

$$k_i = k_{i,B} + k_{i,B^\perp} \; ; \quad \ell_i = \ell_{i,B} + \ell_{i,B^\perp} \; ;$$

$$k_{i,B} , \ell_{i,B} \in B \; ; \quad k_{i,B^\perp} , \ell_{i,B^\perp} \in B^\perp .$$

Let

$$\Phi_\gamma = \{ \phi \in X \mid \langle k_i, \phi \rangle_X = \langle \ell_i, \phi \rangle_X = \gamma_i , \; 1 \leq i \leq N \} .$$

Then

$$\Phi_{k\ell} = \bigcup_{\gamma\in[\eta,\zeta]} \Phi_\gamma \text{ and } \quad \Phi_\gamma = \sum_{i=1}^{N} \{\gamma_i (k_i + \ell_i)\} + A^\perp .$$

Thus

$$\left\langle \sum_{i=1}^{N} \lambda_i k_i + \sum_{i=1}^{N} \mu_i \ell_i , \sum_{i=1}^{N} \gamma_i (k_i + \ell_i) - \sum_{i=1}^{N} \rho_i (k_i + \ell_i) \right\rangle_X \geq 0$$

where

$$\sum_{i=1}^{N} \rho_i (k_i + \ell_i) \in P_A(\tilde{\Delta}) \quad \text{and} \quad \eta_i \leq \gamma_i \leq \zeta_i .$$

Hence

$$\sum_{i=1}^{N} (\lambda_i + \mu_i)(\gamma_i - \rho_i) \geq 0 .$$

Suppose that for an index i_o : $\eta_{i_o} < \rho_{i_o} < \zeta_{i_o}$. We can find numbers γ_i' and γ_i'' such that $\eta_i \leq \gamma_i' \leq \zeta_i$, $\eta_i \leq \gamma_i'' \leq \zeta_i$ and $\gamma_i' = \gamma_i'' = \rho_i$, $i \neq i_o$ with $\rho_{i_o} - \gamma_{i_o}' < 0$ and $\rho_{i_o} - \gamma_{i_o}'' > 0$. Then $\lambda_{i_o} + \mu_{i_o} = 0$. On the other hand, if $\rho_{i_1} = \eta_{i_1}$, for some index $i_1 \in I$ then $\lambda_{i_1} + \mu_{i_1} \geq 0$. Similarly, for an index i_2 such that $\rho_{i_2} = \zeta_{i_2}$, $\lambda_{i_2} + \mu_{i_2} \leq 0$. Thus there exist a unique set of coefficients $\lambda_i^{(1)}$, $\mu_i^{(1)}$, $\lambda_i^{(2)}$, $\mu_i^{(2)}$, $\lambda_i^{(3)}$, $\mu_i^{(3)}$

at least some of which are non-zero, and the set $\tilde{\Delta}$ of splines is characterised as follows:

$$\tau^* \tilde{z} = \sum_{i\in I_1} (\lambda_i^{(1)} k_i + \mu_i^{(1)} \ell_i) + \sum_{i\in I_2} (\lambda_i^{(2)} k_i + \mu_i^{(2)} \ell_i) +$$

$$+ \sum_{i\in I_3} (\lambda_i^{(3)} k_i + \mu_i^{(3)} \ell_i)$$

with

$$\lambda_i^{(1)} + \mu_i^{(1)} \geq 0 , \quad \lambda_i^{(2)} + \mu_i^{(2)} \leq 0 , \quad \lambda_i^{(3)} + \mu_i^{(3)} = 0 ;$$

and

$$\sum_{i \in I_1} (\lambda_i^{(1)} k_{i,B^\perp} + \mu_i^{(1)} l_{i,B^\perp}) + \sum_{i \in I_2} (\lambda_i^{(2)} k_{i,B^\perp} + \mu_i^{(2)} l_{i,B^\perp}) + \\ + \sum_{i \in I_3} (\lambda_i^{(3)} k_{i,B^\perp} + \mu_i^{(3)} l_{i,B^\perp}) = \theta_X ;$$

where

$$I_1 = \{ i \mid 1 \leq i \leq N \text{ and } \langle k_i , s \rangle_X = \langle l_i , s \rangle_X = \eta_i \text{ for all } s \in \tilde{\Delta} \}$$

$$I_2 = \{ i \mid 1 \leq i \leq N \text{ and } \langle k_i , s \rangle_X = \langle l_i , s \rangle_X = b_i \text{ for all } s \in \tilde{\Delta} \}$$

$$I_3 = \{ i \mid 1 \leq i \leq N \text{ and } \eta_i < \langle k_i , s \rangle_X = \langle l_i , s \rangle_X < b_i \text{ for all } s \in \tilde{\Delta} \}$$

Remark.

The interpolating splines of $\Delta = [(s_{a_0} + B^\perp) \cap \Phi_C]$ can be characterised in a similar manner. We have

$$(C) \begin{cases} \tau^* z_0 = \sum_{i \in I'} \lambda_i e_i - \sum_{i \in I''} \mu_i e_i , \quad \lambda_i \geq 0 , \quad \mu_i \geq 0 \\ \sum_{i \in I'} \lambda_i e_{i,B^\perp} - \sum_{i \in I''} \mu_i e_{i,B^\perp} = \theta_X \\ I' = \{ i \in I \mid \langle e_i , s \rangle_X = \langle e_i , c_i^{(1)} \rangle_X \text{ for all } s \in \Delta \} \\ I'' = \{ i \in I \mid \langle e_i , s \rangle_X = \langle e_i , c_i^{(2)} \rangle_X \text{ for all } s \in \Delta \} \end{cases}$$

A characterisation of the type (C) has been obtained by Atteia in [4] when the index set $I = \{1,2, \ldots, N\}$.

We shall now study the minimal problem (* *) in a dual form.

We first define

$$\Psi(x,\lambda) = \frac{1}{2}\|\tau x\|_Z^2 - \sum_{i=1}^{N} m_i^{(1)}\left[\langle k_i, x\rangle_X - \eta_i\right] -$$

$$-\sum_{i=1}^{N} n_i^{(1)}\left[\langle \ell_i, x\rangle_X - \eta_i\right] + \sum_{i=1}^{N} m_i^{(2)}\left[\zeta_i - \langle k_i, x\rangle_X\right] +$$

$$+\sum_{i=1}^{N} n_i^{(2)}\left[\zeta_i - \langle \ell_i, x\rangle_X\right] + \sum_{i=1}^{N} m_i^{(3)}\left[\langle k_i, x\rangle_X - \langle \ell_i, x\rangle_X\right]$$

where

$$\lambda = \left(m_1^{(1)},\dots,m_N^{(1)},n_1^{(1)},\dots,n_N^{(1)},m_1^{(2)},\dots,m_N^{(2)},n_1^{(2)},\dots,n_N^{(2)},m_1^{(3)},\dots,m_N^{(3)}\right)$$

and set

$$\lambda^* = \left(m_1^{(1)*},\dots,m_N^{(1)*},n_1^{(1)*},\dots,n_N^{(1)*},m_1^{(2)*},\dots,m_N^{(2)*},n_1^{(2)*},\dots,n_N^{(2)*},m_1^{(3)*},\dots,m_N^{(3)*}\right)$$

with

$$m_i^{(1)*} = \begin{cases} \lambda_i^{(1)} & \text{if } i\in I_1 \\ 0 & \text{if } i\notin I_1 \end{cases} ;\quad m_i^{(2)*} = \begin{cases} \lambda_i^{(2)} & \text{if } i\in I_2 \\ 0 & \text{if } i\notin I_2 \end{cases} ;\quad m_i^{(3)*} = \begin{cases} \lambda_i^{(3)} & \text{if } i\in I_3 \\ 0 & \text{if } i\notin I_3 \end{cases}$$

$$n_i^{(1)*} = \begin{cases} \mu_i^{(1)} & \text{if } i\in I_1 \\ 0 & \text{if } i\notin I_1 \end{cases} ;\quad n_i^{(2)*} = \begin{cases} \mu_i^{(2)} & \text{if } i\in I_2 \\ 0 & \text{if } i\notin I_2 \end{cases}$$

and

$$\Lambda = \Big\{\lambda \mid \lambda = \left(m_1^{(1)},\dots,m_N^{(1)},n_1^{(1)},\dots,n_N^{(1)},m_1^{(2)},\dots,m_N^{(2)},n_1^{(2)},\dots,n_N^{(2)},m_1^{(3)},\dots,m_N^{(3)}\right),$$

$$m_i^{(1)} + n_i^{(1)} \geq 0,\ m_i^{(2)} + n_i^{(2)} \leq 0,\ m_i^{(3)} \text{ arbitrary},\ 1\leq i\leq N\Big\}$$

Now the problem of minimising $\tau\phi$ for $\phi \in \Phi_{kl}$ is equivalent to finding the value of $\Psi(s, \lambda^*)$ for $s \in \tilde{\Delta}$. Using the techniques of Atteia in [4], we can prove that

THEOREM 3.3.

$$\inf_{x \in X} \sup_{\lambda \in \Lambda} \Psi(x, \lambda) = \sup_{\lambda \in \Lambda} \inf_{x \in X} \Psi(x, \lambda) = \Psi(s, \lambda^*)$$

for all $s \in \tilde{\Delta}$.

We shall now construct a sequence of constraint sets, each of which contains an interpolating spline. From remark 2 of §2, since $B^{\perp} \subset S$, $S^{\perp} \cap B^{\perp} = \{\theta_X\}$ and $S^{\perp}$ and $B^{\perp}$ being orthogonal subspaces, $S^{\perp} + B^{\perp}$ is closed. Thus if we define

$$\Phi_{s_a} = s_a + S^{\perp}$$

and

$$s_a^1 = s_a - s_{s_a}^{\perp} \qquad \text{where} \quad \tau s_{s_a}^{\perp} = P_{\tau S^{\perp}}(\tau s_a)$$

then

$$\| \tau s_a^1 \|_Z = \min_{\phi \in \Phi_{s_a}} \| \tau \phi \|_Z .$$

Corresponding to each $s_a \in S$, there exists the interpolating spline s_a^1 of Φ_{s_a} relative to τ and a class $S_1 = \{ s_a^1 \in X \mid s_a \in S \}$ of interpolating splines relative to τ. Proceeding in this manner, we can construct a sequence Φ_n^a $(n = 1, 2, \ldots)$ of constraint sets such that the existence of the interpolating spline in one set implies the existence of the interpolating spline in the succeeding set. A set of the class $\{ \Phi_n^a \}_{n=1}^{\infty}$ is of the form $\Phi_n^a = s_n + S_n^{\perp}$, $s_n \in S_n$

where each S_n is a class of interpolating splines relative to τ. Further, for any positive integer n, we have

1) The spline of Φ_n^a is the unique element of minimal norm in Φ_{n+1}^a. This implies, in particular, that $\| s_n \|_X$ is a monotonic increasing function,

2) $B^{\perp} \subset \bigcap_{n=1}^{\infty} S_n$,

3) $\tau^*\tau S_{n+1} = S_n \cap B$,

4) For n = 1, 2, ... ,

$$\Phi_n^a \cap S_n = s_n ; \quad \Phi_n^a \cap S_n^{\perp} = \mathcal{N},$$

5) $S_{n+1} = (\tau^* \tau S_n^{\perp})^{\perp}$,

6) At the nth stage,

$$S_n^{\perp} = (\tau^* \tau)^n A^{\perp}$$

where $(\tau^* \tau)^n = (\tau^* \tau) (\tau^* \tau) --- (\tau^*\tau)$ applied n times.

All the results obtained in this section can be extended with suitable modifications to the case when $A^{\perp} \cap B^{\perp} \neq \{\theta_X\}$

If D is a compact convex subset of B, analogs of the results in this section can be obtained when we consider constraint sets of the form

$$\Psi_D = \bigcup_{b \in D} \Psi_b .$$

4. Smoothing spline functions.

Consider the product spaces $G = Z \times A$ and $H = Y \times B$. If $g_1 = (z_1, a_1)$ and $g_2 = (z_2, a_2)$ are any two elements of G and α a real number, we define addition and scalar multiplication in G by

$$g_1 + g_2 = (z_1, a_1) + (z_2, a_2) = (z_1 + z_2 , a_1 + a_2)$$

and

$$\alpha g = \alpha (z, a) = (\alpha z, \alpha a) .$$

We set

$$\langle g_1, g_2\rangle_G = \langle z_1, z_2\rangle_Z + \rho\langle a_1, a_2\rangle_X , \qquad \rho > 0$$

and

$$\|g\|_G = \{\langle g, g\rangle_G\}^{1/2} .$$

With the scalar product and norm thus defined, G is a real Hilbert space. Similarly, H can be made a Hilbert space by defining a suitable inner product in it. Let L and Q be mappings of X into G and H respectively defined by $Lx = (\tau x, P_A x)$ and $Q x = (T x, P_B x)$. Then L and Q are continuous linear transformations and

$$\|Lx\|_G \leq \{\|\tau\|^2 + \rho\}^{1/2}\|x\|_X \quad \text{and} \quad \|Qx\|_H \leq \{\|T\|^2 + \rho\}^{1/2}\|x\|_X .$$

Moreover, L X and Q X are closed subspaces of G and H respectively. The transformations L and Q have the same kernel, namely $A^{\perp} \cap B^{\perp}$ and the adjoint transformations L^* and Q^* exist. L^* is a continuous linear transformation of L X onto $(A^{\perp} \cap B^{\perp})^{\perp}$ and Q^* a continuous linear transformation of Q X onto $(B^{\perp} \cap A^{\perp})^{\perp}$. We now establish two extremal properties in the spaces G and H which are the basis for our definition of smoothing splines.

THEOREM 4.1. <u>Given</u> $g = (\theta_Z, a) \in G$ <u>and</u> $h = (\theta_Y, b) \in H$ <u>there exist two sets</u> $\hat{S}_g$ <u>and</u> $\hat{\Sigma}_h$ <u>of interpolating splines in</u> X <u>such that</u>

$$\|L\hat{s}_g - g\|_G = \min_{x \in X} \|Lx - g\|_G , \qquad \underline{\text{for all}}\ \hat{s}_g \in \hat{S}_g$$

<u>and</u>

$$\|Q\hat{\sigma}_h - h\|_H = \min_{x \in X} \|Qx - h\|_H \qquad \underline{\text{for all}}\ \hat{\sigma}_h \in \hat{\Sigma}_h .$$

<u>Further, the sets $\hat{S}_g$ and $\hat{\Sigma}_h$ reduce to a single element each if and only if $A^\perp \cap B^\perp = \{\theta_X\}$.</u>

<u>Proof</u>. Since L X is a closed subspace of G, given $g = (\theta_Z, a) \in G$, there exists a unique element $u_g \in LX$ such that u_g is the best approximation in L X to g. Similarly if $h = (\theta_Y, b) \in H$ is given we can find a unique $v_g \in QX$ giving the best approximation to h. Hence, there exist two sets $\hat{S}_g$ and $\hat{\Sigma}_h$ of elements in X such that

$$\| L\hat{s}_g - g \|_G = \min_{x \in X} \| Lx - g \|_G \quad \text{for all } \hat{s}_g \in \hat{S}_g$$

and

$$\| Q\hat{\sigma}_h - h \|_H = \min_{x \in X} \| Qx - h \|_H \quad \text{for all } \hat{\sigma}_h \in \hat{\Sigma}_h .$$

We assert that $\hat{S}_g$ and $\hat{\Sigma}_h$ are sets of interpolating splines. We show this only for $\hat{S}_g$, the case of $\hat{\Sigma}_h$ being similar. Now let $\hat{s}_g$ be any element of $\hat{S}_g$. Since $\hat{s}_g \in X$, we have $\hat{s}_g = a_g + a_g^\perp$ with $a_g \in A$ and $a_g^\perp \in A^\perp$. Define $\Phi_g = a_g + A^\perp$. Denote by S_g the set of interpolating splines of Φ_g relative to τ. Then

$$\| \tau s_g \|_Z = \min_{\phi \in \Phi_g} \| \tau \phi \|_Z \quad \text{for all } s_g \in S_g .$$

Consequently $\| \tau s_g \|_Z \leq \| \tau \hat{s}_g \|_Z$.

But $\| L\hat{s}_g - g \|_G \leq \| L s_g - g \|_G$ for all $s_g \in S_g$.

Hence $\| L\hat{s}_g - g \|_G = \| L s_g - g \|_G$ or equivalently,

$$\| \tau \hat{s}_g \|_Z = \| \tau s_g \|_Z .$$

Thus $\hat{s}_g = s_g + b^\perp$ where $b^\perp \in B^\perp$ and

$a_g + a_g^\perp = a_g - a_{a_g}^\perp + b^\perp$ and hence $\hat{s}_g \in S_g$.

But $\hat{S}_g = \hat{s}_g + (A^\perp \cap B^\perp) = s_g + (A^\perp \cap B^\perp) = S_g$ and our assertion is proved.

__Definition 4.1__. An element of $\hat{S}_g$ is called a __smoothing spline relative to__ L __and the point__ $g = (\theta_Z, a)$ and an element of $\hat{\Sigma}_h$ a __smoothing spline relative to__ Q __and the point__ $h = (\theta_Y, b)$.

THEOREM 4.2. __If__ $\hat{S} = \bigcup_{g \in \theta_Z \times A} \hat{S}_g$ __and__ $\hat{\Sigma} = \bigcup_{h \in \theta_Y \times B} \hat{\Sigma}_h$ __then__

1) $\hat{S} \equiv S$ __and__ $\hat{\Sigma} \equiv \Sigma$,

2) $L^* L S = A$ __and__ $Q^* Q \Sigma = B$.

__Proof__. From theorem 4.1, it follows that $\hat{S}_g \subset \hat{S} \Rightarrow \hat{S}_g \subset S$, i.e., $\hat{S} \subset S$. To prove the converse, consider $s_{a_0} \in S$. It is sufficient to find some $a \in A$ such that $g = (\theta_Z, a)$ and $\| L s_{a_0} - g \|_G = \min_{x \in X} \| Lx - g \|_G$. Since $\tau^* \tau S = A \cap B$ (Theorem 2.2), there exists $\hat{a} \in A \cap B$ such that $\tau^* \tau s_{a_0} = \hat{a}$.

Let $\tilde{a} = a_0 + \frac{\hat{a}}{\rho}$. We claim that $\tilde{a}$ is the required element. For, $\langle \rho(\tilde{a} - a_0), a \rangle_X = \langle \hat{a}, a \rangle_X = \langle \hat{a}, a + a^\perp \rangle_X$ for all $a \in A$, for all $a^\perp \in A^\perp$.

In other words,

$$\langle \rho(\tilde{a} - a_0), a \rangle_X = \langle \tau^* \tau s_{a_0}, a + a^\perp \rangle_X$$

for all $a \in A$, for all $a^\perp \in A^\perp$ which in turn implies that

$\langle (-\tau s_{a_0}, \tilde{a} - a_0), (\tau x, P_A x) \rangle_G = 0$ for all $x \in X$ and

so we have

$$(-\tau s_{a_0}, \tilde{a} - a_0) \in (LX)^\perp .$$

Since $(\theta_Z, \tilde{a}) = (-\tau s_{a_0}, \tilde{a} - a_0) + (\tau s_{a_0}, P_A s_{a_0})$,

$$\tilde{g} = (\theta_Z, \tilde{a}) = L s_{a_0} + (-\tau s_{a_0}, \tilde{a} - a_0)$$

or equivalently

$$\| L s_{a_0} - \tilde{g} \|_G = \min_{x \in X} \| Lx - \tilde{g} \|_G$$

and so $S \subset \hat{S}$. Similarly $\hat{\Sigma} \equiv \Sigma$. To prove the second part, observe that for $s \in S$,

$$\langle L a^{\perp}, L s \rangle_G = \langle \tau a^{\perp}, \tau s \rangle_Z = 0 \text{ for all } a^{\perp} \in A^{\perp}$$

and hence $L^* L S \subset A$. Now L^* maps LX onto $(A^{\perp} \cap B^{\perp})^{\perp}$ and so given $a \in A$, there exists $x \in X$ such that $L^* L x = a$. Thus

$$\langle L^* L x , a^{\perp} \rangle_X = 0 \quad \text{for all} \quad a^{\perp} \in A^{\perp}$$

or equivalently,

$$\langle L x , L a^{\perp} \rangle_G = 0 \quad \text{for all} \quad a^{\perp} \in A^{\perp} .$$

We have

$$\langle (\tau x , P_A x) , (\tau a^{\perp}, P_A a^{\perp}) \rangle_G = 0 \quad \text{for all} \quad a^{\perp} \in A^{\perp}$$

and hence

$$\langle \tau x , \tau a^{\perp} \rangle_Z + \rho \langle P_A x , P_A a^{\perp} \rangle_X = 0 .$$

Since for $a^{\perp} \in A^{\perp}$, $P_A a^{\perp} = \theta_X$, this implies

$$\langle \tau x , \tau a^{\perp} \rangle_Z = 0 \quad \text{for all} \quad a^{\perp} \in A^{\perp} .$$

Thus $\tau x \in (\tau A^{\perp})^{\perp}$, i.e., $\tau x \in \tau S$ (from Theorem 2.2) From remark 1 of § 2, $B^{\perp} \subset S$ and hence $x \in S$. Thus $A \subset L^* L S$. Similarly, we can prove that $Q^* Q \Sigma = B$. This completes the proof of the theorem.

5. Some Special Cases.

1) Let $\dim (A) = n$, $Z = L_2 [a,b]$, $\tau = \mathcal{L}$ where $\mathcal{L}$ is a differential operator of the form

$$\mathcal{L} = \sum_{j=0}^{q} a_j \left(\frac{d}{dx}\right)^j, \quad a_q \neq 0 \quad \text{on } [a,b],$$

$$a_j \in C^j[a,b], \quad 0 \leq j \leq q,$$

$X = \mathcal{H}^q[a, b]$, the Hilbert space of real-valued functions $f \in C^{q-1}[a,b]$ such that $f^{(q-1)}$ is absolutely continuous and $\mathcal{L} f \in L_2[a, b]$ with the inner product

$$\langle f, g \rangle_{\mathcal{H}^q} = \sum_{j=0}^{q-1} f^{(j)}(a)\, g^{(j)}(a) + \int_a^b \mathcal{L} f \, \mathcal{L} g .$$

The kernel of $\mathcal{L}$ is spanned by functions $\{u_i\}_1^q$ in $C^q[a,b]$ and is of dimension q. The set $\{u_i\}_1^q$ spans a closed subspace U of $\mathcal{H}^q[a,b]$. Let $\{\ell_i\}_1^n$, $\{m_i\}_1^n$ be two bases for A such that $\langle \ell_i, m_j \rangle_{\mathcal{H}^q} = \delta_{ij}$. Then there exists $r = (r_1, \ldots, r_n) \in \mathbb{R}^n$ such that $a = \sum_{i=1}^{n} r_i \ell_i$

and

$$\Phi_a = \sum_{i=1}^{n} r_i \ell_i + A^{\perp} = \{\phi \in \mathcal{H}^q \mid \langle m_i, \phi \rangle_{\mathcal{H}^q} = r_i, \ 1 \leq i \leq n\}.$$

The set $\{m_i\}_1^n$ are the representors of n linearly independent continuous linear functionals on $\mathcal{H}^q$.

The existence of the class $\mathcal{S}$ of $\mathcal{L}g$-splines introduced by Jerome and Schumaker [10] is an immediate consequence of the finite dimensionality of the kernel of $\mathcal{L}$. Theorem 2.1, its corollary and Theorem 7.1 of [10] can be deduced from our Theorems 2.2 and 3.1. Moreover, if we consider the continuous linear transformation $\mathcal{M}$ of $\mathcal{H}^q$ onto $\mathbb{R}^n$ defined by

$$\mathcal{M}(f) = \left(\langle m_1, f \rangle_{\mathcal{H}^q}, \ldots, \langle m_n, f \rangle_{\mathcal{H}^q}\right), \quad f \in \mathcal{H}^q$$

and set

$$\Psi_{u^{\perp}} = u^{\perp} + U, \quad u^{\perp} \in U^{\perp},$$

then Theorem 2.1 guarantees the existence of a set $\Sigma_{\mathcal{U}^\perp}$ of interpolating splines of $\Psi_{\mathcal{U}^\perp}$ relative to $\mathcal{M}$ satisfying

$$\|\mathcal{M}\sigma_{\mathcal{U}^\perp}\|_{\mathbb{R}^n} = \min_{\psi \in \Psi_{\mathcal{U}^\perp}} \|\mathcal{M}\psi\|_{\mathbb{R}^n} \quad \text{for all } \sigma_{\mathcal{U}^\perp} \in \Sigma_{\mathcal{U}^\perp}$$

and hence a class $\Sigma_{\mathcal{L}}$ of interpolating splines relative to $\mathcal{M}$. All the results of this paper also hold for the two classes $\mathfrak{S}$ and $\Sigma_{\mathcal{L}}$.

2) If dim $(A^\perp) = q$, dim $(B) = N \geq q$ and $B^\perp \cap A^\perp = \{\theta_X\}$, then various known results follow either as special cases or can be deduced from our Theorems 2.1, 2.2, 4.1 and 4.2, in particular, the results of Atteia ([4], Chapters VII and VIII sections 1,2 of [3]) and Anselone and Laurent ([2], Propositions 2.1, 6.1, 8.1 and 8.2).

3) The condition $A^\perp \cap B^\perp = \{\theta_X\}$ which is required for the uniqueness of the spline function together with the finite dimensionality of $B^\perp$ has been extensively used by Atteia ([3], [4], [5]) and Anselone and Laurent ([2]) for the existence of the splines. On the other hand, Jerome and Schumaker [10] used the finite - dimensionality of the kernel of the transformation for the existence of $\mathcal{L}g$-splines. We have in this paper, proved the existence under more general conditions, namely $A^\perp + B^\perp$ is closed. Further, constraint sets of the form $\{\phi \in X \mid T\phi = y\}$, $y \in Y$ considered by Atteia in [3] and [5] are special cases of the constraint sets considered by us since they are translates of subspaces.

The author is deeply grateful to Professor K.R.Unni for his guidance and constant encouragement. Thanks are also due to Professor Joseph W. Jerome for his helpful criticisms.

References

1 Ahlberg, J.H., Nilson, E.N. and Walsh, J.L., The theory of splines and their applications, Academic Press, New York (1967).

2 Anselone, P.M. and Laurent, P.J., A general method for the construction of interpolating or smoothing spline functions. Numer. Math. 12, (1968), 66-82.

3 Attéia, M., Etude de certains noyaux et théorie des fonctions 'spline' en analyse numérique. Thèse, Grenoble (1966).

4 Attéia, M., Fonctions 'spline' avec contraintes linéaires de type inégalité. 6e Congrès de l'AFIRO, Nancy, Mai (1967).

5 Attéia, M., Fonctions 'spline' définies sur un ensemble convexe. Numer. Math. 12 (1968) 192-210.

6 Daniel, J.W. and Schumaker, L.L., On the closedness of the linear image of a set, with applications to generalized spline functions. Applicable Analysis (to appear).

7 Goldstein, A.A., Constructive Real Analysis, Harper and Row. New York (1967).

8 Halmos, P.R., Introduction to Hilbert Space, Chelsea Publishing Company (1951).

9 Halmos, P.R., A Hilbert Space Problem Book, D. Van Nostrand Company, New Jersey (1967).

10 Jerome, J.W. and Schumaker, L.L., On Lg-splines, J.Approx. Th. 2 No.1, (1969), 29-49.

11 Laurent, P.J., Construction of spline functions in a convex set. I.J.Schoenberg (Editor) 'Approximations with special emphasis on spline functions'. Academic Press, New York (1969) 415-446.

12 Meinardus, G., Approximation of functions: theory and numerical methods, Springer-Verlag, Berlin (1967).

13 Schoenberg, I.J., Spline interpolation and the higher derivatives, Proc. Nat. Acad. Sc. 51, No. 1 (1964) 24-28

14 Schoenberg, I.J., Spline functions and the problem of graduation, Proc. Nat. Acad. Sc. 52 No.4 (1964) 947-950.

A SURVEY OF v-INTEGRAL REPRESENTATION THEORY FOR OPERATORS ON FUNCTION SPACES INCLUDING THE TOPOLOGICAL VECTOR SPACE SETTING

S.G.Wayment

In 1909 F.Riesz [20] showed the dual of C[0,1] is BV[0,1] (the functions of bounded variation with $\|g\|_{BV} = V_0^1 g$) via a Stieltjes integral representation theorem. That is, if $T \in C^*[0,1]$, then $T(f) = S\int_0^1 f dg$ for each $f \in C[0,1]$ and where $g \in BV[0,1]$. Subsequent developments have sometimes replaced g by a regular Borel measure μ_g generated by the content formed from differencing g on closed intervals [4].

Briefly, one might capsulize this development as follows: First f is 'approximated' by a step function $h = \sum_{i=1}^n \alpha_i \chi_{E_i}$ where $E_i = (a_i, b_i], i > 1$ and $E_1 = [0, b_1]$ and where α_i is a scalar for each i. Then T(f) should be approximately T(h) which by linearity is $\sum_{i=1}^n \alpha_i T(\chi_{E_i})$. If we define $\mu_g(E) = T(\chi_E)$ and choose $\alpha_i = f(t_i)$ for some $t_i \in E_i$ then our approximation to T(f) is an 'approximating sum' for an integral, and in the limit $T(f) = \lim \sum_{i=1}^n f(t_i)\mu_g(E_i) = \int_0^1 f d\mu_g$. The problem with the preceding is that T does not operate on χ_E because $\chi_E \notin C[0,1]$.

There are many ways to avoid or surmount this problem. One is to use the Hahn-Banach theorem to extend T to the characteristic functions. Another is to extend T to characteristic functions using monotonically pointwise convergent sequences of continuous functions. Still another way is to make an identification of the problem at hand with the Hausdorff Moment Problem, and use Bernstein polynomials. There have been many classical-type integral representation theorem extensions to more general settings including [5] [9] [12] [13]

[16][17][18] [21] [22] [23] [24] [25] (for a more complete list of references, see [3]). In each of these extensions, the problem is more difficult mainly because extending T to operate on characteristic functions (or avoiding this problem by going another route) is more difficult.

In 1969 J.R.Edwards and S.G.Wayment [9] obtained an extended form of the classical Riesz Representation theorem which applied in any setting where the topology on the function space was not stronger than the topology of uniform convergence. This led to the investigation of functionals on BV[0,1] continuous in the BV norm (which is stronger than the topology of uniform convergence) which subsequently led to the development of the v-integral [10] . It is to this theory of the v-integral and extensions thereof that we now turn our attention. For simplicity we shall assume $f(0) = 0$ for each real function in AC[0,1], the set of real absolutely continuous functions on [0,1] where $\| f \|_{BV} = V_0^1 f$, the variation from 0 to 1 of f.

We first note that if $f \in AC[0,1]$ and h is a step function, then $\| f - h\|_{BV} = \| f \|_{BV} + \| h \|_{BV}$. Thus f cannot be approximated in BV-norm by a step function and the process **alluded** to in the second paragraph for obtaining a representation of a continuous linear functional appears doomed for AC[0,1]. However, it turns out that another class of 'elementary' functions can be used to replace step functions, namely polygonal functions, and the procedure of paragraph two can be duplicated almost exactly.

LEMMA 1. [10] Let $\sigma = \{0 = x_0 < x_1 < x_2 < \cdots < x_n = 1\}$ be a partition of [0,1] and let pf_σ be a polygonal function with corners precisely at the points $(x_i, f(x_i))$. If $f \in AC[0,1]$, then pf_σ converges in the BV-norm to f as $|\sigma|$ tends to zero (where $|\sigma| = \max_i |x_i - x_{i-1}|$).

Using this lemma we can proceed as soon as we decide how to decompose pf_σ into a linear combination of building blocks.

Definition 2. [10] Suppose $E = (a,b]$ and let $\psi_E(t) = 0$ if $t \leq a$; $\frac{t-a}{b-a}$ if $a \leq t \leq b$; and 1 if $t \geq b$. (Note that $\psi_E(t) = \int_0^t \chi_E dx$.)

Then $pf_\sigma = \sum_{i=1}^{n} \alpha_i \psi_{E_i}$ where $\alpha_i = f(x_i) - f(x_{i-1}) = \Delta_i f$ and where $E_i = (x_{i-1}, x_i]$. Hence if T is a continuous (in BV-Norm) linear functional on AC[0,1], then for $f \in AC[0,1]$ we conclude that $T(f)$ is approximately $T(pf_\sigma) = \sum_{i=1}^{n} \alpha_i T(\psi_{E_i})$. So if we let the action of T on fundamental functions ψ_E determine a set function K, via $T(\psi_E) = K(E)$, we have that $T(f) = \lim_{|\sigma| \to 0} \sum_{i=1}^{n} K(E_i)\Delta_i f = v\int_0^1 K df$, the last expression being the v-integral.

Definition 3. [10] . If K and μ_f are functions defined on a field of sets over a space S, then the v-integral of K and μ_f is $\lim_{\pi} \sum_{\pi} K(E_i)\, \mu_f(E_i)$.

The special case of the v-integral where $\mu_f(E) = \Delta_i f = f(b)-f(a)$ when $E = (a,b]$ and when $S = (0,1]$ gives immediately the following theorem [10] .

THEOREM 4. If the functional T on AC[0,1] is linear and continuous in the BV-norm, then there exists a unique bounded, convex with respect to length, set function K so that for each $f \in AC[0,1]$ we have $T(f) = v\int_0^1 K df$ and $\|T\| = B$, the least bound on $|K(E)|$ as E ranges over the sets of the form $(a,b]$. Conversely, if K is a bounded and convex with respect to length set function, then $v\int_0^1 K df$ exists for each $f \in AC[0,1]$ and generates a continuous linear functional.

All that remains to be discussed is the concept 'K is convex with respect to length'. Recall that if E_1 and E_2 are disjoint sets and if $E = E_1 \cup E_2$, then $\chi_E = \chi_{E_1} + \chi_{E_2}$. This is why a set function determined by a linear operator on characteristic functions yield an <u>additive</u> set function, viz., $\mu(E) = T(\chi_{E_1} + \chi_{E_2}) = T(\chi_{E_1}) + T(\chi_{E_2}) = \mu(E_1) + \mu(E_2)$ and hence 'measures' are generated in the classical Riesz Representation Theorem. However, for fundamental functions, we have under the same circumstances that $\psi_E = \lambda_1\psi_{E_1} + \lambda_2\psi_{E_2}$ where $\lambda_1 + \lambda_2 = 1$ and hence the set function generated by the action of T on fundamental functions is <u>convexly additive</u>, viz., $K(E) = T(\psi_E) = T(\sum_{i=1}^{n} \lambda_i\psi_{E_i}) = \sum_{i=1}^{n} \lambda_i K(E_i)$ where the $\{E_i\}$ are disjoint and $\bigcup_{i=1}^{n} E_i = E$ and where $\Sigma\lambda_i = 1$. In fact, it is easily seen that λ_i is the ratio of the length of E_i to the length of E.

In some sense then, we have paralleled the classical Riesz Representation Theorem to obtain a v-integral representation theorem for a functional continuous in a norm stronger than the sup-norm. We have had to replace $f(x_i)$ by $\Delta_i f = f(x_i) - f(x_{i-1})$ and we have had to replace an additive set function $\mu(E_i)$ by a convexly additive set function $K(E_i)$ in the approximating sums for the integral. Since this might be viewed as a loss by some, let us next ask what have we gained? The first apparent answer is that we have an integral which converges as nicely as the Riemann Integral, that is, as $|\sigma| \to 0$. The second answer one might give is that ψ_E is still a continuous function, where χ_E is not, and hence it will be simpler to extend our theorem to more complicated settings. Finally, we note that since polygonal functions approximate absolutely continuous functions in the BV-norm, we can use techniques developed in [9] [23] [24] to give v-integral representations on function spaces with norms not stronger than the BV-norm. The first such theorem we give is the v-integral analog of the classical Riesz Representation Theorem. Before stating this analog however, we note that if a functional T is to be continuous on a function space with a given norm, then if V_θ is a neighborhood of the origin in the image space of T, (in the case of functionals the image space is the real numbers), there exists a neighborhood U_θ of the origin in the function space so that if $f \in U_\theta$, then T(f) must be in V_θ. It is precisely this fact applied to the approximating functions (in the case of the classical Riesz Representation Theorem the approximating functions are step functions, and in Theorem 4 of this paper they were polygonal functions) which are built from the 'fundamental building blocks' (in the classical Riesz Theorem characteristic functions and in Theorem 4 fundamental functions) that determine the properties of the set function used to integrate (in the classical Riesz theorem a regular Borel measure and in Theorem 4 a bounded, convexly additive set function). Naturally, as we change the norm or topology on the function space, we change the properties that the set function used to integrate must have since it is determined by the action of T on the building blocks used. These considerations applied to approximating continuous real functions on [0,1] by polygonal functions to represent a functional T which is continuous in the sup-norm lead to the following definition.

Definition 5. The real set function K is said to be convex Gowurin (the name is suggested by analogy to the development in [24]) if there exists a constant M such that for each partition $\{x_i\}_{i=1}^n$ of [0,1] where $E_i = (x_{i-1}, x_i]$ and corresponding collection of scalars $\{\alpha_i\}_{i=1}^n$, it follows that

$$\left| \sum K(E_i)\,\alpha_i \right| \leq M \max_j \left| \sum_{i=1}^{j} \alpha_i \right| .$$

THEOREM 6. *If* T *is a linear operator on* $C_0[0,1]$, *the continuous functions on* $[0,1]$ *which are zero at zero and endowed with the sup-norm topology then there exists a unique convex-Gowurin set function* K *defined on half open intervals which is convex with respect to length so that* $T(f) = v\int_0^1 Kdf$ *for each* $f \in C_0[0,1]$. *Furthermore* $\|T\| = WK$, *the least* M *in definition* 5. *Conversely, if* K *is such a set function then* $v\int_0^1 Kdf$ *exists for each* $f \in C_0[0,1]$ *and generates a continuous linear functional*.

Finally we are in a position to see the advantage of using fundamental functions rather than characteristic functions. Since fundamental functions are still continuous functions, the proof of the previous theorem does not change if we allow the continuous functions on $[0,1]$ to have values in a topological vector space and endow them with the topology of uniform convergence (see [6] for details). This is in stark contrast to the complications which arise when extending via classical techniques to vector valued functions, [5], [23], [24] locally convex topological vector space valued functions [9],[13], and finally topological vector space valued functions [9],[21]. Furthermore, a *characterization* is obtained as opposed to a *representation* using the classical approaches. That is (in the classical approaches referenced) if T took functions into a space Y, then K(E) was an operator from X into Y^{**} and one could not conclude that such K will always generate via integration an operator with range in Y. We summarize with a definition and a theorem.

Definition 7. A set function defined on half open intervals $(a,b]$ in $(0,1]$ and with values in the set $L[X,Y]$ of linear operators from the topological vector space X into the topological vector space Y is said to be quasi-Gowurin provided that given a neighbourhood V of the origin in Y there exists a neighbourhood U of the origin in the space of continuous X-valued functions on [0,1] which are θ (the zero in X) at zero such that if $\Sigma \psi_{E_i} x_i \in U$, then $\Sigma K(E_i)x_i \in V$.

THEOREM 8. [6] Suppose T is a continuous linear operator from C_θ into Y. Then there is a set function K with values in L[X,Y] which is convex with respect to length and quasi-Gowurin such that $T(f) = v\int K df$ for each $f \in C_\theta$. Conversely, if K is such a set function then $v\int K df$ exists for each $f \in C_\theta$ and generates a continuous linear functional.

Utilizing the fact that $\psi_E(t) = \int_0^t \chi_E d\mu$ we can immediately extend theorem 8 to the setting where the domain [0,1] of the functions is replaced by an interval in E^n. For example, if $n = 2$ and E is a rectangle in the unit square $0 \leq t_1 \leq 1$, $0 \leq t_2 \leq 1$, then for each point $p = (p_1, p_2)$ we define $\psi_E(p) = \int_{R_p} \chi_E d\mu$ where R_p is the rectangle with diagonal from (0,0) to (p_1, p_2) and μ is Lebesque measure in the plane. Then if f is a continuous function of the unit square $0 \leq t_1 \leq 1$; $0 \leq t_2 \leq 1$ and if f is zero along $t_1 = 0$ and along $t_2 = 0$, we can approximate f by $\sum \alpha_i \psi_{E_i}$ where now the E_i are half-open rectangles (open on the bottom and left) and come from a grid of horizontal and vertical lines, and where if E_i has lower left corner (a,b) and upper right corner (c,d), then $\alpha_i = f(c,d) - f(a,d) - f(c,b) + f(a,b)$. Of course, the definitions all need to be extended to this setting and some concern needs to be given concerning the order in which the summations are taken in the definition of quasi-Gowurin (for example, one might wish to number the rectangles E_{i_j} where i denotes row and j denotes column that E_{i_j} appears in and then sum on j from 1 to n before incrementing i), but it is straightforward.

Let us return now to the setting where the domain of the functions is again [0,1] but the range is a normed space X, i.e. $f(t) \in X$. If we define bounded variation and absolutely continuous for functions on [0,1] with values in X by replacing absolute value signs by norm signs, then one can investigate the possibility of obtaining the analog of Theorem 4 in this setting. Surprisingly and unfortunately the analog cannot be obtained because Lemma 1 is false [26]. However, if T is a continuous linear operator on the BV-norm closure of the polygonal functions in this setting, and additionally in the setting where the domain [0,1] is replaced by the unit square in Euclidean n space E^n, then the analogous result to Theorem 4 is immediate [8].

We note in passing that one can apply the techniques of [9],[24] to the continuously differentiable vector valued functions to obtain v-integral representation theorems since polygonal functions are 'almost continuously differentiable' in the same sense that step functions are 'almost continuous', i.e. they can be imbedded in the second dual space. Hence, for example, the linear functional $T(f) = f'(1/2)$ can be realized by a v-integral, that is, $f'(1/2) = v\int_0^1 K\,df$ for an appropriate K [7].

We close with the comment that by using the fact that $\psi_E(t) = \frac{1}{\mu(E)}\int_0^t \chi_E\,d\mu$ it is possible to define the analog of polygonal measure in a general measure space [8]. For example if ν is a bounded finitely additive set function, absolutely continuous with respect to a positive bounded set function μ, then define a fundamental measure $\psi_E(F) = \frac{1}{\mu(E)}\int \chi_E\,d\mu$ and a polygonal measure $P\nu_\pi$ approximating ν by $P\nu_\pi(F) = \int_F (\Sigma \frac{\nu(E_i)}{\mu(E_i)}\chi_{E_i})d\mu$ where the $\{E_i\} = \pi$ partition the space. Fefferman [11] has shown that $P\nu_\pi$ converges to ν in the variational sense and so if T is a linear functional (continuous in the BV-norm) on the space $\mathfrak{A}\mathcal{C}$ of finitely additive set functions which are absolutely continuous with respect to μ and $\pi = \{E_i\}$ is a partition of space, then

$$T(\nu) = \lim_{\pi} T(P\nu_{\pi}) = \lim_{\pi} T(\Sigma\, \nu(E_i)\, \psi_{E_i})$$

$$= \lim_{\pi} \Sigma\, \nu(E_i)\, T(\psi_{E_i})$$

$$= \lim_{\pi} \Sigma\, \nu(E_i)\, K(E_i) = v\int K\, d\nu,$$

where $K(E) = T(\psi_E)$ defines the bounded set function K [8]. For further extensions, also see [8], [1]. Also, a derivative to match the v-integral has been developed in [14] as well as a Radon-Nikodyn theorem for the v-integral in [15]. We also note that Alo and deKorvin [2] have extended the representation theorem on finite measure spaces to include vector-valued measures with a measure being AC if it is in the BV closure of the polygonal measures (a necessary definition because of the results in [26]). Finally we note that D.Mauldin [19] has used the integral of Definition 3 to give a representation for the second dual of C[0,1], a problem outstanding since the work of F.Riesz [20].

References

1 Alo, R.A. and deKorvin, A., Functions of bounded variation on idempotent semi-groups, Math. Annalen 194 (1971) 1-11.

2 Alo,R.A. and deKorvin, A., Vector valued absolutely continuous functions on idempotent semi-groups, Trans. Am. Math. Soc., (to appear).

3 Jurgen Batt, Die verallgemeinerungen des darstellungssatzes von grosser F.Riesz und ihre Anwendungen, Jber. Deutsch. Math. -Verein 74 (1973) 147-181.

4 Dunford, N. and Schwartz, Linear operators I: General theory, Pure and appl. Math., Vol.7, Interscience, New York, 1958. MR22#8302.

5 Edwards, J.R. and Wayment, S.G., A unifying integral representation theorem, Math. Ann. 181 (1969) 311-324.

6 Edwards, J.R. and Wayment, S.G., A v-integral representation for linear operators on spaces of continuous functions with values in topological vector spaces, Pac. J. Math., 35 (1970) 327-330.

7 Edwards, J.R. and Wayment, S.G., A v-integral representation for the continuous linear operators on spaces of continuously differentiable vector valued functions, Proc. Am. Math. Soc. 30 (1971) 263-270.

8 Edwards, J.R. and Wayment, S.G., Extensions of the v-integral, submitted June 1970.

9 Edwards, J.R. and Wayment, S.G., Integral representations for continuous linear operators in the setting of convex topological vector spaces, Trans. Am. Math. Soc., 157 (1971) 329-345.

10 Edwards, J.R. and Wayment, S.G., Representations for transformations continuous in the BV-norm, Trans. Am. Math. Soc., 154 (1971) 251-265.

11 Fefferman, C., A Radon-Nikodym theorem for finitely additive set functions, Pac. J. Math., 23 (1967) 35-45.

12 Fihtengol'c, I.G. and Kantorovic, L., Sur les operations lineares dans l'espace des fonctions bornees, Studia Math. 5 (1934) 69-98.

13 Goodrich, R.K., A Riesz representation theorem in the setting of locally convex spaces, Trans. Am. Math. Soc. 131 (1968), 246-258. MR36#5731.

14 Hatte, L. and Wayment, S.G., A derivative to match the v-integral Journal fur die reine und. angewandte Mathematik, 257 (1972)16-28.

15 Hatte, L. and Wayment, S.G., A Radon-Nikodym theorem for the v-integral, Journal fur die reine und angewandte Mathematik, (to appear).

16 Hilderbrandt, T.H., Linear continuous functionals on the space (BV) with weak topologies, Proc. Amer. Math. Soc. 17 (1966) 658-664. MR 33# 1710.

17 Hilderbrandt, T.H., Linear operators on functions of bounded variation, Bull. Amer. Math. Soc. 44 (1938) 75.

18. Leviaton, D., On a representation theorem and application to moment sequences in locally convex spaces, Math. Ann., 182 (251-262) 1969.

19 Mauldin, D., A representation theorem for the second dual of C[0,1], (submitted).

20 Riesz, F., Sur les operations functionelles lineaires, C.R. Acad. Sci. Paris 149 (1909) 974-977.

21 Schuchat, A.H., Integral representation theorem in topological vector spaces, Trans. Am. Math. Soc. 172 (1972) 373-397.

22 Swong, K., A representation theory of continuous linear maps, Math. Ann. 155 (1964) 270-291; errata, ibid 157 (1964) 178. MR 29# 2642.

23 Tucker, D.H., A note on the Riesz representation theorem Proc. Amer. Math. Soc. 14 (1963) 354-258. MR 26# 2865.

24 Tucker, D.H., A representation theorem for a continuous linear transformation on a space of continuous functions, Proc. Am. Math. Soc., 16 (1965) 946-953.

25 Uherka, D.J., Generalized Stieltjes integrals and a strong representation theorem for continuous linear maps on a function space, Math Ann. 182 (1969) 60-66. MR 40 #705.

26 Wayment, S.G., On the BV-norm closure of vector-valued polygonal functions, Revue Roumaine de Mathematiques Pures et Appliquees, 17 (1972) 1123-1126.

ISOTONE MEASURES, 1948-1973

J.H.Williamson

Let G be a locally compact topological group, with left Haar measure. Let $M(G)$ denote as usual the bounded regular Borel measures on G, $L^p(G)$ the equivalence-classes of pth power integrable functions on G (with respect to left Haar measure).

As is well known, measures in $M(G)$ act, by (left-) convolution, as linear operators on a wide variety of spaces of functions on G. Examples are (i) $L^p(G)$ $(1 \leq p < \infty)$. For a full discussion see, for example, [3] , § 20. We note here that if

$$(\mu * f)(x) = \int_G f(y^{-1}x)\, d\mu(y) \tag{1}$$

then μ defines a bounded linear map of L^p to itself, and moreover

$$\| \mu * f \|_p \leq \|\mu\| \, \|f\|_p \tag{2}$$

(although we shall not require this last relation in what follows). The case $p = 1$ is classical ; L^1 is in fact not only a linear space, but a linear algebra ; and measures act as multipliers on L^1:

$$\mu * (f * g) = (\mu * f) * g \tag{3}$$

($\mu \in M(G)$, $f, g \in L^1(G)$). The multipliers can be characterised as being precisely the elements of $M(G)$ (Wendel[8], Theorem 3) but in the present discussion we shall take $M(G)$ itself as the starting-point. (ii) $S^1(G)$, where S^1 is a Segal algebra on G. For details see [7] , § 4. Again we have an inequality of the form

$$\| \mu * f \|_S \leq \|\mu\| \, \| f \|_S \tag{4}$$

(iii) the bounded measurable functions, or bounded continuous functions, or bounded functions vanishing at infinity, or bounded continuous functions vanishing at infinity, with the essential supremum or supremum norm in each case, and convolution as in (1).

We could envisage generalisations of the above situation : there could be two different spaces A and B, with elements of M(G) acting as maps from A to B ; and we could consider spaces of measures as well as spaces of functions. There would be some minor technical modifications required, but the theory would go through in very much the same way. We do not consider such generalisations in the present exposition.

Let A be a space of functions on G, mapped into itself by elements of M(G). There is always a natural order in A, compatible with the natural order in M(G). If A is a space of continuous functions, $\leq$ is to be taken in the 'everywhere' sense : if a space of measurable functions, in the 'almost everywhere' sense. We always have (for $\mu \in M(G)$, $f \in A$)

$$\mu \geq 0, f \geq 0 \Rightarrow \mu * f \geq 0 \ . \tag{5}$$

The basic problem is this : for which measures $\mu \in M(G)$ is it true that for $f \in A$,

$$\mu * f \geq 0 \Longleftrightarrow f \geq 0 \ ? \tag{6}$$

If (6) holds we shall say that μ <u>acts isotonically on A</u>, or is <u>A-isotone</u>. In the classical case $A = L^1(G)$ the term 'isotone' has normally been used without qualification, in the literature. Sometimes if the space A is clear from the context, explicit reference to it may be omitted. If μ is a positive measure with one-point support then the effect of convolution with μ is simply translation and multiplication by a positive scalar. For such measures (6) holds, trivially ; we shall refer to them as the trivial isotone measures. Our problem can now be re-phrased as : do there exist in M(G) any non-trivial A-isotone measures? The original interest of the problem was as a possible means of characterising G in terms of its group algebra $L^1(G)$; it turns out that a characterisation of G in terms of isotone measures is possible in certain cases. However, the problem is evidently of considerable intrinsic interest also, independent of any possible applications to characterisation theorems. We shall for future reference note here the statement

the only A-isotone measures in M(G) are trivial . (7)

Before embarking on a brief account of the history of the problem, we recall one or two facts about topological groups. With any topological group G there are associated two natural uniform structures, the left and the right ; basic vicinities are, respectively, $\{(x,y) : x^{-1}y \in N\}$ and $\{(x,y) : xy^{-1} \in N\}$ where N runs through a basic set of neighbourhoods of the neutral element e of G. For details see [1] , Ch. 3, § 3. If these two structures coincide, G is unistructural, otherwise bistructural (there may well be other structures on G, besides the left and the right, but we are not concerned with them here). A necessary and sufficient condition for G to be unistructural is ([1] ,Ch 3,Ex 3) of § 3)

$$\text{given } N_1 , \exists\ N_2 \text{ such that } x^{-1}N_2\, x \subset N_1, \forall\ x \in G \qquad (8)$$

or, equivalently

$$\text{for each } N, \bigcap_{x\in G} x^{-1}Nx \text{ is a neighbourhood of e.} \qquad (9)$$

In a locally compact group with left Haar measure we have, for appropriate functions f,

$$\int_G f(yx)\,dx = \int_G f(x)\,dx$$

In general

$$\int_G f(xy)\,dx \neq \int_G f(x)\,dx;$$

however we always have

$$\int_G f(xy)\,dx = \Delta(y^{-1}) \int_G f(x)\,dx \qquad (10)$$

where Δ, the <u>modular function</u> of G, is a continuous homomorphism of G into the multiplicative group of strictly positive reals. If $\Delta(x)$ is identically 1, G is <u>unimodular</u>. Unimodular groups include abelian, discrete, and compact groups.

<u>LEMMA 1</u>. <u>If</u> G <u>is unistructural then it is unimodular</u>.

<u>Proof</u>. Immediate from the definitions; note that if Δ is a bounded function then it must be 1 for all $x \in G$.

Not every unimodular group is unistructural. More generally, we have the following situation. Unistructurality is inherited by (closed) subgroups of G; unimodularity is not so inherited. It follows that if G is unistructural then every closed subgroup of G is unimodular. There are examples of groups that are not unistructural, but every closed subgroup of which is unimodular; see e.g. [5].

We now proceed to the promised historical sketch. All work published so far relates to the case $A = L^1(G)$. In [4] Y.Kawada proved that if (6) holds, and if also

$$\mu \text{ is surjective} \qquad (11)$$

(as a map of L^1 to itself, by left convolution) then (7) follows. His original argument can be simplified. In [8], J.G.Wendel proved a rather similar theorem, that if μ acts isometrically on L^1

($\|\mu * f\|_1 = \|f\|_1$, for all $f \in L^1$) then μ must be trivial in the sense that it is a (complex) measure concentrated at a point, with mass equal to 1 in absolute value. There is no surjectivity assumption (11) here. In view of this it is natural to enquire whether (11) is necessary in Kawada's theorem. In [9] some partial results were obtained ; if G is abelian, or discrete, or compact then (11) can be dropped and (7) is still true provided that (6) holds. On the other hand if G is non-unimodular and metrisable, (7) may fail in the absence of (11). In [6] J.S.Pym showed that (7) could fail even in unimodular groups, provided that there were a suitable non-unimodular subgroup. In [2] the positive results of [9] appear as an exercise (Ch. 8, Ex 19) of §4) ; apart from this little attention was paid to the problem for some time. Recently W.Moran has obtained some fresh results [5] . He has proved two things ; if G is unistructural and μ is absolutely continuous then μ cannot be isotone ; and non-trivial isotone measures exist in some groups, all of whose closed subgroups are unimodular. This last result is a rather surprising one, and has caused a recent revival of interest in the problem.

We now present a very simple argument that enables all the known positive results (that is, cases in which (7) is true) to be established easily. First we impose two rather mild restrictions on A : we require that

A contains all continuous functions of compact support (12)

and a localisation condition :

$$\left.\begin{array}{l} \text{given } f \in A,\ f \geq 0 \text{ and } K_1, K_2 \text{ compact in } G \\ \text{with } K_1 \subset \operatorname{int} K_2,\ \exists\ g \in A \text{ with } 0 \leq g \leq f, \\ g = f \text{ on } K_1,\ g = 0 \text{ outside } K_2 \end{array}\right\} \quad (13)$$

Many spaces of interest satisfy a stronger condition ; the slight complication of (13) is due to the need to include spaces of continuous functions.

LEMMA 2 *If* $f \geqslant 0 \Rightarrow \mu * f \geqslant 0$ *then* $\mu \geqslant 0$.

Proof Almost immediate.

The neighbourhoods N_1, N_2 of e will be taken in what follows to be open, with compact closures. This is convenient, but not essential. If $\mu \in M(G)$ the function $\mu(N_1 x N_2)$ is bounded :

$$|\mu(N_1 x N_2)| \leqslant \|\mu\| \qquad \text{for all } x \in G \tag{14}$$

and is lower semi-continuous if $\mu \geqslant 0$, hence in general certainly measurable. The function is not in general continuous, unless μ is an absolutely continuous measure.

THEOREM 1 *If* μ *is an* A-*isotone measure such that for some neighbourhoods* N_1, N_2 *of* e *and some* $f_o \in A$ *we have*

$$\mu(N_1 x N_2) \leqslant f_o(x) \tag{15}$$

then μ *is trivial*.

Proof Suppose μ is non-trivial : then we can write $\mu = \mu_1 + \mu_2 + \mu_3$, where μ_1, μ_2, μ_3 are all $\geqslant 0$, $\mu_1 \neq 0$, $\mu_2 \neq 0$, μ_1 is supported by N_3, μ_2 by $N_3 a$ where N_3 is a compact symmetric neighbourhood of e with $N_3{}^3 \subset N$, and $a \notin N_3{}^5$. Let N_4 be a compact symmetric neighbourhood of e with $N_4{}^2 \subset N_3 \cap N_2$.

Let f_1 be continuous, with $0 \leqslant f_1(x) \leqslant 1$ for all x, $f_1(e) = 1$, $f(x) = 0$ for $x \notin N_4$. Let $f_2 = \dfrac{1}{\|\mu\|} f_o$ $(x \notin N_4{}^2)$, $= 0$ $(x \in N_4)$. The existence of such an $f_2 \in A$ is guaranteed by (13) and the fact that A is a linear space. There exists a continuous function of compact support, g say, such that $g(x) \geqslant 0$ for all x; $g(x) = 1$ for $x \in a^{-1} N_3{}^3$, and $g(x) = 0$ for $x \in N_3$. Write

$f_3 = \dfrac{\| \mu \|}{\| \mu_2 \|} g$; by (12), $f_3 \in A$. Write finally $f = -f_1+f_2+f_3$; then certainly $f \in A$.

We have then

$$\mu * f = -\mu * f_1 + \mu_1 * f_2 + \mu_2 * f_3 + \text{non-negative terms.}$$

Now, in the first place we have

$$(\mu * f_1)(x) = \int f_1(y^{-1}x)\, d\mu(y) \leq \int \chi_{N_4}(y^{-1}x)\, d\mu(y)$$

$$= \int_{xN_4} d\mu(y) = \mu(xN_4).$$

Next

$$(\mu * f_2)(x) = \int f_2(y^{-1}x)\, d\mu_1(y) \geqslant \| \mu_1 \| \inf_{y \in N_3} f_2(y^{-1}x).$$

If $x \notin N_3N_4^{\;2} \subset N_3^{\;2}$, this last expression is

$$\geqslant \inf_{y \in N_3} \mu(N_3y^{-1}xN_4) \geqslant \mu(xN_4).$$

Finally if $x \in N_3^{\;2}$ and $y \in N_3a$ then $y^{-1}x \in a^{-1} N_3^{\;3}$ and $f_3(y^{-1}x) = \| \mu \| / \| \mu_2 \|$, so that $(\mu_2 * f_3)(x) = \int f_3(y^{-1}x)\, d\mu_2(y) = \| \mu \| \geqslant \mu(xN_4)$.

So in any case $(\mu * f)(x) \geqslant 0$. But $f(x)$ is certainly not $\geqslant 0$, since $f(x) = -f_1(x)$ if $x \in N_4$. Hence μ cannot be isotone. This establishes the required result.

<u>COROLLARY 1</u> <u>For any A, a non-trivial A-isotone measure cannot have compact support.</u>

<u>Proof</u> If μ has compact support then $\mu(N_1 \times N_2)$ has compact support, as a function of x, and so is always dominated by a function in A.

COROLLARY 2 *If A is the space of all bounded measurable functions on G, or all bounded continuous functions on G, there are no non-trivial A-isotone measures (whatever G may be)*

Proof $\mu(N_1 \times N_2)$ is bounded.

COROLLARY 3 *If $A = L^P(G)$ and G is unistructural there are no non-trivial A-isotone measures.*

Proof If G is unistructural, and N_0 is given, by (8) there exists N_1 such that $x^{-1}N_1 x \subset N_0$ for all x. But then if μ is positive $\mu(N_1 \times N_2) \leq \mu(xN_0N_2) = (\mu * \chi_N)(x)$ $(N = (N_0N_2)^{-1})$. $\chi_N \in L^P$ so $\mu * \chi_N \in L^P$, hence $\mu(N_1 \times N_2) \in L^P$ for some N_1, N_2 and so μ cannot be non-trivial isotone.

COROLLARY 4 (W.Moran, unpublished) *If $A = L^P(G)$ and G has a compact open normal subgroup then there are no non-trivial A-isotone measures on G.*

Proof. Let H be the subgroup in question : write $N_1 = H$ and then $\mu(N_1 \times N_2) = \mu(H \times N_2) = \mu(xHN_2)$; this last function is in L^P for the same reason as before.

The main interest of Corollary 4 is that it enables us to produce examples of groups that are *not* unistructural but yet have no non-trivial L^P-isotone measures. The following example is due to W.Moran (unpublished : I am much indebted to Dr.Moran for permission to reproduce it here). Let G be the semi-direct product of Z and

$$H = \prod_{-\infty}^{\infty} Z_2 \quad \text{where}$$

$$(n,(h_m))\,(n',(h'_m)) = (n+n',(h_m + h'_{m+n})).$$

Then H is a compact open normal subgroup of G. On the other hand $N = (0,L)$ is a neighbourhood of e, where L is the subset of H consisting of vectors h (regarded as a sequence of 0's and 1's) with 0th component zero.

Then

$$(n,(h_m))\ (0,(h'_m))\ (-n,(h_{m-n})) = (0,h'_{m+n}) \tag{16}$$

and so

$$\bigcap_{x \in G} x^{-1}\ (0,L)\ x = (0,(0)) \tag{17}$$

which is not a neighbourhood of e in G. If G were unistructural, this would be a neighbourhood, by (9). So G is not unistructural, but the condition of Corollary 4 holds, so that G has no non-trivial isotone measures.

The examples that have so far been produced of non-trivial isotone measures have either been absolutely continuous ([9], [5]), or at least absolutely continuous with respect to Haar measure on some closed subgroup [6]. It is not known whether this is the general situation, but the following result shows that some continuity is always present.

<u>THEOREM 2</u> <u>If</u> μ <u>is a non-trivial</u> A-<u>isotone measure and</u> L <u>is any coset of the centre</u> C <u>of</u> G <u>then</u> $\mu(L) = 0$.

<u>Proof</u> The general argument is very similar to that of Theorem 1. We may suppose without loss of generality that $\mu(C) > 0$. If μ is non-trivial, there is a compact symmetric neighbourhood N of e such that $\mu = \mu_1 + \mu_2 + \mu_3$, where $\mu_1,\mu_2,\mu_3 \geq 0$, $\mu_1 \neq 0, \mu_2 \neq 0$, and μ_1 is supported by $N \cap C$, μ_2 by Na, where $a \notin N^6$.

Choose f_o to be a continuous function of compact support such that $0 \leq f_o(x) \leq 1$ everywhere, $f_o(e) = 1, f_o(x) = 0$ for $x \notin N$. Choose g_o also continuous and of compact support such that $0 \leq g_o(x) \leq 1$ for all x, $g_o(x) = 1$ if $x \in a^{-1}N^4$ and $g_o(x) = 0$ if $x \in N$. Write $f_1(x) = (\mu_1 * f_o)(x)$, and choose $f_2 \in A$ such that $0 \leq f_2(x) \leq (\mu * f_o)(x)$ everywhere, $f_2(x) = (\mu * f_o)(x)$ if $x \notin N^2$,

$f_2(x) = 0$ if $x \in N$. Let $f_3(x) = \|\mu\| \; \|\mu_1\| \; g_o(x) / \|\mu_2\|$, and write $f = -f_1 + f_2 + f_3$; then $f \in A$.

We have, as in Theorem 1,

$$\mu * f = -\mu * f_1 + \mu_1 * f_2 + \mu_2 * f_3 + \text{non-negative terms.}$$

In the first place, for all x,

$$(\mu * f_1)(x) = (\mu * \mu_1 * f_o)(x) = (\mu_1 * \mu * f_o)(x) \text{ since}$$

μ_1 and μ commute.

Next, $(\mu_1 * f_2)(x) = \int f_2(y^{-1}x)\, d\mu_1(y)$ and if $y \in N$ and $x \notin N^3$ then $y^{-1}x \notin N^2$ so that

$$f_2(y^{-1}x) = (\mu * f_o)(y^{-1}x)$$

and hence $(\mu_1 * f_2)\,(x) = \int (\mu * f_o)\,(y^{-1}x)\, d\mu_1(y)$

$$= (\mu_1 * \mu * f_o)(x) \; .$$

Finally, if $x \in N^3$ then if also $y \in Na$ we have $y^{-1}x \in a^{-1}N^4$, $f_3(y^{-1}x) = \|\mu\| \; \|\mu_1\| / \|\mu_2\|$ and so $(\mu_2 * f_3)(x) = \int f_3(y^{-1}x)\, d\mu_2(x) = \|\mu\| \; \|\mu_1\|$; also

$$(\mu * \mu_1 * f_o)\,(x) \leq \int (\sup f_o)\, d(\mu * \mu_1)$$

and $\sup f_o = 1$, so $(\mu * \mu_1 * f_o)(x) \leq \|\mu * \mu_1\| \leq \|\mu\| \; \|\mu_1\| \leq (\mu_2 * f_3)\,(x)$.

Hence, whether $x \in N^3$ or not, $(\mu * f)(x) \geqslant 0$, and f is not $\geqslant 0$ since $f(x) = -f_1(x)$ if $x \in N$ and $f_1(x) > 0$ for any x in the support of μ_1. It follows that μ is not isotone.

COROLLARY *If μ is non-trivial and A-isotone then μ must be continuous.*

Proof. If not, $\mu(a) > 0$ for some $a \in G$. Since $e \in C$, $\mu(h) > 0$ for some coset of C.

<u>Remark</u> In Theorem 2 it is sufficient if L is a coset of the centre of the subgroup of G generated by the support of μ, not of G itself.

<u>References</u>.

1 Bourbaki, N., Topologie générale, Chh 3,4 (3me éd). (Actualités Sci et Ind 1143; Hermann, Paris, 1960)

2 Bourbaki, N., Intégration, Chh 7, 8. (Actualités Sci et Ind 1306; Hermann, Paris, 1963)

3 Hewitt, E. and Ross, K., Abstract harmonic analysis, vol 1. (Grundlehren Math Wiss 115; Springer-Verlag, Berlin, Göttingen, Heidelberg 1963)

4 Kawada, Y., On the group ring of a topological group. Math Jap <u>1</u> (1948) 1-5

5 Moran, W., Isotone measures on locally compact groups. J London Math Soc (2nd series) <u>5</u> (1972) 347-355

6 Pym, J.S., A note on the Kawada-into theorem; Proc Edinburgh Math Soc <u>13</u> (1963) 295-296

7 Reiter, H., L^1-algebras and Segal algebras (Lecture notes in Mathematics, no 231; Springer-Verlag, Berlin, Göttingen, Heidelberg, 1971)

8 Wendel, J., Left centralizers and isomorphisms of group algebras; Pacific J Math <u>2</u> (1952) 251-261

9 Williamson, J.H., On theorems of Kawada and Wendel; Proc Edinburgh Math Soc <u>11</u> (1958) 71-77

Unsolved Problems

The following open problems were suggested by the partipants at the Conference

PROBLEM 1. Concerning the spectrum of a topological algebra and that of its completion : by A. Mallios.

Suppose we are given a topological algebra E (we assume the multiplication in E to be separately continuous) and let E be the completion of the respective topological vector space E (we assume the topology of E to be Hausdorff). Very little is known about criteria ensuring the extension of the ring multiplication of E to $\tilde{E}$ in the general case.

Now, suppose that E has a jointly continuous multiplication, so that regarding the spectra $\mathfrak{M}(E)$ and $\mathfrak{M}(\tilde{E})$ of the respective topological algebras one has the relation

$$\mathfrak{M}(\tilde{E}) = \mathfrak{M}(E), \tag{1}$$

within a continuous bijection. On the other hand, the preceding relation does not, in general, subsist within a homeomorphism: Cf. B. Guennebaud, Bull. Sci. math. 91 (1967), p.90; W. E. Dietrich, Jr., Math. Ann. 183 (1969), p.210; A. Mallios, Math. Ann. 154 (1964), p.173. A sufficient condition in order (1) to hold true within a homeomorphism is $\mathfrak{M}(E)$ to be a locally equicontinuous subset of E_s' (: weak topological dual of E) (cf. the paper of this author included in these Proceedings). The situation includes the case of commutative Banach algebras, but the condition is not always necessary: Cf. A. Mallios, J. Funct. Anal. 6 (1970), p.478.

Now, it would be of interest to know the class of topological algebras (: continuous multiplication) for which (1) is a homeomorphism.

On the other hand, within the same flavor of ideas seems to be the finding of that class of topological algebras, which behave relative to the 'topological tensor product functor' $\hat{\otimes}_\tau$ (τ is an appropriate 'admissible' tensorial topology; Cf.A.Mallios, Math. Ann. 162 (1966), p.247) in such a way, that one has the relation:

(2) $$\mathfrak{M}(E \hat{\otimes}_\tau F) = \mathfrak{M}(E) \times \mathfrak{M}(F),$$

within a homeomorphism (cf. the author's paper in this volume: Part I, (2.7) and Part II, (2.6)).

PROBLEM 2. Proposed by Sz.-Nagy.

Every contraction T on Hilbert space whose spectrum lies on the unit circle and which satisfies

$$\| (T-\lambda I)^{-1} \| \leq \frac{M}{1-|\lambda|}$$

for $|\lambda| < 1$, is similar to a unitary operator. This theorem was established by unitary dilation techniques(B.Sz.-Nagy - C.Foiaş and I.C.Gohberg - M.G.Krein). Try to find a straightforward proof.

PROBLEM 3. Proposed by G.Freud.

Let $1 \leq p \leq \infty$, let $f(x)$ be 2π-periodic and $f \in \mathcal{L}_p$. By $\|\cdot\|_p$ we denote the usual $\mathcal{L}_p$-norm, and $T_h f(x) = f(x+h)$. We consider the $\mathcal{L}_p$-continuity modulus

$$\omega_1(\mathcal{L}_p ; f ; \delta) = \inf_{0 \leq h \leq \delta} \| T_h f - f \|_p$$

and the '$\mathcal{L}_p$ Zygmund modulus'

$$\omega_2(\mathcal{L}_p ; f ; \delta) = \frac{1}{2} \inf_{0 \leq h \leq \delta} \| T_{h/2} f + T_{-h/2} f - 2f \|_p .$$

We proved in a recent paper (Freud, G., On the $\mathcal{L}_2$-continuity moduli of functions. Periodica Mathematica Hungarica, 3 (1973) 27-35, Theorem 3. In the statement '$r < s$' must be corrected to '$r > s$'.)

that for every $f \in \mathcal{L}_2$ there exist absolute constants C_1 and C_2 so that

$$C_1 h^2 \int_h^\infty u^{-3} [\omega_2(\mathcal{L}_2 ; f ; u)]^2 du \leq [\omega_1(\mathcal{L}_2 ; f ; h)]^2$$

$$\leq C_2 h^2 \int_h^\infty u^{-3} [\omega_2(\mathcal{L}_2 ; f ; u)]^2 du .$$

This means that, in space $\mathcal{L}_2$, if the Zygmund moduli of two functions f_1 and f_2 have the same order of decrease then also their continuity moduli have the same order of decrease. It is easy to see that this is no more the case in the space $\mathcal{L}_\infty$ (i.e. C).

Problem. For which values of p is the situation in $\mathcal{L}_p$ similar to $\mathcal{L}_2$ and for which values of p is it similar to $\mathcal{L}_\infty$?

Vol. 215: P. Antonelli, D. Burghelea and P. J. Kahn, The Concordance-Homotopy Groups of Geometric Automorphism Groups. X, 140 pages. 1971. DM 16,–

Vol. 216: H. Maaß, Siegel's Modular Forms and Dirichlet Series. VII, 328 pages. 1971. DM 20,–

Vol. 217: T. J. Jech, Lectures in Set Theory with Particular Emphasis on the Method of Forcing. V, 137 pages. 1971. DM 16,–

Vol. 218: C. P. Schnorr, Zufälligkeit und Wahrscheinlichkeit. IV, 212 Seiten. 1971. DM 20,–

Vol. 219: N. L. Alling and N. Greenleaf, Foundations of the Theory of Klein Surfaces. IX, 117 pages. 1971. DM 16,–

Vol. 220: W. A. Coppel, Disconjugacy. V, 148 pages. 1971. DM 16,–

Vol. 221: P. Gabriel und F. Ulmer, Lokal präsentierbare Kategorien. V, 200 Seiten. 1971. DM 18,–

Vol. 222: C. Meghea, Compactification des Espaces Harmoniques. III, 108 pages. 1971. DM 16,–

Vol. 223: U. Felgner, Models of ZF-Set Theory. VI, 173 pages. 1971. DM 16,–

Vol. 224: Revètements Etales et Groupe Fondamental. (SGA 1). Dirigé par A. Grothendieck XXII, 447 pages. 1971. DM 30,–

Vol. 225: Théorie des Intersections et Théorème de Riemann-Roch. (SGA 6). Dirigé par P. Berthelot, A. Grothendieck et L. Illusie. XII, 700 pages. 1971. DM 40,–

Vol. 226: Seminar on Potential Theory, II. Edited by H. Bauer. IV, 170 pages. 1971. DM 18,–

Vol. 227: H. L. Montgomery, Topics in Multiplicative Number Theory. IX, 178 pages. 1971. DM 18,–

Vol. 228: Conference on Applications of Numerical Analysis. Edited by J. Ll. Morris. X, 358 pages. 1971. DM 26,–

Vol. 229: J. Väisälä, Lectures on n-Dimensional Quasiconformal Mappings. XIV, 144 pages. 1971. DM 16,–

Vol. 230: L. Waelbroeck, Topological Vector Spaces and Algebras. VII, 158 pages. 1971. DM 16,–

Vol. 231: H. Reiter, L^1-Algebras and Segal Algebras. XI, 113 pages. 1971. DM 16,–

Vol. 232: T. H. Ganelius, Tauberian Remainder Theorems. VI, 75 pages. 1971. DM 16,–

Vol. 233: C. P. Tsokos and W. J. Padgett. Random Integral Equations with Applications to stochastic Systems. VII, 174 pages. 1971. DM 18,–

Vol. 234: A. Andreotti and W. Stoll. Analytic and Algebraic Dependence of Meromorphic Functions. III, 390 pages. 1971. DM 26,–

Vol. 235: Global Differentiable Dynamics. Edited by O. Hájek, A. J. Lohwater, and R. McCann. X, 140 pages. 1971. DM 16,–

Vol. 236: M. Barr, P. A. Grillet, and D. H. van Osdol. Exact Categories and Categories of Sheaves. VII, 239 pages. 1971. DM 20,–

Vol. 237: B. Stenström, Rings and Modules of Quotients. VII, 136 pages. 1971. DM 16,–

Vol. 238: Der kanonische Modul eines Cohen-Macaulay-Rings. Herausgegeben von Jürgen Herzog und Ernst Kunz. VI, 103 Seiten. 1971. DM 16,–

Vol. 239: L. Illusie, Complexe Cotangent et Déformations I. XV, 355 pages. 1971. DM 26,–

Vol. 240: A. Kerber, Representations of Permutation Groups I. VII, 192 pages. 1971. DM 18,–

Vol. 241: S. Kaneyuki, Homogeneous Bounded Domains and Siegel Domains. V, 89 pages. 1971. DM 16,–

Vol. 242: R. R. Coifman et G. Weiss, Analyse Harmonique Non-Commutative sur Certains Espaces. V, 160 pages. 1971. DM 16,–

Vol. 243: Japan-United States Seminar on Ordinary Differential and Functional Equations. Edited by M. Urabe. VIII, 332 pages. 1971. DM 26,–

Vol. 244: Séminaire Bourbaki – vol. 1970/71. Exposés 382–399. IV, 356 pages. 1971. DM 26,–

Vol. 245: D. E. Cohen, Groups of Cohomological Dimension One. V, 99 pages. 1972. DM 16,–

Vol. 246: Lectures on Rings and Modules. Tulane University Ring and Operator Theory Year, 1970–1971. Volume I. X, 661 pages. 1972. DM 40,–

Vol. 247: Lectures on Operator Algebras. Tulane University Ring and Operator Theory Year, 1970–1971. Volume II. XI, 786 pages. 1972. DM 40,–

Vol. 248: Lectures on the Applications of Sheaves to Ring Theory. Tulane University Ring and Operator Theory Year, 1970–1971. Volume III. VIII, 315 pages. 1971. DM 26,–

Vol. 249: Symposium on Algebraic Topology. Edited by P. J. Hilton. VII, 111 pages. 1971. DM 16,–

Vol. 250: B. Jónsson, Topics in Universal Algebra. VI, 220 pages. 1972. DM 20,–

Vol. 251: The Theory of Arithmetic Functions. Edited by A. A. Gioia and D. L. Goldsmith VI, 287 pages. 1972. DM 24,–

Vol. 252: D. A. Stone, Stratified Polyhedra. IX, 193 pages. 1972. DM 18,–

Vol. 253: V. Komkov, Optimal Control Theory for the Damping of Vibrations of Simple Elastic Systems. V, 240 pages. 1972. DM 20,–

Vol. 254: C. U. Jensen, Les Foncteurs Dérivés de $\varprojlim$ et leurs Applications en Théorie des Modules. V, 103 pages. 1972. DM 16,–

Vol. 255: Conference in Mathematical Logic – London '70. Edited by W. Hodges. VIII, 351 pages. 1972. DM 26,–

Vol. 256: C. A. Berenstein and M. A. Dostal, Analytically Uniform Spaces and their Applications to Convolution Equations. VII, 130 pages. 1972. DM 16,–

Vol. 257: R. B. Holmes, A Course on Optimization and Best Approximation. VIII, 233 pages. 1972. DM 20,–

Vol. 258: Séminaire de Probabilités VI. Edited by P. A. Meyer. VI, 253 pages. 1972. DM 22,–

Vol. 259: N. Moulis, Structures de Fredholm sur les Variétés Hilbertiennes. V, 123 pages. 1972. DM 16,–

Vol. 260: R. Godement and H. Jacquet, Zeta Functions of Simple Algebras. IX, 188 pages. 1972. DM 18,–

Vol. 261: A. Guichardet, Symmetric Hilbert Spaces and Related Topics. V, 197 pages. 1972. DM 18,–

Vol. 262: H. G. Zimmer, Computational Problems, Methods, and Results in Algebraic Number Theory. V, 103 pages. 1972. DM 16,–

Vol. 263: T. Parthasarathy, Selection Theorems and their Applications. VII, 101 pages. 1972. DM 16,–

Vol. 264: W. Messing, The Crystals Associated to Barsotti-Tate Groups: With Applications to Abelian Schemes. III, 190 pages. 1972. DM 18,–

Vol. 265: N. Saavedra Rivano, Catégories Tannakiennes. II, 418 pages. 1972. DM 26,–

Vol. 266: Conference on Harmonic Analysis. Edited by D. Gulick and R. L. Lipsman. VI, 323 pages. 1972. DM 24,–

Vol. 267: Numerische Lösung nichtlinearer partieller Differential- und Integro-Differentialgleichungen. Herausgegeben von R. Ansorge und W. Törnig, VI, 339 Seiten. 1972. DM 26,–

Vol. 268: C. G. Simader, On Dirichlet's Boundary Value Problem. IV, 238 pages. 1972. DM 20,–

Vol. 269: Théorie des Topos et Cohomologie Etale des Schémas. (SGA 4). Dirigé par M. Artin, A. Grothendieck et J. L. Verdier. XIX, 525 pages. 1972. DM 50,–

Vol. 270: Théorie des Topos et Cohomologie Etale des Schémas. Tome 2. (SGA 4). Dirigé par M. Artin, A. Grothendieck et J. L. Verdier. V, 418 pages. 1972. DM 50,–

Vol. 271: J. P. May, The Geometry of Iterated Loop Spaces. IX, 175 pages. 1972. DM 18,–

Vol. 272: K. R. Parthasarathy and K. Schmidt, Positive Definite Kernels, Continuous Tensor Products, and Central Limit Theorems of Probability Theory. VI, 107 pages. 1972. DM 16,–

Vol. 273: U. Seip, Kompakt erzeugte Vektorräume und Analysis. IX, 119 Seiten. 1972. DM 16,–

Vol. 274: Toposes, Algebraic Geometry and Logic. Edited by. F. W. Lawvere. VI, 189 pages. 1972. DM 18,–

Vol. 275: Séminaire Pierre Lelong (Analyse) Année 1970–1971. VI, 181 pages. 1972. DM 18,–

Vol. 276: A. Borel, Représentations de Groupes Localement Compacts. V, 98 pages. 1972. DM 16,–

Vol. 277: Séminaire Banach. Edité par C. Houzel. VII, 229 pages. 1972. DM 20,–

Vol. 278: H. Jacquet, Automorphic Forms on GL(2). Part II. XIII, 142 pages. 1972. DM 16,–

Vol. 279: R. Bott, S. Gitler and I. M. James, Lectures on Algebraic and Differential Topology. V, 174 pages. 1972. DM 18,–

Vol. 280: Conference on the Theory of Ordinary and Partial Differential Equations. Edited by W. N. Everitt and B. D. Sleeman. XV, 367 pages. 1972. DM 26,–

Vol. 281: Coherence in Categories. Edited by S. Mac Lane. VII, 235 pages. 1972. DM 20,–

Vol. 282: W. Klingenberg und P. Flaschel, Riemannsche Hilbertmannigfaltigkeiten. Periodische Geodätische. VII, 211 Seiten. 1972. DM 20,–

Vol. 283: L. Illusie, Complexe Cotangent et Déformations II. VII, 304 pages. 1972. DM 24,–

Vol. 284: P. A. Meyer, Martingales and Stochastic Integrals I. VI, 89 pages. 1972. DM 16,–

Vol. 285: P. de la Harpe, Classical Banach-Lie Algebras and Banach-Lie Groups of Operators in Hilbert Space. III, 160 pages. 1972. DM 16,–

Vol. 286: S. Murakami, On Automorphisms of Siegel Domains. V, 95 pages. 1972. DM 16,–

Vol. 287: Hyperfunctions and Pseudo-Differential Equations. Edited by H. Komatsu. VII, 529 pages. 1973. DM 36,–

Vol. 288: Groupes de Monodromie en Géométrie Algébrique. (SGA 7 I). Dirigé par A. Grothendieck. IX, 523 pages. 1972. DM 50,–

Vol. 289: B. Fuglede, Finely Harmonic Functions. III, 188. 1972. DM 18,–

Vol. 290: D. B. Zagier, Equivariant Pontrjagin Classes and Applications to Orbit Spaces. IX, 130 pages. 1972. DM 16,–

Vol. 291: P. Orlik, Seifert Manifolds. VIII, 155 pages. 1972. DM 16,–

Vol. 292: W. D. Wallis, A. P. Street and J. S. Wallis, Combinatorics: Room Squares, Sum-Free Sets, Hadamard Matrices. V, 508 pages. 1972. DM 50,–

Vol. 293: R. A. DeVore, The Approximation of Continuous Functions by Positive Linear Operators. VIII, 289 pages. 1972. DM 24,–

Vol. 294: Stability of Stochastic Dynamical Systems. Edited by R. F. Curtain. IX, 332 pages. 1972. DM 26,–

Vol. 295: C. Dellacherie, Ensembles Analytiques, Capacités, Mesures de Hausdorff. XII, 123 pages. 1972. DM 16,–

Vol. 296: Probability and Information Theory II. Edited by M. Behara, K. Krickeberg and J. Wolfowitz. V, 223 pages. 1973. DM 20,–

Vol. 297: J. Garnett, Analytic Capacity and Measure. IV, 138 pages. 1972. DM 16,–

Vol. 298: Proceedings of the Second Conference on Compact Transformation Groups. Part 1. XIII, 453 pages. 1972. DM 32,–

Vol. 299: Proceedings of the Second Conference on Compact Transformation Groups. Part 2. XIV, 327 pages. 1972. DM 26,–

Vol. 300: P. Eymard, Moyennes Invariantes et Représentations Unitaires. II. 113 pages. 1972. DM 16,–

Vol. 301: F. Pittnauer, Vorlesungen über asymptotische Reihen. VI, 186 Seiten. 1972. DM 18,–

Vol. 302: M. Demazure, Lectures on p-Divisible Groups. V, 98 pages. 1972. DM 16,–

Vol. 303: Graph Theory and Applications. Edited by Y. Alavi, D. R. Lick and A. T. White. IX, 329 pages. 1972. DM 26,–

Vol. 304: A. K. Bousfield and D. M. Kan, Homotopy Limits, Completions and Localizations. V, 348 pages. 1972. DM 26,–

Vol. 305: Théorie des Topos et Cohomologie Etale des Schémas. Tome 3. (SGA 4). Dirigé par M. Artin, A. Grothendieck et J. L. Verdier. VI, 640 pages. 1973. DM 50,–

Vol. 306: H. Luckhardt, Extensional Gödel Functional Interpretation. VI, 161 pages. 1973. DM 18,–

Vol. 307: J. L. Bretagnolle, S. D. Chatterji et P.-A. Meyer, Ecole d'été de Probabilités: Processus Stochastiques. VI, 198 pages. 1973. DM 20,–

Vol. 308: D. Knutson, λ-Rings and the Representation Theory of the Symmetric Group. IV, 203 pages. 1973. DM 20,–

Vol. 309: D. H. Sattinger, Topics in Stability and Bifurcation Theory. VI, 190 pages. 1973. DM 18,–

Vol. 310: B. Iversen, Generic Local Structure of the Morphisms in Commutative Algebra. IV, 108 pages. 1973. DM 16,–

Vol. 311: Conference on Commutative Algebra. Edited by J. W. Brewer and E. A. Rutter. VII, 251 pages. 1973. DM 22,–

Vol. 312: Symposium on Ordinary Differential Equations. Edited by W. A. Harris, Jr. and Y. Sibuya. VIII, 204 pages. 1973. DM 22,–

Vol. 313: K. Jörgens and J. Weidmann, Spectral Properties of Hamiltonian Operators. III, 140 pages. 1973. DM 16,–

Vol. 314: M. Deuring, Lectures on the Theory of Algebraic Functions of One Variable. VI, 151 pages. 1973. DM 16,–

Vol. 315: K. Bichteler, Integration Theory (with Special Attention to Vector Measures). VI, 357 pages. 1973. DM 26,–

Vol. 316: Symposium on Non-Well-Posed Problems and Logarithmic Convexity. Edited by R. J. Knops. V, 176 pages. 1973. DM 18,–

Vol. 317: Séminaire Bourbaki – vol. 1971/72. Exposés 400–417. IV, 361 pages. 1973. DM 26,–

Vol. 318: Recent Advances in Topological Dynamics. Edited by A. Beck, VIII, 285 pages. 1973. DM 24,–

Vol. 319: Conference on Group Theory. Edited by R. W. Gatterdam and K. W. Weston. V, 188 pages. 1973. DM 18,–

Vol. 320: Modular Functions of One Variable I. Edited by W. Kuyk. V, 195 pages. 1973. DM 18,–

Vol. 321: Séminaire de Probabilités VII. Edité par P. A. Meyer. VI, 322 pages. 1973. DM 26,–

Vol. 322: Nonlinear Problems in the Physical Sciences and Biology. Edited by I. Stakgold, D. D. Joseph and D. H. Sattinger. VIII, 357 pages. 1973. DM 26,–

Vol. 323: J. L. Lions, Perturbations Singulières dans les Problèmes aux Limites et en Contrôle Optimal. XII, 645 pages. 1973. DM 42,–

Vol. 324: K. Kreith, Oscillation Theory. VI, 109 pages. 1973. DM 16,–

Vol. 325: Ch.-Ch. Chou, La Transformation de Fourier Complexe et L'Equation de Convolution. IX, 137 pages. 1973. DM 16,–

Vol. 326: A. Robert, Elliptic Curves. VIII, 264 pages. 1973. DM 22,–

Vol. 327: E. Matlis, 1-Dimensional Cohen-Macaulay Rings. XII, 157 pages. 1973. DM 18,–

Vol. 328: J. R. Büchi and D. Siefkes, The Monadic Second Order Theory of All Countable Ordinals. VI, 217 pages. 1973. DM 20,–

Vol. 329: W. Trebels, Multipliers for (C, α)-Bounded Fourier Expansions in Banach Spaces and Approximation Theory. VII, 103 pages. 1973. DM 16,–

Vol. 330: Proceedings of the Second Japan-USSR Symposium on Probability Theory. Edited by G. Maruyama and Yu. V. Prokhorov. VI, 550 pages. 1973. DM 36,–

Vol. 331: Summer School on Topological Vector Spaces. Edited by L. Waelbroeck. VI, 226 pages. 1973. DM 20,–

Vol. 332: Séminaire Pierre Lelong (Analyse) Année 1971-1972. V, 131 pages. 1973. DM 16,–

Vol. 333: Numerische, insbesondere approximationstheoretische Behandlung von Funktionalgleichungen. Herausgegeben von R. Ansorge und W. Törnig. VI, 296 Seiten. 1973. DM 24,–

Vol. 334: F. Schweiger, The Metrical Theory of Jacobi-Perron Algorithm. V, 111 pages. 1973. DM 16,–

Vol. 335: H. Huck, R. Roitzsch, U. Simon, W. Vortisch, R. Walden, B. Wegner und W. Wendland, Beweismethoden der Differentialgeometrie im Großen. IX, 159 Seiten. 1973. DM 18,–

Vol. 336: L'Analyse Harmonique dans le Domaine Complexe. Edité par E. J. Akutowicz. VIII, 169 pages. 1973. DM 18,–

Vol. 337: Cambridge Summer School in Mathematical Logic. Edited by A. R. D. Mathias and H. Rogers. IX, 660 pages. 1973. DM 42,–

Vol: 338: J. Lindenstrauss and L. Tzafriri, Classical Banach Spaces. IX, 243 pages. 1973. DM 22,–

Vol. 339: G. Kempf, F. Knudsen, D. Mumford and B. Saint-Donat, Toroidal Embeddings I. VIII, 209 pages. 1973. DM 20,–

Vol. 340: Groupes de Monodromie en Géométrie Algébrique. (SGA 7 II). Par P. Deligne et N. Katz. X, 438 pages. 1973. DM 40,–

Vol. 341: Algebraic K-Theory I, Higher K-Theories. Edited by H. Bass. XV, 335 pages. 1973. DM 26,–

Vol. 342: Algebraic K-Theory II, "Classical" Algebraic K-Theory, and Connections with Arithmetic. Edited by H. Bass. XV, 527 pages. 1973. DM 36,–